Archimedes

New Studies in the History and Philosophy of Science and Technology

Volume 75

Archimedes has three fundamental goals: to further the integration of the histories of science and technology with one another; to investigate the technical, social and practical histories of specific developments in science and technology; and finally, where possible and desirable, to bring the histories of science and technology into closer contact with the philosophy of science.

The series is interested in receiving book proposals that treat the history of any of the sciences, ranging from biology through physics, all aspects of the history of technology, broadly construed, as well as historically-engaged philosophy of science or technology. Taken as a whole, Archimedes will be of interest to historians, philosophers, and scientists, as well as to those in business and industry who seek to understand how science and industry have come to be so strongly linked.

Submission / Instructions for Authors and Editors: The series editors aim to make a first decision within one month of submission. In case of a positive first decision the work will be provisionally contracted: the final decision about publication will depend upon the result of the anonymous peer-review of the complete manuscript. The series editors aim to have the work peer-reviewed within 3 months after submission of the complete manuscript.

The series editors discourage the submission of manuscripts that contain reprints of previously published material and of manuscripts that are below 150 printed pages (75,000 words). For inquiries and submission of proposals prospective authors can contact one of the editors:

Please find on the top right side of our webpage a link toour *Book Proposal Form.*

Stefano Gattei
Editor

The Incomparable Historian

Essays in Memory of John L. Heilbron

Springer

Editor
Stefano Gattei
Department of Sociology and Social Research
University of Trento
Trento, Italy

ISSN 1385-0180 ISSN 2215-0064 (electronic)
Archimedes
ISBN 978-3-032-25677-5 ISBN 978-3-032-25678-2 (eBook)
https://doi.org/10.1007/978-3-032-25678-2

This Springer imprint is published by the registered company Springer Nature Switzerland AG
The registered company address is: Gewerbestrasse 11, 6330 Cham, Switzerland

Preface

Magnam vim habet ad amicitias inter homines eruditos conjungendas morum studiorumque similitudo. Majorem etiam, si alter alteri ad virtutem aut scientiam viam præeundo monstraverit. Maximam denique, si studia, quorum inter eos communio est, minimè sint vulgaria.

Johannes Clauberg[1]

Seldom do projects turn out as they were initially conceived. Sadly, this collection of essays is no exception: originally planned as a *Festschrift* volume in honor of John L. Heilbron on the occasion of his 90th birthday, it turned into a volume in his memory, after John's unexpected death in Padua, on 5 November 2023, aged 89.[2]

A scholar of infinite curiosity, vast interests, and exceptional learning, a master wordsmith with a rare ability to range with extreme competence and absolute proficiency from the sixteenth to the twentieth century, John was trained at the University of California, Berkeley, earning bachelor's and master's degrees in physics (1955,

[1] "The similarity of character and interests has great power in forming friendships between learned individuals. Even greater is the power if one of them shows the other, by leading the way, the path to virtue or knowledge. Finally, the power is at its highest if the studies they share are far from commonplace": Johannes Clauberg, *Physica, Quibus rerum corporearum Vis & natura, Mentis ad Corpus relatæ proprietates, denique Corporis ac Mentis arcta & admirabilis in Homine conjunctio explicantur*, Amsterdam: Daniel Elzevir, 1664, f. *3*r*.

[2] Heilbron's last talk, "*The Assayer* Assayed", was delivered 2 months earlier at the opening of the XLIII SISFA (Italian Society for the History of Science and Astronomy) annual conference; it is available online at https://www.youtube.com/watch?v=L13T4ERlsco&t=1796s. In the months that followed Heilbron's passing, a few commemorative events were held to honour his life and scholarly contributions. On 5–6 July 2024, the Ludwig-Maximilians-Universität hosted "Recollections and New Directions: Symposium in Memory of John Heilbron (1934–2023)", organized by Kärin Nickelsen and Johannes-Geert Hagmann; on 10 November 2024, the HSS Annual Meeting included a special session on "The Life and Work of the Late John L. Heilbron", organized by Daniel L. Kevles; on 20 March 2025, the Center for Science, Technology, Medicine & Society of the University of California, Berkeley, organized the symposium "Remembering John Heilbron"; and on 11 September 2024, at its XLV National Congress, the Italian Society for the History of Physics and Astronomy held a memorial session in his honor. Vol. 17 of *The Collected Papers of Albert Einstein* (Princeton: Princeton University Press, 2024) was dedicated to his memory.

1958) and a doctorate in history (1964). In 1962–1964, as assistant director, he joined Thomas S. Kuhn on a major project—promoted jointly by the American Physical Society and the American Philosophical Society—to collect materials useful in reconstructing the history of quantum physics, cataloging all available primary sources and conducting, with Kuhn but more frequently alone, dozens of interviews with physicists who contributed to the field in the crucial decades (1900–1930) that shaped our picture of the world.[3] This project formed the basis of his doctoral dissertation, *A History of the Problem of Atomic Structure from the Discovery of the Electron to the Beginning of Quantum Mechanics* (1964), written under Kuhn.[4]

Heilbron came to the history of science in the postwar decades, when the discipline was primarily engaged in setting its own standards and terms. He was a driving force in the transformation of the history of science into an independent academic discipline (not only in the USA), shaping the way scholars approach the history of science. "Not hocus pocus but *hoc opus est, hic labor:* the business of trying to understand scientists on their own often arcane terms, to account for and evaluate their beliefs, practices, and broader place within society, and to make those findings not just legible but also engaging to a broader and perhaps indifferent public would occupy Heilbron for his entire career as teacher, writer, editor, and administrator."[5]

As documented in the amazing list of his publications,[6] he was prodigiously productive: across some six decades, he authored more than twenty books, dozens of studies and reviews, as well as edited collections of sources of fundamental importance to the history of science. His major books alone deal with a stunning variety of subjects, from early modern studies to quantum physics. To name but a few: five major biographies (H. G. J. Moseley, 1974; Max Planck, 1986 and 2000;

---

[3]Thomas S. Kuhn, John L. Heilbron, Paul Forman and Lini Allen, *Sources for History of Quantum Physics: An Inventory and Report*, Philadelphia: The American Philosophical Society, 1967. Later, Heilbron would collaborate with Kuhn on "The Genesis of the Bohr Atom", *Historical Studies in the Physical Sciences*, 1, 1969, pp. 211–290, and with Forman (and Spencer Weart) on *Physics* circa *1900: Personnel, Funding, and Productivity of the Academic Establishments*, Princeton-London: Princeton University Press, 1975. Despite their differing focus and scope, these works remain foundational to much of the scholarship in the history of quantum physics.

[4]Heilbron worked closely with Kuhn in preparing *The Structure of Scientific Revolutions* (Chicago-London: The University of Chicago Press, 1962); he is the only student of Kuhn's—together with philosophers of science Paul K. Feyerabend (1924–1994) of Berkeley and Ernest Nagel (1901–1985) of Columbia, and theoretical physicist H. Pierre Noyes (1923–2016) of Lawrence Radiation Laboratory—acknowledged in the book (p. xii).

[5]Eileen Reeves, "John Lewis Heilbron (1934–2023)", *Galilæana*, XXI, 1, 2024, pp. 255–270: pp. 257–258.

[6]See *infra*, pp. xix–xliii. Heilbron was keenly aware of the importance of bibliographies as key tools for historians, himself authoring and promoting a few that were included in the "Berkeley Papers in History of Science" series, which he founded and edited for some twenty years (1977–1994).

Ernest Rutherford, 2003; Galileo, 2010; Francesco Bianchini, 2022[7]), a groundbreaking study of the history of electricity in the seventeenth and eighteenth centuries (1979), a history of the Lawrence Berkeley Laboratory (with Robert W. Seidel, 1989), studies in the history of physics from the seventeenth to the twentieth century and astronomy (among them, the magisterial *The Sun in the Church*, 1999, which was awarded the Pfizer Prize in 2001), as well as his longstanding research project on scientific academies and learned societies, which he carried on until the very end. To these, two remarkable works need to be added, namely, *Geometry Civilized* (1998) and *The Ghost of Galileo in a Forgotten Painting from the English Civil War* (2021), uniquely testifying to Heilbron's incredible curiosity, attention to detail and taste for the uncommon. All his works are uniformly of outstanding quality and display an ability to deal with the technical aspects of science as well as the complex social, political, and institutional contexts (not to mention funding regimes) in which science has been pursued in the past. Nor can be overlooked his editorial work, as documented, for example, in such volumes as *The Quantifying Spirit in the 18th Century* (edited with Tore Frängsmyr and Robin E. Rider, 1990), and *The Oxford Companion to the History of Modern Science* (2003). For his scholarly production he received the discipline's most prestigious awards, from the History of Science Society's Sarton Medal (1993) to the Médaille Alexandre-Koyré of the Académie internationale d'Histoire des Sciences (2000) and the Pais Prize for the History of Physics from the American Physical Society (2006). A member of the Académie internationale d'Histoire des Sciences (1981), the Royal Swedish Academy of Sciences (1987), the American Academy of Arts and Sciences (1988) and the American Philosophical Society (1990), Heilbron received honorary doctorates from the universities of Bologna (1988), Pavia (2000), Uppsala (2000), and Yale (2011).

His writing was entertaining and literate, often witty.[8] He did not refrain from expressing sharp views, both in his essays and in his many reviews (not to mention in his rejoinders to the authors' occasional replies), but was always willing to help, and showed affection and care for his students. Working at the intersection of people, scientific ideas, and institutions, Heilbron enthusiastically and effectively taught many undergraduate and graduate courses, as well as directed a variety of doctoral dissertations. "[A] man of institutions, even though he preferred to work on his own

---

[7] He wished to write a biography of Niels Bohr, too, but was denied full access to family and personal documents in Bohr's archive. He then collaborated with Finn Aaserud, Director of the Niels Bohr Archive in Copenhagen, with whom he published *Love, Literature, and the Quantum Atom: Niels Bohr's 1913 Trilogy Revisited* (Oxford: Oxford University Press, 2013).

[8] And even more so were his talks: see, to offer but one brilliant example, "Historical Myths About Science", the keynote address Heilbron gave at the Washington and Lee University, Lexington VA, on May 9, 2014, available at https://www.youtube.com/watch?v=PxkxCSyP60Y&t=652s (a shorter and substantially revised version of this talk appeared in print as "Myth 5: That Galileo Publicly Refuted Aristotle's Conclusions about Motion by Repeated Experiments Made from the Campanile of Pisa", in Ronald L. Numbers, Kostas Kampourakis (eds.), *Newton's Apple and Other Myths about Science*, Cambridge, MA-London: Harvard University Press, 2015, pp. 40–47.

terms," as Cathryn Carson aptly described him,[9] "he worked within and alongside existing institutions–journals, museums, libraries, encyclopedias, research centers, departments–and created new ones,"[10] weaving together networks of scholars across a few generations.

In 1964–1967, he worked as an assistant professor to establish the Program in the History and Philosophy of Science, now the Department of History and Sociology of Science, at the University of Pennsylvania. Having been appointed to the faculty of the Department of History at the University of California, Berkeley, in 1967, he advanced from the rank of assistant to associate professor of History in 1971, and to full professor in 1973, when he founded and directed the Office for History of Science and Technology. In 1980, he took on the editorship of *Historical Studies in the Physical Sciences* (subsequently *Historical Studies in the Physical and Biological Sciences*, now *Historical Studies in the Natural Sciences*—to many, "John's journal"), moving it on to the University of California Press as it transitioned from its founder Russell McCormmach: he would go on to edit 56 issues (Vols. 11–37, 1980–2007), turning it into one of the leading journals in the discipline. Across three decades, the journal had a profound influence on the historiography of science, both because it published only work that met Heilbron's exacting standards, and because of his legendary critical and clarifying editorial comments and revisions of the submitted papers. A demanding and challenging incubator for all authors, junior or senior, which would produce excellent works that went well beyond the volumes of the journal itself, contributing to mold a growing interest in the history of modern physical sciences and to shape the way history of science (and, more broadly, history) should be written. The words inaugurating his editorship read as a manifesto:

> Discussion of philosophical or other ideas that do not relate directly to scientific work or practice falls outside the scope of the journal. So does fascination with precursors. Pedantry should be kept to the minimum needed for good scholarship. Words like "Newtonian" and "Cartesian" should be defined and used univocally or not at all. The organization of knowledge current in 1900 or 1950 or 1980 should not be imposed on earlier periods. [...] Conflation of quite different meanings of "physics" has caused historians to miss a fundamental problem in the history of physical science: identifying the forces responsible for the constitution of classical physics and the definition of neighboring sciences.
>
> Periodization has not received enough attention. Division of history by hundred-year intervals is an unsatisfactory expedient [...]. One pernicious result is that the joins between centuries receive too little attention. Continuity between the 18th and 19th centuries has suffered severely; much that is usually taken by specialists to be new in the later period can be found in embryo in the Ancien Régime. Similar organic connections have been recognized by general historians, from whom historians of science still have much to learn. Historians of very recent physics face an additional hazard in the implicit periodization of the subject established by living physicists. We must arrive at an independent estimate of the relative importance of the science they made.

---

[9] Cathryn Carson, "John L. Heilbron (1934–2023)", *Nature*, 626, 1 February 2024, p. 25.

[10] Cathryn Carson, "John L. Heilbron (1934–2023): In Memoriam", *Historical Studies in the Natural Sciences*, 54, 4, 2024, pp. 521–526: p. 523.

> Science is polyglot and international; its historians tend to stick to one or two languages or cultures. [...] Even the most active historical imagination cannot generate the wealth of possibilities and connections that appear from cross-national comparisons.[11]

Some ten years earlier, in the opening issue of the journal, founding editor Russell McCormmach outlined his goals and set out how he intended to pursue them: "There are facets of scientific activity that can only be understood when related sciences are treated as a group of interacting disciplines rather than as isolated specialties," he wrote. "The vision of the history of science that relegates the historiographic traditions of internal and external history to mutually exclusive roles is sterile," he continued, "obstructing the synthesis of the intellectual and social history of science that must come. I will be strongly sympathetic to studies that implicitly challenge the cogency of the commonly held view of the internalist-externalist distinction. I would like, for example, to encourage articles exposing the intimate relation of institutions and ideas."[12] From the very outset, McCormmach made it clear that he intended the journal to break new methodological and substantive ground in the history of science. When he took over as new editor, Heilbron followed suit, reiterating several of McCormmach's goals—especially those his predecessor only partially achieved, as *HSPS* still lacked truly interdisciplinary studies and studies outside the fields of physics and chemistry—modifying others and adding to them a call for articles on historiography, bibliography, historical demography, and the international character of modern science. And if McCormmach called for literary quality of the articles, urging authors "to submit writing that is clear, direct, and free of jargon and pedantry,"[13] Heilbron complemented such desiderata with "economy of expression and an occasional smile. The ridiculous in the human condition touches scientists as well as their historians."[14]

As Cathryn Carson noted,

> John was determined to cultivate scholarship on the conceptual history of the recent physical sciences (since the Scientific Revolution) in their social, cultural, and institutional contexts. Articulating a call whose power today is sometimes domesticated or taken for granted, he insisted that historians recognize the stakes of scientific ideas exactly while breaking through the internal-external divide. Along with more studies of less well-trodden sciences,

[11] John L. Heilbron, "Foreword", *Historical Studies in the Physical Sciences*, 11, 1, 1980, pp. I–III: pp. II–III.

[12] Russell McCormmach, "Editor's Foreword", *Historical Studies in the Physical Sciences*, 1, 1969, pp. VII–IX: pp. VII, VIII.

[13] Ibid., p. IX.

[14] Heilbron, "Foreword" (cited, footnote 11), p. III. In the "Foreword" to the first issue of the journal after its change of name into *Historical Studies in the Physical and Biological Sciences*, Heilbron clarified and extended his editorial priorities: the journal was not meant to be limited to the twentieth century (1700 was established as a rough *terminus a quo*; articles dealing with the seventeenth century were welcome, too, provided they brought their topics into the eighteenth) or to the history of physics, and physical sciences were broadly defined to include, besides physics and chemistry, meteorology, geology, astronomy and those areas of biological sciences that employ the tools and techniques of the physical sciences. See Heilbron, "Foreword", *Historical Studies in the Physical and Biological Sciences*, 16, 1, 1986, pp. III–IV.

> he pointed to the urgent need to illuminate institutional settings, funding landscapes, and other forms of social support. He wanted to see histories of experimental techniques along with ideas. Above all, he spurred colleagues' ambition to attend rigorously, not just gesturally, to connections between the content of science and its societal setting.[15]

Heilbron came to emphasize, especially in his later years, the historian's role as both an interpretive and creative endeavor: the historian's craft, he said in an interview, is "a funny combination of patience and boldness," where consideration for evidence and respect for the sources do not only mean taking them at face value, insofar as possible, but also adopting principles of selection and organization, and taking into account what others have written about the subject. Distancing himself from what Kuhn famously claimed—historians have "to climb into other people's heads"[16]—Heilbron argued that whatever persona historians construct, based on the available remains (which are not necessarily confined to literary sources), this is inevitably artificial: their success depends on how effectively they can use what is known to construct a plausible picture. Historians, in other words, are somewhat like novelists, but more constrained—their narrative, that is, necessarily involves conjecture and artistry. "One of the great treats of history," he continued, is that "it frees you from time and place."[17]

To quote again from Eileen Reeves' compelling portrait of John's life and work,

> Heilbron's increasing emphasis on the relevance of literary techniques to historiography is complemented by greater attention to literary works themselves within his later monographs, both as an effective means to reach a broader public, and as part of his conviction that a scientist's aesthetic choices correlated—sometimes—with other and less obvious intellectual dispositions.[18]

This volume brings together essays from former students, colleagues, and friends. Reflecting the wide variety of John's interests and works, their subjects range widely—from the Scientific Revolution to quantum physics through Lamarck and Darwin, from styles of historiography and applied history of science to contemporary narratives of science—yet each contribution reflects something of Heilbron's method: an insistence on technical competence, a sensitivity to context, a distaste

---

[15] Carson, "John L. Heilbron (1934–2023): In Memoriam" (cited, footnote 10), p. 523.

[16] "Probably the thing I do best and certainly the one to which I have devoted most time is climbing from the writings into the minds of dead scientists, figuring out how they thought, why they believed what they did, and how they came to change their minds": Thomas S. Kuhn, "Reflections on Receiving the John Desmond Bernal Award", *4S Review*, 1, 4, 1983, pp. 26–30: p. 27. See also Thomas S. Kuhn, "History of Science", in Peter D. Asquith and Henry E. Kyburg, jr. (eds.), *Current Research in Philosophy of Science: Proceedings of the P.S.A. Critical Research Problems Conference*, East Lansing, MI: Philosophy of Science Association, 1979, pp. 121–127: p. 122; and "A Discussion with Thomas S. Kuhn", in Thomas S. Kuhn, *The Road Since* Structure*: Philosophical Essays 1970–1993, with an Autobiographical Interview*, edited by James Conant and John Haugeland, Chicago-London: The University of Chicago Press, 2000, pp. 255–323: p. 280.

[17] Quotes in this paragraph are from "Science and History", an interview to John Heilbron by Harry Kreisler (Berkeley, 8 November 2007), available at https://iis.berkeley.edu/publications/john-heilbron-science-and-history.

[18] Reeves, "John Lewis Heilbron (1934–2023)" (cited, footnote 5), p. 266.

for cliché, and an appreciation for the eccentricities of scientific life. Across more than seven decades of relentless, passionate, and rigorous activity, John Heilbron transformed a discipline and marked the lives of those who had the privilege and pleasure of working with him, encouraging them to "follow where your subject leads you, into any area of history—institutional as well as intellectual, practical as well as philosophical, technical as well as theoretical. [...] The demise of the so-called internalist view of science may have resulted less from methodological revolutionizing than from empirical overwhelming: it burst at the seams due to vast and mounting evidence that science is inseparably woven into the rest of life."[19]

Whether dealing with Galileo, Planck, or Bohr, the development of quantum physics or the Copernican Revolution, an enigmatic painting from the English Civil War or an Italian polymath of the late seventeenth century, the multi-faceted transformations of geometry, the practice of science in a major laboratory or European scientific academies and learned societies, his works always reflect a thorough mastery of the subject, an original approach, and an engaging narrative.

All of us whose lives and work were touched by his breathtaking intellect and wit will miss him terribly. He leaves the memory of his uncompromising dedication to intellectual work, carried through to his very last days. He made it clear that, no matter how very different our contributions were from his and no matter how unattainable he often seemed to us, we were all engaged in a fundamental and exciting activity worth devoting our lives to.

Pisa-Trento, Italy
June 2025

Stefano Gattei

---

[19] Jessica Riskin, "Introduction: A Mingled Yarn", in *Nature Engaged: Science in Practice from the Renaissance to the Present*, edited by Mario Biagioli and Jessica Riskin, New York: Palgrave Macmillan, 2012, p. 2.

Acknowledgments

My gratitude to Jed Buchwald, who kindly welcomed this volume in the "Archimedes" series he edits. My heartfelt thanks to him and to all other authors, both for sending their contributions and for bearing with me along the long path that eventually led to its publication. Special thanks to John's wife, Alison Browning, who has worked "behind the scenes" from the very outset of the project, and Concetta Luna, who helped me bring it to completion.

Heilbron did not keep a full list of his publications, but only included a selected list of his major works in his cv. The bibliography published here—which, despite the editor's efforts, is likely incomplete—is based on two such partial lists and was compiled by appealing to online resources. It also greatly benefited from the help of many friends and colleagues who kindly supplied me with material and information: besides those already mentioned, I am indebted to Cathryn Carson, Matthias Dörries, Paula Fass, Mordechai Feingold, Yuanlin Guo, Malachi H. Hacohen, Jack Lesch, Massimo Mazzotti, Giuliano Pancaldi, Markku Peltonen, John M. Preston, Nicolas Roudet, Michael B. Shank, Ignacio Suay Matallana, James R. Voelkel and Cristiano Zanetti.

Jim Bennett, distinguished scholar and a close friend of Heilbron's for many years, passed away on 28 October 2023, a week before John. Despite his illness, he welcomed my invitation and promptly sent me his contribution, which now appears thanks to his wife, Sylvia Sumira.

Another good friend of John's, Mario Biagioli (1955–2025), was unfortunately unable to submit his contribution, as he had wished to do, owing to the illness that regrettably led to his passing on 17 May 2025.

Finally, I wish to thank Gigi Scully, who kindly granted me permission to include her portrait of John working in the garden of his house in Shilton. (*SG*)

In his study. Shilton (UK), September 2011

In front of April House. Shilton (UK), 11 June 2021

On a short trip to Italy. Lucca, 1 October 2022

Gigi Scully, *John in his garden at Shilton* (2022). Oil on canvas, 40′ × 40′

A Bibliography of the Writings of John L. Heilbron, 1964–2025

Non omnis moriar multaque pars mei
vitabit Libitinam.
Horace[1]

Monographs

1. *A History of the Problem of Atomic Structure from the Discovery of the Electron to the Beginning of Quantum Mechanics*, submitted in partial satisfaction of the requirements for the degree of Doctor of Philosophy in History in the Graduate Division of the University of California, Berkeley, 1964; v + 420 pp. (Committee in charge: Thomas S. Kuhn, A. Hunter Dupree, Walter D. Knight).
2. (Thomas S. Kuhn, JLH, Paul Forman, Lini Allen) *Sources for History of Quantum Physics: An Inventory and Report* (Memoirs of the American Philosophical Society, 68), Philadelphia: The American Philosophical Society, 1967; xi + 176 pp.
3. *H. G. J. Moseley: The Life and Letters of an English Physicist, 1887–1915*, Berkeley-Los Angeles-London: University of California Press, 1974; xiii + 312 pp.

   Reprinted in the "UC Press Voices Revived" series, 2024.

4. (Paul Forman, JLH, Spencer Weart) *Physics* circa *1900: Personnel, Funding, and Productivity of the Academic Establishments* (*Historical Studies in the Physical Sciences*, 5), Princeton-London: Princeton University Press, 1975; 185 pp.
5. (Wayne Shumaker, JLH) *John Dee on Astronomy:* Propaedeumata aphoristica *(1558 and 1568)*, Berkeley-Los Angeles-London: University of California Press, 1978; xii + 264 pp.
6. *Electricity in the 17th and 18th Centuries: A Study of Early Modern Physics*, Berkeley-Los Angeles-London: University of California Press, 1979; xv + 606 pp.

[1] "I shall not wholly die: a large part of me will elude the Goddess of death": *Odes*, III 30, 6–7.

Reprinted, New York: Dover, 1999; xxx + 606 pp., with "Preface to the Dover Edition", pp. xi-xxv.
Reprinted in the "UC Press Voices Revived" series, 2023.

7. *Historical Studies in the Theory of Atomic Structure* (The Development of Science), New York: Arno Press, 1981; ix + 302 pp.
8. (JLH, Bruce R. Wheaton, with the assistance of John G. May, Robin E. Rider, David Robinson) *Literature on the History of Physics in the 20th Century* (Berkeley Papers in History of Science, 5), Berkeley: Office for History of Science and Technology, 1981; ix + 485 pp.
9. *Elements of Early Modern Physics*, Berkeley-Los Angeles-London: University of California Press, 1982; xii + 301 pp.

   Reprinted in the "UC Press Voices Revived" series, 2024.
   Italian translation by Ferdinando Abbri, *Alle origini della fisica moderna. Il caso dell'elettricità* (Collezione di testi e studi. Storia e critica della scienza), Bologna: il Mulino, 1984; 454 pp.

10. (Bruce R. Wheaton, JLH) *An Inventory of Published Letters to and from Physicists, 1900–1950* (Berkeley Papers in History of Science, 6), Berkeley: Office for History of Science and Technology, 1982; vii + 102 pp., with four microfiches.
11. *Physics at the Royal Society during Newton's Presidency*, Los Angeles: William Andrews Clark Memorial Library, 1983; xiv + 123 pp.
12. *The Dilemmas of an Upright Man: Max Planck as Spokesman for German Science*, Berkeley-Los Angeles-London: University of California Press, 1986; xiii + 238 pp.

    Revised edition, *The Dilemmas of an Upright Man: Max Planck and the Fortunes of German Science*, Cambridge, MA-London: Harvard University Press, 2000; xiii + 254 pp., with "Afterword, 2000", pp. 205–219.
    German translation by Norma von Ragenfeld-Feldman, *Max Planck: Ein Leben für die Wissenschaft, 1858–1947*, mit einer Auswahl der allgemein verständlichen Schriften von Max Planck, Stuttgart: S. Hirzel, 1988; 474 pp.

    Second, revised and augmented edition (translation of additional texts by Carsten Heinisch), Stuttgart: S. Hirzel, 2006; xiii + 647 pp., with "Vorwort zur 2. Auflage", p. xiii, and "Nachwort", pp. 300–325.

    Chinese translation [of the "Nachwort" to the revised German edition] by Zaiqing Fang, Keke Chen, "重评纳粹时期的普朗克 [Max Planck in the Nazi Period – A Reappraisal]", 科學文化評論 (*Science & Culture Review*), Vol. 5, No. 6, December 2008, pp. 31–47.

    Italian translation [of the first German edition] by Riccardo Valla, *I dilemmi di Max Planck. Portavoce della scienza tedesca*, a cura di Ivana Gambaro, Turin: Bollati Boringhieri, 1988; 222 pp.

French translation [of the first German edition] by Nicole Dhombres, *Planck, 1858-1947: une conscience déchirée*, Préface de Pierre Gilles de Gennes, Paris: Belin, 1988; 255 pp.

Excerpted as “Planck. La révolution quantique”, *Les génies de la science*, No. 27, Mai-Juillet 2026, pp. 30, 32–120.

Korean translation [of the first German edition] by Jeong Myeong-sik and Kim Yeong-sik, 막스 플랑크: 한 양심적 과학자의 딜레마, Seoul: Minumsa, 1992; 220 pp.

Chinese translation [of the first English edition] by Bing Liu, 正直者的困境——作为德国科学发言人的马克斯·普朗克, Shanghai: Orient Publishing Center, 1997; 242 pp.

Japanese translation [of the first German edition] by Shinichi Muraoka, マックス・プランクの生涯:ドイツ物理学のディレンマ, Tokyo: Hosei University Press, 2000; x + 288 pp.

Hebrew translation [of the second German edition, with the addition of “Afterword, 2000”] by Tal Golan, התלבטויותיו של אדם ישר: מקס פלנק וגורל המדע הגרמני, Tel Aviv: Kamin, 2014; 222 pp.

13. (Elisabeth Crawford, JLH, Rebecca Ullrich) *The Nobel Population, 1901–1937: A Census of the Nominators and Nominees for the Prizes in Physics and Chemistry* (Berkeley Papers in History of Science, 11; Uppsala Studies in History of Science, 4), Berkeley: Office for the History of Science and Technology, University of California, and Uppsala: Office for History of Science, Uppsala University, 1987; [v] + 337 pp.
14. (JLH, Robert W. Seidel) *Lawrence and His Laboratory: A History of the Lawrence Berkeley Laboratory*, Vol. 1 (California Studies in the History of Science), Berkeley-Los Angeles-Oxford: University of California Press, 1989; xxvii + 586 pp.
15. *Weighing Imponderables and Other Quantitative Science Around 1800*, Berkeley: University of California Press, 1993 (*Historical Studies in the Physical and Biological Sciences*, Supplement to Vol. 24, No. 1; Berkeley Papers in History of Science, 13; Uppsala Studies in History of Science, 11; Bologna Studies in History of Science, 2); [x] + 337 pp.
16. *Geometry Civilized: History, Culture, and Technique*, Oxford-New York: Clarendon Press, 1998; viii + 309 pp.
17. *The Sun in the Church: Cathedrals as Solar Observatories*, Cambridge, MA-London: Harvard University Press, 1999; ix + 366 pp.

French translation by Maurice Mashaal, *Astronomie et églises* (Bibliothèque Scientifique), Paris: Belin, 2003; 367 pp.

Italian translation by Giulia Bancheri and Piero Ranfagni, *Il Sole nella Chiesa: Le grandi chiese come osservatori astronomici*, Bologna: Compositori, 2005; iv + 425 pp.

18. *Ernest Rutherford and the Explosion of Atoms* (Oxfords Portraits in Science), Oxford-New York: Oxford University Press, 2003; 144 pp.

    Korean translation by Go Moon-ju, 핵물리학과 러더퍼드, Seoul: Badapublishing, 2006; 179 pp.
    Japanese translation by Haruo Nashimoto, アーネスト・ラザフォード：原子の宇宙の核心へ, Tokyo: Otsuki Shoten, 2009; 17 + 184 pp.

19. *Coming to Terms with the Scientific Revolution* (Salvia Småskrifter, 6), Uppsala: Tryck Wikströms, 2006; 36 pp.

    Reprinted, in revised form, as "Coming to Terms with the Scientific Revolution", *European Review*, Vol. 15, No. 4, October 2007, pp. 473–489.
    See also [Essays, 109].

20. *Galileo*, Oxford-New York: Oxford University Press, 2010; xiv + 508 pp.

    Corrected paperback reprint, 2010; xiv + 508 pp.
    Italian translation by Stefano Gattei, *Galileo. Scienziato e umanista*, Turin: Giulio Einaudi, 2013; pp. xvi + 543, with "Prefazione all'edizione italiana", p. ix.

21. (Finn Aaserud, JLH) *Love, Literature, and the Quantum Atom: Niels Bohr's 1913 Trilogy Revisited*, Oxford: Oxford University Press, 2013; x + 284 pp.
22. *Physics: A Short History from Quintessence to Quarks*, Oxford: Oxford University Press, 2015; ix + 228 pp.

    Reprinted as *The History of Physics: A Very Short Introduction* (Very Short Introductions), Oxford: Oxford University Press, 2018; xvi + 175 pp.
    Turkish translation by Toros Öztürk, *Fizik: Beşinci Elementten Kuarklara Kısa Bir Tarih*, Istanbul: Koç Üniversitesi Yayınları, 2017; 191 pp.

23. *Niels Bohr: A Very Short Introduction* (Very Short Introductions), Oxford: Oxford University Press, 2020; xxii + 132 pp.

    Chinese translation by Lü Yingqi and An Junhong, 尼尔斯·玻尔传, Lanzhou: Lanzhou University Press, 2024; xi + 141 pp.

24. *The Ghost of Galileo in a Forgotten Painting from the English Civil War*, Oxford: Oxford University Press, 2021; vii + 501 pp.
25. *The Incomparable Monsignor: Francesco Bianchini's World of Science, History, and Court Intrigue*, Oxford: Oxford University Press, 2022; vi + 327 pp.
26. (Jim Baggott, JLH) *Quantum Drama: From the Bohr-Einstein Debate to the Riddle of Entanglement*, Oxford: Oxford University Press, 2024; x + 336 pp.

    French translation by Guillaume Merle, L'odyssée quantique. De la controverse entre Einstein et Bohr à l'énigme de l'intrication, Préface de Franck Laloë, Louvain-la-Neuve: De Boeck Supérieur, 2025; 336 pp.

Edited Volumes

1. Benjamin Franklin, *Briefe von der Elektrizität*, übersetzt und mit Anmerkungen versehen von Carl Wilcke, Braunschweig: Vieweg & Sohn, 1984; LI + 136 pp.
   "Einleitung", pp. XV-LI.
2. (Tore Frängsmyr, JLH, Robin E. Rider, eds.) *The Quantifying Spirit in the 18th Century*, Berkeley-Los Angeles-Oxford: University of California Press, 1990; VIII + 411 pp.
   (Tore Frängsmyr, JLH, Robin E. Rider) "Preface", pp. VII-VIII.
   Reprinted in the "UC Press Voices Revived" series, 2023.
3. *The Oxford Companion to the History of Modern Science*, Oxford-New York: Oxford University Press, 2003; XXX + 941 pp.
   "Preface", pp. VII-X.

   Japanese translation by Kuniyuki Sakurai, Hisashi Yamazaki, Ryuji Kimura, Masatoshi Mori and Noriko Kumura, supervised by Jirō Ōta, 現代科学史大百科事典, Tokyo: Asakura Shoten, 2014; XXIV + 909 pp.

4. *The Oxford Guide to the History of Physics and Astronomy*, Oxford-New York: Oxford University Press, 2005.
   "Preface", pp. VII-IX.
5. *Advancements of Learning: Essays in Honour of Paolo Rossi* (Biblioteca di *Nuncius*. Studi e testi, 62), Florence: Leo S. Olschki, 2007; VIII + 275 pp.
   "Preface", pp. VII-VIII.
6. *The Oxford Illustrated Companion to the History of Modern Science*, Black Dog & Leventhal, 2008.
   "Preface to the Illustrated Edition", pp. V-VI.
7. Niels Bohr, *L'estructura i la filosofia dels àtoms* (Clàssics de la Ciència, 10), Catalan translation by Xavier Roqué, Barcelona: Institut d'Estudis Catalans / Eumo Editorial, 2010; LXXXII + 177 pp.
   "Part primera", in four sections: "Prefaci", pp. IX-XI; "La paradoxa inicial", pp. XIII-XXXI; "Èxits i limitacions", pp. XXXIII-LI; "La paradoxa final", pp. LIII-LXX; "Bibliografia", pp. LXXI-LXXXII.
8. (JLH, René Sigrist, eds.) *Jean-André Deluc: Historian of Earth and Man*, Geneva: Éditions Slatkine, 2011; 350 pp.
   (JLH, René Sigrist) "Introduction", pp. 11–18.

Other Editorial Capacities

1. Founder and Editor, "Berkeley Papers in the History of Science" series, Berkeley: Office for History of Science and Technology, 1977–1994.
2. Editor, *Historical Studies in the Physical Sciences*, Vol. 11, No. 1, 1980 – Vol. 15, No. 2, 1985.
3. Editor, *Historical Studies in the Physical and Biological Sciences*, Vol. 16, No. 1, 1986 – Vol. 37, No. 2, March 2007.
4. Subject Editor (physics), *Dictionary of the History of Science*, edited by William F. Bynum, E. Janet Browne, Roy Porter, London-Basingstoke: The Macmillan Press, 1981; revised edition, 1983.

5. Member of the Editorial Board, *Dictionary of Scientific Biography: Supplement II*, Frederic L. Holmes editor in chief, Vol. XVII: *Leason Heberling Adams–Fritz H. Laves*, and Vol. XVIII: *Aleksandr Nikolaevich Lebedev–Fritz Zwicky*, New York: Charles Scribner's Sons, 1990.
6. Editor, "California Studies in the History of Science" series, Berkeley-Los Angeles-London: University of California Press, 1989–1996.
7. Scientific Coordinator, *Storia della scienza*, Sandro Petruccioli editor in chief, Vol. 6: *L'Età dei Lumi*, Rome: Istituto della Enciclopedia Italiana, 2002.
8. Scientific Coordinator (physics, chemistry, technology), *Storia della scienza*, Sandro Petruccioli editor in chief, Vol. 8: *La seconda rivoluzione scientifica*, Rome: Istituto della Enciclopedia Italiana, 2004.

Essays

1. "A propos de l'invention de la bouteille de Leyde", *Revue d'histoire des sciences et de leurs applications*, Vol. 19, No. 2, Avril-Juin 1966, pp. 133–142.
2. "G. M. Bose: The Prime Mover in the Invention of the Leyden Jar?", *Isis*, Vol. 57, No. 2, Summer 1966, pp. 264–267.
3. "The Work of H. G. J. Moseley", *Isis*, Vol. 57, No. 3, Autumn 1966, pp. 336–364.

   Reprinted in [Monographs, 7 (1981)], pp. 230–258.
   See also [Notes, comments and replies, 1].

4. "The Kossel-Sommerfeld Theory and the Ring Atom", *Isis*, Vol. 58, No. 4, Winter 1967, pp. 450–485.

   Reprinted in [Monographs, 7 (1981)], pp. 260–295.

5. "The Scattering of α and β Particles and Rutherford's Atom", *Archive for History of Exact Sciences*, Vol. 4, No. 4, 21 February 1968, pp. 247–307.

   Reprinted in [Monographs, 7 (1981)], pp. 85–145.

6. "Quantum Historiography and the Archive for History of Quantum Physics", *History of science*, Vol. 7, No. 1, March 1968, pp. 90–111.
7. (JLH, Thomas S. Kuhn) "The Genesis of the Bohr Atom", *Historical Studies in the Physical Sciences*, Vol. 1, 1969, pp. vi, 211–290.

   Reprinted in Bobbs-Merrill Reprint Series in the History of Science (Indianapolis: The Bobbs-Merrill Company, 1969), HS-28.
   Reprinted in [Monographs, 7 (1981)], pp. 149–228.

8. "Honoré Fabri, S.J., and the Accademia del Cimento", in *XIIe Congrès International d'Histoire des Sciences, Paris 1968: Actes*, Vol. III B: *Science et Philosophie: XVIIe et XVIIIe siècles*, Paris: Librairie Scientifique et Technique Albert Blanchard, 1971, pp. 45–49.
9. "Robert Symmer and the Two Electricities", *Isis*, Vol. 67, No. 1, March 1976, pp. 7–20.

10. "Franklin's Physics", *Physics Today*, Vol. 29, No. 7, July 1976, pp. 32–37.

    Reprinted in Spencer R. Weart, Melba Phillips (eds.), *History of Physics: Readings from Physics Today, Number Two*, New York: American History of Physics, 1985, pp. 12–17.

11. "Lectures on the History of Atomic Physics, 1900–1922", in Charles Weiner (ed.), *History of Twentieth Century Physics / Storia della fisica del XX secolo: Proceedings of the International School of Physics 'Enrico Fermi'. Course LVII*, New York etc.: Academic Press, 1977, pp. 40–108.

    Reprinted in [Monographs, 7 (1981)], pp. 14–82.

12. "J. J. Thomson and the Bohr Atom", *Physics Today*, Vol. 30, No. 4, April 1977, pp. 23–30.

    Reprinted in Spencer R. Weart, Melba Phillips (eds.), *History of Physics: Readings from Physics Today, Number Two*, New York: American Institute of Physics, 1985, pp. 303–309.

13. "Franklin, Haller, and Franklinist History", *Isis*, Vol. 68, No. 4, December 1977, pp. 539–549.
14. "Éloge: Victor Fritz Lenzen, 1890–1975", *Isis*, Vol. 68, No. 4, December 1977, pp. 598–600.
15. "Introductory Essay", in [Monographs, 5 (1978)], pp. 1–99.

    Excerpted in *Literature Criticism from 1400 to 1800*, Vol. 20, edited by James E. Person, Jr., Detroit: Gale Research, 1993, pp. 30–36.

16. "Volta's Path to the Battery", in George Dubpernell, Jack H. Westbrook (eds.), *Proceedings of the Symposium on Selected Topics in the History of Electrochemistry*, Princeton, NJ: The Electrochemical Society, 1978, pp. 39–65.
17. "Physics at McGill in Rutherford's Time", in Mario Bunge, William R. Shea (eds.), *Rutherford and Physics at the Turn of the Century*, New York: Dawson and Science History Publications, 1979, pp. 42–73.
18. "La fisica negli Stati Uniti subito prima della meccanica quantistica", Italian translation by Elisabetta Donini, in *Fisica & società negli anni '20*, Milan: CLUP-CLUED, 1980, pp. 135–158.
19. "Experimental Natural Philosophy", in George S. Rousseau, Roy Porter (eds.), *The Ferment of Knowledge: Studies in the Historiography of Eighteenth-Century Science*, Cambridge-New York: Cambridge University Press, 1980, pp. 357–387.
20. "Rutherford-Bohr Atom", *American Journal of Physics*, Vol. 49, No. 3, March 1981, pp. 223–231.
21. (JLH, Robert W. Seidel, Bruce R. Wheaton) "A Strong Interaction between Science and Society: 1931–1981", *CERN Courier*, Vol. 21, No. 8, October 1981, pp. 335–346.

22. (JLH, Robert W. Seidel, Bruce R. Wheaton) "Lawrence and His Laboratory: Nuclear science at Berkeley, 1931–1961", *LBL Newsmagazine*, Vol. 6, No. 3, Fall 1981, pp. 2–106.

    Reprinted in book form, Berkeley: Office for History of Science and Technology, 1981; 106 pp.

23. "The Electrical Field Before Faraday", in Geoffrey N. Cantor, Michael J. S. Hodge (eds.), *Conceptions of Ether: Studies in the History of Ether Theories, 1740-1900*, Cambridge etc.: Cambridge University Press, 1981, pp. 187–213.
24. "*Fin-de-siècle* Physics", in Carl G. Bernhard, Elisabeth Crawford, Per Sörbom (eds.), *Science, Technology, and Society in the Time of Alfred Nobel*, Oxford etc.: Pergamon Press, 1982, pp. 51–73.
25. "The Origins of the Exclusion Principle", *Historical Studies in the Physical Sciences*, Vol. 13, No. 2, 1983, pp. 261–310.
26. "The Virtual Oscillator as a Guide to Physics Students Lost in Plato's Cave", in Fabio Bevilacqua, Peter J. Kennedy (eds.), *Using History of Physics in Innovatory Physics Education*, Pavia: Centro Studi per la Didattica della Facoltà di Scienze Matematiche, Fisiche e Naturali, 1983, pp. 162–182.

    Reprinted in *Science & Education*, Vol. 3, No. 2, April 1994, pp. 177–188.

27. "La radio e il ciclotrone", Italian translation by Lanfranco Belloni, *Museoscienza*, Vol. 22, 1983, pp. 13–24.
28. "La ristrutturazione della fisica fra le due guerre", Italian translation by Silvio Bergia, in Giovanni Battimelli, Michelangelo De Maria, Arcangelo Rossi (eds.), *La ristrutturazione delle scienze tra le due guerre mondiali*, Vol. 1: *L'Europa*, Rome: La Goliardica, 1984, pp. 21–56.
29. "Bohr's First Theories of the Atom", *Physics Today*, Vol. 38, No. 10, October 1985, pp. 28–36.

    Reprinted in Anthony P. French, Peter J. Kennedy (eds.), *Niels Bohr: A Centenary Volume*, Cambridge, MA-London: Harvard University Press, 1985, pp. 33–49.

30. "The Earliest Missionaries of the Copenhagen Spirit", *Revue d'histoire des sciences*, Vol. 38, Nos. 3-4, Juillet-Décembre 1985, pp. 195–230.

    Reprinted in Edna Ullmann-Margalit (ed.), *Science in Reflection – The Israel Colloquium: Studies in History, Philosophy, and Sociology of Science*, Vol. 3, Dordrecht-Boston-London: Kluwer Academic Publishers, 1988, pp. 201–233.

31. (Elisabeth Crawford, JLH, John G. May, Rebecca Ullrich) "Les élites scientifiques internationales: Physique et chimie dans le premier tiers du 20e siècle", *Vingtième Siècle: Revue d'histoire*, No. 12, Octobre-Décembre 1986, pp. 97–100.

32. "The First European Cyclotrons", *Rivista di storia della scienza*, Vol. 3, No. 1, March 1986, pp. 1–44.
33. "Applied History of Science", *Isis*, Vol. 78, No. 4, December 1987, pp. 552–563.

    Italian translation by Annamaria Tagliavini, "Storia applicata della scienza", *Intersezioni*, Vol. 8, No. 1, Aprile 1988, pp. 5–21.

34. (JLH, Daniel J. Kevles) "Science and Technology in U. S. History Textbooks: What's There–and What Ought to Be There", *Reviews in American History,* Vol. 16, No. 2, June 1988, pp. 173–185.

    Reprinted in *The History Teacher*, Vol. 21, No. 4, August 1988, pp. 425–438.

35. (JLH, Daniel J. Kevles) "Finding a Policy for Mapping and Sequencing the Human Genome: Lessons from the History of Particle Physics", *Minerva*, Vol. 26, No. 3, September 1988, pp. 299–314.
36. "Science in the Church", *Science in Context*, Vol. 3, No. 1, Spring 1989, pp. 9–28.
37. "The Politics of the Meter Stick", *American Journal of Physics*, Vol. 57, No. 11, November 1989, pp. 988–992.
38. (Gad Freudenthal, JLH) "Éloge: Joseph Ben-David, 19 August 1920–12 January 1986", *Isis*, Vol. 80, No. 4, December 1989, pp. 659–663.
39. (Gian-Paolo Biasin, JLH, Owen Chamberlain, A. Carl Helmholz, Eugene D. Commins) "Emilio G. Segrè, 1905–1989. Professor of Physics, Emeritus", *University of California: In Memoriam*, 1989, pp. 145–151.
40. "The Detection of the Anti-Proton", in Michelangelo De Maria, Mario Grilli, Fabio Sebastiani (eds.), *The Restructuring of Physical Sciences in Europe and the United States 1945–1960*, Singapore etc.: World Scientific, 1989, pp. 161–217.
41. "An Historian's Interest in Particle Physics", in Laurie M. Brown, Max Dresden, Lillian Hoddeson, with the assistance of May West (eds.), *Pions to Quarks: Particle Physics in the 1950s*, Cambridge-New York: Cambridge University Press, 1989, pp. 47–54.
42. (Elisabeth Crawford, JLH) "Die Kaiser-Wilhelm-Institute für Grundlagenforschung und die Nobelinstitution", German translation by Norma Feldman-von Ragenfeld, in Rudolf Vierhaus, Bernhard vom Brocke (eds.), *Forschung im Spannungsfeld von Politik und Gesellschaft. Geschichte und Struktur der Kaiser-Wilhelm-/Max-Planck-Gesellschaft. Aus Anlaß 75-jährigen Bestehens*, Stuttgart: Deutsche Verlags-Anstalt, 1990, pp. 835–857.
43. "Introductory Essay", in [Edited volumes, 2 (1990)], pp. 1–23.
44. "The Measure of Enlightenment", in [Edited volumes, 2 (1990)], pp. 207–242.
45. "The Contributions of Bologna to Galvanism", *Historical Studies in the Physical and Biological Sciences*, Vol. 22, No. 1, 1991, pp. 57–85.

46. "Fisica e astronomia nel Settecento", Italian translation by Libero Sosio, in William R. Shea (ed.), *Storia delle scienze*, Vol. 2: *Le scienze fisiche e astronomiche*, Milan: Arnoldo Mondadori, 1991, pp. 318–443.

    Reprinted, Turin: Giulio Einaudi, 1992.

47. "Creativity and Big Science", *Physics Today*, Vol. 45, No. 11, November 1992, pp. 42–47.

    See also [Notes, comments and replies, 15].

48. "A Mathematicians' Mutiny, with Morals", in Paul Horwich (ed.), *World Changes: Thomas Kuhn and the Nature of Science*, Cambridge, MA: The MIT Press, 1993, pp. 81–129.

    Reprinted, Pittsburgh, PA: University of Pittsburgh Press, 2010.

49. "Franklin As an Enlightened Natural Philosopher", in J. A. Leo Lemay (ed.), *Reappraising Benjamin Franklin: A Bicentennial Perspective*, Newark: University of Delaware Press, and London-Toronto: Associated University Presses, 1993, pp. 196–220.
50. "Two Previous Standard Models", *Beam Line*, Vol. 23, No. 2, Summer 1993, pp. 8–14.

    Reprinted in slightly revised form in Lillian Hoddeson, Laurie Brown, Michael Riordan, Max Dresden (eds.), *The Rise of the Standard Model: Particle Physics in the 1960s and 1970s*, Cambridge-New York: Cambridge University Press, 1997, pp. 45–54.

51. "Meridiane and Meridians in Early Modern Science", in Piers Bursill-Hall, *R. J. Boscovich: Vita e attività scientifica / His Life and Scientific Work*, Rome: Istituto della Enciclopedia Italiana, 1993, pp. 385–406.
52. "Science in the American Research University", in Giuliano Pancaldi (ed.), *Le università e le scienze. Prospettive storiche e attuali / Universities and the Sciences: Historical and Contemporary Perspectives*, Bologna: Centro internazionale per la storia delle università, della scienza e delle istituzioni di alta cultura / International Centre for the History of Universities and Science, Università di Bologna, 1993, pp. 121–131.

    Italian translation by Paola Govoni, "La scienza nell'università di ricerca americana", *ibidem*, pp. 132–141.

53. "Il mercanteggiamento della scienza", Italian translation by Heinz Winter and Elio Sindoni, in Domenico Galbiati, Padre Eligio, Renato A. Ricci, Gioachino Rigamonti, Elio Sindoni (eds.), *Scienza ed etica alle soglie del terzo millennio*, Bologna: Società Italiana di Fisica, 1993, pp. 105–114.
54. "Some Uses for Catalogues of Old Scientific Instruments", in Robert G. W. Anderson, James A. Bennett, William F. Ryan (eds.), *Making Instruments Count: Essays on Historical Scientific Instruments presented to Gerard*

L'Estrange Turner, Aldershot (UK): Variorum, and Brookfield, VT: Ashgate, 1993, pp. 1–16.

55. "The Affair of the Countess Görlitz", *Proceedings of the American Philosophical Society*, Vol. 138, No. 2, June 1994, pp. 284–316.
56. "[Untitled Address at Opening]", *Annual Report, Max-Planck-Institut für Wissenschaftsgeschichte / Max Planck Institute for the History of Science*, 1995, pp. 183–201.
57. "Remarks on the Writing of Biography", in Ramesh S. Krishnamurthy, in association with Clifford S. Mead, Mary Jo Nye, assisted by Sean C. Goodlett, Mavin E. Kirk (eds.), *The Pauling Symposium: A Discourse on the Art of Biography*, Corvallis, OR: Oregon State University Libraries, 1996, pp. 230–238.
58. "X-Rays in the History of Science", *Physica Scripta*, Vol. T61, January 1996, pp. 60–66.

    Reprinted in *Kosmos*, 73, 1996, pp. 17–30.

59. "Churches as Scientific Instruments", *Bulletin of the Scientific Instrument Society*, No. 48, March 1996, pp. 4–9.

    Reprinted in *Universitas: Newsletter of the International Centre for the History of Universities and Science*, No. 9, April 1996, pp. 1–12.
    French translation by Julie Brumberg-Chaumont, "Des églises, instruments de science", *Alliage*, no. 32, automne 1997, pp. 31–40.

60. "An Institute for the History of Science", *Lychnos: Årsbok för Idé- och Lärdomshistoria*, 1996, pp. 9–24.
61. "Civility in Physics California between the Wars", *La Lettre de la Maison Française d'Oxford*, No. 7, Trinity Term, 1997, pp. 150–158.
62. "I detriti e la scienza. Leggi della natura e leggi dell'uomo nell'Inghilterra del XVIII secolo", Italian translation by Marco Bresadola, in Antonio Santucci (ed.), *L'età dei Lumi. Saggi sulla cultura settecentesca*, Bologna: il Mulino, 1998, pp. 49–67.
63. "Thomas Samuel Kuhn, 18 July 1922–17 June 1996", *Isis*, Vol. 89, No. 3, September 1998, pp. 505–515.

    Abridged version, "Thomas Samuel Kuhn (18 July 1922–17 June 1996)", *Proceedings of the American Philosophical Society*, Vol. 142, No. 4, December 1998, pp. 677–686.
    Catalan translation by Josep Batalla, "Thomas Samuel Kuhn. 18 de juliol del 1922 – 17 de juny del 1996", in Thomas S. Kuhn, *L'estructura de les revolucions científiques*, Santa Coloma de Queralt: Obrador Edèndum, 2007, pp. 21–40.

64. "Galvani, Volta, and the Uses of Centennials", in Marco Bresadola, Giuliano Pancaldi (eds.), *Luigi Galvani International Workshop: Proceedings*, Bologna: Dipartimento di Filosofia / Centro Internazionale per la Storia delle Università e della Scienza, 1999, pp. 17–32.

65. "Lessons from Teaching History of Science", in Claude Debru (ed.), *History of Science and Technology in Education and Training in Europe*, Luxembourg: Office for Official Publications of the European Communities, 1999, pp. 175–190.

 Czech translation by Jaroslav Folta and Hana Barvíková, "Ponaučení z výuky dějinám věd", *Dějiny věd a techniky*, Vol. 32, No. 1, 1999, pp. 5–18.

66. "The Sun in the Church", *The Sciences*, Vol. 39, No. 5, September-October 1999, pp. 29–35.
67. "Analogy in Volta's Exact Natural Philosophy", in Fabio Bevilacqua, Lucio Fregonese (eds.), *Nuova Voltiana: Studies on Volta and His Times*, Vol. 1, Milan: Ulrico Hoepli, 2000, pp. 1–23.
68. "Domesticating Science in the Eighteenth Century", in William R. Shea (ed.), *Science and the Visual Image in the Enlightenment*, Canton, MA: Science History Publications, 2000, pp. 1–24.
69. "Some Connections among the Heroes", *Revue d'histoire des sciences*, Vol. 54, No. 1: *L'électricité dans ses premières grandeurs (1760–1820)*, Janvier-Mars 2001, pp. 11–28.
70. "Physik und Physikgeschichte in Göttingen um 1800" (German translation by Wolfgang Böker), *Georgia Augusta: Nachrichten aus der Universität Göttingen*, Vol. 74, Mai 2001, pp. 45–58.

 Original English version, "Physics and its History at Göttingen around 1800", in Nicolaas Rupke (ed.), *Göttingen and the Development of the Natural Sciences*, Göttingen: Wallstein, 2002, pp. 50–71.

71. "The Duomo as a Big Science Facility", in Timothy Verdon, Annalisa Innocenti (eds.), *Atti del VII centenario del Duomo di Firenze*, Vol. 2: *La cattedrale come spazio sacro. Saggi sul Duomo di Firenze*, Part 2, Florence: EDIFIR, 2001, pp. 643–660.
72. "History in Science Education, with Cautionary Tales about the Agreement of Measurement and Theory", in Fabio Bevilacqua, Enrico Giannetto, Michael R. Matthews (eds.), *Science Education and Culture: The Contribution of History and Philosophy of Science*, Dordrecht-Boston-London: Kluwer Academic Publishers, 2001, pp. 5–15.

 Reprinted in *Science & Education*, Vol. 11, No. 4, July 2002, pp. 321–331.

73. "The Nobel Science Prizes of World War I", in Elisabeth Crawford (ed.), *Historical Studies in the Nobel Archives: The Prizes in Science and Medicine*, Tokyo: Universal Academy Press, 2002, pp. 19–38.
74. "L'Età dei Lumi: tempi, luoghi e modi" (Introduction), in *Storia della scienza*, Sandro Petruccioli editor in chief, Vol. VI: *L'Età dei Lumi*, Rome: Istituto della Enciclopedia Italiana, 2002, pp. 1–28, 785–786.
75. "La fisica matematica" (Ch. XIII), *ibidem*, pp. 199–216, 792.
76. "'*Naturphilosophie*'" (Ch. XV.2), *ibidem*, pp. 237–240, 793.

77. "Figuras sobre un fondo romántico. Representantes de las ciencias físicas en Göttingen en la década de 1790" (Spanish translation by José Ferreirós), in José Montesinos, Javier Ordóñez, Sergio Toledo (eds.), *Ciencia y Romanticismo*, La Orotava: Fundación Canaria Orotava de historia de la ciencia, 2003, pp. 185–206.
78. "Experimental Nuclear Physics in the Thirties and Forties", in *Proceedings of the International Conference "Enrico Fermi and the Universe of Physics", Rome, September 29 – October 2, 2001*, Rome: Accademia Nazionale dei Lincei/Istituto Nazionale di Fisica Nucleare, 2003, pp. 341–360.
79. "Science as a Subject of History", in *Actes de la VII trobada d'història de la ciència i de la tècnica*, Barcelona: SCHCT, 2003, pp. 33–42.
80. "La fisica e la chimica" (Ch. XLI), in *Storia della scienza*, Sandro Petruccioli editor in chief, Vol. VIII: *La seconda rivoluzione scientifica*, Rome: Istituto della Enciclopedia Italiana, 2004, pp. 313–333, 931–932.
81. "Fisica e chimica dei raggi" (Ch. XLVIII), *ibid*em, pp. 400–408, 934.
82. "La struttura atomica" (Ch. L), *ibid*em, pp. 413–422, 934.
83. "La guerra e i fisici" (Ch. LV), *ibid*em, pp. 456–463, 935.
84. "Censorship of Astronomy in Italy after Galileo", in Ernan McMullin (ed.), *The Church and Galileo*, Notre Dame, IN: University of Notre Dame Press, 2005, pp. 279–322.
85. "Bianchini as an Astronomer", in Valentin Kockel, Brigitte Sölch (eds.), *Francesco Bianchini (1662–1729) und die europäische gelehrte Welt um 1700*, Berlin: Akademie Verlag, 2005, pp. 57–82.
86. "Citoyen de Genève and Philosopher to the Queen of England", *Archives des sciences*, Vol. 58, No. 1, Mai 2005, pp. 75–92.
87. "Frederic Lawrence Holmes, 6 February 1932–27 March 2003", *Proceedings of the American Philosophical Society*, Vol. 149, No. 2, June 2005, pp. 255–260.
88. "Oppenheimer's Guru", in Cathryn Carson, David A. Hollinger (eds.), *Reappraising Oppenheimer: Centennial Studies and Reflections*, Berkeley: Office for History of Science and Technology, 2005, pp. 275–291.
89. "La meridiana di Bianchini e la politica estera di Clemente XI" (Italian translation by Piero Ranfagni), *Giornale di Astronomia*, Vol. 32, No. 1, Marzo 2006 (Atti del Convegno "Il Sole nella Chiesa: Cassini e le grandi meridiane come strumenti di indagina scientifica", Bologna, Archiginnasio, 22–23 Settembre 2005, edited by Fabrizio Bònoli, Gianluigi Parmeggiani, Francesco Poppi), pp. 46–53.
90. "Plus and Minus: Franklin's Zero-Sum Way of Thinking", *Proceedings of the American Philosophical Society*, Vol. 150, No. 4, December 2006, pp. 607–617.
91. "Historical Pointers to the Future of Academies", in Edoardo Vesentini, Leopoldo Mazzarolli (eds.), *L'esperienza delle Accademie e la vita morale e civile dell'Europa*, Venezia: Istituto Veneto di Scienze, Lettere ed Arti, 2006, pp. 123–136.

92. "Max Planck's Compromises on the Way to and from the Absolute", in James Evans, Alan S. Thorndike (eds.), *Quantum Mechanics at the Crossroads: New Perspectives from History, Philosophy and Physics*. Berlin-Heidelberg: Springer, 2007, pp. 21–37.
93. "On Connoisseurship", in Kostas Gavroglu, Jürgen Renn (eds.), *Positioning the History of Science*, Dordrecht: Springer, 2007, pp. 73–76.
94. "Jean-André Deluc and the Fight for Bacon around 1800", in [Edited volumes, 5 (2007)], pp. 77–99.
95. "Benjamin Franklin in Europe: Electrician, Academician, Politician", *Notes & Records of the Royal Society*, Vol. 61, No. 3, 22 September 2007, pp. 353–373.
96. "Ciencia a la sombra del Vaticano", Spanish translation by Lara García Quintás, in José Montesinos, Sergio Toledo (eds.), *Ciencia y religión en la edad moderna*, La Orotava: Fundación Canaria Orotava de historia de la ciencia, 2007, pp. 195–215.
97. "Francesco Bianchini, Historian: In Memory of Amos Funkenstein", in Robert S. Westman, David Biale (eds.), *Thinking Impossibilities: The Intellectual Legacy of Amos Funkenstein*, Toronto: University of Toronto Press, 2008, pp. 227–277.
98. "Solomon's Houses: Francesco Bianchini's Academic Places", in Marco Beretta, Karl Grandin, Svante Lindqvist (eds.), *Aurora Torealis: Studies in the History of Science and Ideas in Honor of Tore Frängsmyr*, Sagamore Beach, MA: Science History Publications, 2008, pp. 51–70.
99. "Bianchini and Natural Philosophy", in Luca Ciancio, Gian Paolo Romagnani (eds.), *Unità del sapere molteplicità dei saperi: Francesco Bianchini (1662–1729) tra natura, storia e religione*, Verona: QuiEdit, 2010, pp. 33–73.
100. "A Short History of Light in the Western World", in Raymond Y. Chiao, Marvin L. Cohen, Anthony J. Leggett, William D. Phillips, Charles L. Harper, Jr. (eds.), *Visions of Discovery: New Light on Physics, Cosmology, and Consciousness*, Cambridge-New York: Cambridge University Press, 2011, pp. 3–23.
101. "Natural Philosophy", in Peter Harrison, Ronald L. Numbers, Michael H. Shank (eds.), *Wrestling with Nature: From Omens to Science*, Chicago-London: The University of Chicago Press, 2011, pp. 173–199.
102. "Cold War Culture, History of Science and Postmodernity: Engagement of an Intellectual in a Hostile Academic Environment", in Cathryn Carson, Alexei Kojevnikov, Helmuth Trischler (eds.), *Weimar Culture and Quantum Mechanics: Selected Papers by Paul Forman and Contemporary Perspectives on the Forman Thesis*, London: Imperial College Press and Singapore: World Scientific Publishing, 2011, pp. 7–20.
103. "Historian of Nature and Mankind", in [Edited volumes, 8 (2011)], pp. 82–103.
104. "Savior of Science and Society", in [Edited volumes, 8 (2011)], pp. 186–240.
105. "The *Bizzarrie* of the *Dialogo*: Myth, Marvel, and Make-Believe in Galileo's Force-Free Physics", *Galilæana*, Vol. 9, 2012, pp. 29–64.

106. "The First Solvay Council, 'A Sort of Private Conference'", in David J. Gross, Marc Henneaux, Alexander Sevrin (eds.), *The Theory of the Quantum World: Proceedings of the 25th Solvay Conference on Physics*, Singapore: World Scientific, 2013, pp. 1–16.
107. "Galileo Puppeteer", in Andrea Albrecht, Giovanna Cordibella, Volker R. Remmert (eds.), *Tintenfass und Teleskop. Galileo Galilei im Schnittpunkt wissenschaftlicher, literarischer und visueller Kulturen im 17. Jahrhundert*, Berlin-Boston: Walter de Gruyter, 2014, pp. 33–49.
108. "The Mind that Created the Bohr Atom", in Olivier Darrigol, Bertrand Duplantier, Jean-Michel Raimond, Vincent Rivasseau (eds.), *Niels Bohr, 1913–2013: Poincaré Seminar 2013*, Cham (Switzerland): Springer, 2013, pp. 55–101.
109. "Was There a Scientific Revolution?", in Jed Z. Buchwald, Robert Fox (eds.), *The Oxford Handbook of the History of Physics*, Oxford: Oxford University Press, 2013, pp. 7–24.

See also [Monographs, 19]

110. "British Participation in the First Solvay Councils on Physics", *The European Physical Journal. Special Topics*, Vol. 224, No. 10, September 2015: *The Early Solvay Councils and the Advent of the Quantum Era*, pp. 2041–2055.
111. "Boscovich in Britain", in Theodore Arabatzis, Jürgen Renn, Ana Simões (eds.), *Relocating the History of Science: Essays in Honor of Kostas Gavroglu*, Cham (Switzerland) etc.: Springer, 2015, pp. 99–116.
112. "Myth 5: That Galileo Publicly Refuted Aristotle's Conclusions about Motion by Repeated Experiments Made from the Campanile of Pisa", in Ronald L. Numbers, Kostas Kampourakis (eds.), *Newton's Apple and Other Myths about Science*, Cambridge, MA-London: Harvard University Press, 2015, pp. 40–47.
113. "'My Courage is Ablaze so Wildly': Niels Bohr *en route* to His Quantum Atom", in Finn Aaserud, Helge Kragh (eds.), *One Hundred Years of the Bohr Atom: Proceedings from a Conference*, Copenhagen: Det Kongelige Danske Videnskabernes Selskab, 2015, pp. 27–50.
114. "A proposito di *Libertà di filosofare* in naturalibus. *I mondi paralleli di Cesi e Galileo* di Paolo Galluzzi", Part I: "Galileo and the Tides", *Quaderni storici*, Nuova Serie, Vol. 150, No. 3, dicembre 2015, pp. 873–884.
115. "Where to Start?", in Alexander Blum, Kostas Gavroglu, Christian Joas, Jürgen Renn (eds.), *Shifting Paradigms: Thomas S. Kuhn and the History of Science*, Berlin: Max Planck Institute for the History of Science, 2016, pp. 3–13.
116. "Benedict XIV and the Natural Sciences", in Rebecca Messbarger, Christopher M. S. Johns, Philip Gavitt (eds.), *Benedict XIV and the Enlightenment. Art, Science, and Spirituality*, Toronto: University of Toronto Press, 2016, pp. 177–205.

117. "Accounts of Moseley and Versions of His Laws", in Roy MacLeod, Russell G. Egdell, Elizabeth Bruton (eds.), *For Science, King & Country: The Life and Legacy of Henry Moseley*, London: Unicorn, 2018, pp. 224–243.
118. "Tore Frängsmyr (1938–2017)", *Isis*, Vol. 109, No. 2, June 2018, pp. 356–359.

    Reprinted as "Tore Frängsmyr, 8 July 1938–28 August 2017", *Proceedings of the American Philosophical Society*, Vol. 162, No. 3, September 2018, pp. 313–319.

119. "History of Science or History of Learning", *Berichte zur Wissenschaftsgeschichte*, Vol. 42, Nos. 2–3, September 2019 (special issue: *History of Science or History of Knowledge?*, edited by Christian Joas, Dominik Knaupp, Fabian Krämer, Kärin Nickelsen), pp. 200–219.
120. "Myth 1: That Myths Are Simple Falsehoods", in Kostas Kampourakis (ed.), *Darwin Mythology: Debunking Myths, Correcting Falsehoods*, Cambridge-New York: Cambridge University Press, 2023, pp. 14–23.
121. (JLH, Carlo Rovelli) "Matrix Mechanics Mis-Prized: Max Born's Belated Nobelization", *The European Physical Journal H*, Vol. 48, No. 11, 2023 (online).
122. "Feyerabend, Bohr and Quantum Physics", in Stefano Gattei, Roberta Corvi (eds.), *Feyerabend in Dialogue: Critical Essays*, Cham (Switzerland): Springer, 2024, pp. 109–129.
123. "Music and Motion", in Natacha Fabbri, Ferdinando Abbri (eds.), *Vincenzo Galilei: The Renaissance Dialogue between Music and Science*, Florence: Leo S. Olschki, 2025, pp. 3–24.

Popular Articles

1. "Millenarister och hundraårsfirare [Millenarians and Centenarians]", Swedish translation by Caj Lundgren, *Svenska Dagbladet*, 8 January 1995, p. 4.
2. "Vetenskapsmän som tankepoliser [Scientists as Thought Police]", *Svenska Dagbladet*, 28 May 1995, p. 4.
3. "Strålarna som satte fart på [The Rays that Set Things in Motion]", *Forskning och framsteg*, Vol. 30, No. 7, 1995, pp. 36–41.
4. "La douleur en 'moins' et le plaisir en 'plus'", *Les Cahiers de Science & Vie*, hors série, No. 28, août 1995, pp. 38–47.
5. "Laddad upptäckt [Charged Discovery]", *Forskning & framsteg*, Vol. 32, No. 4, 1997, pp. 46–50.
6. "Les églises, instruments de science", *La Recherche*, no. 307, 1 mars 1998, pp. 78–83.

    Reprinted in *Dieu: la science et la religion*, special issue of *La Recherche*, hors série, no. 14, janvier-mars 2004, pp. 68–73.

7. "Marie Curie: 100 år sedan upptäckten av radium [Marie Curie: 100 years since the Discovery of Radium]", *Forskning & framsteg*, Vol. 34, No. 7, 1999, pp. 46–50.

8. “La découverte du micromonde”, *Les Cahiers de Science & Vie*, No. 52, août 1999, pp. 22–30.
9. “Konflikt med Galvani tände Volta [Conflict with Galvani sparked Volta’s Interest]”, *Forskning & framsteg*, Vol. 35, No. 4, 2000, pp. 32–36.
10. “Mathématiques pour dames… et militaires”, French translation by Nicholas Witkowski, *La Recherche*, no. exceptionnel “Jeux mathématiques”, mai-juin 2000, pp. 8–10.

    Reprinted in *Tangente*, no. 74, 2000, pp. 8–10.

11. “Un prix «Haberrant»”, French translation by Guillaume Orvas, *Les Cahiers de Science et Vie*, No. 60, décembre 2000, pp. 52–55.
12. “Quand la science investit les cathédrales”, French translation by Alice Rolland, *Les Cahiers de Science & Vie*, No. 62, avril 2001, pp. 24–30.
13. “Max Planck – en motvillig revolutionär [Max Planck – A Reluctant Revolutionary]”, *Forskning & framsteg*, Vol. 37, No. 3, 2002, pp. 42–47.
14. “John Dalton & atomen [John Dalton and the Atom]”, *Forskning & framsteg*, Vol. 3, No. 7, 2003, pp. 40–43.
15. “Benjamin Franklin: Elektriker och revolutionär [Benjamin Franklin: Electrician and Revolutionary]”, *Forskning & framsteg*, Vol. 41, No. 1, 2006, pp. 50–55.

Reviews and Essay Reviews

1. Review of J. A. Leo Lemay, *Ebenezer Kinnersley, Franklin’s Friend* (Philadelphia: University of Pennsylvania Press, 1964), *Isis*, Vol. 56, No. 3, Autumn 1965, pp. 387–388.
2. “Gilbert in Mundane Garb”, review of William Gilbert, *De mundo nostro sub-lunari philosophia nova*, facsimile of the 1651 edition (Amsterdam: Menno Hertzberger, 1965); Suzanne Kelly, The *De mundo* of William Gilbert (Amsterdam: Menno Hertzberger, 1965), *Isis*, Vol. 57, No. 2, Summer 1966, pp. 274–275.
3. Review of René Taton (ed.), *History of Science: Science in the Twentieth Century*, English translation by Arnold J. Pomerans (New York: Basic Books, 1966), *Isis*, Vol. 58, No. 2, Summer 1967, pp. 283–284.
4. Review of William T. Scott, *Erwin Schrödinger: An Introduction to His Writings* (Amherst: University of Massachusetts Press, 1967), *Isis*, Vol. 59, No. 2, Summer 1968, pp. 242–243.
5. Review of Angelo Crescini, *Le origini del metodo analitico: Il Cinquecento* (Udine: Del Bianco, 1965), *Isis*, Vol. 59, No. 3, Autumn 1968, pp. 343–344.
6. Review of Basil Schonland, *The Atomists (1805–1933)* (New York: Oxford University Press, 1968), *The American Historical Review*, Vol. 74, No. 2, December 1968, pp. 545–546.
7. Review of Gerald J. Whitrow (ed.), *Einstein: The Man and His Achievement* (London: British Broadcasting Corporation, 1967), *The British Journal for the History of Science*, Vol. 4, No. 2, December 1968, pp. 195–196.

8. Review of David P. Wheatland, with the assistance of Barbara Carson, *The Apparatus of Science at Harvard 1765–1800* (Cambridge, MA: Harvard University Press, 1968), *Isis*, Vol. 60, No. 3, Autumn 1969, pp. 412–413.
9. "Yankee Science", review of George H. Daniels, *American Science in the Age of Jackson* (New York-London: Columbia University Press, 1968), *Trans action: Social Science and Modern Society*, Vol. 7, No. 5, March 1970, pp. 58–60.
10. Review of Maurice Clavelin, *La philosophie naturelle de Galilée. Essai sur les origines et la formation de la méchanique classique* (Paris: Armand Colin, 1968), *Journal of the History of Philosophy*, Vol. 8, No. 3, July 1970, pp. 341–343.
11. Review of Lawrence Badash (ed.), *Rutherford and Boltwood: Letters on Radioactivity* (New Haven: Yale University Press, 1969), *Technology and Culture*, Vol. 11, No. 3, July 1970, pp. 460–462.
12. "Contributions of a Controversialist", review of Conor Reilly, *Francis Line, S.J.: An Exiled English Scientist 1595–1675* (Rome: Institutum historicum Societatis Jesu, 1969), *Isis*, Vol. 61, No. 3, Autumn 1970, pp. 397–398.
13. Review of Robert J. Forbes (ed.), *Martinus Van Marum: Life and Work*, Vol. I (Haarlem: H. D. Tjeenk Willink & Zoon, 1971), *Isis*, Vol. 61, No. 4, Winter 1970, pp. 544–545.
14. "Ehrenfest's Earlier Years", review of Martin J. Klein, *Paul Ehrenfest*, Vol. 1: *The Making of a Theoretical Physicist* (Amsterdam: North-Holland, and New York: Elsevier, 1970), *Science*, New Series, Vol. 171, No. 3971, 12 February 1971, pp. 558–559.
15. Review of Max Born, *Physics in My Generation* (New York: Springer, 1969); Friedrich Herneck, *Bahnbrecher des Atomzeitalters: Grosse Naturforscher von Maxwell bis Heisenberg* (Berlin: Buchverlag der Morgen, 1970); Bernard Jaffe, *Moseley and the Numbering of the Elements* (New York: Doubleday, 1971); Robert Jungk, *The Big Machine*, English translation by Grace Marmor Spruch and Traude Wess (New York: Scribner's, 1970) *Isis*, Vol. 63, No. 1, March 1972, pp. 111–112.
16. Review of John Jewkes, David Sawers, Richard Stillerman, *The Sources of Invention: A Study of the Causes and Consequences of Industrial Innovation through the Inventions of the Nineteenth and Twentieth Centuries* (New York: W. W. Norton, 1971^2); Norman A. F. Smith, *Victorian Technology and Its Preservation in Modern Britain* (Leicester: Leicester University Press, and New York: Humanities Press, 1970); Illinois Institute of Technology Research Institute, *Technology in Retrospect and Critical Events in Science*, Vol. I (Chicago: Illinois Institute of Technology, 1968), *Isis*, Vol. 63, No. 1, March 1972, p. 115.
17. "Physics: Historico-criticism", review of John Bradley, *Mach's Philosophy of Science* (London: Athlone Press, 1971), *Science*, Vol. 176, No. 4032, 21 April 1972, pp. 271–272.

18. Review of Ronald W. Clark, *Einstein: The Life and Times* (New York: World publishing Company, 1971), *The American Historical Review*, Vol. 77, No. 3, June 1972, pp. 748–749.
19. Review of Herbert W. Meyer, *A History of Electricity and Magnetism* (Cambridge, MA: The MIT Press, 1971), *Isis*, Vol. 63, No. 4, December 1972, p. 572.
20. Review of Wolfgang Yourgrau, Alwayn van der Merwe (eds.), *Perspectives in Quantum Theory: Essays in Honor of Alfred Landé* (Cambridge, MA: The MIT Press, 1971), *Isis*, Vol. 63, No. 4, December 1972, p. 596.
21. Review of Loyd S. Swenson Jr., *The Ethereal Aether: A History of the Michelson-Morley-Miller Aether-Drift Experiments, 1880–1930*, Foreword by Gerald Holton (Austin: University of Texas Press, 1972), *The American Historical Review*, Vol. 78, No. 1, February 1973, p. 70.
22. Review of Armin Hermann, *The Genesis of Quantum Theory (1899–1913)*, English translation by Claude W. Nash (Cambridge, MA: The MIT Press, 1971), *Isis*, Vol. 64, No. 1, March 1973, pp. 132–133.
23. Review of Max Jammer, *Concepts of Space: The History of Theories of Space in Physics*, Foreword by Albert Einstein (Cambridge, MA: Harvard University Press, 1969^{2}), *Isis*, Vol. 64, No. 2, June 1973, p. 251.
24. Review of Michael S. Mahoney, *The Mathematical Career of Pierre de Fermat (1601–1665)* (Princeton, NJ: Princeton University Press, 1973), *History: Review of New Books*, Vol. 1, No 8, July 1973, p. 180.
25. Review of Victor F. Weisskopf, *Physics in the Twentieth Century: Selected Essays*, Foreword by Hans A. Bethe (Cambridge, MA: The MIT Press, 1972), *Isis*, Vol. 64, No. 4, December 1973, pp. 572–573.
26. Review of Alexandre Koyré, *The Astronomical Revolution: Copernicus—Kepler—Borelli*, translated by Robert E. W. Maddison (Ithaca, NY: Cornell University Press, 1973), *History: Reviews of New Books*, Vol. 2, No 3, January 1974, p. 73.
27. Review of Alan G. R. Smith, *Science and Society in the 16th and 17th Centuries* (New York: Science History Publications, 1972), *American Scientist*, Vol. 62, No. 1, January-February 1974, p. 122.
28. Review of Banesh Hoffmann, with the collaboration of Helen Dukas, *Albert Einstein Creator and Rebel* (New York: Viking, 1972), *Isis*, Vol. 65, No. 1, March 1974, pp. 128–129.
29. Review of Nathan Reingold (ed.), Stuart Pierson, Arthur P. Molella (ass. eds.), with the assistance of James M. Hobbins, John R. Kerwood, *The Papers of Joseph Henry*, Vol. I: *December 1797–October 1832: The Albany Years* (Washington, DC: Smithsonian Institution Press, 1972), *Journal of the History of Medicine and Allied Sciences*, Vol. 29, No. 2, April 1974, pp. 244–245.
30. Review of Ernst Benz, *Theologie der Elektrizität: Zur Begegnung und Auseinandersetzung von Theologie und Naturwissenschaft im 17. und 18. Jahrundert* (Wiesbaden: Franz Steiner, 1971), *Isis*, Vol. 65, No. 2, June 1974, p. 275.

31. Review of Abdus Salam, Eugene Wigner (eds.), *Aspects of Quantum Theory* (Cambridge-New York: Cambridge University Press, 1972), *Isis*, Vol. 65, No. 2, June 1974, pp. 292–293.
32. "Emotional Lines in Modern Physics", review of Lewis S. Feuer, *Einstein and the Generations of Science* (New York: Basic Books, 1974), *Science*, New Series, Vol. 185, No. 4153, 30 August 1974, pp. 777–779.
    See also [Notes, comments and replies, 3].
33. Review of William H. Brock, *H. E. Armstrong and the Teaching of Science, 1880–1930* (Cambridge-New York: Cambridge University Press, 1973), *Isis*, Vol. 66, No. 1, March 1975, p. 142.
34. Review of Dorothy Michelson Livingston, *The Master of Light: A Biography of Albert A. Michelson* (New York: Charles Scribner's Sons, 1973), *The British Journal for the History of Science*, Vol. 8, No. 2, July 1975, pp. 187–188.
35. "Arthur Holly Compton's Work", review of Roger H. Stuewer, *The Compton Effect: Turning Point in Physics* (New York: Science History Publications, 1975); *Scientific Paper of Arthur Holly Compton: X-Ray and Other Studies*, edited by Robert S. Shankland (Chicago: The University of Chicago Press, 1974), *Science*, New Series, Vol. 189, No. 4204, 29 August 1975, p. 711.
36. Review of Robert J. Forbes (ed.), *Martinus Van Marum: Life and Work*, Vol. III (Haarlem: H. D. Tjeenk Willink & Zoon, 1971), *Isis*, Vol. 66, No. 3, September 1975, p. 427.
37. Review of Frederick Soddy, *Radioactivity and Atomic Theory: Presenting Facsimile Reproduction of the Annual Progress Reports on Radioactivity 1904–1920 to the Chemical Society*, edited with commentary by Thaddeus J. Trenn (London: Taylor & Francis, 1975), *Ambix*, Vol. 22, No. 3, November 1975, p. 222.
38. Review of E. Scott Barr, *An Index to Biographical Fragments in Unspecialized Scientific Journals* (University, AL: The University of Alabama Press, 1973), *Isis*, Vol. 66, No. 4, December 1975, pp. 569–570.
39. Review of Silvio A. Bedini, *Thinkers and Tinkers: Early American Men of Science* (New York: Charles Scribner's Sons, 1975), *History: Reviews of New Books*, Vol. 4, No. 4, February 1976, pp. 75–76.
40. Review of Yehuda Elkana (ed.), *The Interaction between Science and Philosophy* (Atlantic Highlands, NJ: Humanities Press, 1974), *Annals of Science*, Vol. 33, No. 2, March 1976, pp. 213–215.
41. Review of David Knight, *Sources for the History of Science, 1660–1914* (Ithaca NY: Cornell University Press, 1975), *The Library Quarterly*, Vol. 46, No. 2, April 1976, p. 218.
42. Review of Maria Luisa Righini Bonelli, William R. Shea (eds.), *Reason, Experiment, and Mysticism in the Scientific Revolution* (New York: Science History Publications, 1975), *History: Reviews of New Books*, Vol. 4, No. 7, May-June 1976, pp. 149–150.
43. "Richard Courant", review of Constance Reid, *Courant in Göttingen and New York: The Story of an Improbable Mathematician* (New York: Springer,

1976), *Science*, New Series, Vol. 195, No. 4280, 25 February 1977, pp. 775–776.

44. Review of Ellen G. Gartrell, *Electricity, Magnetism and Animal Magnetism: A Checklist of Printed Sources, 1600–1850* (Wilmington, DE: Scholarly Resources, 1975), *Isis*, Vol. 68, No. 1, March 1977, pp. 124–125.
45. "The Upward Path", review of Harriet Zuckerman, *Scientific Elite: Nobel Laureates in the United States* (New York: The Free Press, and London: Collier Macmillan, 1977), *Science*, New Series, Vol. 196, No. 4291, 13 May 1977, pp. 754–755.
46. Review of Charles Webster, *The Great Instauration: Science, Medicine and Reform, 1626–1660* (London: Duckworth, and New York: Holmes & Meier, 1975), *Isis*, Vol. 68, No. 3, September 1977, pp. 485–487.
47. Review of Lawrence Badash, *Rutherford Correspondence Catalog* (New York: American Institute of Physics, 1974); Albert F. Gunns, Judith Goodstein, *Guide to the Robert Andrews Millikan Collection at the California Institute of Technology* (New York: American Institute of Physics, 1975), *Technology and Culture*, Vol. 18, No. 4, October 1977, pp. 749–750.
48. Review of Hans Kangro, *Early History of Planck's Radiation Law*, English translation by Ralph E. W. Maddison (London: Taylor & Francis, 1976), *Annals of Science*, Vol. 35, No. 2, March 1978, pp. 214–216.
49. "Researchers on Radioactivity", review of Thaddeus J. Trenn, *The Self-Splitting Atom: The History of the Rutherford-Soddy Collaboration* (London: Taylor & Francis, 1977), *Science*, New Series, Vol. 200, No. 4346, 9 June 1978, pp. 1143–1144.
50. "The Senior Bragg", review of Gwendolen M. Caroe, *William Henry Bragg, 1862–1942: Man and Scientist* (New York: Cambridge University Press, 1978), *Science*, New Series, Vol. 202, No. 4369, 17 November 1978, pp. 740–741.
51. "A Harvest of Mechanical Invention", review of Lynn White, *Medieval Religion and Technology: Collected Essays* (Berkeley: University of California Press, 1978), University Publishing.
52. "A Public Institution", review of Morris Berman, *Social Change and Scientific Organization: The Royal Institution, 1799–1844* (Ithaca, NY: Cornell University Press, 1978), *Science*, New Series, Vol. 203, No. 4380, 9 February 1979, pp. 537–538.
53. "Max Born", review of Max Born, *My Life: Recollections of a Nobel Laureate* (New York: Scribner, 1979), *Science*, New Series, Vol. 204, No. 4394, 18 May 1979, pp. 740–741.
54. Review of Willem D. Hackmann, *Electricity from Glass: The History of the Frictional Electrical Machine, 1600–1850*, Vol. 4 (Alphen aan der Rijn: Sijthoff & Noordhoff, 1978), *Annals of Science*, Vol. 37, No. 1, January 1980, pp. 116–117.
55. Review of Leonard G. Wilson, *Benjamin Silliman and His Circle: Studies on the Influence of Benjamin Silliman on Science in America* (New York: Science

History Publications, 1979), *History: Reviews of New Books*, Vol. 8, No. 4, February 1980, pp. 80–81.

56. "Letters Between Physicists", review of *Wolfgang Pauli: Wissenschaftlicher Briefwechsel mit Bohr, Einstein, Heisenberg, u.a.*, Vol. 1: *1919–1929*, edited by Armin Hermann, Karl von Meyenn, Victor F. Weisskopf (New York: Springer, 1979), *Science*, New Series, Vol. 208, No. 4445, 16 May 1980, pp. 724–726.
57. Review of John M. Spalek, in collaboration with Adrienn Ash, Sandra H. Hawrylchak, *Guide to the Archival Materials of the German-Speaking Emigration to the United States after 1933* (Charlottesville: University of Virginia Press, 1978), *Philosophy of Science*, Vol. 48, No. 1, March 1981, p. 161.
58. "The Publishing Enterprise", review of Arthur J. Meadows (ed.), *Development of Science Publishing in Europe* (New York: Elsevier, 1980), *Science*, New Series, Vol. 212, No. 4496, 15 May 1981, pp. 776–777.
59. Review of Brian Easlea, *Witch Hunting, Magic & the New Philosophy: An Introduction to Debates of the Scientific Revolution, 1450–1750* (Atlantic Highlands, NJ: Humanities Press, 1981), *History: Reviews of New Books*, Vol. 9, No. 8, July 1981, p. 197.
60. "Nineteenth-Century Shifts", review of Hans N. Jahnke, Michael Otte (eds.), *Epistemological and Social Problems of the Sciences in the Early Nineteenth Century* (Boston: Reidel, 1979), *Science*, New Series, Vol. 215, No. 4530, 15 January 1982, pp. 281–282.
61. Review of *Heralds of Science (As Represented by Two Hundred Epochal Books and Pamphlets in the Dibner Library, Smithsonian Institution)*, revised edition, Preface and notes by Bern Dibner, Introduction by Robert P. Multhauf (New York: Neale Watson Academic Publications), *The Quarterly Review of Biology*, Vol. 57, No. 1, March 1982, p. 54.
62. "Products of a Centennial", review of Peter C. Aichelburg, Roman U. Sexl (eds.), *Albert Einstein: His Influence on Physics, Philosophy and Politics* (Braunschweig: Vieweg, 1979); Engelbert Broda, *Einstein und Österreich* (Vienna: Verlag der Österreichischen Akademie der Wissenschaften, 1980); Consejo Nacional de Ciencia y Tecnologia, *Einstein* (Mexico City: Consejo Nacional de Ciencia y Tecnologia, 1979); Helen Dukas, Banesh Hoffmann (eds.), *Albert Einstein: The Human Side* (Princeton, NJ: Princeton University Press, 1979); Albert Einstein, *Autobiographical Notes: A Centennial Edition*, edited and translated by Paul A. Schilpp (La Salle, IL: Open Court, 1979); Anthony P. French, *Einstein: A Centenary Volume* (Cambridge, MA: Harvard University Press, 1979); Gerald Holton, Yehuda Elkana (eds.), *Albert Einstein: Historical and Cultural Perspectives*, Princeton, NJ: Princeton University Press, 1982); Colette M. Kinnon, Anatolii N. Kholodilin, Jacques G. Richardson (eds.), *The Impact of Modern Scientific Ideas on Society: In Commemoration of Einstein* (Boston: Reidel, 1981); Heinz Nelkowski, Armin Hermann, Hans Poser, Rolf Schrader, Reinhard Seiler (eds.), *Einstein Symposium* (New York: Springer, 1979); Ilse Rosenthal-Schneider, *Reality*

and Scientific Truth: Discussions with Einstein, von Laue, and Planck, edited by Thomas Braun (Detroit: Wayne State University Press, 1980); Loyd S. Swenson, Charles P. Snow, Howard Stein, Ilya Prigogine, *Albert Einstein: Four Commemorative Lectures* (Austin: University of Texas Humanities Research Center, 1979); Gerald E. Tauber (ed.), *Albert Einstein's Theory of General Relativity* (New York: Crown, 1979); Harry Woolf (ed.), *Some Strangeness in the Proportion: A Centennial Symposium to Celebrate the Achievements of Albert Einstein* (Reading, MA: Addison-Wesley Advanced Book Program, 1980), *Science*, New Series, Vol. 216, No. 4548, 21 May 1982, pp. 846–849.

63. "An Era at the Cavendish", review of John Hendry (ed.), *Cambridge Physics in the Thirties* (Bristol: Adam Hilger, 1984), *Science*, New Series, Vol. 225, No. 4664, 24 August 1984, p. 825.
64. Review of William R. Shea (ed.), *Otto Hahn and the Rise of Nuclear Physics* (Dordrecht-Boston-Lancaster: D. Reidel, 1983), *Annals of Science*, Vol. 42, No. 2, March 1985, pp. 181–182.
65. "Artes compilationis", essay review of Jagdish Mehra, Helmut Rechenberg, *The Historical Development of Quantum Theory*, 4 Volumes in 6 Parts (New York-Heidelberg-Berlin: Springer, 1982), *Isis*, Vol. 76, No. 3, September 1985, pp. 388–393.
   See also no. 94.
66. Review of Andrew Pickering, *Constructing Quarks: A Sociological History of Particle Physics* (Chicago: The University of Chicago Press, 1984), *American Journal of Sociology*, Vol. 91, No. 6, May 1986, pp. 1479–1481.
67. Review of Werner Heisenberg, *Gesammelte Werke / Collected Works*, Series A, Part 1: *Original scientific papers / Wissenschaftlichen Originalarbeiten*, edited by Walter Blum, Hans-Peter Dürr, Helmut Rechenberg (Berlin-Heidelberg-New York-Tokyo: Springer, 1984); Series B: *Scientific Review Papers, Talks, and Books / Wissenschaftliche Übersichtsartikel, Vorträge und Bücher*, edited by Walter Blum, Hans-Peter Dürr, Helmut Rechenberg (Berlin-Heidelberg-New York-Tokyo: Springer, 1984), *Historical Studies in the Physical and Biological Sciences*, Vol. 16, No. 2, 1986, pp. 372–373.
68. Review of Rudolf Peierls, *Bird of Passage: Recollections of a Physicist* (Princeton, NJ: Princeton University Press, 1985), *Historical Studies in the Physical and Biological Sciences*, Vol. 16, No. 2, 1986, pp. 374–375.
69. Review of Hilde Levi, *George de Hevesy: Life and Work* (Copenhagen: Rhodes, and Bristol: Adam Hilger, 1985), *Historical Studies in the Physical and Biological Sciences*, Vol. 17, No. 1, 1986, pp. 185–186.
70. "The Rise of Theoretical Physics", review of Christa Jungnickel, Russell McCormmach, *Theoretical Physics from Ohm to Einstein*, 2 Vols. (Chicago: The University of Chicago Press, 1986), *The Scientist*, 15 December 1986, p. 22.
71. Review of Margarita Bowen, *Empiricism and Geographical Thought from Francis Bacon to Alexander von Humboldt* (Cambridge: Cambridge University

Press, 1981), *Historical Studies in the Physical and Biological Sciences*, Vol. 17, No. 2, 1987, p. 370.

72. Review of Marcello Pera, *La rana ambigua. La controversia sull'elettricità animale tra Galvani e Volta* (Turin: Giulio Einaudi, 1986), *Historical Studies in the Physical and Biological Sciences*, Vol. 17, No. 2, 1987, p. 377.
73. Review of Ronald E. Zupko, *A Dictionary of Weights and Measures for the British Isles: The Middle Ages to the Twentieth Century* (Philadelphia: American Philosophical Society, 1985), *Historical Studies in the Physical and Biological Sciences*, Vol. 17, No. 2, 1987, pp. 380–381.
74. "Einstein's Early Endeavours", review of *The Collected Papers of Albert Einstein*, Vol. 1: *The Early Years, 1879–1902*, edited by John Stachel, with English annotations and a translation book by Anna Beck (Princeton: Princeton University Press, 1987), *Physics Bulletin*, Vol. 39, No. 2, February 1988, p. 74.
75. Review of Maristella Casciato, Maria Grazia Ianniello, Maria Vitale (eds.), *Enciclopedismo in Roma barocca. Athanasius Kircher e il Museo del Collegio Romano tra Wunderkammer e museo scientifico* (Venice: Marsilio, 1986), *Historical Studies in the Physical and Biological Sciences*, Vol. 18, No. 2, 1988, p. 404.
76. Review of Klaus Grewe, *Bibliographie zur Geschichte des Vermessungswesens* (Stuttgart: Konrad Wittwer, 1984), *Historical Studies in the Physical and Biological Sciences*, Vol. 18, No. 2, 1988, pp. 406–407.
77. Review of Peter J. Kuznick, *Beyond the Laboratory: Scientists as Political Activists in 1930s America* (Chicago: The University of Chicago Press, 1987), *Historical Studies in the Physical and Biological Sciences*, Vol. 18, No. 2, 1988, p. 409.
78. Review of Herbert F. York, *Making Weapons, Talking Peace: A Physicist's Odyssey from Hiroshima to Geneva* (New York: Basic Books, 1987), *Historical Studies in the Physical and Biological Sciences*, Vol. 18, No. 2, 1988, p. 414.
79. Review of Lanfranco Belloni, *Da Fermi a Rubbia. Storia e politica di un successo mondiale della scienza italiana* (Milan: Rizzoli, 1988), *Historical Studies in the Physical and Biological Sciences*, Vol. 19, No. 1, 1988, p. 199.
80. Review of Hans A. Bethe, Robert F. Bacher, M. Stanley Livingston, *Basic Bethe: Seminal Articles on Nuclear Physics, 1936–1937*, Introduction by Roger H. Stuewer (s.l.: Tomash Publishers/American Institute of Physics, 1986), *Historical Studies in the Physical and Biological Sciences*, Vol. 19, No. 1, 1988, p. 200.
81. Review of Alfred Landé, *Selected Scientific Papers*, edited by Asim O. Barut, Alwyn Van der Merwe (Dordrecht: D. Reidel, 1988), *Historical Studies in the Physical and Biological Sciences*, Vol. 19, No. 1, 1988, p. 205.
82. Review of *L'électricité dans l'histoire. Problèmes et méthodes* (Paris: Presses Universitaires de France, 1985); Louis Figuier, *Les merveilles de l'électricité. Textes choisis*, edited by Fabienne Cardot (Paris: Association pour l'histoire de l'électricité en France, 1985), *Archives internationales d'histoire des sciences*, Vol. 38, No. 121, Décembre 1988, pp. 307–308.

83. Review of Thomas F. Glick (ed.), *The Comparative Reception of Relativity* (Dordrecht-Boston: D. Reidel, 1987), *Historical Studies in the Physical and Biological Sciences*, Vol. 19, No. 2, 1989, p. 406.
84. Review of Thomas F. Glick, *Einstein in Spain: Relativity and the Recovery of Science* (Princeton, NJ: Princeton University Press, 1988), *Historical Studies in the Physical and Biological Sciences*, Vol. 19, No. 2, 1989, p. 406.
85. Review of Hans Mark, Lowell Wood (eds.), *Energy in Physics, War and Peace: A Festschrift Celebrating Edward Teller's 80th Birthday* (Dordrecht: Kluwer, 1988), *Historical Studies in the Physical and Biological Sciences*, Vol. 19, No. 2, 1989, p. 408.
86. Review of Joan L. Richards, *Mathematical Visions: The Pursuit of Geometry in Victorian England* (Boston: Academic Press, 1988), *Historical Studies in the Physical and Biological Sciences*, Vol. 19, No. 2, 1989, p. 411.
87. Review of Jürgen Teichmann, *Zur Geschichte der Festkörperphysik: Farbzentrenforschung bis 1940* (Stuttgart: Franz Steiner, 1988), *Historical Studies in the Physical and Biological Sciences*, Vol. 19, No. 2, 1989, p. 413.
88. Review of *The History of Science and Technology: A Narrative Chronology*, Vol. 1: *Prehistory–1900*, Vol. 2: *1900–1970* (New York: Facts on File, 1988), *Historical Studies in the Physical and Biological Sciences*, Vol. 20, No. 1, 1989, p. 189.
89. Review of Franz Stulhofer, *Lohn und Strafe in der Wissenschaft: Naturforscher im Urteil der Geschichte* (Vienna: Böhlau, 1987), *Historical Studies in the Physical and Biological Sciences*, Vol. 20, No. 1, 1989, p. 193.
90. Review of Johannes D. van der Waals, *On the Continuity of the Gaseous and Liquid States*, edited with an introductory essay by John S. Rowlinson (Amsterdam: North Holland, 1988), *Historical Studies in the Physical and Biological Sciences*, Vol. 20, No. 1, 1989, p. 194.
91. Review of Steven Shapin, Simon Schaffer, *Leviathan and the Air-Pump: Hobbes, Boyle and the Experimental Life*, including a translation of Thomas Hobbes, *Dialogus physicus de natura aeris*, by Simon Schaffer (Princeton: Princeton University Press, 1985), *Medical History*, Vol. 33, No. 2, 1 April 1989, pp. 256–257.
    See also [Notes, comments and replies, 44].
92. "Honesty's the Best Policy", review of Tony Rothman, *Science à la Mode: Physical Fashions and Fictions* (Princeton: Princeton University Press, 1989), *Nature*, Vol. 340, No. 6235, 24 August 1989, p. 608.
93. "Careful Mismanagement", review of Catherine Caufield, *Multiple Exposures: Chronicles of the Radiation Age* (London: Secker & Warburg, 1989); Joseph C. Morone, Edward J. Woodhouse, *The Demise of Nuclear Energy? Lessons for Democratic Control of Technology* (New Haven-London: Yale University Press, 1999), *London Review of Books*, Vol. 12, No. 1, 11 January 1990, p. 23.
94. Review of Jagdish Mehra, Helmut Rechenberg, *The Historical Development of Quantum Theory*, Vol. 5 in 2 Parts (New York: Springer, 1987), *Annals of Science*, Vol. 47, No. 1, January 1990, pp. 99–100.
    See also no. 65.

95. "Shadows of the Past", review of Mark Walker, *German National Socialism and the Quest for Nuclear Power, 1939–1949* (Cambridge University Press, 1989), *Nature*, Vol. 343, No. 6257, 1 February 1990, pp. 421–422.
96. Essay review of David Gooding, Trevor Pinch, Simon Schaffer (eds.), *The Uses of Experiment: Studies in the Natural Sciences* (Cambridge-New York: Cambridge University Press, 1989); Frank A. J. L. James (ed.), *The Development of the Laboratory: Essays on the Place of Experiment in Industrial Civilization* (London: Macmillan, 1989), *Medical History*, Vol. 34, No. 3, 1 July 1990, pp. 335–336.
97. "Relatively Significant", review of *The Collected Papers of Albert Einstein*, Vol. 2: *The Swiss Years: Writings, 1900–1909*, edited by John Stachel, with an English translation book by Anna Beck (Princeton: Princeton University Press, 1989), *Physics World*, Vol. 3, No. 7, July 1990, p. 53.
98. "Technical Targets", review of John Gimbel, *Science, Technology, and Reparations: Exploitations and Plunder in Postwar Germany* (Stanford: Stanford University Press, 1990), *Nature*, Vol. 348, No. 6301, 6 December 1990, pp. 496–497.
99. Review of John R. Millburn in collaboration with Henry C. King, *Wheelwright of the Heavens: The Life & Work of James Ferguson, FRS* (London: Vade-Mecum Press, 1988), *Historical Studies in the Physical and Biological Sciences*, Vol. 20, No. 2, 1990, p. 432.
100. Review of Christophe Charle, Eva Telkes, *Les professeurs de la Faculté des sciences de Paris. Dictionnaire biographique (1901–1939)* (Paris: Editions du CNRS, 1989), *Historical Studies in the Physical and Biological Sciences*, Vol. 21, No. 1, 1990, p. 200.
101. Review of Sabina Fleitmann, *Walter Charleton (1620–1707), "Virtuoso". Leben und Werk* (Frankfurt am Main: Peter Lang, 1986), *Historical Studies in the Physical and Biological Sciences*, Vol. 21, No. 1, 1990, p. 201.
102. Review of John Gascoigne, *Cambridge in the Age of the Enlightenment: Science, Religion, and Politics from the Restoration to the French Revolution* (Cambridge-New York: Cambridge University Press, 1989), *Historical Studies in the Physical and Biological Sciences*, Vol. 21, No. 1, 1990, p. 202.
103. Review of Derek Howse, *Nevil Maskelyne: The Seaman's Astronomer*, Foreword by Francis Graham-Smith (Cambridge: Cambridge University Press, 1989), *Historical Studies in the Physical and Biological Sciences*, Vol. 21, No. 1, 1990, p. 204.
104. Review of Desmond G. King-Hele, A. Rupert Hall (eds.), *Newton's* Principia *and Its Legacy: Proceedings of a Royal Society Discussion Meeting Held on 30 June 1987* (London: The Royal Society, 1988), *Historical Studies in the Physical and Biological Sciences*, Vol. 21, No. 1, 1990, p. 205.
105. Review of Burghard Weiss, *Wie finde ich Literatur zur Geschichte der Naturwissenschaft und der Tecknik?* (Berlin: Arno Spitz, 1990^2), *Historical Studies in the Physical and Biological Sciences*, Vol. 21, No. 1, 1990, p. 210.

106. Review of Michael Hunter, Simon Schaffer (eds.), *Robert Hooke: New Studies* (Woodbridge [UK], and Wolfeboro, NH: Boydell Press, 1989), *Historical Studies in the Physical and Biological Sciences*, Vol. 21, No. 2, 1991, p. 401.
    See also no. 109.
107. Review of René Sigrist, *Les origines de le Société de Physique et d'Histoire Naturelle (1790–1822). La science genevoise face au modèle français* (Geneva: ATAR, 1990), *Historical Studies in the Physical and Biological Sciences*, Vol. 21, No. 2, 1991, pp. 405–406.
108. Review of Trevor I. Williams, *Robert Robinson: Chemist Extraordinary* (Oxford: Clarendon Press, 1990), *Historical Studies in the Physical and Biological Sciences*, Vol. 21, No. 2, 1991, p. 407.
109. Review of Michael Hunter, Simon Schaffer (eds.), *Robert Hooke: New Studies* (Woodbridge, NH: Boydell Press, 1989), *Medical History*, Vol. 35, No. 3, 1 July 1991, pp. 359–360.
    See also no. 106.
110. Review of Isaac Benguigui, *Trois physiciens genevois et l'Europe savante. Les De la Rive, 1800–1920* (Geneva: Georg, 1990), *Historical Studies in the Physical and Biological Sciences*, Vol. 22, No. 1, 1991, pp. 194–195.
111. Review of Alberto Elena, Ana Martínez-Albertos, *Bibliografía española de historia de la ciencia y de tecnología*, Vol. 1 (1988); *Addenda* (1988); Vol. 2 (1989) (Madrid: Ediciones de la Universidad Autónoma), *Historical Studies in the Physical and Biological Sciences*, Vol. 22, No. 1, 1991, p. 197.
112. Review of Joachim Fischer, *Napoleon und die Naturwissenschaften* (Stuttgart: Franz Steiner, 1988), *Historical Studies in the Physical and Biological Sciences*, Vol. 22, No. 1, 1991, p. 198.
113. "The Truth about Physics", review of Joseph Schwartz, *The Creative Moment: How Science Made Itself Alien to Modern Culture* (New York: Jonathan Cape/HarperCollins, 1992), *Nature*, Vol. 360, No. 6401, 19 November 1992, pp. 217–218.
114. Review of Nicholas Copernicus, *Complete Works*, translation and commentary by Edward Rosen, Vol. 1: *On the Revolutions*; Vol. 2 (with the assistance of Enra Hilfstein): *Minor Works* (Baltimore: The Johns Hopkins University Press, 1992), *Historical Studies in the Physical and Biological Sciences*, Vol. 23, No. 2, 1993, p. 367.
115. Review of Evan M. Melhado, Tore Frängsmyr (eds.), *Enlightenment Science in the Romantic Era: The Chemistry of Berzelius and Its Cultural Setting* (Cambridge: Cambridge University Press, 1992), *Historical Studies in the Physical and Biological Sciences*, Vol. 23, No. 2, 1993, p. 371.
116. Review of Kenne Fant, *Alfred Nobel: A Biography*, English translation by Marianne Ruuth (New York: Arcade Publishing, 1993), *Historical Studies in the Physical and Biological Sciences*, Vol. 25, No. 1, 1994, pp. 177–178.
    See also no. 117.

117. "A Misantropic Philantropist", review of Kenne Fant, *Alfred Noble: A Biography*, English translation by Marianne Ruuth (New York: Arcade Publishing, 1993), *Nature*, Vol. 370, No. 6489, 11 August 1994, p. 426.
   See also no. 116.
118. Review of Steven Shapin, *A Social History of Truth: Civility and Science in Seventeenth-Century England* (Chicago: The University of Chicago Press, 1994), *Annals of Science*, Vol. 52, No. 5, September 1995, pp. 521–523.
119. Review of Peter Damerow, Gideon Freudenthal, Peter McLaughlin, Jürgen Renn, *Exploring the Limits of Preclassical Mechanics* (New York: Springer, 1992), *Historical Studies in the Physical and Biological Sciences*, Vol. 25, No. 2, 1995, pp. 380–381.
120. Review of Brian J. Ford, *Images of science: A History of Scientific Illustration* (New York: Oxford University Press, 1993), *Historical Studies in the Physical and Biological Sciences*, Vol. 25, No. 2, 1995, pp. 382–383.
121. Review of Marelene F. Rayner-Canham, Geoffrey W. Rayner-Canham, *Harriet Brooks: Pioneer Nuclear Scientist* (Montreal-Kingston: McGill-Queen's University Press, 1992), *Historical Studies in the Physical and Biological Sciences*, Vol. 25, No. 2, 1995, p. 385.
122. Review of Giambattista Guglielmini, *Carteggio «De diurno terrae motu»: Canterzani, Isolani, Matteucci, Bonfioli Malvezzi, Caldani, Calandrelli, Bonati*, edited by Maria Teresa Borgato, Alessandra Fiocca (Florence: Leo S. Olschki, 1994), *Isis*, Vol. 87, No. 1, March 1996, p. 177.
   See also no. 126.
123. Review of *A Guide to the Archival Collections in the Niels Bohr Library at the American Institute of Physics* (College Park, MD: American Institute of Physics, 1994), *Isis*, Vol. 87, No. 2, June 1996, pp. 384–385.
124. Review of Walter Brandmüller, Egon J. Greipl, *Copernico, Galileo e la Chiesa. Fine della controversia (1820). Gli atti del Sant'Uffizio* (Florence: Leo S. Olschki, 1992), *Historical Studies in the Physical and Biological Sciences*, Vol. 27, No. 1, 1996, p. 178.
125. Review of Vincenzo Ferrone, *The Intellectual Roots of the Italian Enlightenment: Newtonian Science, Religion, and Politics in the Early Eighteenth Century*, English translation by Sue Brotherton (Atlantic Highlands, NJ: Humanities Press, 1995), *Historical Studies in the Physical and Biological Sciences*, Vol. 27, No. 1, 1996, p. 181.
126. Review of Giambattista Guglielmini, *Carteggio «De diurno terrae motu». Canterzani, Isolani, Matteucci, Bonfioli Malvezzi, Caldani, Calandrelli, Bonati*, edited by Maria Teresa Borgato, Alessandra Fiocca (Florence: Leo S. Olschki, 1994), *Historical Studies in the Physical and Biological Sciences*, Vol. 27, No. 1, 1996, pp. 182–183.
   See also no. 122.
127. Review of Luigi Pepe, *Copernico e la questione Copernicana in Italia dal XVI al XIX secolo* (Florence: Leo S. Olschki, 1996), *Historical Studies in the Physical and Biological Sciences*, Vol. 27, No. 1, 1996, pp. 187–188.

128. Review of Geoffrey V. Sutton, *Science for a Polite Society: Gender, Culture & the Demonstration of Enlightenment* (Oxford: Westview, 1995), *Annals of Science*, Vol. 53, No. 6, November 1996, pp. 638–639.
129. Review of Albrecht Dürer, *Géométrie*, edited and translated by Jeanne Peiffer (Paris: Seuil, 1995), *Annals of Science*, Vol. 53, No. 6, November 1996, pp. 645–646.
130. "The Magnet Wars", review of Patricia Fara, *Sympathetic Attractions: Magnetic Practices, Beliefs, and Symbolism in Eighteenth-Century England* (Princeton: Princeton University Press, 1996), *The Times Literary Supplement*, no. 4934, 24 October 1997, p. 32.
131. "The Data Day", review of Raymond L. Francis, *The Illustrated Almanac of Science, Technology, and Invention: Facts, Figures, and the Fanciful* (New York: Plenum, 1997), *Nature*, Vol. 389, No. 6654, 30 October 1997, p. 926.
132. Review of Gerard L'E. Turner, *The Practice of Science in the Nineteenth Century: Teaching and Research Apparatus in the Teyler Museum* (Haarlem: The Teyler Museum, 1996), *Historical Studies in the Physical and Biological Sciences*, Vol. 27, No. 2, 1997, pp. 369–370.
133. Review of Peter de Clercq, *At the Sign of the Oriental Lamp: The Musschenbroek Workshop in Leiden, 1660–1750* (Rotterdam: Erasmus Publishing, 1997), *Historical Studies in the Physical and Biological Sciences*, Vol. 28, No. 1, 1997, pp. 197–198.
134. Review of Edward G. Ruestow, *The Microscope in the Dutch Republic: The Shaping of Discovery* (Cambridge: Cambridge University Press, 1996), *Historical Studies in the Physical and Biological Sciences*, Vol. 28, No. 1, 1997, pp. 206–207.
135. Review of Martin Harwit, *An Exhibit Denied: Lobbying the History of* Enola Gay (New York: Copernicus, 1996), *Physics Today*, Vol. 50, No. 6, June 1997, p. 79.
136. Review of Beate Ceranski, *"Und sie fürchtet sich vor niemandem". Die Physikerin Laura Bassi (1711–1778)* (Frankfurt etc.: Campus, 1996), *Archives internationales d'histoire des sciences*, Vol. 47, No. 139 bis, 1997, pp. 188*-189*.
137. Review of Neil A. Porter, *Physicists in Conflict: From Antiquity to the New Millennium* (Bristol: Institute of Physics Publishing, 1998), *Historical Studies in the Physical and Biological Sciences*, Vol. 29, No. 1, 1998, p. 187.
138. "Beware Scientists Writing History", review of Bernard Pullman, *The Atom in the History of Human Thought* (Oxford-New York: Oxford University Press, 1998), *Nature*, Vol. 395, No. 6700, 24 September 1998, pp. 339–340.
139. Review of Clifford D. Connor, *Jean Paul Marat: Scientist and Revolutionary* (Atlantic Highlands, NJ: Humanities Press, 1997), *Annals of Science*, Vol. 55, No. 4, October 1998, pp. 423–424.
140. Review of Andrea A. Rusnock (ed.), *The Correspondence of James Jurin (1684–1750): Physician and Secretary to the Royal Society* (Amsterdam-Atlanta: Rodopi, 1996), *Medical History*, Vol. 42, No. 4, 1 October 1998, pp. 536–537.

141. "The *DSB* on Laplace Reprinted", review of Charles C. Gillispie, with the assistance of Robert Fox and Ivor Grattan-Guinness, *Pierre-Simon Laplace, 1749–1827: A Life in Exact Science* (Princeton: Princeton University Press, 1997), *Journal for the History of Astronomy*, Vol. 30, No. 1, February 1999, p. 83.
142. "Why We Can Have It Now", review of Paul Israel, *Edison: A Life of Invention* (New York-Chichester: Wiley & Sons, 1998); and of Victor K. McElheny, *Insisting on the Impossible: The Life of Edwin Land* (Reading, MA: Perseus Books, 1998), *The Times Literary Supplement*, No. 5009, 2 April 1999, p. 31.
143. "Science in Culture", review of Kristen Lippincott (ed.), *The Story of Time* (London: Merrell Holberton, 1999), *Nature*, Vol. 402, No. 6761, 2 December 1999, p. 462.
144. Review of Iwan Rhys Morus, *Frankenstein's Children: Electricity, Exhibition and Experiment in Early-Nineteenth Century London* (Princeton: Princeton University Press, 1998), *Medical History*, Vol. 44, No. 1, 1 January 2000, pp. 138–139.
145. "Around the World via Greenwich", review of Elly Dekker (ed.), *Globes at Greenwich: A Catalogue of the Globes and Armillary Spheres at the National Maritime Museum, Greenwich* (Oxford: Oxford University Press, 1999), *Nature*, Vol. 404, No. 6773, 2 March 2000, pp. 16–17.
146. Review of Paul L. Rose, *Heisenberg and the Nazi Atomic Bomb Project: A Study in German Culture*, (Berkeley-Los Angeles: University of California Press, 1998), *The American Historical Review*, Vol. 105, No. 2, April 2000, pp. 638–639.
147. "Surfing the Nebulae, Churning the Waters", review of W. Garrett Scaife, *From Galaxies to Turbines: Science, Technology and the Parsons Family* (New York: Taylor & Francis, 2000), *Nature*, Vol. 407, No. 6801, 14 September 2000, pp. 133–134.
148. "Hven's Gate", review of John R. Christianson, *On Tycho's Island: Tycho Brahe and His Assistants, 1570–1601* (Cambridge: Cambridge University Press, 2000), *London Review of Books*, Vol. 22, No. 21, 22 November 2000, p. 24.
149. Review of Scott L. Montgomery, *The Moon and the Western Imagination* (Tucson: University of Arizona Press, 1999), *International Journal of the Classical Tradition*, Vol. 8, No. 1, Summer 2001, pp. 127–129.
150. Review of Nigel Hiscock, *The Wise Master Builder: Platonic Geometry in Plans of Medieval Abbeys and Cathedrals* (Aldershot [UK]–Brookfield, VT: Ashgate, 2000), *Isis*, Vol. 93, No. 1, March 2002, pp. 111–112.
151. Review of Paul Brouzeng, Suzanne Débarbat (eds.), *Sur les traces des Cassini: Astronomes et observatoires du sud de la France* (Paris: Éditions du Comité des Travaux Historiques et Scientifiques, 2001), *Isis*, Vol. 93, No. 2, June 2002, pp. 286–287.
152. Review of Kenneth J. Howell, *God's Two Books: Copernican Cosmology and Biblical Interpretation in Early Modern Science* (Notre Dame, IN: University

of Notre Dame Press, 2002), *The American Historical Review*, Vol. 107, No. 5, December 2002, pp. 1626–1627.

153. Review of William T. Lynch, *Solomon's Child: Method in the Early Royal Society of London* (Stanford: Stanford University Press, 2001), *Isis*, Vol. 93, No. 4, December 2002, p. 693.
154. "Information Technology", review of Richard Yeo, *Encyclopaedic Visions: Scientific Dictionaries and Enlightenment Culture* (Cambridge: Cambridge University Press, 2001), *Notes and Records of the Royal Society of London*, Vol. 57, No. 1, January 2003, pp. 111–112.
155. "Lab Boy to Professor", review of James Hamilton, *Faraday: The Life* (London: HarperCollins, 2002), *The Times Literary Supplement*, No. 5205, 3 January 2003, p. 23.
156. Review of Stephen Hawking (ed.), *On the Shoulders of Giants: The Great Works of Physics and Astronomy* (Philadelphia: Running Press, 2002), *Astronomy*, Vol. 31, No. 8, August 2003, p. 105.
157. Review of Peter Galison, Emily Thompson (eds.), *The Architecture of Science* (Cambridge, MA-London: The MIT Press, 1999), *Annals of Science*, Vol. 60, No. 4, October 2003, pp. 454–455.
158. Review of Mordechai Feingold (ed.), *Jesuit Science and the Republic of Letters* (Cambridge, MA-London: The MIT Press, 2003), *The American Historical Review*, Vol. 109, No. 1, February 2004, pp. 242–244.
159. "Careful Readers", review of Owen Gingerich, *The Book Nobody Read: Chasing the Revolutions of Nicolaus Copernicus* (London: Arrow, 2005), *London Review of Books*, Vol. 27, No. 18, 22 September 2005, pp. 28–29.
160. Review of Joyce E. Chaplin, *The First Scientific American: Benjamin Franklin and the Pursuit of Genius* (New York: Basic Books, 2006); Philip Dray, *Stealing God's Thunder: Benjamin Franklin's Lightning Rod and the Invention of America* (New York: Random House, 2005), *Physics Today*, Vol. 59, N. 9, September 2006, pp. 59–60.
161. Review of Peter J. Bowler, Iwan Rhys Morus, *Making Modern Science: A Historical Survey* (Chicago-London: The University of Chicago Press, 2005), *The British Journal for the History of Science*, Vol. 40, No. 1, March 2007, pp. 118–119.
162. "Around the Sun", review of William T. Vollmann, *Uncentering the Earth: Copernicus and* On the Revolutions of the Heavenly Spheres (London: Weidenfeld & Nicolson, 2006), *The Times Literary Supplement*, No. 5437, 15 June 2007, p. 23.
163. Review of Peter Dear, *The Intelligibility of Nature: How Science Makes Sense of the World* (Chicago-London: The University of Chicago Press, 2006), *The British Journal for the History of Science*, Vol. 40, No. 4, December 2007, pp. 599–601.
164. Review of Johann S. Pütter, Jacob C. Saalfeld, Georg H. Oesterley, *Versuch einer academischen Gelehrten-Geschichte von der Georg-August-Universität*

zu Göttingen (Vols. 1–2, Göttingen: Vandenhoek, 1765 and 1788; Vol. 3, Hanover: Helmwing, 1820; Vol. 4, Göttingen: Vandenhoek & Ruprecht, 1838; reprint, Hildesheim-Zürich-New York: Olms-Weidmann, 2006, 4 Vols.), *Göttingische Gelehrte Anzeigen*, Jahrgang 260, Heft 1–2, 2008, pp. 10–20.

165. Review of William R. Shea, Mariano Artigas, *Galileo Observed: Science and the Politics of Belief* (Sagamore Beach, MA: Science History Publications, 2006); Richard J. Blackwell, *Behind the Scenes at Galileo's Trial: Including the First English Translation of Melchior Inchofer's* Tractatus Syllepticus (Notre Dame, IN: Notre Dame University Press, 2006), *The British Journal for the History of Science*, Vol. 41, No. 2, June 2008, pp. 294–296.
166. Review of Adam Mosley, *Bearing the Heavens: Tycho Brahe and the Astronomical Community of the Late Sixteenth Century* (New York: Cambridge University Press, 2007), *The Journal of Interdisciplinary History*, Vol. 39, No. 1, Summer 2008, pp. 124–125.
167. "The Invention of the Telescope", review of Eileen Reeves, *Galileo's Glassworks: The Telescope and the Mirror* (Cambridge, MA: Harvard University Press, 2008), *Journal for the History of Astronomy*, Vol. 39, No. 4, November 2008, pp. 530–531.
168. "Vinum novum in utres veteres", essay review of *New Dictionary of Scientific Biography*, Noretta Koertge editor in chief, 8 Vols. (Farmington Mills, MI: Charles Scribner's Sons/Thomson Gale, 2008), *Isis*, Vol. 101, No. 4, December 2010, pp. 839–842.
169. "*Sidereus Nuncius* Debated", review of Horst Bredekamp (ed.), *Galileo's O*, Vol. I: *Galileo's* Sidereus Nuncius*: A Comparison of the Proof Copy (New York) with Other Paradigmatic Copies*, edited by Irene Brückle, Oliver Hahn; Vol. II: Paul Needham, *Galileo Makes a Book: The First Edition of the* Sidereus Nuncius*, Venice 1610* (Berlin: Akademie Verlag, 2011), *Journal for the History of Astronomy*, Vol. 43, No. 2, May 2012, pp. 244–246.
170. "Robert Westman on Galileo and Related Matters", review of Robert S. Westman, *The Copernican Question: Prognostication, Skepticism, and Celestial Order* (Berkeley: University of California Press, 2011), *Perspectives on Science*, Vol. 20, No. 3, Fall 2012, pp. 379–388.

    See also [Notes, comments and replies, no. 42].
171. "Fireworks and the Origin of Modern Science", review of Simon Werrett, *Fireworks: Pyrotechnic Arts and Sciences in European History* (Chicago-London: The University of Chicago Press, 2010), *Metascience*, Vol. 21, No. 3, November 2012, pp. 685–686.
172. Review of Alexander Marr, *Between Raphael and Galileo: Mutio Oddi and the Mathematical Culture of Late Renaissance Italy* (Chicago-London: The University of Chicago Press, 2011), *Annals of Science*, Vol. 71, No. 1, January 2014, pp. 140–142.
173. "Visualizing Astronomical Knowledge in Rome", review of Ulrike Feist, *Sonne, Mond und Venus: Visualisierungen astronomischen Wissens im frühneuzeitlichen Rom* (Berlin: Akademie Verlag, 2013), *Journal for the History of Astronomy*, Vol. 45, No. 4, November 2014, pp. 485–487.

174. "Roman Thought Police and Early-Modern Astrology", essay review of Monica Azzolini, *The Duke and the Stars: Astrology and Politics in Renaissance Milan* (Cambridge, MA: Harvard University Press, 2013); Ugo Baldini, Leen Spruit (eds.), *Catholic Church and Modern Science: Documents from the Archives of the Roman Congregations of the Holy Office and the Index*, Vol. I: *Sixteenth-Century Documents*, in 4 Tomes (Rome: Libreria Editrice Vaticana, 2009); Fabrizio Bònoli, Giuseppe Bezza, Salvo de Meis, Cinzia Colarita, *I pronostici di Domenico Maria Novara* (Florence: Leo S. Olschki, 2012); Mary Quinlan-McGrath, *Influences: Art, Optics, and Astrology in the Italian Renaissance* (Chicago-London: The University of Chicago Press, 2013), *Perspectives on Science*, Vol. 23, No. 2, Summer 2015, pp. 228–240.
175. "Logic and the Sizes of Stars", review of Christopher M Graney, *Setting Aside All Authority: Giovanni Battista Riccioli and the Science against Copernicus in the Age of Galileo* (Notre Dame, IN: University of Notre Dame Press, 2015), *Journal for the History of Astronomy*, Vol. 47, No. 2, May 2016, pp. 222–224.
176. Review of Thomas F. Mayer, *The Roman Inquisition: Trying Galileo* (Philadelphia: University of Pennsylvania Press, 2015), *The American Historical Review*, Vol. 121, No. 3, June 2016, pp. 1034–1035.
177. "Deep Narrow Territories", review of Peter Harrison, *The Territories of Science and Religion* (Chicago-London: The University of Chicago Press, 2015), *Metascience*, Vol. 25, No. 3, November 2016, pp. 349–353.
178. "Universal Joint", review of Peter Watson, *Convergence: The Idea at the Heart of Science* (New York: Simon & Schuster, 2017), *The New York Times Book Review*, Vol. 122, No. 11, 12 March 2017, p. 22.
179. "A Choice of Companions", essay review of Bernard V. Lightman (ed.), *A Companion to the History of Science* (Chichester: Wiley-Blackwell, 2016); Georgina M. Montgomery, Mark A. Largent (eds.), *A Companion to the History of American Science* (Chichester: Wiley-Blackwell, 2015), *Isis*, Vol. 108, No. 3, September 2017, pp. 660–663.

    See also [Notes, comments and replies, 47].
180. Essay review of Richard Yeo, *Notebooks, English Virtuosi, and Early Modern Science* (Chicago-London: The University of Chicago Press, 2014); Renée Raphael, *Reading Galileo: Scribal Technologies and the* Two New Sciences (Baltimore: Johns Hopkins University Press, 2017), *History of Universities*, Vol. 31, Part 1, 2018, pp. 213–220.
181. Review of Klaus Hentschel, *Photons: The History and Mental Models of Light Quanta* (Cham, CH: Springer International, 2018), *Annals of Science*, Vol. 76, Nos. 3–4, July-October 2019, pp. 387–389.
182. Review of Alfredo Dinis, *A Jesuit against Galileo? The Strange Case of Giovanni Battista Riccioli Cosmology*, edited by Álvaro Balsas, Ricardo Barroso Batista (Braga: Universidade Católica Portuguesa, 2017); Aviva Rothman, *The Pursuit of Harmony: Kepler on Cosmos, Confession, and*

Community (Chicago-London: The University of Chicago Press, 2017), *History of Universities*, Vol. 33, Part 1, 2020, pp. 225–231.

183. Review of Per Pippin Aspass, László Kontler, *Maximilian Hell (1720–92) and the Ends of Jesuit Science in Enlightenment Europe* (Leiden: Brill, 2020), *Church History*, Vol. 89, No. 4, December 2020, pp. 953–955.

Prefaces and Introductions

1. ["Preface"], in *Max Planck: A Bibliography of His Non-Technical Writings*, Berkeley: Office for History of Science and Technology, 1977, pp. I-IV.
2. ["Preface"], in *William Henry Bragg and William Lawrence Bragg: A Bibliography of Their Non-Technical Writings*, Berkeley: Office for History of Science and Technology, 1978, pp. I-III.
3. ["Preface"], in *Ernest Rutherford: A Bibliography of His Non-Technical Writings*, Berkeley: Office for History of Science and Technology, 1979, pp. I-III.
4. "Preface", in Frederic L. Holmes, *Eighteenth-Century Chemistry as an Investigative Enterprise*, Berkeley: Office for History of Science and Technology, 1989, pp. I-II.
5. "Preface", in Frederic L. Holmes, *Between Biology and Medicine: The Formation of Intermediary Metabolism*, Berkeley: Office for History of Science and Technology, 1992, pp. I-III.
6. "Introduction", in Galileo Galilei, *Dialogue Concerning the Two Chief World Systems: Ptolemaic and Copernican*, translated with revised notes by Stillman Drake, Foreword by Albert Einstein, New York: The Modern Library, 2001, pp. XIII-XXII.

Notes, Comments and Replies

1. "Reply" to Philip M. Heimann, "The Work of H. G. J. Moseley" (comment on [Essays, 3]), *Isis*, Vol. 58, No. 2, Summer 1967, pp. 232–233: *ibidem*, p. 234.
2. "Commentary: Duhem and Donahue", in Robert S. Westman (ed.), *The Copernican Achievement*, Berkeley-Los Angeles-London, 1975, pp. 276–284.
3. "Clarification", on [Reviews and essay reviews, 32], *Science*, New Series, Vol. 187, No. 4179, 7 March 1975, p. 792.
4. "Moseley Letters", on Thaddeus J. Trenn, "Moseley and More Moseleyana" (essay review of [Monographs, 3]), *Annals of Science*, Vol. 33, No. 1, January 1976, pp. 105–109: *ibid*em, No. 3, May 1976, p. 309.
5. "Foreword", *Historical Studies in the Physical Sciences*, Vol. 11, No. 1, 1980, pp. I-III.
6. "Appendix" to Patricia Fortini Brown, "*Laetentur Caeli*: The Council of Florence and the Astronomical Fresco in the Old Sacristy", *Journal of the Warburg and Courtauld Institutes*, Vol. 44, 1981, pp. 176–180: *ibid*em, p. 180.
7. "Foreword" [unsigned], *Historical Studies in the Physical and Biological Sciences*, Vol. 16, No. 1, 1986, pp. [III-IV].
8. "H. Moseley and the Nobel Prize", *Nature*, Vol. 330, No. 6150, 24/31 December 1987, p. 694.
9. (William F. Bynum, JLH) "Eighteen Eighty Eight and All That',' *Nature*, Vol. 331, No. 6151, 7 January 1988, pp. 27–30.

10. (William F. Bynum, JLH) "Eighteen Eighty Nine and All That", *Nature*, Vol. 337, No. 6202, 5 January 1989, pp. 29–32.
11. (JLH, Daniel J. Kevles) "By Failing to Discuss the 'Civics' of Science and Technology, History Textbooks Distort the Past and Endanger the Future", *The Chronicle of Higher Education*, Vol. 35, No. 23, 15 February 1989, p. A48.

    Reprinted, in revised form, as "The 'Civics' of Technology in History Textbooks", in *The Education Digest*, Vol. 55, No. 1, September 1989, pp. 42–44.

12. (William F. Bynum, JLH) "Eighteen Ninety and All That", *Nature*, Vol. 343, No. 6253, 4 January 1990, pp. 11–14.
13. "Why Revisit Solomon's House? A Commentary on Sessions V and VI", in Tore Frängsmyr (ed.), *Solomon's House Revisited: The Organization and Institutionalization of Science*, Canton, MA: Science History Publications, 1990, pp. 331–342.
14. (William F. Bynum, JLH) "Eighteen Ninety One and All That", *Nature*, Vol. 349, No. 6304, 3 January 1991, pp. 9–12.
15. (William F. Bynum, JLH) "Eighteen-Ninety Two and All That", *Nature*, Vol. 355, No. 6355, 2 January 1992, pp. 11–14.
16. ["Reply"] to Patricio A. A. Laura, "Jesuits' Role in Revolutionary Science", *Physics Today*, Vol. 47, No. 1, January 1994, p. 70 (comment on [Essays, 47]): *ibidem*.
17. (William F. Bynum, JLH) "Eighteen-Ninety Three and All That", *Nature*, Vol. 361, No. 6407, 7 January 1993, pp. 9–12.
18. (William F. Bynum, JLH) "Eighteen-Ninety Four and All That", *Nature*, Vol. 367, No. 6458, 6 January 1994, pp. 11–14.
19. "On Coulomb's Electrostatics Balance", in Christine Blondel, Matthias Dörries (eds.), *Restaging Coulomb: usages, controverses et réplications autour de la balance de torsion*, Florence: Leo S. Olschki, 1994, pp. 151–161.
20. (William F. Bynum, JLH) "Eighteen Ninety Five and All That", *Nature*, Vol. 373, No. 6509, 5 January 1995, pp. 11–14.
21. (William F. Bynum, JLH) "Eighteen-Ninety-Six and All That", *Nature*, Vol. 379, No. 6560, 4 January 1996, pp. 15–18.
22. (William F. Bynum, JLH), "Eighteen-Ninety-Seven and All That", *Nature*, Vol. 385, No. 6611, 2 January 1997, pp. 13–16.
23. (William F. Bynum, JLH), "1397 bis 1997", *Naturwissenschaftliche Rundschau*, Vol. 50, No. 10, Oktober 1997, pp. 397–402.
24. (William F. Bynum, JLH) "Eighteen-Ninety-Eight and All That", *Nature*, Vol. 391, No. 6662, 1 January 1998, pp. 13–16.
25. (William F. Bynum, JLH) "Eighteen-Ninety-Nine and All That", *Nature*, Vol. 397, No. 6714, 7 January 1999, pp. 15–18.
26. "From Horsehair to Lightning Rods", *Nature*, Vol. 401, No. 6751, 23 September 1999, p. 329.
27. "Copernicus in the Cathedral", *The Wall Street Journal*, Vol. 234, No. 80, 22 October 1999, p. W15.

28. "Science in Culture", *Nature*, Vol. 402, No. 6761, 2 December 1999, p. 462.
29. (William F. Bynum, JLH), "Plus ça change", *Nature*, Vol. 402, No. 6761 Suppl., 2 December 1999, pp. C86–88.
30. (William F. Bynum, JLH) "Millennial Highlights… from Gerbert d'Aurillac to Watson and Crick", *Nature*, Vol. 403, No. 6765, 6 January 2000, pp. 13–16.
31. "Concluding Remarks", in Svante Lindqvist (ed.), *Museums of Modern Science*, Canton, MA: Science History Publications, 2000, pp. 189–192.
32. (William F. Bynum, JLH) "1901 and All That", *Nature*, Vol. 409, No. 6816, 4 January 2001, pp. 13–16.
33. "That Quantum Leap…", *The Times Higher Education Supplement*, No. 1468, 5 January 2001, p. 19.
34. (William F. Bynum, JLH) "1902 and All That", *Nature*, Vol. 415, No. 6867, 3 January 2002, pp. 15–18.
35. "Coming to Terms: Caloric, Cathode, Curium and Quark – Coinage from the Mint of Science", *Nature*, Vol. 415, No. 6872, 7 February 2002, p. 585.
36. "American Optimism set Franklin's Work Apart", *The Philadelphia Inquirer*, 9 June 2002, p. A C5.
37. (William F. Bynum, JLH) "1903 and All That", *Nature*, Vol. 421, No. 6918, 2 January 2003, pp. 15–18.
38. (William F. Bynum, JLH) "1904 and All That", *Nature*, Vol. 426, No. 6968, 18 December 2003, pp. 761–764.
39. "Wit and Wisdom", *Nature*, Vol. 438, No. 7064, 3 November 2005, p. 29.
40. "Swansong", *Historical Studies in the Physical and Biological Sciences*, Vol. 37, No. 2, March 2007, pp. 185–187.
41. "The Celestial Message", *Nature*, Vol. 467, No. 7314, 23 September 2010, pp. 398–399.
42. "Reply to Westman" (on Robert S. Westman, "The Copernican Question Revisited: A Reply to Noel Swerdlow and John Heilbron", *Perspectives on Science*, Vol. 21, No. 1, Spring 2013, pp. 100–136): *ibidem*, No. 3, Fall 2013, p. 386.
43. "Vielfache Wahrheiten. Niels Bohr und sein Quantenmodell des Atoms", German translation by Michaela Grabinger, *Kultur & Technik*, No. 3: *Bohrs Modell. Vor hundert Jahren begründete Niels Bohr die moderne Atomtheorie*, 2013, pp. 6–11.
44. "The Path to the Quantum Atom", *Nature*, Vol. 498, No. 7452, 6 June 2013, pp. 27–30.

Spanish translation, "La ruta hacia el átomo cuántico", *Investigación y ciencia*, No. 447, Diciembre 2013, pp. 52–55.

45. "Are Historians Fit to Rule?", *Isis*, Vol. 107, No. 2, June 2016, pp. 350–352.
46. "A Last Judgment", *Isis*, Vol. 108, No. 1, March 2017, pp. 122–123.
47. "In Reply" to Bernard Lightman, "To the Editor", *Isis*, Vol. 108, No. 4, December 2017, p. 865; and Mark Largent, Georgina Montgomery, "To the Editor", *ibidem*, p. 866 (on [Reviews and essay reviews, 179]): *ibidem*, p. 867.

48. “Have I Explained Anything?”, *Isis*, Vol. 110, No. 2, June 2019, pp. 305–307.
49. “David Michael Heilbron (November 25, 1936 – March 8, 2020)” [unsigned], *San Francisco Chronicle*, 15 March 2020, p. B6.

Entries

1. *Dictionary of Scientific Biography*, Charles C. Gillispie editor in chief, New York: Charles Scribner’s Sons, 1970–1981, 16 Vols.

   Vol. I: *Pierre Abailard–L. S. Berg* (1970)

   “Aepinus, Franz Ulrich Theodosius”, pp. 66–68.
   “Beccaria, Giambatista”, pp. 546–548.

   Vol. II: *Hans Berger–Christoph Buys Ballot* (1970)

   “Bose, Georg Matthias”, pp. 324–325.
   (Adrienne R. Weill-Brunschvicg, JLH) “Broglie, Louis-César-Victor-Maurice de”, pp. 487–489.

   Vol. III: *Pierre Cabanis–Heinrich von Dechen* (1971)

   “Canton, John”, pp. 51–52.
   “Cavallo, Tiberius”, pp. 153–154.

   Vol. IV: *Richard Dedekind–Firmicus Maternus* (1971)

   “Dufay (Du Fay), Charles-François de Cisternai”, pp. 214–217.

   Vol. V: *Emil Fischer–Gottlieb Haberlandt* (1972)

   “Gray, Stephen”, pp. 515–517.

   Vol. VII: *Iamblichus–Karl Landsteiner* (1973)

   “Kinnersley, Ebenezer”, pp. 371–372.
   “Kleist, Ewald Georg von”, p. 403.

   Vol. IX: *A. T. Macrobius–K. F. Naumann* (1974)

   “Moseley, Henry Gwyn Jeffreys”, pp. 542–545.

   Vol. X: *S. G. Navashin–W. Piso* (1974)

   “Nollet, Jean-Antoine”, pp. 145–148.

   Vol. XI: *A. Pitcairn–B. Rush* (1975)

   “Richman, Georg Wilhelm”, pp. 432–434.

   Vol. XIII: *Hermann Staudinger–Giuseppe Veronese* (1976)

   “Symmer, Robert”, pp. 224–225.
   “Thomson, Joseph John”, pp. 362–372.

Vol. XIV: *Addison Emery Verrill–Johann Zwelfer* (1976)

"Volta, Alessandro Giuseppe Antonio Anastasio", pp. 69–82.
"Watson, William", pp. 193–196.
"Wilcke, John Carl", pp. 352–356.

2. *Academic American Encyclopedia*, Princeton, NJ: Aretê, 1979–1980, 21 Vols.

Vol. 15 (1980)

"Physics, History of", pp. 282–286.

3. *Dictionary of the History of Science*, edited by William F. Bynum, E. Janet Browne, Roy Porter, London-Basingstoke: The Macmillan Press, 1981; revised edition, 1983.

Italian translation, *Dizionario di storia della scienza*, edited by Mauro La Forgia and Sandro Petruccioli, Rome-Naples: Theoria, 1988; 566 pp.

"Aether", pp. 7–8 (It. tr., "etere", pp. 183–184).
"Aristotelian Physics", pp. 25–26 (It. tr., "fisica aristotelica", pp. 202–203).
"Atomic Structure", pp. 34–36 (It. tr., "atomo, struttura dell'", pp. 60–62).
"Attraction", pp. 36–37 (It. tr., "attrazione", pp. 62–63).
"Cambridge School", pp. 50–51 (missing in the It. tr.).
"Cartesian", p. 53 (It. tr., "cartesiano", p. 87).
"Charge", p. 64 (It. tr., "carica", p. 86).
"Classical Physics", p. 68 (It. tr., "fisica classica", p. 203).
"Cosmic Rays", pp. 83–84 (It. tr., "raggi cosmici", pp. 407–408).
"Effluvium", pp. 110–111 (It. tr., "effluvio", pp. 153–154).
"Electricity and Magnetism", pp. 113–115 (It. tr., "elettricità e magnetismo", pp. 158–161).
"Force", pp. 151–152 (It. tr., "forza", pp. 209–210).
"Gravity", pp. 170–171 (It. tr., "gravità", pp. 235–237).
"Hydrodynamics and Hydrostatics", pp. 193–195 (It. tr., "idrodinamica e idrostatica", pp. 244–246).
"Latitude of Forms", p. 229 (It. tr., "latitudine delle forme", pp. 275–276).
"Leyden Jar", p. 233 (It. tr., "bottiglia di Leyda", pp. 72–73).
"Lightning", p. 237 (It. tr., "fulmine", p. 215).
"Magneto-Optics", p. 244 (It. tr., "magnetoottica", pp. 293–294).
"Mechanical Reduction", pp. 252–253 (It. tr., "riduzionismo meccanico", pp. 425–426).
"Mechanics", pp. 253–256 (It. tr., "meccanica", pp. 305–308).
"Multiplication of Species", p. 281 (It. tr., "specie, moltiplicazione delle", p. 475).
"Newtonian", p. 299 (It. tr., "newtoniano", pp. 351–352).
"Physics", p. 326 (It. tr., "fisica", pp. 201–202).
"Quality", p. 353 (It. tr., "qualità", pp. 403–404).

4. *The Oxford Companion to the History of Modern Science*, John L. Heilbron editor in chief, Oxford-New York: Oxford University Press, 2003.

   "Academies and Learned Societies", pp. 1–5.
   "Atomic Structure", pp. 70–71.
   "Cathedral", p. 128.
   "Cavendish, Henry (1731–1810) and Charles Augustin Coulomb (1736–1806)", pp. 131–133.
   "Cosmic Rays", pp. 181–182.
   "Curie, Marie Sklodowska (1867–1934) and Pierre Curie (1859–1906)", pp. 192–193.
   "Dynasty", pp. 221–222.
   "Ether", pp. 273–274.
   "Experimental Philosophy", pp. 286–288.
   "Fire and Heat", pp. 303–305.
   "Geodesy", pp. 333–335.
   "History of Science", pp. 370–374.
   "Imponderables", pp. 396–398.
   "Leyden Jar and Electrophore", pp. 459–460.
   "Lightning Conductor", pp. 466–467.
   "Magneto-Optics", p. 484.
   "Nobel Prize", pp. 576–578.
   "Noble Gases", pp. 578–579.
   "Physics", pp. 643–646.
   "Planck, Max (1858–1947)", pp. 650–651.
   "Pneumatics", pp. 658–659.
   "Putti in Science", pp. 685–686.
   "Quantum Physics", pp. 691–694.
   "Rainbow", pp. 706–707.
   "Relativity", pp. 711–713.
   "Röntgen, Wilhelm Conrad (1845–1923)", pp. 723–724.
   "Rutherford, Ernest (1871–1937), pp. 724–726.
   "Scientist", pp. 744–745.
   "Slogans from Science", pp. 755–756.
   "Spontaneous Combustion", p. 772.
   "Standard Model", pp. 776–777.
   "Sympathy and Occult Quality", p. 786.
   "Terminology", p. 794.
   "X Rays", pp. 835–836.

5. *Encyclopedia of the Enlightenment*, Alan C. Kors editor in chief, Oxford-New York: Oxford University Press, 2003, 4 Vols.

   Vol. 3: *Mably-Ruysch*

      "Natural Philosophy and Science", pp. 134–142.

6. *Oxford Dictionary of National Biography: From the Earliest Times to the Year 2000*, edited by Henry C. G. Matthew, Brian Harrison, Oxford-New York: Oxford University Press, 2004, 60 Vols.

   Vol. 39: *Morant–Murray*

   (Ernest Rutherford, revised by JLH) "Moseley, Henry Gwyn Jeffreys (1887–1915)", pp. 455–456.

7. *The Oxford Guide to the History of Physics and Astronomy*, edited by John L. Heilbron, Oxford-New York: Oxford University Press, 2005.[2]

   "Atomic Structure", pp. 23–25.
   "Cassini Family: Gian Domenico (1625–1712); Jacques (1677–1756); César-François Cassini de Thury (1714–84); Jean-Dominique (1748–1845)", pp. 45–47.
   "Cavendih, Henry (1731–1810), and Charles Augustin Coulomb (1736–1806)", pp. 48–51.
   "Cosmic Rays", pp. 74–76.
   "Curie, Marie Slodowska (1867–1934), and Pierre Curie (1859–1906)", pp. 81–83.
   "Deluc, Jean André (1727–1817)", pp. 85–86.
   "Ether", pp. 108–110.
   "Experimental Philosophy", pp. 111–113.
   "Fire and Heat", pp. 121–122.
   "Geodesy", pp. 131–133.
   "Imponderables", pp. 168–170.
   "Lightning Conductor", pp. 193–194.
   "Magneto-Optics", pp. 201–202.
   "Noble Gases", pp. 230–231.
   "Physics", pp. 248–252.
   "Planck, Max", pp. 252–253.
   "Pneumatics", pp. 260–261.
   "Quantum Physics", pp. 267–271.
   "Rainbow", pp. 277–278.
   "Relativity", pp. 280–283.
   "Röntgen, Wilhelm Conrad (1845–1923)", pp. 283–284.
   "Rutherford, Ernest (1871–1937)", pp. 284–285.
   "Standard Model", pp. 309–310.
   "Sympathy and Occult Quality", pp. 318–319.
   "Terminology", pp. 321–322.
   "X Rays", pp. 344–345.

[2] This volume offers a coherent selection of articles from *The Oxford Companion to the History of Modern Science* (2003). If some of the original entries may have been slightly revised or shortened, two entries—namely, "Cassini Family: Gian Domenico (1625–1712); Jacques (1677–1756); César-François Cassini de Thury (1714–84); Jean-Dominique (1748–1845)" and "Deluc, Jean André (1727–1817)"—are included here for the first time.

8. *The Oxford Illustrated Companion to the History of Modern Science*, John L. Heilbron editor in chief: Black Dog & Leventhal, 2008.[3]

   "Academies and Learned Societies", pp. 1–3.
   "Atomic Structure", pp. 31–33.
   "Cosmic Rays", pp. 75–76.
   "Ether", pp. 113–114.
   "Imponderables", pp. 152–153.
   "Magneto-Optics", pp. 179–180.
   "Nobel Prize", pp. 212–213.
   "Noble Gases", pp. 213–214.
   "Physics", pp. 239–242.
   "Pneumatics", pp. 249–250.
   "Quantum Physics", pp. 263–266.
   "Relativity", pp. 270–272.
   "Slogans from Science", p. 281.
   "Standard Model", p. 292.
   "Sympathy and Occult Quality", p. 297.
   "X Rays", pp. 311–312.

9. *Europe 1789 to 1914: Encyclopedia of the Age of Industry and Empire*, John Merriman, Jay Winter editors in chief, Farmington Hills, MI: Thomson Gale, 2006, 5 Vols.

   Vol. 4: *Pacifism to Syphilis*

   "Planck, Max (1858–1947)", pp. 1798–1800.

10. *Compendium of Quantum Physics: Concepts, Experiments, History and Philosophy*, edited by Daniel Greenberger, Klaus Hentschel, Friedel Weinert, Dordrecht etc.: Springer, 2009.

    "Rutherford Atom", pp. 671–675.

11. *Oxford Dictionary of National Biography: 2005–2008*, edited by Lawrence Goldman, Oxford: Oxford University Press, 2013.

    "North, John David (1934–2008)", pp. 844–845.

Translations

1. Alfred de Musset, "Never Swear to Anything: A Comedy in Three Acts", English translation by Charles B. Paul and John L. Heilbron, *Bulletin of the New York Public Library*, Vol. 75, No. 10, December 1971, pp. 543–572.
2. Giovanni Paltrinieri, *The* Meridiana *in the Basilica of San Petronio*, English translation by John L. Heilbron, Bologna: Basilica di San Petronio, 2007; 31 pp.

[3] This volume includes simplified versions of a selection (some 200 out of 600) of the same entries to the *The Oxford Companion to the History of Modern Science* (2003).

Interviews

1. "«J'ai été captivé par cette découverte»", interview by Aline Kiner, *Sciences et avenir*, avril 2005, p. 87.
2. "物理学革命如何影响20世纪的世界文化和政府 [How Did the Revolution in Physics Impact the Global Culture and Politics in the 20th Centuary?]", interview by Li Hu, *Sanlian Life Weekly*, No. 9, 28 February 2022 (= No. 1178), pp. 60–63.

Online Video and Audio Recordings[4]

1. "Science and History"
   https://www.youtube.com/watch?v=AXy3dhfXYZ8&t=831s
   A conversation with Harry Kreisler, University of California, Berkeley, 15 November 2007.
2. "Physics and History: Links Between Two Cultures Forged in the Baroque".
   https://www.youtube.com/watch?v=IWd84FyDxmM
   Charles and Martha Hitchcock Lecture, University of California, Berkeley, 8 March 2008.
3. "Galileo's Life and Work"
   https://www.youtube.com/watch?v=-NvJtopQ4uM
   Oxford University Press promotional video, 17 February 2011.
   See [Monographs, 20].
4. "Bohr's Creation of his Quantum Atom"
   https://www.youtube.com/watch?v=K-elU0hylBs
   Meeting of American Physical Society, Denver CO, 16 April 2013.
5. "My Courage is Ablaze so Wildly"
   https://www.youtube.com/watch?v=xstWDB5NjQQ
   Regia Academia Scientiarum Danica, Copenhagen, 11 June 2013.
   See [Essays, 113].
6. "H.G.J. Moseley in interaction with Bohr and Rutherford"
   https://www.youtube.com/watch?v=EYIhx5YFNg0
   Institute of Physics History of Physics Group, Manchester, October 2013.
7. "The Mind that created the Bohr Atom"
   https://www.youtube.com/watch?v=34pXSdDaBtY&t=1147s
   "Niels Bohr 1913–2013", Séminaire Poincaré, Paris, 7 December 2013.
   See [Essays, 102].
8. "Historical Myths About Science"
   https://www.youtube.com/watch?v=PxkxCSyP60Y
   Washington and Lee University, Lexington VA, 9 May 2014.
   See [Essays, 112].

[4] This list comprises Heilbron's video and audio recordings that were accessible online at the time of this volume's submission for press (10 July 2025). By their very nature, such items prove more ephemeral than printed works: they may appear on websites with varying URLs, be segmented and issued in separate parts, or indeed be entirely withdrawn from the internet. Furthermore, novel audio or video content might be uploaded in the future. I have decided to include this list—likely the most provisional of all—as further attestation to Heilbron's multifaceted scholarly activity, not least in the public understanding of science and its history.

9. "Wartime Activities of Participants in the Prewar Solvay Institutes for Physics and Chemistry"
   https://www.youtube.com/watch?v=39jl9GtPwaU
   Académie Royale de Belgique, Brussels, 14 November 2014.
10. "John Heilbron's Lightning Tour of the Big Questions in Physics"
    https://www.youtube.com/watch?v=7xhH_iFx18g&t=11s
    Oxford University Press promotional video, 26 November 2015.
    See [Monographs, 22].
11. "Epistemology: How Do We Know What We Know?"
    https://closertotruth.com/video/heijo-005/?referrer=7670
    "What is Consonance in Science and Theology?"
    https://closertotruth.com/video/heijo-002/?referrer=7681
    "Challenging Theological Doctrines"
    https://closertotruth.com/video/heijo-003/?referrer=7689
    "Critical Realism in Science and Theology"
    https://closertotruth.com/video/heijo-001/?referrer=7695
    "What Are the Scope and Limits of Science?"
    https://closertotruth.com/video/heijo-004/?referrer=8357
    Parts of a conversation with Robert L. Kuhn, for the "Closer to Truth" series, 22 March 2018.
12. "Welcome"
    https://www.youtube.com/watch?v=QnPjfSSiGaM
    "Looking Back as We Move Forward: The Past, Present and Future of the History of Science", opening remarks by Jean-Laurent Rosenthal and JLH, California Institute of Technology, Pasadena, 26 April 2019.
13. "Paradigm Shifts Across the Ages"
    https://www.youtube.com/watch?v=i6s5qqHz8uU
    St Cross Centre for the History and Philosophy of Physics, Oxford University, 19 February 2020.
14. "On the Ghost of Galileo"
    https://www.youtube.com/watch?v=k8Nt8R4OzYk
    A conversation with Katelyn Phillips for "The Side Comment" series, Episode 12, 27 January 2021.
    See [Monographs, 24].
15. "1643: The Ghost of Galileo"
    https://www.youtube.com/watch?v=NT1wYefRf9w
    Conversation with Violet Moller, introduced by John Hillman, 2 February 2021.
    See [Monographs, 24].
16. "The Assayer Assayed"
    https://www.youtube.com/watch?v=L13T4ERlsco
    XLIII National Congress, Italian Society for the History of Physics and Astronomy, Padua, 5 September 2023.

17. "Albert Einstein"
https://www.bbc.co.uk/sounds/play/m001qdx1
Melvyn Bragg in conversation with Richard Staley, Diana Kormos Buchwald and JLH, released on 14 September 2023.

On John L. Heilbron

Baggott, Jim. "Obituary of John Heilbron", *The Guardian*, 26 December 2023, p. 7.
Carson, Cathryn. "John L. Heilbron (1934–2023)", *Nature*, Vol. 626, No. 7997, 1 February 2024, p. 25.
Carson, Cathryn. "John L. Heilbron (1934–2023): In Memoriam", *Historical Studies in the Natural Sciences*, Vol. 54, No. 4, September 2024, pp. 521–526.
Gattei, Stefano. "John L. Heilbron", *Physis: International Journal for the History of Science*, Vol. 59, No. 1, 2024, pp. 287–295.
Gattei, Stefano. "John L. Heilbron, 1934–2023", *Journal for the History of Astronomy*, Vol. 55, No. 2, 2024, pp. 279–284.
Reeves, Eileen A. "John Lewis Heilbron (1934–2023)", *Galilæana*, Vol. 21, No. 1, 2024, pp. 255–270.

An extended *Éloge* (by Stefano Gattei, Daniel Kevles, Jessica Riskin and Peter Westwick) is forthcoming in *Isis*, Vol. 117, No. 1, 2026, pp. 171–177.

Online notices

Max Planck Institute for the History of Science, "In Memoriam, John L. Heilbron, 1934–2023", 5 November 2023.
https://www.mpiwg-berlin.mpg.de/de/aktuelles/memoriam-john-l-heilbron-1934-2023
Department of History, University of California, Berkeley. "In Memoriam, John L. Heilbron, 1934–2023", 6 November 2023.
https://history.berkeley.edu/news/memoriam-john-l-heilbron-1934-2023
Gambaro, Ivana (SISFA, Società Italiana per la Storia della Fisica e dell'Astronomia). "Addio a John Heilbron", 6 November 2023.
https://www.sisfa.org/notizie-eventi/addio-a-john-heilbron/
Baggott, Jim. "John Heilbron: A Distinguished Mind", 7 November 2023.
https://jimbaggott.substack.com/p/coming-soon
Museo Galileo, "John Lewis Heilbron, in memoriam", 7 November 2023.
https://www2.museogalileo.it/it/archivio-news/165-archivio-news-2023/2557-john-lewis-heilbron-in-memoriam.html
Castel-Branco, Nuno. "John Heilbron, 17 March 1934 – 05 November 2023", 29 November 2023.
https://pubs.aip.org/physicstoday/online/42911/John-Heilbron
American Academy of Arts and Sciences. "John L. Heilbron (1934–2023)", November 2023.
https://www.amacad.org/person/john-l-heilbron

A *Festschrift* volume was published in 2012: *Nature Engaged: Science in Practice from the Renaissance to the Present*, edited by Mario Biagioli and Jessica Riskin, New York: Palgrave Macmillan.

Stefano Gattei

Contents

Part II The Nineteenth and Twentieth Centuries

Part III Styles in the Historiography of Science

My Autobiography for John: An Open Letter to a Dear Friend

Dear John:

I submitted the first version of this letter to a *Festschrift* intended to celebrate your 90th birthday on 17 March 2024. Tragically and unexpectedly, you died at the end of 2023, after which the *Festschrift* had to be transformed into the present memorial volume. Nevertheless, I still want to formulate my greeting to you as a personal letter addressed to you with appropriate revisions, thus seeking to underscore in the best possible way the close friendship we developed over many years and to show the readers how I intended you to read my message. My wife, Anne, and I are happy that you accepted our dinner invitation when you last visited Copenhagen in August 2023, which we fortunately did not know then was soon going to be a sweet memory of our last meeting together. At that time, you were your old vigorous self, without any hint of bad health. That is how we wish to remember you, and this is the spirit in which I have sought to revise my letter.

This is not the contribution I first intended to write for the planned *Festschrift*. When I was asked to contribute to it, I considered it most appropriate to write a continuation of the first part of our co-authored book, *Love, Literature and the Quantum Atom*, which marked the hundredth anniversary of Niels Bohr's path-breaking atomic model from 1913. As readers of this open letter may not know or recall, that first part of our book constituted an article by me based on the early correspondence between Niels Bohr and his fiancée Margrethe Nørlund, which the Bohr family had generously given me special permission to consult for the purpose. My main argument there, which you agreed with and supported in your own contribution, was that Margrethe played an immeasurably important role not only for Niels's emotional—as opposed to professional—life, but indeed also for his ability to do physics research, which to him also involved his emotions. The letters add substantially to the understanding of Bohr the person, documenting, among many other things, the couple's shared strong interest in literature and the enormous debt that Bohr felt he owed his fiancée, which he wrote to her that he would never be able to pay back. Your own contribution to our book starts with the words: "The correspondence excerpted in the first part of this volume prompted a reconsideration of the account I published over forty years ago in collaboration with the late Thomas Kuhn."[1] In other words, the few letters between Niels and Margrethe that you and Kuhn had been allowed to see in preparation for your article had not been sufficient to allow such a reconsideration.[2]

The letters in the correspondence that I was given access to were written from the time when Niels and Margrethe first met and fell in love, through Bohr's postdoctoral stay in Cambridge and Manchester, until just after Niels and Margrethe married in 1913. The marriage took place shortly after Bohr's stay with Ernest Rutherford in Manchester, where his first thoughts about the atomic model were generated leading to the publication in 1913 of Bohr's famous "Trilogy," the first part of which introduced the model. The correspondence did not stop after they moved together, mostly because Niels for significant periods traveled without Margrethe when she was unable to join him on his journeys. My original intention was to contribute to the planned *Festschrift* with a continuation of my earlier article in order to investigate to what extent, based on this later correspondence, Margrethe continued to play a similar role in Niels's life and work. Unfortunately, I discovered somewhat late in my planning that the Bohr family is not yet ready to make the entire Niels–Margrethe correspondence available to researchers. The collection is likely to be a treasure, and the Bohr family intends to release it in due time. Under the circumstances, however, I could not wait for this, as I had to finish my article for you in time for it to appear in the *Festschrift*.

Now my problem was: how could I invent another subject for a tribute to you in order to finish my article in time? Considering the fast-approaching deadline, I decided that a genuine scholarly contribution was out of the question. The only alternative I found realistic was to make some autobiographical notes concentrating on matters that I deem to have some relation to what you, John, have meant to my own life and career. So the positive side to the Bohr family's negative decision is that it has inspired me to have some thoughts on this aspect of my life that I might never otherwise have had. My contribution is hence more of a personal greeting than what I had originally intended. To what extent it is appropriate for this memorial volume, readers must decide for themselves; at any rate, I am glad that my piece has been accepted for inclusion. From my own point of view, I feel both honored and obliged to contribute, and am anxious to join the other contributors in making a tribute to your memory.

[1] Finn Aaserud and J.L. Heilbron, *Love, Literature and the Quantum Atom* (Oxford: Oxford University Press, 2013), p. 105.

[2] John L. Heilbron and Thomas S. Kuhn, "The Genesis of the Bohr Atom" in *Historical Studies in the Physical Sciences 1* (1969), 211–289.

Intellectual Influence

I entered the world of the history of science only slowly, in a country with little experience in the field and with a native language spoken by only a small number of individuals around the world. Norwegian, needless to say, is unreadable to most people on the planet, including scholars. As a physics student at the University of Oslo, I was kindly allowed by my unconventional thesis adviser, Kristoffer Gjøtterud, to write a historical piece as my physics dissertation.[3] Apart from the realization that I would never become a great physicist and my reluctance to spend my entire life in the Norwegian high school system, my decision to turn into some kind of historian was motivated by my observation that historical introductions in physics textbooks were often not based on facts but were rather rational reconstructions for the teaching of physics. Such introductions struck me as utterly misleading. At this time, I also read the still influential book *The Structure of Scientific Revolutions* by Thomas Kuhn, which, with its presentation of paradigms and revolutions, introduced a model in agreement with my own amateurish view of the history of science.[4] In the process I discovered that many history of science journals were readily available at one of the university libraries close to the physics department. I devoured them, enjoying in particular some of your own articles, which you had published at this time. The discovery that history of science was a discipline with its own positions and publications was a great revelation to me.

An exemplary case of the misrepresentations of the history of science in textbooks is the treatment of the origin of quantum physics through Max Planck's introduction of the quantum of action, h, at the turn of the twentieth century.[5] This was made clear to me by the history of science literature on the matter that I had access to. From these readings I was motivated to dig into the prehistory of Planck's contribution, reading with great appetite the original publications of the Austrian physicist Ludwig Boltzmann and others in old physics journals easily available in the physics library, many of which nobody else seemed ever to have opened. As a consequence of these extensive studies, which took much longer than that of an average physics student at the university, I was able to conclude that even Martin J. Klein and other historians writing on the matter had not got the story quite right.[6] To me, it turned out that the question was not whether Planck had understood or had not

---

[3] Finn Aaserud, *Plancks energikvantebegrep: Et forsøk på å avklare hvordan det ble til* [*Planck's energy quantum concept: An attempt to clarify how it came about*], unpublished dissertation in physics, Institute of Physics, University of Oslo, May 1976.

[4] Thomas S. Kuhn, *The Structure of Scientific Revolutions* (Chicago: Chicago University Press, 1962).

[5] See, for example, Arthur Beiser, *Concepts of Modern Physics* (London: McGraw-Hill, 1967), p. 253. A search on the internet for, e.g, "Planck" and "energy quantum" confirms that this misrepresentation is still prevalent today.

[6] Klein has published extensively on the topic. See, for example, Martin J. Klein, "Max Planck and the Beginnings of Quantum Theory" in *Archive for the History of Exact Sciences 1* (1962), 459–479.

understood statistical mechanics, but that he had understood it on his own premises at the time. Because I wrote the dissertation in Norwegian, I was protected from getting into a discussion with all the historians of physics in many countries who had also written on the topic. They probably never heard about my dissertation criticizing their work. And had they heard about it, they would not have been able to read it anyway. There is one important exception, though, as I did send the dissertation to you, John, when I subsequently applied to Berkeley among other universities in North America to become a graduate student in the history of science. You (who were not among the historians criticized in my dissertation) claimed to be able to read Norwegian, and were certainly the only proper historian of science who may have understood my rather convoluted argument from reading my dissertation. However, we were never to discuss these matters in any detail.

My decision to begin an entirely new education abroad in the history of science was helped by my friendship with Robert Marc Friedman, a graduate student at the Department of the History of Science of the Johns Hopkins University, who had traveled the opposite way to study the biography of Vilhelm Bjerknes, the internationally renowned doyen of Norwegian meteorology.[7] In addition, a Norwegian friend from high school, who had gone on to study mathematics at the University of Oslo—where we continued to spend time together until he continued as a postdoc at the University of Chicago—strongly encouraged me to follow in his footsteps. For the same reasons that I decided to write a historical dissertation at the University of Oslo, I followed their advice, applying to graduate programs in the history of science at American universities. It was also helpful that the Norwegian Research Council for Science and the Humanities (*Norges Almenvitenskapelige Forskningsråd, NAVF*) was seeking to develop history of science in Norway and for this reason accepted my application for a grant over several years to study the field abroad. For a student, I lived quite well on the grant for as long as it lasted. In fact, because of the low US dollar rate, I was able to live more comfortably than the average American graduate student.

My main reason for accepting the offer from Johns Hopkins was my great admiration for Russell McCormmach, whose writings relating to my own dissertation topic in Oslo I had already learned to respect. Russ (as we called him) was the originator and editor of the leading-edge journal *Historical Studies in the Physical Sciences* (*HSPS*), the annual volumes of which included impressive "Editor's Forewords" providing context for the specialized articles in the rest of the volumes. His Editor's Foreword to Volume 3 of the journal from 1971 sought to place Planck's discovery within the context of the newly developing subdiscipline of theoretical physics, basing the argument on the general view that "the prevailing institutions and culture affect the scientist's thought and career largely through the mediation of the discipline."[8] I felt greatly privileged to have Russ as my dissertation adviser, and

[7] Robert Marc Friedman, *Appropriating the Weather: Vilhelm Bjerknes and the Construction of a Modern Meteorology* (Ithaca and London: Cornell University Press, 1989).

[8] Russell McCormmach, "Editor's Foreword" in *Historical Studies in the Physical Sciences, 3* (1971), Ix–xxiv, on p. x.

it was quite a learning experience to be able to discuss matters pertaining to the history of physics with him on a one-on-one basis. However, Russ left Hopkins before I had finished there, and I was lucky to receive Robert Kargon as my second adviser, not least because he was more insistent than Russ that I complete my degree promptly. I still remember Kargon's reaction to the seminar I held on my Oslo dissertation at the end of my first year at Hopkins: "Now you've shown that you can handle the physics. That's the easy part. Next you will turn to the real challenge, the history of it!" That's something I would never hear from the physicists at the University of Oslo!

At this time there was an exchange program for graduate students between the Hopkins Department of the History of Science and the *Institut d'histoire et de sociopolitique des sciences* at the French-speaking *Université de Montréal*. I was lucky enough to spend my third semester at this quite unique institution, which, unfortunately, has since folded. My main teacher there was the American historian of science, Lew Pyenson, who as a graduate student had obtained his Ph.D. with Russ at Johns Hopkins.

In Montreal I had been able to have a conversation with your dissertation adviser, Thomas Kuhn. Upon returning to Baltimore, by March 1978 I learned through the grapevine that Kuhn was about to finish a book on the history of Planck's energy quantum concept.[9] I soon wrote to him about my Oslo dissertation on the same topic, and he answered immediately that he would be happy to continue our conversation. From my letter he gathered that we may have come to similar conclusions, and wrote that he looked forward to talking with me about the matter at the annual event of the Joint Atlantic Seminar in the Physical Sciences for graduate students to be held later in the spring of 1978 at the Rensselaer Institute of Technology, where it just so happened that he was invited to give the senior lecture. I recall that the other students at the seminar were quite impressed that I was able to entertain the great Kuhn during the entire conference dinner and beyond. Although it was quite by chance that Kuhn and I shared the same interest, it felt like quite a victory for the fresh and unknown Norwegian student. It even seemed that Kuhn agreed with my argument with regard to whether or not Planck understood statistical mechanics at the time when he introduced the quantum of action. I never followed up on the matter, and Kuhn's book has since come to constitute the main contribution to the topic of how and why Planck introduced the energy quantum and how Planck was later to be seen as the originator of quantum physics. My encounter with your dissertation adviser, John, was an experience which helped bring my understanding of your own writing into context.

In the early 1960s you had taken part in the project *Archive for the History of Quantum Physics* (*AHQP*), with headquarters hosted by Bohr in Copenhagen. *AHQP* was created with the ambitious purpose of documenting the history of quantum physics by collecting correspondence and manuscripts from the physicists

[9] Thomas S. Kuhn, *Black-Body Theory and the Quantum Discontinuity, 1894–1912* (New York: Oxford University Press, 1978).

involved and by conducting oral history interviews with them.[10] Here Kuhn was the leader, assisted by his students—you and Paul Forman. *AHQP* has since been a crucial source for historians writing on the history of quantum physics, as it was also for me after I decided to write my doctoral dissertation on the history of Bohr's Institute for Theoretical Physics at Copenhagen University in the 1930s, in the hope that this topic someday might bring me back to Scandinavia. We are many who are in debt to you for your part in conducting the *AHQP* project.

The very first issue of *HSPS* from 1969 contained the important article, referred to in your above quotation from our book, by you and Kuhn on the genesis of Bohr's atomic model.[11] It was based to a great extent on the archival material collected for *AHQP*. A major point of that article is that Bohr's atomic model arose within a quite different physics paradigm than the present one and that its historical origin could hardly be understood from the point of view of physics at the much later time when *AHQP* was pursued. Toward the end of his life, even Bohr himself had difficulty appreciating that his discovery was made in an intellectual physics environment quite different from what he later helped to create, paradoxically enough, partly by means of his own discovery.[12] As such, although it is not stated clearly in your article, your own and Kuhn's contribution may in my opinion be seen as an earlier confirmation of Kuhn's view in *The Structure of Scientific Revolutions*[13] than Kuhn's own later book on Planck's energy quantum.[14] If my presumption is correct, Kuhn's thoughts have had an important effect on the early work of you and me.

Baltimore being close to Washington, D.C., I also had the opportunity to befriend Paul Forman, Curator at the National Museum of American History at the Smithsonian Institution. Paul became an important informal adviser to and influence on me for many years thereafter. When I first met him, his seminal article in the third volume of *HSPS*, arguing that the social environment in Weimar Germany was crucial for the germination of quantum mechanics, was hotly debated among both historians of science and physicists; his argument has subsequently become known as the "Forman thesis."[15] Unlike many of my physics friends, I greatly admired the article. The originally isolated Norwegian was definitely in the process of coming out of his hiding place and joining the international history of science community,

---

[10] Thomas S. Kuhn, J.L. Heilbron, Paul Forman and Lini Allen, *Sources for History Physics: An Inventory and Report* (Philadelphia: American Philosophical Society, 1967).

[11] Heilbron and Kuhn 1969 (note 2).

[12] See the description of the oral history interview with Bohr as part of the *AHQP* project, held shortly before Bohr's death: Thomas S. Kuhn, "Revisiting Planck" in *Historical Studies in the Physical Sciences 14* (2, 1984), 231–252, on 247–248.

[13] Kuhn 1962 (note 4).

[14] In 1984 Kuhn was compelled to point explicitly to the relationship between his philosophical and historical contributions and why he hardly ever pointed to this relationship in the latter. Kuhn 1984 (note 12).

[15] Paul Forman, "Weimar Culture, Causality, and Quantum Theory, 1918–1927: Adaptation by German Physicists and Mathematicians to a Hostile Intellectual Environment" in *Historical Studies in the Physical Sciences 3* (1971), 1–115.

which at this time was clearly centered in the USA. Later on, I continued to follow Paul's important work, first on the relationship between physics and national security in the USA,[16] and then his contribution to the understanding of the origins of postmodernism.[17] Indeed, the work of Kuhn and his two students has had a great influence on me!

I arrived in Copenhagen in 1980 to start research for my dissertation for Johns Hopkins at the Niels Bohr Archive (NBA), which had been in existence by name since Bohr's death in 1962 and which held the substantial collection of Bohr's scientific letters and manuscripts. It was established essentially to publish the *Niels Bohr Collected Works* and as such received limited funds from the Carlsberg Foundation. While there, I developed a close personal and working relationship with the director, Erik Rüdinger, who introduced me to the archival collections while he was busy preparing the publication of the *Collected Works*, his main task as director. Furthermore, I arrived just in time to meet the Australian Peter Robertson, who had just finished his excellent book on Bohr's institute in the 1920s.[18] Although our approaches to the institute's history were rather different, his well-researched book has been important for my own work, and we have kept in touch in spite of the long distance between us.

At Johns Hopkins I was educated to become a professional rather than an amateur historian, as I had been at Oslo. For one thing I learned to handle archival material. When writing my dissertation at Johns Hopkins, not only did I use the sources made available by the *AHQP* project, in particular Bohr's scientific correspondence and manuscripts available at NBA, but also employed other archival material, notably resources at the Rockefeller Archive Center in New York, which made possible a view of Bohr and his institute from the external point of view of a funding agency. The latter experience, in particular, confirmed in practice the realization that Paul Forman and others had already alerted me to, namely how the seemingly independent development of science may depend on circumstances entirely outside the field itself. In other words, science is a cultural entity to be considered as part of general history.[19]

Not long after completing my doctoral dissertation at Hopkins, I was accepted in 1985 in the temporary position as Associate Historian at the Center for History of Physics of the American Institute of Physics (AIP), then on Manhattan in New York. Although at the time I was helping at NBA to prepare the celebration of the

---

[16] See, in particular, Paul Forman, "Behind Quantum Electronics: National Security as Basis for Physical Research in the United States, 1940-1960" in *Historical Studies in the Physical Sciences 18* (1, 1987), 149–229.

[17] See, in particular, "The Primacy of Science in Modernity, of Technology in Postmodernity, and of Ideology in the History of Technology" in *History and Technology 23* (2007), 1–152.

[18] Peter Robertson, *The Early Years: The Niels Bohr Institute 1921–1930* (Copenhagen: Akademisk Forlag, 1979).

[19] My doctoral dissertation at Hopkins was later published in a much revised form as a book: Finn Aaserud, *Redirecting Science: Niels Bohr, Philanthropy and the Rise of Nuclear Physics* (Cambridge, etc.: Cambridge University Press, 1990).

centennial of Bohr's birth to take place in the fall of that year, NBA's director of the board, the late Aage Winther, strongly encouraged me to accept the position. He was going to be a great friend and supporter after I returned to Copenhagen four years later, eventually inviting the entire family every year for coffee on the day of Christmas Eve. When I applied for the position at AIP, my girlfriend from the University of Oslo from the mid-1970s—archaeology student Gro Synnøve Næs—promised immediately to join me in New York, feeling secure that someone else would get the job. We were both positively surprised when I was chosen for the position. Fortunately, Gro kept her promise.

In connection with the Bohr centennial in 1985, NBA was formalized as an institution at about the same time as I moved to the USA. The setup of the small institution was quite complex in spite of consisting of only two employees in addition to the director. It was established as an independent institution with its own statutes and its own board of directors chosen from relevant Danish institutions as well as from the Bohr family. It received partial funding from the Danish Ministry of Education, with the right to apply for support elsewhere. Its finances were administered centrally by Copenhagen University, while its location at the Niels Bohr Institute was determined according to a contract with that institution.

The stay at the AIP Center proved a great learning experience, where I picked up on the late Joan Warnow's innovative approaches to archival work, and where the director, Spencer Weart, introduced me to new ways of doing the history of science. As I was expected to, I chose an oral history and a historical research project, consisting, respectively, of interviewing American physicists regarding their involvement in postwar science policy and conducting a case study on the JASON group of academic physicists established in 1959 in the wake of Sputnik to advise the US government on national security questions. I was also expected to apply to foundations for support for my project and was able to acquire funding in addition to my salary provided by AIP, which allowed for research trips. Incidentally, the choice of JASON as a research project was suggested to me by one of your students, Robert Seidel.

My new project brought me to many places, including the University of California at Berkeley, where, if I recall correctly, I met you for the first time. It was during this period that you wrote and published your wonderful biography of Max Planck, with an emphasis on Planck's personal life, which, as a physicist dabbling in history, I had considered irrelevant when I wrote about Planck myself.[20] This was an eye-opener for me with regard to how a historian can, and often should, involve him- or herself in the personal life of the individuals dealt with. During my stay in New York, the American government compelled me to marry Gro, threatening that she would otherwise be expelled from the country. Our first child, Andreas, was conceived and born in God's own country. Because of this American intervention, he was born, morally correct, a good nine months after our marriage.

---

[20] John L. Heilbron, *The Dilemmas of an Upright Man: Max Planck as Spokesman for German Science* (Berkeley and London: University of California Press, 1986).

After four years, from 1985 to 1989, in New York City, I received an international call at my office from Erik Rüdinger, who told me that he was retiring from NBA to become involved with the Danish government in teaching physics in developing countries. Since physics and developing countries had long been his two main concerns, he simply could not forgo such an opportunity to combine the two interests. At first, I could not understand why he took the trouble to call me internationally to tell me this, so he had to bite the bullet and ask me directly whether I was interested in becoming his successor at NBA. I told him I would need a week to think about it. When I told Spencer and Joan about the call, they reproached me for not having accepted this unique offer immediately. Fortunately, the offer still stood when I accepted the position the week after. My strategy to return to Scandinavia had worked better than I could have imagined! While I hoped to exploit a continued Norwegian effort to build up the history of science there, not only would I now end up in Scandinavia, but even in the very lion's den of my Hopkins dissertation!

A Friendship Begins

While my story has so far been written mainly from memory, the rest will be based to a considerable extent on my correspondence with you from Copenhagen, which has been retained at NBA. It is so extensive that it has taken me days just to quickly go through it. That has indeed been an emotional as well as an intellectual experience, constituting a necessary reminder of the scope of your help and friendship over the last several years. Reading it has strongly added to my motivation for contributing to this memorial volume for you.

One of the other two employees at NBA I inherited from Erik was Felicity Pors. As you have experienced yourself, Felicity has an extraordinary memory regarding NBA's archival collections. In a very positive sense, she has been a challenge to work with, not least with regard to translating Bohr's writings from Danish into English. Coming from the USA, I was used to writing American-English, while Felicity, having grown up in northern England, insisted on proper English-English, which had indeed been the language used by previous editors of the *Bohr Collected Works*. Felicity has a great sense for language, and usually she won out. The other employee who stayed with me for most of my tenure at NBA was Anne Lis Rasmussen, whom I hired myself not long after being employed. Being educated as a librarian, she was responsible for the library and, as I am sure you agree, took pride in taking good care of you and other research visitors. She was also my much-needed teacher of written Danish, and I greatly profited from having her read proofs of my letters and other writings in the language.

The personal relationship between you and me took some time to develop. True enough, my pregnant Norwegian wife Gro and I did visit you and your wife Pat in Berkeley on a short trip to the USA in 1991, where I remember that you took us to an exquisite fish restaurant. But when Pat sadly died two years later, I learned about it not from you, but from your secretary, whom I also knew personally. The relationship between the two of us was still rather formal.

When I moved to Copenhagen, I was still hoping to finish a book on JASON, and when in 1992 Gro showed me a newspaper announcement of a grant from the American Council of Learned Societies, I applied for it. The grant enabled me to do research during the spring semester of 1993 at the University of California at San Diego (UCSD), where JASON still met regularly for its summer meetings; in the summer of 1993, I was able to conduct oral history interviews with several members of the group. Unfortunately, although I was in California, I did not find the time to see you during my time there. Owing to my heavy duties at NBA—which included the editorship of the *Bohr Collected Works*, obtaining necessary funding in addition to the insufficient amount provided by the Danish state, and doing my best to service researchers from all over the world seeking to use NBA's rich archival material—I was never able to finish the book. To compensate, I have recently donated the substantial archival material that I collected for the project to the AIP Center, now located in College Park, Maryland, in the expectation that other historians will have use for it.

My half-year at UCSD did, however, lead to an article in *HSPS*,[21] which Russ had left as editor at this time, being replaced by no other than you, John. In this way I was going to experience you as an editor when you reduced my article to half the size of my submitted manuscript. My immediate reaction was to consider your action a massacre, but I soon came to realize that it was a sign of your conscientiousness as well as your incredible capacity for work. It certainly improved the article, and I am grateful to this day for your great endeavor, which must have taken quite some time from your many other duties.

In 1995, actually a year after you had formally resigned from your Berkeley professorship, the two of us were modern enough to start corresponding by email. The following year I spent spring with the family at Churchill College in Cambridge, England, on a grant from the Carlsberg Foundation, studying documents pertaining to Bohr's political activities for an "Open World," which I was obliged to document in volume 11 of the *Bohr Collected Works*, and which has since been a major research interest of mine.[22] At the end of my stay I was happy to be introduced to your new companion, Alison Browning. I was soon to realize that in spite of your formal retirement, this was rather the beginning of a new life for you, in which I was going to be happily included. By then, you and I had become close enough for you to tell me about your failed attempt to buy a house in Oxford, which was followed by your buying April House in the village of Shilton near Oxford. I had hoped to meet you at the international history of science conference in Liège, Belgium, but you told me that it coincided with your father's 90th birthday, which of course was first priority for you. In 2000 my family of four, which nine years earlier had been

[21] Finn Aaserud, "Sputnik and the 'Princeton Three:' The National Security Laboratory That Was Not to Be" in *Historical Studies in the Physical and Biological Sciences 25* (2, 1985), 185–239.

[22] My most recent publication on this topic is Finn Aaserud, "Niels Bohr's Diplomatic Mission during and after World War Two" in *Berichte der Wissenschaftsgeschichte 43* (4, 2020), 493–520.

Fig. 1 You doing gardening at April House, summer 2010

supplemented with our daughter Karen, visited April House for the first time, where we experienced how at home you felt with the villagers, not least in the local pub (Figs. 1 and 2). On the academic side, this was the year when you invited me to write a couple of articles for the *Oxford Companion to the History of Modern Science*, of which you were the General Editor.[23] At the end of the year we met in Berlin, appropriately enough at the centennial for Planck's introduction of the energy quantum.

[23] Finn Aaserud, "Bohr, Niels (1885–1962);" *idem.*, "Bohr Institute;" *idem.* "Complementarity and Uncertainty." *Oxford Companion to the History of Modern Science*, J.L. Heilbron, gen. ed. (Oxford: Oxford University Press, 2003).

Fig. 2 You and I with our wives in your pub in Shilton

In 1998, the British playwright Michael Frayn's play *Copenhagen* premiered in London to great acclaim. As the play described a fictional conversation in the afterlife between Niels Bohr, his wife Margrethe, and Bohr's German physics colleague Werner Heisenberg about why Heisenberg had visited Bohr in Copenhagen in 1941 during the German occupation of Denmark, it was to be expected that Frayn had visited NBA to seek documentation for the Bohr–Heisenberg relationship. Unfortunately, I was away at the time. Of course, it did not take long for the play to be reenacted in the city carrying its name. So I found it only natural to organize a seminar in 1999 at the Niels Bohr Institute in which physicists, historians, dramatists, and actors were encouraged to interact, with the play *Copenhagen* as the point of departure.[24] The seminar proved a success, even though you were not present, John!

Actually, the seminar was so successful that I decided to organize a second and more ambitious colloquium two years later on the same theme and with the same name, "*Copenhagen* and beyond," in which you *did* participate actively.[25] The event took place in the Carlsberg Mansion, a house of honor that served as Bohr's residence from 1934 until his death in 1962. Naturally, Frayn himself was the main

[24] Transcriptions of the lectures and discussions can be seen on https://www.nbarchive.dk/nba_calendar/1999-2011/copenhagen_and_beyond_1/

[25] https://www.nbarchive.dk/nba_calendar/1999-2011/copenhagensymposium/

Fig. 3 Me and you at the second "*Copenhagen* and beyond" colloquium, in 2001

attraction of both events, but your contribution to the second helped making it into what I consider myself a high point of my directorship at NBA (Fig. 3). At the end of it, I was able to announce that the Bohr family had agreed to make public Bohr's drafts of his letters to Werner Heisenberg, written late in life, regarding the latter's Copenhagen visit in 1941. They were released on NBA's website, in facsimile, transcription, and English translation, shortly thereafter.[26] As far as I can tell, you genuinely enjoyed the conference and certainly appreciated the value of making the unsent letters available.

My main activity at the time, however, was the completion of the *Niels Bohr Collected Works*, the first volume of which had been published as early as 1972, and which it was still the NBA director's task to see through to publication. Even here, you contributed generous help and advice, John, not least improving both the language and content of the introductions that as editor I was obliged to write for each subsection of the last two volumes.[27] Your help has certainly added to the quality of the volumes that I was solely responsible for.

[26] The letters can be read at https://www.nbarchive.dk/collections/bohr-heisenberg/

[27] *Niels Bohr Collected Works*, Vol. 11, *The Political Arena (1934–1961)* and Vol. 12, *Popularization and People (1911–1962)* (Amsterdam, etc.: Elsevier, 2005 and 2007).

The Friendship Strengthens

Toward the end of 2005 I told you that I was diagnosed with chronic lymphatic leukemia (CLL), as well as diabetes, type 2, which has limited my activities, though certainly not stopped them, ever since. John, you have exhibited remarkable concern, understanding, support, and comfort throughout, showing your cordial friendship. I have been lucky, as almost 20 years later I have been able to quit medication for the CLL.

Shortly after my diagnosis, I was grateful to have the opportunity to write a small article in an issue of *HSPS* celebrating the journal's first editor, Russell McCormmach.[28] Naturally, I showed you the manuscript before submitting it. The special issue was edited by Lew Pyenson. Lew wrote an extensive Editor's Foreword, emulating our shared teacher's "Editor's Forewords" in the first issues of *HSPS*, in which he illuminated Russ's editorship, writings and personality through long quotations from Russ's personal letters to him.[29] Likewise, my own contribution included quotations from my correspondence with Russ during my Hopkins studies, which, like Lew, I felt obliged to ask Russ for permission to quote. In this connection it may be relevant to mention that the editors of the current volume insisted that the planned *Festschrift* should be a surprise for you, thus discouraging me from asking your permission to quote from the correspondence between the two of us, which I trust you would not have taken offense with. In the issue celebrating Russ, which included contributions to the journal by its contributors during the first ten years, when Russ edited it, you delivered your "swansong," announcing your retirement as editor of the journal and, of course, giving praise to its first editor.[30]

In January 2007 you told me that your father had died just before Christmas the year before. He was close to 100 years old.

In 2007 I was invited to take part in the celebration of a major contributor to *HSPS* over the years, namely Paul Forman, and in particular his revolutionary Forman Thesis.[31] The conference was to take place at the University of British Columbia in Vancouver, Canada,[32] and I looked forward to presenting an overview of Paul's academic work. Paul even sent me a box containing all his publications, from which I could prepare my talk. But this time my health failed me, as my immune system proved delinquent, and I was simply compelled to cancel my participation. Naturally, being so close to Paul, you took part in the conference. In your

[28] Finn Aaserud, "Russell McCormmach as a teacher" in *Historical Studies in the Physical and Biological Sciences 37* (2, 2007), 453–462.

[29] Lewis Pyenson, "Editor's Foreword" in *ibid.*, pp. 189–204.

[30] John Heilbron, "Swansong" in *ibid.*, pp. 185–187.

[31] Forman 1971 (note 15).

[32] A book based on the conference has been published as Weimar Culture and Quantum Mechanics: Selected Papers by Paul Forman and Contemporary Perspectives on the Forman Thesis, Alexei B Kojevnikov, Cathryn Carson and Helmuth Trischler, eds. (Singapore: World Scientific Publishing, 2011).

talk you admiringly referred to your three-years-younger former fellow student as "my old teacher," emphasizing Paul's unflinching independence as a historian, which did not always bring him friends, and illustrating your points with personal letters between the two of you.[33] You showed your understanding of my situation by writing to me on 15 January 2007 after I told you about my decision not to come: "I'm very sorry that you will not be able to go to Vancouver, but of course you cannot travel with a depleted immune system. You need a cast-iron stomach and an empty head to fly."

Your contribution to the *HSPS* issue honoring Russ was not your "swansong" in any other way, either concerning your own work or concerning your willingness and ability to help me and no doubt many others. Thus on 14 July 2005, when there was light at the end of the tunnel with regard to completing the *Niels Bohr Collected Works*, you wrote to me: "I'll be glad to be a sounding board for plans about the future activities of the NB Archive." Over the next several years you were going to live up to this promise to the fullest extent. For what should NBA represent after the publication of the *Bohr Collected Works* was completed, when many people understandably identified the institution with this publication? Both you and I strongly believed in a life for NBA after the *Collected Works*, not just as a continuation, but as a substantial expansion of activities. After all, there was still more to be written regarding Bohr—a biography of all aspects of his life, his physics, his thoughts on philosophy, his administrative accomplishments, his political involvement, his milieu, etc.—on the basis of the extensive archival collections held at NBA, which ought to be the natural center for such an enterprise. I have been grateful for your unflinching support in this regard against all odds of my unstable health as well as doubts and disagreements from the outside, in part as a consequence of NBA's complex setup. As you encouragingly wrote to me on 21 November 2007:

You have a number of promising pieces—the Ministry, the Bohr family, the existing Archive, Bohr's name, the centennial in 2013 [of the publication of the Bohr atomic model], the new edition [of the *Collected Works*], international interest, and, possibly, some connection with the university (as a source of graduate students in association with research projects etc.).

The new edition of the *Collected Works* referred to in your above quotation was realized as an electronic version and a reprint under new cover together with a new cumulative index. It was celebrated with a formal bestowal of the new edition from the publisher, Elsevier, at the Niels Bohr Institute on 6 June 2008.[34]

On 3 March 2011 you wrote in an email: "Alison and I are working on a curious project I'll tell you about in June that requires us to know everything we can about Cleyn." The Cleyn in question was the early seventeenth-century painter, Franz Cleyn, who, you wrote, "spent many years working for [the Danish King] Christian IV before going in 1623/4 to England." "So": you continued, "can you find out who

[33] J.L. Heilbron, "Cold War Culture, History of Science and Postmodernity: Engagement of an Intellectual in a Hostile Academic Environment" in *ibid.*, pp. 7–20.

[34] See https://www.nbarchive.dk/nba_calendar/1999-2011/bcw-newedition/

Fig. 4 You speaking at the 25th anniversary of NBA, 9 June 2011

the reigning Danish expert on Cleyn is?" I was of course glad to be trusted with this task and to become at least a little acquainted with your work on an entirely different century than the one that I have limited myself to. It was also the least I could do after all the help you had provided me with. You and I continued to correspond on and off about this for nearly ten years before your efforts resulted in a

Fig. 5 You and Alison enjoying summer in Copenhagen before celebrating the silver wedding anniversary of Gro and me, 10 June 2011

much-acclaimed book.[35] During this long period I collaborated first with Gro and after Gro died with my current wife Anne in order best to accommodate your wish to find out about Cleyn. I got the feeling that we were of some help.

On 9 June 2011, the 25-year anniversary of NBA's formal establishment was celebrated, one year late. At this event, you gave a talk on the history of NBA, introducing it with comments on the history and importance of archives in general, in particular, the background of *AHQP*. More importantly for me, you ended with a suitably ambitious vision for NBA's future. A revised transcript of this visionary talk, presented with your characteristic flair, is available on NBA's website.[36] It corresponded fully with my own vision for NBA's future at the time (Fig. 4). The following day happened to be the silver wedding anniversary of Gro and me, and my family of four was very happy to have you and Alison with us celebrating at our favorite Italian restaurant together with some of our other closest friends (Fig. 5).

Fortunately, your help has not been limited to administration and institution building. In fact, our most extensive and satisfying collaboration has been, as

---

[35] J.L. Heilbron, *The Ghost of Galileo: In a forgotten painting from the English Civil War* (Oxford: Oxford University Press, 2021).

[36] John Heilbron, "Talk at the Niels Bohr Archive, June 2011." Reproduced at https://www.nbarchive.dk/doc/heilbron.pdf

mentioned at the outset, the preparation of our book, *Love, Literature and the Quantum Atom*, written and published for the celebration in 2013 of the centennial of Bohr's atomic model.[37] The decision that my article on Niels and Margrethe should be included in a book co-authored by the two of us came about only slowly, when it became clear that I was the only one permitted to see the letters themselves, and that you would be allowed to see only the quotations that I chose from them in my forthcoming article. I still admire your acceptance of this condition and am grateful for the opportunity it gave me to prepare and publish a book together with you. Our discussions over an extensive period of the historical implications of the letters were the greatest possible pleasure and learning experience for me. I find it revealing to note that the tables had turned, and that it was now *I* who dealt with personal biography and *you* who took care of the scientific content considered in your chapter of our book. My approach to the history of science had certainly changed since I wrote my dissertation at the University of Oslo.

In mid-2011 the Danish Ministry decided to provide full support for NBA's day-to-day expenses, which meant that NBA would no longer need to apply to foundations simply in order to survive. This was of course a great relief, as I could from now on concentrate on applying for new projects rather than just keeping NBA above water. At the same time, NBA ceased being an independent institution under the Ministry, with Copenhagen University taking over this role. These decisions were made in connection with the grand preparations for the 2013 celebrations, in which NBA would naturally play a considerable role. While most of the celebratory events were organized by the Niels Bohr Institute, NBA organized an ambitious history of science conference, with contributions from a large number of some of the foremost historians of science all over the world, dealing with Bohr's atomic model and its historical consequences. The conference was held at the premises of the prestigious Royal Danish Academy of Sciences and Letters, with you—by now unsurprisingly—giving the public lecture introducing the conference. The proceedings from the conference were edited by me and my esteemed colleague, Helge Kragh, who at his retirement from his professorship at the University of Aarhus was then moving into an office at NBA, which he still occupies, adding much to the milieu at the small institution. These proceedings, comprising 560 pages, were published in 2015 by the Royal Danish Academy as the first volume of its new series *Scientia Danica. Series M, Mathematica et physica*.[38]

For a number of months in 2013 and early 2014, I had been informing you about Gro's condition after she had been struck with cancer. Then, on 17 April 2014, I told you and Alison that she had passed away the day before. Your response the day after you received my message gave me great comfort:

---

[37] Aaserud and Heilbron 2013 (note 1).

[38] *One Hundred Years of the Bohr Atom: Proceedings from a Conference*, Finn Aaserud and Helge Kragh, eds. (Copenhagen: The Royal Danish Academy of Sciences and Letters, 2015). Your lecture, "'My courage is ablaze so wildly' – Niels Bohr *en route* to his quantum atom," is reproduced on pp. 27–50.

I know something about the way you feel: it is devastating, the end of a way of life. You will need a change of scene. When you feel up to it, and would like to spend a little time here [in April House], you are very welcome.

We had come a long way in our personal relationship since I was told by your secretary in 1993 that your wife, Pat, had died. Although I did not take you up on your generous offer, it moved me greatly and helped me on my return to normal life.

That "normal life" consisted to a great extent of our continued shared effort to expand the activities of NBA. Already on 5 May I sent you an improved version of my detailed argument to establish a "Niels Bohr Archive Research Unit," to be submitted to the university. You responded the day after that "[t]his organization could be clarified by dividing NBA into a Collections Unit (CU) and a R[esearch] U[nit]. …" When the university representative responded that the document had the character of a status report rather than a vision for NBA's future, you gave me some comfort by interpreting this response as "not negative about prospects, but (correctly in my view) insistent on a better defined and more salable project or set of projects." After further thought and a meeting with the university's Vice Chancellor, the Research Unit was proposed to investigate "the inner dynamics of creative environments," a vision with which you expressed agreement in an email of 23 November. This is just one example of our extensive correspondence about the possible future of NBA.

In the meantime, on 4 June 2014 you had accepted the formal invitation to become a member of the NBA's Scientific Advisory Board (SAB), which was being established at this time. The first meeting of SAB took place in Copenhagen on 21 June the following year. It resulted in a document, received by NBA on 14 August, in which SAB gave advice regarding the future of NBA and in which your formulations undoubtedly played a considerable part. One of the several recommendations was to appoint a professor connected to NBA in order to establish and run the "Research Unit"; the university offered to establish the professorship not long after. You warned about not acting too fast, writing to me on 4 January 2016 that "care and caution are required. The wrong person can kill the edifice that has been so painfully built up." You need not have worried. The appointment of the professor, actually two half-time professors, was going to take place many years later, long after I had retired as director of NBA. The wheels do not always turn fast in Denmark!

This has also been the case for opening to researchers the private correspondence between Niels and Margrethe Bohr. Not surprisingly, after our work on our co-authored book, you had an interest yourself in the matter. On 22 June 2015 you wrote to the pertinent representative of the Bohr family and chairman of the NBA board of directors, Vilhelm Bohr, with a copy to me:

I'm grateful to you for taking the trouble to [make the family] release the documents that any serious intellectual biography of Bohr will have to consider. But it appears that the time is not ripe, and as I have other projects in hand and do not want to enter into one where there are any reservations about my use of documents, I ask you not to carry this matter further on my behalf. At some point the documents will be available and an appropriate biography will be written. Let's leave some tasks to posterity.

As stated at the outset, more than eight years later, the material is still not available to researchers.

The year 2016 saw further progress when NBA employed an archivist. As you wrote to me: "Very good news about the archivist. The hiring will keep up the momentum."

In early 2017, I decided to retire from my position, partly because I felt tired, and partly because I thought it might be advantageous to introduce fresh blood, showing to the powers that be that there might be a future for NBA. At that time I had also found a new companion, Anne Brockdorff Kristensen from the town of Odense in Denmark, whose first husband had died the same year as Gro. The relationship started when out of the blue she called me in order to offer a painting to NBA, and developed into marriage in September 2018. So there is hope for a future also for me. Retired from a long career as a teacher of Danish and English in the Danish high school (*gymnasium*) system, Anne has taken over Felicity's role in improving my English, in this article as well as in other writings. I have also become infected with her interest in and knowledge of literature.

On 17 March 2017, you accepted membership of a committee of five, set up to recommend my successor. For once, the procedure proved efficient, for on 8 September I was able to write to you that my successor "Christian Joas has just started here, and I'm not the boss any more." The salutation of my retirement on 2 October 2017 was quite a splash,[39] with the auditorium filled with colleagues from many places, several of whom gave flattering talks. On 25 April you had written to me with your characteristic sense of humor that you accepted "enthusiastically the invitation to ruin your retirement ceremony." Your lecture for me on that occasion is something that I still savor as a moving expression of our long-time friendship and collaboration. The lectures were followed up by a dinner for a select group at a nearby restaurant.

I am happy to say that my retirement has not meant that we have retired from keeping in touch, although the frequency of our correspondence has decreased considerably from the time when we sometimes wrote several emails a day. After you were introduced to my wife Anne, you invited both of us to April House, where we enjoyed our stay in late August 2019, as my family had done several years earlier.

After my retirement I have even been able to pay back another very small part of the debt that I owe you for your dependability and patience over the many years. On 17 December 2018, you wrote to me: "I shall make your life miserable in real time by sending you the chapters as I finish the final editing before sending the manuscript to the Press." These were chapters of your little book *Niels Bohr: A Very*

[39] The program for the conference can be read at https://www.nbarchive.dk/nba_calendar/2017/celebration-marking-finn-aaseruds-and-felicity-porss-retirements-from-the-niels-bohr-archive-nba/

Short Introduction, published by Oxford University Press in 2020.[40] I was proud to get the assignment to criticize the manuscript, and you even accepted some of my suggestions for improvement.

During your vacation in Italy, Alison got a stroke on 8 September, and some weeks later you too were taken ill. You died there on 8 November, while Alison sometime later was allowed by Italian doctors to return to England after being held back in Italy because of her health situation. She is only slowly recovering after her return. I learned of this tragic situation through my correspondence with Stefano Gattei, the editor of what was originally intended as a *Festschrift* for you, but has now become the memorial volume in which this article is a contribution. I am truly grateful to Stefano for keeping me informed and for helping you and Alison during these difficult times. He has furthermore alerted me to the Italian history of science milieu, of which you were already a part. Stefano has encouraged me to make contact with Alison, with whom I have phoned in Italy as well as at a nursing home in Shilton, where she now lives. We will keep in touch and I greatly hope that she will recover fully.

Concluding Remarks

I have called this rambling greeting "My autobiography for John," which may imply that my story is more about me than you. However, when you were still alive and this volume was still intended as a *Festschrift*, I found it most reasonable to place you in the context of my own life and career in the form of a personal letter addressed to you. I still consider this the most natural form in view of our close friendship. Moreover, as the first part of this essay is written from memory and not from written sources, I am sure that it suffers from lapses of memory similar to those you have pointed out yourself in those of your own writings that include personal recollections.[41] But my intentions have been good, and I hope that such lapses do not take away from my main point of remembering you.

Having been prevented from writing the scholarly piece I expected to celebrate you with, then, I have chosen instead to say something personal about what you have meant to *me*. True enough, our personal relationship came late in life, but from the very time I became interested in the history of science, your work has provided an important background, consciously and unconsciously. Developing a friendship and collaboration with you has been a greater pleasure and privilege for that reason. You have accomplished so incredibly much in your life, both administratively and intellectually, and have even been able to cover different centuries in your writings. I have admired most of these accomplishments at a distance without participating in

---

[40] John L. Heilbron, *Niels Bohr: A Very Short Introduction* (Oxford, etc.: Oxford University Press, 2020).

[41] See, for example, Heilbron 2011 (note 36), pp. 6–7.

them. My great privilege has been to be part at all of only a few of your activities, to receive your help and advice in my own work, and to be able to regard you as a close friend. I am sure that many other contributors to this memorial volume have had similar experiences with regard to other aspects of your life and career and could have written quite different autobiographies describing them.

I have greeted you on your birthday many times before and sometimes asked you how you were able to celebrate it. Once as an answer to this question you said something like: "The celebration went well. I was allowed to spend the whole day writing." Writing scholarly articles and books was your major enjoyment. It is tragic that it cannot be continued, but it is a solace that your writings and example will continue to provide inspiration and knowledge for others for generations to come.

With my very warmest greetings,
Finn

Finn Aaserud

Encomium

J.L. Heilbron has passed these to his Doktortochter, who has done her best to pass them to his many Enkel who have come through her hands: the utility of an armillary sphere, fidelity to truth, a scrupulous sense of justice, scholarly exactitude, ruthless editing, the distinction between restrictive and nonrestrictive clauses, insistence on the serial comma, and love of brevity.

A. J. Lustig

Part I
Early Modern Studies

Chapter 1
Vincenzo Galilei on Zarlino's Mathematical Demonstrations: Towards an Account of Science vs. Music in the Scientific Revolution

Maurice A. Finocchiaro

Abstract This essay begins with some autobiographical remarks intended to indicate the author's indebtedness to his former teacher John Heilbron with regard to the ideal of historical accuracy. It then sketches one of the author's current research projects, dealing with the contributions of Galileo's father, Vincenzo Galilei, directly to the sixteenth century revolution in music and musicology, and indirectly to the seventeenth century scientific revolution. As a step in this direction, the paper then gives an English translation of a 12-page passage in Vincenzo's *Discorso intorno all'opere di messer Gioseffo Zarlino* (1589); this is important because this passage contains a critique from the point of view of mathematical demonstration of the works of Zarlino, who was a leading exponent of traditional musicology. Then I undertake a provisional analysis of this passage, meant both to understand and to evaluate Vincenzo's critique. It turns out that the critique contains some merits and many faults, but there is much more to be done for a full account.

Keywords Galileo · Gioseffo Zarlino · Mathematical demonstration · Musicology · Science-vs.-music · Scientific revolution · Vincenzo Galilei

1.1 Introduction

In 1967, I was a graduate student in philosophy at the University of California-Berkeley. My specialization was the philosophy of science, but I was developing an increasing interest in the history of science, primarily as a result of my taking courses with Paul Feyerabend and studying his works (Feyerabend 1963, 1965).

M. A. Finocchiaro (✉)
Department of Philosophy, University of Nevada-Las Vegas, Las Vegas, NV, USA
e-mail: maurice.finocchiaro@unlv.edu

S. Gattei (ed.), *The Incomparable Historian*, Archimedes 75,
https://doi.org/10.1007/978-3-032-25678-2_1

The connection was that Feyerabend preached and practiced an historical approach to scientific methodology and philosophy of science; the key point was to learn about the nature and methods of science by studying the works of great scientists (such as Galileo, Newton, and Einstein) and the details of crucial historical developments (such as the seventeenth century Scientific Revolution and the transition from classical to twentieth century physics).

That year, John Heilbron joined the faculty of the history department at Berkeley. I had never met or heard about this new professor. However, I quickly learned that his specialty was the history of physics; that he had recently received his undergraduate and graduate education at Berkeley; and that his research had focused on early twentieth century atomic and nuclear physics, especially the work of Ernest Rutherford (Heilbron 1968). Heilbron's background impressed me because I myself had studied physics and philosophy as an undergraduate, and indeed I had written a bachelor thesis on the history and philosophy of quantum mechanics. Thus, I started taking some history-department seminars with Heilbron.

To be sure, at first I worried that my involvement with Heilbron's historical teaching and scholarship might distract me from my philosophical focus. That is, although (as mentioned) I had come to accept the historical approach to methodology and philosophy of science, the aim of such an approach is to develop general principles about the methods and nature of science; thus there might be some tension between such a general aim and the focus on specific and particular historical facts. Moreover, besides learning and accepting the historical approach to philosophy of science from Feyerabend, in my graduate studies I had also come under the influence of another one of my professors, namely Michael Scriven. From the latter, I had come to accept a focus on a cluster of fields described by such labels as applied logic, practical logic, informal logic, argumentation, reasoning, and critical thinking (Scriven 1968, 1976). It was already enough of a challenge to be torn between Feyerabend's historicist methodology of science and Scriven's applied-logical approach to argumentation and reasoning. Adding Heilbron's history of science to these two things increased the challenge even more.

Eventually, however, I was able to overcome such a challenge as follows. First, from Heilbron's work and teaching I derived something that may be labeled historical accuracy and conceptualized as a general aim which one can try to pursue in any field. Then I conceived a dissertation project that would combine my three influences. My dissertation followed Feyerabend's historicist methodology by focusing on Joseph Agassi's (1963) book *Towards an Historiography of Science*, which may be regarded as the first work dealing explicitly and systematically with the history of the history of science. It followed Scriven's applied logic by focusing on the logical interpretation and evaluation of Agassi's own arguments supporting his claims and of the arguments by his followers and critics; and here the focus was on explanatory claims, namely claims by historians of science trying to understand historical facts after they have been established, and claims by Agassi aiming to understand facts in the history of the history of science. And it followed Heilbron's ideal of historical accuracy by a careful examination of the accuracy of the claims about historical facts and figures by Agassi, his supporters, and his critics. The completed dissertation was accepted by these three scholars in the Spring of 1969, and it was later published as a book (Finocchiaro 1973).

Heilbron's ideal of historical accuracy with the rest of my tripartite approach was fruitful enough that I was able to apply and refine it in my next book-length project (Finocchiaro 1980): a critical examination of Galileo's *Dialogue on the Two Chief World Systems, Ptolemaic and Copernican*. This is the 1632 work which contains Galileo's synthesis of physics, astronomy, and methodology, and which was the immediate cause of his trial and condemnation by the Inquisition the following year. My book thus contains a meticulous analysis of the content, structure, and strength of the many arguments for or against the earth's motion discussed by Galileo; and this is meant to provide a critical understanding of Galileo's own reasoning and argumentation, as well as a data basis for theorizing in the style of applied logic. My book also contains an exhaustive account of the methodological and philosophical discussions elaborated by Galileo, together with a theoretical systematization of such discussions, and with an elaboration of the interconnections between these theoretical principles and Galileo's own scientific practice. And thirdly, my book carries to an unprecedented degree the textual analysis of what Galileo is saying, and in the process I evaluate the analyses advanced by other scholars. Finally, I am pleased to report that the many authoritative reviews of my book can testify to the success of my multi-faceted approach.

I am inclined to believe that I continued following and refining such an approach during my scholarly career. However, here my primary aim is neither autobiographical nor biographical vis-à-vis Heilbron.[1] Rather, it is to present my latest scholarly effort which hopefully follows such an approach, but which is also valuable in its own right. It deals with Galileo's father, Vincenzo Galilei, to whom I now turn.

1.2 Galileo's Father: Vincenzo Galilei

There is no need to elaborate the fact that Galileo's life, science, trial, and legacy can be studied and have been widely studied in order to learn more about such key topics as the Scientific Revolution of the seventeenth century, the nature of science in general, the relationship between science and religion, and his interdisciplinary aftermath in the last four centuries.[2] On the other hand, there would be more of a need to elaborate the fact that Galileo can be profitably studied from more original points of view, such as the theory and practice of logical reasoning; I have already mentioned such a possibility, but this is not what I want to elaborate here. Instead, what I want to do here is to point out that anyone seriously interested in Galileo will sooner or later come across his father, the relationship between father and son, and the relationship between science and music.

---

[1] However, elsewhere (Finocchiaro 2015) I have elaborated a thesis along both of these lines.

[2] Cf. Heilbron (1999, 2010), Finocchiaro (2005, 2019), Bucciantini 2021.

Vincenzo Galilei was born around 1520 in Tuscany and died in Florence in 1591. He was a musician in several senses of this word: performer (with the lute), composer, historian, critic, and musicologist. When he was about 40, he was able to study with a famous musician in Venice, Gioseffo Zarlino (1517–1590), a priest and the chapel master of St. Mark's Basilica. Zarlino was widely renowned for his 1558 synthetic book *Le Istitutioni harmoniche*. Despite their teacher-pupil relationship, Vincenzo did not become a follower of Zarlino. Indeed, their relationship soon soured, and they became engaged in a bitter controversy, recorded in their later and most famous books: Zarlino's *Harmonic Demonstrations (*1571*)*; Vincenzo's *Dialogue on Ancient and Modern Music* (1581, 2003); Zarlino's *Musical Supplements* (1588); and Vincenzo's *Discourse on the Works of Mr. Gioseffo Zarlino* (1589).[3]

There are several reasons why Vincenzo is an important figure in cultural history. One reason is that he pioneered the art of experimentation, which he practiced in the field of musical acoustics.[4] That is, he investigated how the musical sounds of vibrating strings changed by changing their length, tension, diameter, and material; this is something that had not been done systematically since the time of Pythagoras, whose alleged results had been uncritically accepted for two thousand years. Thus, Vincenzo had a pivotal role in making the science of music a more empirical discipline. Moreover, he influenced his son Galileo, who however applied the experimental method in mechanics and kinematics; that is, Galileo investigated whether and how the speed of falling bodies changed as a result of changing such variables as their weight, material, distance fallen, time elapsed, and angle in an inclined plane.

Although, as just mentioned, both father and son were thus revolutionizing their respective fields of research by making them more empirical, their empiricism did not consist primarily of qualitative and passive observation, which had been commonly practiced since the time of Aristotle. Rather, their experimentation was both active and quantitative; that is, it involved an intervention into the physical world and a manipulation of physical objects, with the aim of formulating mathematical formulas that interrelated the quantities being manipulated. This involved a new appreciation of mathematics, which could be contrasted to the uncritical mathematicism and number mysticism that had become common in both the science of music and the science of motion.[5]

In previous work (Finocchiaro 2020), I have already examined Vincenzo's *Dialogue*. Thus, now I feel that I should examine his other main work, the *Discourse*, which contains his critique of Zarlino's views. As many scholars have emphasized (e.g., Walker 1978, 14–26), the Galilei-Zarlino controversy was only partly intellectual, but also partly personal, and partly socio-political; moreover, the two thinkers were often contradicting themselves, often over-critical of each other, and

[3] Cf. Goldberg 2011, Kelleher 1993.

[4] Palisca (1961, 1992), Drake (1975, 1992), Finocchiaro 2020.

[5] Cf. Cohen 1984, Heller-Roazen 2011, Kelleher 1993, Peterson (2011, 149–173, 255–258), Walker 1978.

sometimes insincere in their criticism. However, in the present context, what impressed me most, and what I found most relevant, was a passage in the *Discourse* that explicitly criticizes Zarlino from the point of view of mathematics. I first learned about this passage from two scholarly works dealing comprehensively with the Galilei-Zarlino controversy, by Kelleher (1993) and Goldberg (2011); the latter even includes an annotated translation of Galilei's *Discourse*. I am much indebted to these two works; for example, I found Goldberg's notes very informative and useful. However, I found the translation itself unsatisfactory, partly because it contains many errors, and partly because the translation tends to be too literal, which has the consequence that the meaning is often hard to grasp. Thus, I decided to translate the passage myself, and this is done in the next section.

1.3 Vincenzo Galilei's *Discourse*: The Passage on Zarlino's Mathematical Demonstrations[6]

[§1.1] In accordance with the previously mentioned plan, I will now respond to Zarlino's complaints concerning mathematics; and I will do so with the few principles which I learned as a child. First of all, I say that in my *Dialogue* all calculations and computations are most correct and very easily explained. Admittedly, most of them are easy since the content did not require greater complication, which I thus avoided as best I could. Instead, whenever something could be done with simple words, I did not want to do it with difficult schemes and demonstrations; for the latter would not be understood by everyone, while the former would not be available everywhere or could not be used by everyone.

[§1.2] Now, going back to Zarlino, [p.57] in his book which he entitles *Harmonic Demonstrations* I do not see what he is trying to say, nor what the stories with which the book is filled have to do with the required demonstrations. To be specific, since this is what he wants me to do (although I myself do not like doing it), in Chap. 8 of Book I of his *Musical Supplements*, he writes this beautiful sentence: there can be no man of fame, reputation, or worth who is not well versed in mathematics. However, if one is to judge the worth of men from their understanding of mathematics, I will show how much of it he understands.

[§2.1] Let us begin with this. In the first discussion, he treats the fourth postulate[7] as very well known. But its obscurity has led to efforts to demonstrate it on the part of very great men, such as Eudoxus, Pappus, and Theon, leaving aside the fact that Euclid treated it as a definition. And he does this in all the following claims which he labels axioms, even though they are treated by Euclid as propositions worthy of

[6]Galilei 1589, 56–68. Numbers in square brackets are my own additions; some refer to my own subdivisions into sections, some to my own subdivisions into paragraphs, and some to the original page numbers.

[7]"A proportion between two extremes is said to consist of its mean proportionals as its parts" (Zarlino 1571, 27).

being demonstrated because of their difficulty; examples are the first, the fourth, the sixth, the seventh, and others. Now, this is common practice for commentators about easy things: these commentators are silent about difficult things because they do not understand them, and they excuse themselves by saying that they want to be brief and concise; but then about those things that are easy they have very long discussions. I let stand [p.58] the disorder in which he deals with them, when he lists some physical ones (such as the second) among the others that are mathematical; moreover, he lists them by taking them at random from the first and the seventh books of Euclid.

[§2.2] Then, with regard to demonstrations, besides their not being presented for the most part in the manner appropriate to them, there are many that are false. One example is the eighth,[8] in which he claims that, if one wants to place any number as the difference of any proportion, one should multiply that number by the terms of the proportion, whether or not they are root; he does not notice that this is not true except when the given proportion is between terms that are lowest and different from 1. To show that this is true, suppose that one wants to place 3 as the difference of the proportion 6:4; then one multiplies these terms by 3, yielding the numbers 18 and 12; but their difference is 6, not 3. Or suppose that the proportion is in its lowest terms, such as 5:2; the same difficulty arises; for if one wants to change its terms by 4, one gets 20 and 8; but their difference is 12, not 4. Therefore, this rule of his, which he advances as absolutely universal, is not true except when the proportion is of a specific one of its five kinds (namely the superparticular[9]), and when it is also in its lowest terms.

[§3.1] A ridiculous manner of arguing is used regarding proposition 36[10] of the same first discussion: he treats proportions as if they were quantities rather than [p.59] relations of quantities, by attributing to them those properties that belong to quantities, such as that the whole is greater than any of its parts. Thus he argues that, since the proportion of the first to the third consists of the proportion of the first to the second and of the proportion of the second to the third, the former is greater than each of the latter by being the whole consisting of the latter as parts. How true this is may be judged from this example: in *6:8:4* it is absolutely true that the proportion *6:4* consists of the two *6:8* and *8:4* as its parts (according to him). Nevertheless, whether the proportion *6:4* is greater that *8:4,* I shall leave it to the judgment of those who understand the eighth proposition of the fifth book of Euclid; he does not seem to understand it, as far as I can tell from what he writes here as well as in the next proposition. In that next proposition, based on the preceding false one, he writes these words: "Thus, by the preceding, it will be a larger proportion to have *8* compared to the composite number *12 and 417/512* than to have it compared to the

[8] "Any number can be placed as the difference between the terms of any proportion" (Zarlino 1571, 43).

[9] The Webster Unabridged Dictionary defines "superparticular" as "relating to a ratio in which the greater term exceeds the less by a unit (… 4 to 3 … 8 to 7)."

[10] "If we place three terms in sequence, the proportion that holds between the extremes is greater than each of those that hold between one of the extremes and the middle term" (Zarlino 1571, 76).

simple number *12*."[11] This is contrary to the right manner of demonstration: indeed, it's his misfortune that he never demonstrates anything; he always leaves in his pen all that is right in mathematics, which is the necessary demonstration of its conclusions.

[§4.1] How true this is may also be seen from proposition 14 of the second discussion, regarding Ptolemy's square:[12]

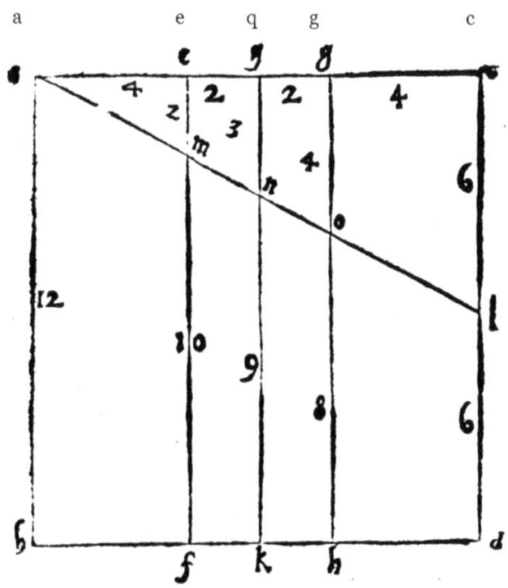

From this he derives the musical consonances. [p.60] In order to prove what needs to be proved, he writes these words: "subdivide *gh* into *go* and *oh* such that *go* is one-third of *ac*" (a beautiful conclusion indeed!). Next: "subdivide *qk* into *qn* and *nk* such that *qn* corresponds to one-fourth of *ac*" (indeed a concise way of saying what this man does not understand!). Next further: "finally, subdivide *ef* into *em* and *mf* such that *em* corresponds to one-sixth of *ac*" (something which is likewise left undemonstrated). Here, any person of judgment will know that in this proposition to which he gives so much space there is nothing needing proof except what he leaves undemonstrated; he does not provide a demonstration because he does not know how to do it; consequently, he has not understood Ptolemy from whom he takes it wholesale. He follows the same style in all the other things which he calls demonstrations.

[§4.2] However, using the few principles I know, I would have demonstrated it as follows. Since *cal* and *gao* are similar triangles, *ca* to *ag* has the same proportion as

[11] Zarlino 1571, 77.

[12] Here Galilei does not reproduce Zarlino's diagram, which is the following (Zarlino 1571, 116).

cl to *go*; but *ca* is 1 ½ times *ag*, and *cl* is 6 and *cd* 12; hence *go* is 4 and thus 1/3 of *ac*. For the same reason, *ca* to *aq* has the same proportion as *cl* to *qn*; but by hypothesis *ca* is double *aq*; and so *cl* is double *qn* and *cd* four times bigger. Similarly, *ca* to *ae* is as *cl* to *em*; and *ca* is triple of *ae*; and so *cl* is triple of *em* and *cd* is six times bigger. Q.E.D.

[§5.1] [p.61] This makes me think of the most kind Mr. Claudio da Coreggio; although he is the impersonation of modesty itself, I cannot believe that he would have heard some of these oversimplifications without laughing at them together with the others whom Zarlino introduces in his discussions. Indeed, he has done them the greatest harm by depicting them as men in the predicament of having to learn by demonstration things that are very well known; for example, that by drawing the fifth from the octave the fourth remains, which is an idea that would really bore even idle or obtuse persons. I wish that we would read only one such thing in his *Demonstrations*; instead the book is filled with them from beginning to end.

[§5.2] Now, please tell me, Mr. Gioseffo, from which mathematicians you have learned that we can formulate definitions and at the same time try to demonstrate them. To do this is precisely to want to litigate what has been agreed upon. In the eighth definition of the second discussion, he says that the octave is defined by the double proportion; then, in the fifth demonstration, he wants to demonstrate that the interval of the octave is multiple. In the ninth definition, he says that the [consonance of the] fifth is defined by the 3:2 proportion; and in the tenth definition, that the [consonance of the] fourth is defined by the 4:3 proportion; then, in the third proposition, he wants to demonstrate that the [consonances of the] fifth and the fourth are located in the largest superparticulars. [p.62] In the eleventh and twelfth definitions, he says that the ditone is defined by the 5:4 proportion, and the semiditone by the 6:5 proportion; then, in the seventh proposition, he demonstrates that the ditone and semiditone are superparticular (which is no small accomplishment!). In proposition 15 of the same discussion, he continues, the interval of the major semitone is defined by the 16:15 proportion; then, in the following proposition, he wants to demonstrate that the interval of the major semitone is superparticular (a beautiful accomplishment!); in proposition 21 of the third discussion, he demonstrates that the major semitone consists of a proportion larger than 17:16; in the following proposition, he says that the major semitone consists of a proportion larger than 18:17 (a beautiful novelty!); then, in the following proposition, he adds that the 15:14 proportion is larger than the major semitone; and then, in the following proposition, he shows that the 17:16 proportion is smaller than the major semitone.

[§5.3] As I said, the book of his *Demonstrations* is full of such subtlety. This is not surprising because in Venice there are so many printers that books are sold in large quantities. And this was the reason that induced him to expand his *Supplements* by so many pages, filling them as he pleased with words from my *Dialogue*. And besides doing what I just said, he was silent about my name, not out of charity, as he says in the preface [p.63] to his *Supplements*, but because of what I am about to say here.

[§6.1] He was silent about my name first because his many slanders would have prevented his book from being printed, and also to make sure that I and my *Dialogue* would not become known. With such a snare, he was trying to deceive the world into believing that the words he ascribes to me correspond to my thoughts; and so, without consulting my *Dialogue* and without further research, readers would attribute to me what he was alleging. Thus, if you want to see directly how deceptive this man is, even more than I have mentioned so far, and how wrongly he complains about me, you should place my *Dialogue* in front of you without believing what Zarlino writes in his *Supplements* and elsewhere; you will see that in these works he reports my clauses, sentences, and paragraphs incorrectly, truncated, and broken up.

[§6.2] Let me mention just one example. In the third chapter of the first of his *Supplements*, he quotes some of my words thus: "consider whether an instrument made from the shinbone of a crane, vulture, or eagle can act to strike men and take away their life."[13] In quoting these few words, he makes three kinds of errors: first, he translates my Florentine mother tongue into that of Bergamo; second, he shows that orthography had not been born in his time; and third, which is more important, he ruins the sentence by interpreting [p.64] the topic in reverse and attributing to himself my concept and to me his blunder. To explain this, one would need many words, so that for brevity's sake I will say only that one should see the passage directly; the same applies to the passage about "the quantity of the comma," which he quotes in the eighth chapter of the fourth book.[14] Finally, to the last-mentioned error, he adds one of a different kind, that is, to cite incorrectly chapters and passages; thus, when he says that I cite chapter 35 of his *Institutions*, in my *Dialogue* I write the fourth.[15] And it is very often that he plays these tricks to the reader.

[§7.1] However, it's now time to wrap up my comments on mathematics; and to be less tedious, I will discuss only some of the most famous things which he writes about. For example, regarding the fifth[16] proposition of the third discussion, he advances a very universal claim but a particular demonstration; that is, he proposes to demonstrate something about any space subdivided into many parts, and in what he calls the demonstration he limits himself to a particular space subdivided into nine parts. Moreover, what manner of demonstrating is it to use numbers? Perhaps he learned it from the eighth proposition of the fifth book of Euclid, which speaks of larger and smaller proportions.

[§7.2] Here, I want to teach him how to demonstrate it.[17]

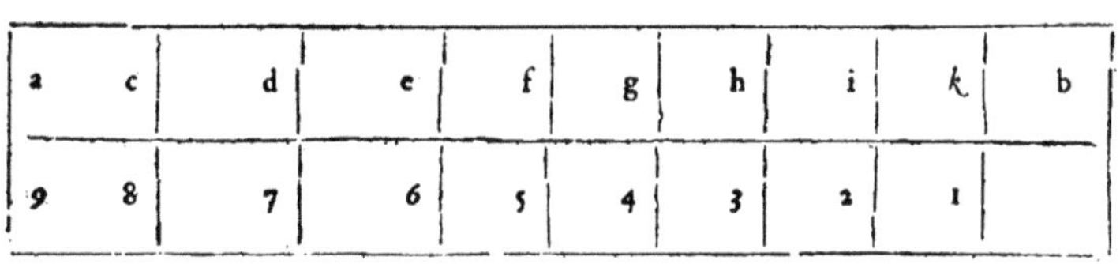

[13] Cf. Zarlino 1588, 14; Galilei 1581, 100–101.

[14] Cf. Zarlino 1588, 147; Galilei 1581, 29.

[15] Cf. Zarlino 1588,15; Galilei (1581, 80; 2003, 199, n. 388).

[16] The proposition Galilei is talking about is the one discussed in Zarlino (1571, 154–155), which is mislabeled no. 7, but which is actually no. 5 in the third discussion.

[17] Again, Galilei does not reproduce Zarlino's diagram, which is the following (Zarlino 1571, 155).

Ac being equal to *cd*, then by the seventh proposition of the fifth book, *ac* will have the same proportion to *cb* as *cd* has to *cb*; but by the eighth proposition of the same book, *cd* has a [p.65] smaller proportion to *cb*[18] than *cd* has to *db*; thus, combining by the 28th proposition of the same book, *ab* will have a smaller proportion to *bc* than *cb* has to *bd*. Q.E.D.

[§7.3] What's more, in his *Supplements*, in support of the truth of everything he says, he refers to his novel and ingenious *Demonstrations*. Let the above suffice for my comments about them.

[§8.1] Now, I will discuss one or two demonstrations which I read during a very quick look at his *Supplements*, since this is what he wants me to do (against my own inclination).

[§8.2] In Chap. 21 of the fourth book,[19] while elaborating things that are very superficial and of no importance, he overlooks the demonstration of what he proposes to demonstrate; that is, to multiply by adding whatever interval one chooses. First of all, he says that he understood this from the proof given for the preceding proposition of his same book; but the latter has nothing to do with what he wants to say. Thus, out of charity, I want to teach it to him.

[§8.3] Now, place in front of you the figure constructed as it is in the mentioned passage.[20]

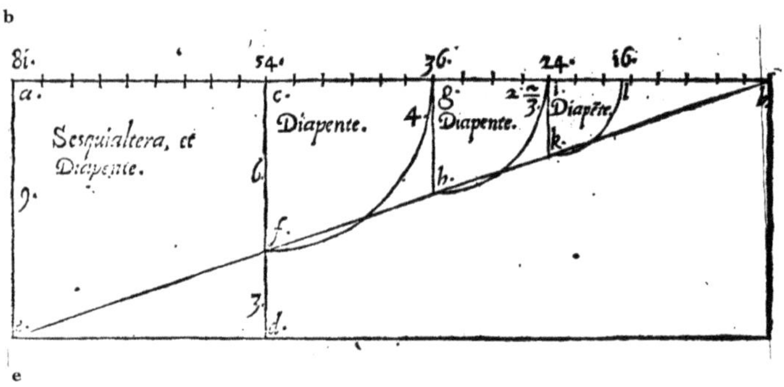

[18] Here, the text has 'ch', which is obvious a typographical error (Galilei 1589, 64).

[19] Here, the proposition which Zarlino is trying to prove reads: "How one can multiply by adding whatever interval one chooses; and some cautions about how to measure or divide a given quantity" (Zarlino 1588, 183).

[20] That is, in Chap. 21 of the fourth book of the *Supplements*; Galilei does not reproduce Zarlino's diagram, which is the following (Zarlino 1588, 183).

There, he says that *ae* is 1½ *cf*, by the definitions[21] of the first discussion of the *Demonstrations*. And he should say that this is true by the second proposition of the third book of Euclid. And because he did not know what one should say, he used his bow and arrow without taking aim, as usual; and so he did not specify which definition or which demonstration he was citing. Then he takes *cg* to be equal to *cf*, concerning which he composes a very tedious speech. [p.66] Then to demonstrate that *cb* is 1½ *bg*, since this is not meant for his teeth, he makes a series of blunders which is the most terrible in the world. That is, he first says that this is true because of the second part of definition 15 of the first book of Euclid, which reason has as much to do with the conclusion as I have to do with Peru. And to strengthen his reason he adds this: by the fourth proposition of the second book of the same Euclid, the two triangles *abe* and *cbf* are proportional. Now, in saying this he does not realize that he does not understand the terminology; for he is saying that Euclid is comparing two triangles, calling them proportional, and thus applying the concept of proportionality to three terms; he should have said that the two triangles are equiangular and their sides are proportional. But even if he had said so, it would not have been relevant; for the line *bg*, concerning which he intends to demonstrate that line *bc* is 1½ longer, is not a side of the two triangles mentioned by him. But let's not waste time on this, and instead let us teach him how to demonstrate that *cb* is 1½ *bg*.

[§8.4] He should say: since *cf* is parallel to *ae*, the two triangles *abe* and *cbf* are similar; and by the fourth proposition of the sixth book of Euclid, *ab:bc* as *ae:cf*, that is, as *ca:cg* since *cg* = *cf* [and *ae* = *ca*]. Now, since the whole *ab* is to the whole *bc* as the part *ac* is to the part *cg*, the remainder *cb* is to the remainder *bg* as the whole *ab* is to the whole *bc*; this is in accordance with proposition 19 of the fifth book of Euclid; therefore, *cb* is 1½ *bg*. [p.67] Q.E.D. And all other propositions are demonstrated in the same manner.

[§9.1] Now, it will seem impossible to some of his clique that, since he was able to demonstrate the preceding proposition[22] which is more difficult than this one,[23] he would have rambled in this manner in demonstrating a less difficult one; to remove such a thought from their minds, I will tell them the reason, which is this. Regarding the demonstration he gives to find two mean proportionals between two given lines, he has translated it verbatim from Eudoxus or Vitruvius; and again in this translation one can see how little he understands this topic. In fact, he shows that he does not know the difference between a rectangle and a square, when he calls the rectangle *bacg* a square:[24]

---

[21] Here Galilei speaks of "definition" in the singular; but there are several (14) in the first discussion of the *Demonstrations* (Zarlino 1571, 19–26).

[22] The proposition which Zarlino tries to prove in Chap. 20 reads: "How one can find two straight lines that are mean proportionals between two given lines, without the help of the Mesolabio" (Zarlino 1588, 181).

[23] In Chap. 21 (Zarlino 1588, 183).

[24] As usual, Galilei does not reproduce Zarlino's (1588, 182) diagram, which is this.

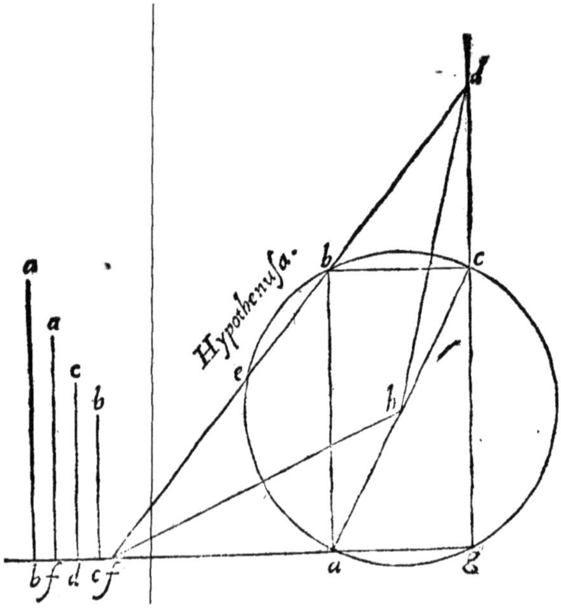

and yet their difference is learned from the definitions in the first book of Euclid. And, in order not to be accused of pedantry, I won't say much about his translation of the phrase "*Quod autem continentur sub et cetera*" which occurs several times in this demonstration, and which he translates as "all that is found below." From this one can see that he does not understand the point of such words, which refer specifically to the rectangle found below.

[§9.2] Now let him learn how much better it would have been if he had followed the advice which his friend gave him, according to what he says at the beginning of the fifth discussion; with as much modesty as possible, his friend tried to dissuade him from the project of publishing his demonstrations. From this one can see that the friend knew much better than he did what they were worth; but if he had followed [p.68] the advice he would not have learned what I have written so far for his instruction, assuming he is capable of it. Perhaps he is not, given that in his *Supplements* he misuses the poor Archimedes,[25] thus showing that he has mastered the latter's work as much as he has the Chaldean language.

[§9.3] Moreover, if he had understood the mentioned proposition, he would not have mourned as dead the duplication of the cube; and the same holds for his understanding of the proposition from which he derived the use of the Mesolabio, in accordance with Philo of Byzantium. But if he had read and understood the

[25] In Chap. 29 of book IV (Zarlino 1588, 207).

demonstration of Archytas regarding the construction of the two mean proportionals, he would have known that the demonstration is purely and simply geometrical, and not mechanical, like the different ones found by many others. Now, I believe he knows that the duplication of the cube involves nothing but the following: given two lines the first of which is ½ of the other, one finds the two mean proportionals between them; then the cube of the first mean is double the cube of the first (shorter) line; thus, once one demonstrates the construction of these mean proportionals, one has found the duplication of the cube. I am surprised that this man had the impudence of publishing such superficialities, for I am sure that in Venice this particular competence is not only appreciated among the nobles, but also that there are many who have it to an excellent degree.

1.4 Analysis and Commentary

This translated passage from Vincenzo Galilei's *Discourse* of 1589 certainly deserves some analysis and commentary. To begin with, it should be mentioned that my translation already embodies a considerable amount of analysis. In fact, Vincenzo's original book displays no breaks of chapters, sections, or paragraphs, but is a continuous script of 128 pages(!); moreover, the punctuation within the text is archaic and idiosyncratic in its use of commas, semicolons, colons, and periods. On the other hand, my translation obviously embodies and displays a subdivision into sections and paragraphs and a present-day more or less standard use of punctuation.

Another thing that makes this passage very difficult reading is the fact that it is full of specific references to Euclid's *Elements* and to Zarlino's *Harmonic Demonstrations* and *Musical Supplements*. I have not added notes quoting any of Euclid's statements, partly because they are too numerous, and partly because they can be checked and identified relatively easily from the standard translation and edition by Heath (1956). However, I have added some notes quoting and translating some of Zarlino's statements, namely those which I thought were more important for understanding or evaluating Vincenzo's critique, at least as a first approximation. Needless to say, a full reconstruction and evaluation of this critique would require the identification of all of Vincenzo's references to Zarlino, both statements of his claims and details of his demonstrations.

Here it may be useful to point out explicitly some features of Zarlino's books which are already implicit in Vincenzo's passage translated above. Zarlino's *Demonstrations* consists of five parts, each of which he labels "Ragionamento," which literally means "reasoning" or "argument," but which I have translated as "discussion." In so doing, I was partly following previous scholarly practice (Kelleher 1993; Goldberg 2011); but also I felt that labeling such parts "arguments" might detract from, or be confused with, the argumentative structure of what Zarlino calls "demonstrations." In fact, each part (discussion) of his book is structured basically like the various "books" of Euclid's *Elements*. That is, each of the five discussions consists of a series of numbered definitions, axioms, postulates, propositions, and demonstrations, all explicitly so labeled. On the other hand, Zarlino's *Supplements* has eight parts each labeled "book," and the only other subdivision is that each such book has many chapters.

A final general comment is that one should not miss the fact that Vincenzo's focus in this part of his book is mathematics and mathematical proof and argumentation. Referring to the sections and paragraphs which I have inserted in the passage, his focus is clear and explicit in paragraphs 1.1, 1.2, 2.1, 3.1, and 7.1.

Vincenzo's first criticism (in §2.1) concerns Zarlino's postulates and axioms: many such claims are problematic and deserve to be demonstrated, as indeed Euclid and other great mathematicians regarded them; others are not mathematical claims, but claims about the physical world. If and insofar as this is true, the criticism is valid, since Vincenzo is correct in assuming the principle that mathematical demonstration requires clarity about what needs or does not need demonstration. On the other hand, it would require a more detailed textual analysis to determine whether Zarlino's postulates and axioms have these characteristics. Moreover, according to Vincenzo, the source of this flaw is the common tendency to pretend that difficult things are easy and thus require little or no demonstration, and that easy things are difficult and thus can become the subjects of lengthy demonstrations; however, this explanation of Zarlino's alleged flaw is highly speculative and need not be accepted.

The next criticism (in §2.2) regards Zarlino's demonstrations, which Vincenzo claims are often false; that is, their validity is limited to special cases, whereas they pretend to prove universal generalizations. Vincenzo gives an example from Zarlino, but I confess that the details of Zarlino's demonstration and of Galilei's critique make no sense to me. On the other hand, I can acknowledge that one scholar (Goldberg 2011, 315) claims that Vincenzo's criticism is itself false because Zarlino's demonstration explicitly contains qualifications about the relevant special cases. In any case, Vincenzo is also implicitly assuming a principle which is correct, namely that a mathematical demonstration is invalid if its conclusion is formulated more generally than what is warranted by the special cases it examines.

Vincenzo's criticism in the next section (§3.1) is relatively clear and compelling. Zarlino is charged with a misunderstanding and misapplication of the Euclidean axiom that the whole is greater than any of its parts. This happens in Zarlino's proof of proposition 34, where his error is to treat relations of quantities as if they were quantities. And in the immediately following demonstration, this error leads him to the absurdity of claiming that *8:12 and 417/512* is greater than *8:12.*

In Sect. 4, the best sense I can make of Vincenzo's criticism is to conceptualize it in terms of what I would call elegance. That is, he is objecting to (indeed, making fun of) the fact that Zarlino makes it sound as if the three quoted sentences are disconnected assumptions resulting from how the diagram is constructed. By contrast, in his own substitute demonstration, Vincenzo focuses on the similarity of the triangles involved, and thereby derives the relationships described in those three sentences. Thus, Vincenzo's suggestion would yield a more systematic and inter-connected (i.e., more "elegant") demonstration. However, this does not mean that Zarlino's more roundabout demonstration is invalid, and that he does not know what he is doing.

In Sect. 5, Vincenzo's complaint is that Zarlino's *Demonstrations* is full of what he (Vincenzo) calls oversimplifications, redundant demonstrations, demonstrations that are essentially definitions, and subtleties. Many examples are given, all of which presuppose what at the time was the standard manner of defining musical consonances.[26] That is, consider the situation in which one produces sounds by plucking stretched strings whose tension is constant but whose length varies. When the length is shortened by a factor of two, the pitch goes up by what is called an "octave"; thus, the octave is defined in terms of the mathematical ratio 2:1. When the length is shortened to two-thirds of the original, the pitch goes up by what is called a "fifth"; thus the fifth is defined in terms of the ratio 3:2. When the length is reduced to three-quarters of the original, the pitch goes up by what is called a "fourth"; and so the fourth is defined in terms of the ratio 4:3.

It is difficult not to sympathize with Vincenzo if we rephrase his examples of what Zarlino is doing as follows: (1) from the definition of the octave in terms of 2:1, Zarlino demonstrates that the octave is defined in terms of a multiple ratio; (2) from the definitions of the fifth and the fourth in terms of the ratios 3:2 and 4:3, Zarlino demonstrates that they are defined in terms of the largest "superparticulars" (namely, ratios where the numerator exceeds the denominator by 1); (3) from the definitions of the ditone and semiditone in terms of the ratios 5:4 and 6:5, he demonstrates that they are defined in terms of superparticulars; (4) after defining the major semitone in terms of the ratio 16:15, he demonstrates in turn that this 16:15 is superparticular, that it is larger than 17:16, that it is larger than 18:17, that the ratio 15:14 is larger than it, and that the ratio 17:16 is smaller than it.

In Sect. 6, Vincenzo begins by complaining that, although Zarlino's *Supplements* is full of criticism of his *Dialogue*, he is never mentioned by name; this allegedly made it easier for Zarlino to have his own book published, but harder for readers to check the accuracy of what was being attributed to Galilei. In fact, in his quotations from the *Dialogue*, Zarlino takes liberties and commits errors such as the following: Zarlino tends to modify Galilei's Florentine language into a dialect from the city of Bergamo; he is careless about reproducing Vincenzo's original spelling and punctuation; he sometimes rephrases a quotation in such a way that the reader is led to confuse Galilei's thought with Zarlino's, and vice-versa; and he sometimes gives incorrect references to Galilei's works.

These objections, if and to the extent that they are true, show that Zarlino's practice of historical accuracy leaves much to be desired. However, even if true, they are irrelevant to the logical validity of Zarlino's mathematical demonstrations, which is what Vincenzo is supposed to be evaluating in this passage.

[26] Cf. Goldberg 2011, 1–23; Peterson 2011, 154–160.

Regarding the seventh section, let us begin by trying to understand Vincenzo's own demonstration. Referring to Zarlino's diagram (in note 17 above), Vincenzo argues as follows:

(A1) *ac* = *cd*; this is an assumption.
(A2) *ac/cb* = *cd/cb*; this is derived from (1) and justified by Euclid's proposition 7 of book V; and this amounts to saying that if we start with two equal quantities and then divide them by the same quantity, the results will be equal.
(A3) *cd/cb* < *cd/db*; Vincenzo justifies this only by referring to Euclid's proposition 8 of book V; but he should have added that this Euclidean proposition can be applied given that we assume that *cb* > *db*, which is true by construction of the diagram.
(A4) ∴ *ab/bc* < *cb/bd*; Vincenzo justifies this step by referring to Euclid's proposition 28 of book V, but unfortunately there is no such proposition in Euclid's book V, which has only 25 propositions; in any case, I have been unable to find a way of deriving (4) from (3).

On the other hand, I examined Zarlino's (1571, 154–155) own demonstration, and it seems to make sense. With some very minor rephrasing and some modernized notation, it's the following:

(B1) *ac* = *cd* = *de* = *ef* = *fg* = *gh* = *hi* = *ik* = *kb*; assumption, by construction of the diagram.
(B2) ∴ *ab* = *9 s*; where *s* is my symbol, not Zarlino's, for each segment of the diagram.
(B3) and *cb* = *8 s*.
(B4) *ab = cb + ac = cb + 1/8 cb = cb x (1 + 1/8) = 9/8 cb*.
(B5) *db = cb—cd = 8 s—1 s = 7 s*.
(B6) *cb = db + cd = db + 1/7 db = db x (1 + 1/7) = 8/7 db*.
(B7) *ab/cb = (9/8 cb) x (1/cb) = 9/8 = 1 1/8*; from line B4, by division.
(B8) *cb/db = (8/7 db) x (1/db) = 8/7 = 1 1/7*; from line B6, by division.
(B9) but, *1 1/8 < 1 1/7*.
(B10) ∴ *ab/cb* < *cb/db*; from lines B7, B8, and B9, by substitution.

My main reservation about this demonstration concerns the manner in which Zarlino formulates its conclusion, the proposition to be proved. His wording is this: "Consider any given distance subdivided into many equal segments: the proportion of the whole distance to the part in the nearby division is smaller than the proportion of this part to the remaining parts closest to it" (Zarlino 1571, 154). The two problematic phrases are "the part in the nearby division" and "the remaining parts closest to it", which strike me as meaningless; of course, they refer, respectively, to the "*cb*" and "*db*" in the proof, but I don't see that they can be described the way Zarlino describes them.

On the other hand, I don't see that there is a difficulty of over-generalization, which Vincenzo raises when he says that here Zarlino "advances a very universal claim but a particular demonstration." It seems to me that the logic of the demonstration would also apply if one was considering a different distance or a different subdivision into parts. Still less do I see any logical difficulty with Zarlino's use of described numbers, which is the other objection raised by Vincenzo.

In Sect. 8, Vincenzo begins to move away from Zarlino's *Demonstrations*, which is explicitly demonstrative, and to focus on the *Supplements*, where there is less of an emphasis on demonstrations. He criticizes the chapter dealing with the proposition, "How one can multiply by adding whatever interval one chooses" (Zarlino 1588, 183). He accuses Zarlino of an assortment of flaws such as superficiality, irrelevance, unsatisfactory references to Euclid's *Elements*, and misunderstanding of the concepts of proportionality and similarity of triangles. Then he concentrates on a particular strand of Zarlino's analysis, involving the demonstration that *cb* is 1½ *bg* in the diagram Zarlino uses;[27] Vincenzo complains that Zarlino's cannot prove this particular claim, and constructs his own proof. With some minor additions, rewording, and more modern symbolization, Vincenzo's demonstration (§8.4) is the following:

(C1) *cf* is parallel to *ae*; assumption, by construction of the diagram.
(C2) ∴ triangles *abe* and *cbf* are similar.
(C3) *ab/bc* = *ae/cf*; from C2, by Euclid, book VI, proposition 4.[28]
(C4) *cg* = *cf*; assumption, by construction of the diagram.
(C5) *ae* = *ca*; assumption, by construction of the diagram.
(C6) *ab/bc* = *ca/cg*; from C3, C4, and C5, by substitution.
(C7) *cb/bg* = *ab/bc*; from C6, by Euclid, book V, proposition 19.[29]
(C8) *ae* = *1½ cf*; assumption, by construction of the diagram.
(C9) *ab* = *1½ bc*; from C8, by Euclid, book VI, proposition 4.
(C10) *cb/bg* = *(1½ bc) x (1/bc)* = *1 ½*; from C7 and C9.
(C11) *cb* = *1 ½ bg*; from C10.

Vincenzo's criticism in this section is reminiscent of the one in Sect. 4, in which he provided a more elegant demonstration than the one given by Zarlino. However, there is one important difference: in that earlier critique Vincenzo was providing a more inter-connected and systematic demonstration than the one which he himself was attributing to Zarlino, whereas now he is providing a demonstration for something which he himself claims Zarlino does not or cannot provide; so, here the notion of elegance does not really apply. On the other hand, the more basic notion of demonstration does apply, in the sense that in the present context Zarlino presumably should provide a demonstration, but does not.

In the last section (§9), Vincenzo criticizes Zarlino's demonstration in Chap. 20 of the *Supplements*, which deals with the proposition, "How to find, without the help of the Mesolabio, two straight lines that are the mean proportionals between two given lines" (Zarlino 1588, 181). His motivation for doing so is understandable:

[27] Cf. the diagram in note 20 above.

[28] "In equiangular triangles the sides about the equal angles are proportional, and those are corresponding sides which subtend the equal angles" (Heath 1956, 2: 200).

[29] "If, as a whole is to a whole, so is a part subtracted to a part subtracted, the remainder will also be to the remainder as whole to whole" (Heath 1956, 2: 174).

at the beginning of Chap. 21, Zarlino states that he was led to the demonstration in that chapter by his demonstration in Chap. 20; thus, someone (especially among Zarlino's followers) might think that Vincenzo's criticism of Chap. 21 can't be correct unless there is something wrong with Zarlino's demonstration in Chap. 20.

Vincenzo begins this additional criticism by denying that the previous proposition (Chap. 20) has anything to do with the subsequent one (Chap. 21). Then he points out two things he finds wrong with that previous demonstration: at one point Zarlino calls "square" a geometrical figure which is a rectangle; and in several places he uses an imprecise Italian translation for a Latin phrase found in the text by Eudoxus or Vitruvius from which Zarlino is adopting his demonstration. Now, these attributions by Vincenzo are basically true. However, these criticisms are so secondary and unimportant that I think here Vincenzo is engaged in nitpicking.

Next, it should be noted that the mention of Archimedes reminds us of Vincenzo's aim in this passage to stress issues of mathematical demonstration. The background of these references is as follows. In the *Dialogue*, Vincenzo has one of the interlocutors end a discussion (that need not otherwise concern us here) by saying, as a kind of afterthought: "these considerations could provide, to a beautiful intellect skilled in mathematics, an effective instrument to find the way to the squaring to the circle, now mourned as dead."[30] In the *Supplements*, besides some substantive criticism, Zarlino tried to make fun of this claim with the following remark: "Oh blessed and happy Archimedes! If only in your own time you could have known someone who could have given you this advice and shown you the way! With such help, easily and to your greater glory, perhaps you could have discovered the squaring of the circle, not yet found by anyone and now mourned as dead, as my disciple hyperbolically puts it" (Zarlino 1588, 207).

To these relatively rhetorical exchanges, the next and last paragraph introduces a deeply mathematical issue, namely the duplication of the cube. The problem is that of constructing a cube whose volume is twice that of a given cube. Now, to begin with, it is obvious that the answer is not a cube whose side is twice as long as that of the given cube, for such a new cube would have a volume eight times that of the original cube; however, the side of the constructed cube must be longer than the side of the original cube but shorter than twice that length. This problem was solved in the fifth and fourth centuries B.C.by ancient Greek mathematicians such as Hippocrates and Archytas (see Huffman 2020). The solution involves two main steps regarding two lines the second of which is twice the first: one step is to show that if we find the two mean proportionals between these two lines, the cube of the first mean proportional equals twice the cube of the first line; the second step is to discover a method for finding or constructing two mean proportionals between two given lines.

Vincenzo complains that Zarlino mourns the duplication of the cube as an unsolved or insoluble problem. Supposedly, Zarlino does so because he does not understand the proposition he is talking about (in Chap. 20 of the *Supplements*). However, Zarlino does not even mention the duplication of the cube, let alone claim

[30] Galilei 1581, 135; cf. Galilei 2003, 331–333.

that it is an issue to be mourned as dead. So the relevance of this criticism by Vincenzo is questionable. Perhaps he is confusing this mathematical problem with the squaring of the circle, although that is unlikely. Or perhaps his criticism is just attacking a straw man.

1.5 Epilogue

It is obvious that Vincenzo's critique of Zarlino's mathematical demonstrations does not always involve mathematical or demonstrative issues, nor always questions of logical validity of reasoning, nor always the truth of Zarlino's claims. Instead, it often involves questions of relevance, clarity, understanding, elegance, rhetorical propriety, and fair reporting. However, this is not to say that the critique does not provide material for further theorizing about such concepts.

In this regard, I already had occasion to briefly discuss the concept of demonstrative elegance, and to illustrate it with two potential examples. I connected elegance to systematicity and simplicity; I pointed out that the only genuine example was the one where Vincenzo was providing a better demonstration than the one provided by Zarlino; and I pointed out that the second potential example was not a real one because there was no demonstration by Zarlino to be contrasted with the one suggested by Vincenzo.

Another item for further theorizing involves Vincenzo's criticism which I have labeled "nitpicking" and the one I have labeled "attacking a straw man." In my commentary above, I did not label these as "fallacies" committed by Vincenzo, but in fact I think they are good examples of such fallacies. Indeed, they would provide instructive case studies for practical logicians engaged in the theory of fallacies.[31]

It is also obvious that this passage has almost nothing directly or explicitly dealing with music and the science of music. Admittedly, if it had, the passage would have been much more difficult to understand, evaluate, and analyze. Thus, the present essay merely scratches the surface of the topic of the relationship between science and music in the scientific revolution.

References

Agassi, Joseph. 1963. *Towards an Historiography of Science*. The Hague: Mouton.

Bucciantini, Massimo, ed. 2021. *The Science and Myth of Galileo Between the Seventeenth and the Nineteenth Centuries in Europe*. Florence: Olschki.

Cohen, H. Floris. 1984. *Quantifying Music: The Science of Music at the First Stage of the Scientific Revolution, 1580–1650*. Dordrecht: Reidel.

Drake, Stillman. 1975. The Role of Music in Galileo's Experiments. *Scientific American* 232 (6): 98–104.

[31] Cf. Scriven 1976, 71–73; Finocchiaro 1980, 332–342.

Drake, Stillman. 1992. Music and Philosophy in Early Modern Science. In *Music and Science in the Age of Galileo*, ed. Victor Coelho, 3–16. Dordrecht: Springer Netherlands.

Feyerabend, Paul K. 1963. How to Be a Good Empiricist. In *Philosophy of Science: The Delaware Seminar*, ed. B. Baumrin, vol. 2, 3–39. New York: Interscience.

Feyerabend, Paul K. 1965. Problems of Empiricism. In *Beyond the Edge of Certainty*, ed. R. Colodny, 145–260. Englewood Cliffs: Prentice-Hall.

Finocchiaro, Maurice A. 1973. *History of Science as Explanation*. Detroit: Wayne State University Press.

Finocchiaro, Maurice A. 1980. *Galileo and the Art of Reasoning: Rhetorical Foundations of Logic and Scientific Method*. Dordrecht: Reidel.

Finocchiaro, Maurice A. 2005. *Retrying Galileo, 1633–1992*. Berkeley: University of California Press.

Finocchiaro, Maurice A. 2015. A Para-clerical Approach to the Galileo Affair and to Science vs. Religion. *Science, Religion and Culture* 2 (4): 125–142. https://doi.org/10.17582/journal.src/2015/2.4.125.142.

Finocchiaro, Maurice A. 2019. *On Trial for Reason: Science, Religion, and Culture in the Galileo Affair*. Oxford: Oxford University Press.

Finocchiaro, Maurice A. 2020. Vincenzo Galilei's Musicology and Galileo's Science: Methodological Comparison and Contrast. *Isis* 111:740–758.

Galilei, Vincenzo. 1581. *Dialogo della musica antica et della moderna*. Florence.

Galilei, Vincenzo. 1589. *Discorso intorno all'opere di messer Gioseffo Zarlino*. Florence.

Galilei, Vincenzo. 2003. *Dialogue on Ancient and Modern Music*. Trans. and Edited by Claude V. Palisca. New Haven: Yale University Press.

Goldberg, Randall E. 2011. "Where Nature and Art Adjoin: Investigations into the Zarlino-Galilei Dispute, Including an Annotated Translation of Vincenzo Galilei's *Discorso intorno all'opere di messer Gioseffo Zarlino*." Indiana University: Ph.D. Dissertation in Musicology.

Heath, Thomas L., trans. and ed. 1956. *The Thirteen Books of Euclid's Elements*, 3 vols. New York: Dover.

Heilbron, John L. 1968. The Scattering of α and β Particles and Rutherford's Atom. *Archive for History of Exact Sciences* 4:247–307.

Heilbron, John L. 1999. *The Sun in the Church: Cathedrals as Solar Observatories*. Cambridge, MA: Harvard University Press.

Heilbron, John L. 2010. *Galileo*. Oxford: Oxford University Press.

Heller-Roazen, Daniel. 2011. *The Fifth Hammer: Pythagoras and the Disharmony of the World*. New York: Zone.

Huffman, Carl. 2020. Archytas. In *The Stanford Encyclopedia of Philosophy (Winter 2020 Edition)*, ed. Edward N. Zalta. https://plato.stanford.edu/archives/win2020/entries/archytas/.

Kelleher, John Emil. 1993. Zarlino's *Dimostrationi harmoniche* and Demonstrative Methodologies in the Sixteenth Century. Columbia University; Ph.D. Dissertation.

Palisca, Claude V. 1961. Scientific Empiricism in Musical Thought. In *Seventeenth Century Science and the Arts*, ed. Hedley Howell Rhys, 91–137. Princeton: Princeton University Press.

Palisca, Claude V. 1992. Was Galileo's Father an Experimental Scientist? In *Music and Science in the Age of Galileo*, ed. Victor Coelho, 143–152. Dordrecht: Springer Netherlands.

Peterson, Mark A. 2011. *Galileo's Muse: Renaissance Mathematics and the Arts*. Cambridge, MA: Harvard University Press.

Scriven, Michael. 1968. Science: The Philosophy of Science. *International Encyclopedia of the Social Sciences* 14:83–92.

Scriven, Michael. 1976. *Reasoning*. New York: McGraw-Hill.

Walker, Daniel Pickering. 1978. *Studies in Musical Science in the Late Renaissance*. London: Warburg Institute.

Zarlino, Gioseffo. 1558. *Le Istitutioni harmoniche*. Venice.

Zarlino, Gioseffo. 1571. *Dimostrationi harmoniche*. Venice.

Zarlino, Gioseffo. 1588. *Sopplimenti musicali*. Venice.

Maurice A. Finocchiaro, majored in physics and philosophy at M.I.T, and then earned a Ph.D. at the University of California, Berkeley, with a dissertation in the logic of the historiography of science. He has published 17 books and about 400 essays on such topics as the concept of scientific explanation, the science vs. religion problem, Galileo, and informal logic and critical thinking. Among these, worth mentioning here are *The Galileo Affair* (Berkeley, 1989), *Arguments about Arguments* (Cambridge, 2005), and *On Trial for Reason* (Oxford, 2019).

Chapter 2
Reading Copernicus in Padua: The Case of Paolo Sarpi

Stefano Gattei

Abstract This article reconstructs Paolo Sarpi's engagement with Copernican astronomy and his gradual, informed acceptance of the heliocentric hypothesis. Through a close analysis of Sarpi's *Pensieri naturali, metafisici e matematici* (Natural, Metaphysical, and Mathematical Thoughts), it traces his intellectual trajectory from Aristotelian conceptions of motion and causality to a reasoned endorsement of the new, Sun-centred world system. Drawing on Sarpi's early reflections on mechanics, optics, and celestial motion, the study situates his evolving thought within the dynamic cultural and philosophical milieu of late sixteenth-century Padua—a context shared with Galileo and sustained by the intellectual freedom of the Venetian Republic. The paper examines Sarpi's original interpretations of Copernican arguments, including his responses to problems of parallax, impetus, and the tides, and highlights the depth of his understanding of the later, technical books of *De revolutionibus*. Finally, it considers Sarpi's nuanced reaction to the 1616 condemnation of Copernicus, showing how his position as *consultore in iure* of Venice shaped his public stance.

Keywords Paolo Sarpi · Galileo · *De revolutionibus* · *Pensieri naturali, metafisici e matematici* · Padua · Astronomy · Natural philosophy

> Accordingly, I have no desire to conceal from your Holiness that I was prompted to consider a different system of deducing the motions of the universe's spheres for no other reason than the realization that astronomers do not agree among themselvces in their investigation of this subject.
>
> Nicholas Copernicus[1]

---

[1] "Itaque nolo Sanctitatem tuam latere, me nihil aliud mouisse, ad cogitandum de alia ratione subducendorum motuum sphærarum mundi, quam quod intellexi, Mathematicos sibi ipsis non con-

---

S. Gattei (✉)
Department of Sociology and Social Research, University of Trento, Trento, Italy
e-mail: stefano.gattei@unitn.it

S. Gattei (ed.), *The Incomparable Historian*, Archimedes 75,
https://doi.org/10.1007/978-3-032-25678-2_2

The ten months spanning June 1609 to March 1610, marked by the advent of the telescope and the publication of Galileo's *Sidereus nuncius*, represent a pivotal moment in the history of scientific thought. While the Copernican hypothesis, set forth some 70 years prior in *De revolutionibus* (1543), furnished the mathematical groundwork for a novel conception of the world system, Galileo's telescopic observations imbued it with palpable reality, thereby contributing decisively to removing it from the traditional realm of mathematical models and cementing its place within natural philosophy. They also proved critical in embedding the heliocentric hypothesis within a tumultuous cultural, political and social milieu, thus restoring the revolutionary impetus its author had intended.

2.1 Sarpi's (and Galileo's) Padua

Galileo's groundbreaking telescopic discoveries emerged from an exceptionally vibrant and intellectually fertile historical and cultural setting. Padua stood as one of Europe's preeminent universities between the sixteenth and seventeenth centuries. Esteemed figures such as Agostino Nifo, Pietro Pomponazzi and Jacopo Zabarella had both studied and taught there, consolidating the University of Padua's reputation as a foremost hub for the development of the scientific method. Andreas Vesalius, Gabriele Falloppio and Nicholas Copernicus (who, albeit briefly, was a been a pupil of Girolamo Fracastoro and Luca Gaurico) pursued their studies in Padua, and William Harvey would follow suit in 1599. At the time of Galileo's arrival from Pisa in 1592, the faculty in Padua included Girolamo Fabrici d'Acquapendente, Cesare Cremonini and Santorio Santorio.

Moreover, the city boasted Gian Vincenzo Pinelli's very rich research library, which, over time, became a celebrated meeting place for intellectuals from across the Continent. It was within this stimulating environment that Galileo encountered several scholars and scientists who sojourned in Padua for varying periods of time, thereby forging significant connections with leading figures of the Venetian milieu. Finally, Padua benefited immensely from its proximity to Venice. Though Venice had by then surpassed its zenith of prosperity, it retained its formidable significance as a geographical, political and economic center, fostering a rich tapestry of cultural

stare in illis perquirendis": Nicholas Copernicus, *De revolutionibus orbium coelestium, Libri VI*, Nuremberg: Johannes Petreius, 1543; critical edition, French translation and notes by Michel-Pierre Lerner, Alain-Philippe Segonds and Jean-Pierre Verdet, with the collaboration of Concetta Luna, in Nicolas Copernic, *De revolutionibus orbivm cœlestium / Des révolutions des orbes célestes*, Paris: Les Belles Lettres, 2015 (henceforth, *De revolutionibus*; English translation by Edward Rosen, *On the Revolutions*, Baltimore-London: The Johns Hopkins University Press, 1978, 1992^2), Vol. II, pp. 6.32–7.4 (from the dedicatory letter to Pope Paul III; Engl. transl., p. 4).

and commercial exchange. Crucially, the Venertian Republic could guarantee its citizens, and the University of Padua in particular, an invaluable measure of freedom and intellectual autonomy.

An integral, and in many respects central, figure within the intellectual landscape that nurtured Galileo's discoveries was Paolo Sarpi (1552–1623). Most likely, Galileo first encountered Sarpi within Pinelli's scholarly circle, alongside Martino Sandelli, Lorenzo Pignoria, Paolo Gualdo (who would later write Pinelli's biography), and Antonio Querenghi, to name but a few.[2] Indeed, Sarpi was arguably Galileo's principal interlocutor during his Paduan years and undoubtedly played a pivotal role in Galileo's physical and astronomical investigations undertaken in Padua.

Nonetheless, in light of the surviving documentation, precisely assessing Sarpi's actual contribution to the burgeoning new science proves challenging, as his influence remains elusive and not readily quantifiable. Of paramount importance for any endeavour to evaluate Sarpi's input is his compilation of *Pensieri naturali, metafisici e matematici (Natural, Metaphysical and Mathematical Thoughts)*. This collection offers a rare glimpse into the scientific intellect of a complex author, whose intricate reasonings may be tentatively reconstructed today, some four centuries later, from scattered elements — much as the true scale of a floating iceberg can only be surmised by observing its visible tip.[3]

[2] The first document attesting to their friendship and the commonality of their scientific interests is a letter from Sarpi to Galileo on September 2, 1602: see *Le Opere di Galileo Galilei*, National Edition edited by Antonio Favaro, Florence: G. Barbèra, 1890–1909 (henceforth, *OG*), Vol. 10, no. 83*, pp. 91–93.

[3] In *Pensieri naturali* — a sort of notebook, and at the same time a concise intellectual diary — Sarpi gathers his reflections on a diverse range of subjects, dating from approximately 1578 to 1597, though not all the annotations precisely are dated. Regrettably, the autograph text, along with numerous other manuscripts by Sarpi, was lost to a fire that consumed the library of the Venetian convent of the Order of the Servants of Mary on the night of 17–18 October 1769. Fortunately, a complete copy and a partial copy of the text, both dating from the mid-eighteenth century, have survived. A comprehensive edition of Sarpi's 'thoughts' is now available, supplemented by additional materials: Paolo Sarpi, *Pensieri naturali, metafisici e matematici. Manoscritto dell'iride e del calore. Arte di ben pensare. Pensieri medico-morali. Pensieri sulla religione. Fabulae. Massime e altri scritti*, edited by Luisa Cozzi and Libero Sosio, Milan-Naples: Ricciardi, 1996 (henceforth, *Pensieri naturali*). Prior to this, portions of Sarpi's 'thoughts' were edited by Romano Amerio in Paolo Sarpi, *Scritti filosofici e teologici editi ed inediti*, Bari: Laterza, 1951 (see also Romano Amerio, *Il Sarpi dei pensieri filosofici inediti*, Turin: Edizioni di "Filosofia", 1950). An earlier, largely unreliable transcription was provided by Giovanni Papini in Paolo Sarpi, *Scritti filosofici inediti*, Lanciano: R. Carabba, 1910. See also Giovanni Da Pozzo, "Per il testo dei *Pensieri* del Sarpi", *Bollettino dell'Istituto di Storia della Società e dello Stato*, 3, 1961, pp. 139–196; Libero Sosio, "Nota al testo dei *Pensieri* di Paolo Sarpi", *Differenze*, 9, 1970, pp. 285–329; Luisa Cozzi, "La tradizione settecentesca dei *Pensieri* sarpiani", *Studi veneziani*, 13, 1971, pp. 393–450; and Libero Sosio, "I *Pensieri* di Sarpi: breve storia del testo", *Studi storici dell'Ordine dei Servi di Maria*, 52, 2002, pp. 7–69. On Sarpi and Copernicus, see Libero Sosio, "Il copernicanesimo del Sarpi nei *Pensieri*", in Pacifico M. Branchesi and Corrado Pin (eds.), *Fra Paolo Sarpi dei Servi di Maria*, Venice: Comune di Venezia, 1983, pp. 153–186; this work was later re-elaborated as "Fra Paolo Sarpi e la cosmologia", in *Pensieri naturali*, pp. LXXXIX-CXCIV. I am indebted to Libero

2.2 Between Rome and Venice

"Upon completing his duties as provincial superior [1582], and once relieved of his charge, he entered a period of tranquillity, which he regarded as the most profound repose he had ever known in his life […]. And just as rest is more gratifying and enjoyable to one who is wearied, so he wholly immersed himself in the contemplation of natural phenomena for three years".[4] Thus wrote Fulgenzio Micanzio, Sarpi's successor as consultant to the Venetian Republic and biographer, as well as diligent and devoted custodian of his memory and manuscripts. Although deeply rooted in the years of his youth, Sarpi's reflections on scientific and natural matters received a renewed impetus during the three years preceding his departure for Rome. There, in 1585, he was dispatched to serve as Procurator General to the Apostolic See, a role he fulfilled until 1588. During this time, he garnered the esteem of Cardinal Giulio Antonio Santori, the protector of the Servite Order, and Pope Sixtus V. Concurrently, he forged associations with notable figures from the Curia, including the Jesuits Nicolás de Bobadilla (one of the Society's early members) and Roberto Bellarmino (then a professor at the Roman College, who had contended that the pope was not entitled to exercise temporal jurisdiction), as well as Giovannni Battista Castagna, who would ascend to the papacy as Urban VII in 1590.

In Rome, Sarpi assiduously cultivated his mathematical and speculative interests, and whilst frequenting Jesuit circles, he likely encountered Cristopher Clavius, who was lectuiring at the Roman College. It is perhaps through Clavius that Sarpi became acquainted with the unpublished optical works of Francesco Maurolico (1494–1575), arguably the foremost Italian scholar of optics and mathematics in the first half of the sixteenth century.[5] Furthermore, during a sojourn in Naples whilst residing in Rome, Sarpi reconnected with Giovanni Battista Della Porta (1435–1515),

Sosio (1935–2023) for providing copies of these works of his, and for his assistance in understanding Sarpi's scientific thought.

[4] "Col finir il carico di provinciale, deposto il peso, entrò in una quiete ch'egli chiamava tutto il riposo che godesse nella sua vita […]. E come a chi è stanco è più soave il riposo e più lo gusta, così per tre anni s'immerse tutto nelle speculazioni delle cose naturali": Fulgenzio Micanzio, *Vita del padre Paolo* [1646], in Paolo Sarpi, *Istoria del Concilio Tridentino seguita dalla "Vita del padre Paolo" di Fulgenzio Micanzio*, edited by Corrado Vivanti, Turin: Einaudi, 1974, 2011^2, Vol. 2, pp. 1291–1433: p. 1309.

[5] Clavius, who had been a friend of Maurolico, received from the latter manuscripts of his unpublished works on optics for publication 1 year prior to Maurolico's demise. *Photismi de lumine et umbra* was indeed published in 1611, accompanied by notes from an unnamed author, which were in fact penned by Clavius himself (see Francesco Maurolico, *Optica*, edited by Riccardo Bellé and Ken'ichi Takahashi, Pisa-Rome: Fabrizio Serra, 2017). Sarpi's expertise in optics is also acknowledged by Girolamo Fabrici d'Acquapendente (ca. 1533–1619), who wrote: "The secret was observed and revealed to me by the Reverend Father Master Paolo Veneto, of the Order of the so-called Servants, a distinguished theologian and philosopher, as well as a great scholar of the mathematical disciplines, and especially of Optics, whose name I mention here to honor him" ("Quod arcanum obseruatum est, & mihi significatum à Reu. Patre Magistro Paulo Veneto, Ordinis vt appellant Seruorum, Theologo, Philosophoque insigni, sed Mathematicarum disciplinarum, præcipueque Optices maximè studioso, quem hoc loco honoris gratia nomino": *Liber de visione*

whom he had previously met in Venice in 1579 (where Della Porta had travelled on behalf of Cardinal Luigi d'Este to oversee the production of certain optical instruments). Years later, in the Proem to Book VII of the *Magia Naturalis*, which was dedicated to magnetic phenomena, Della Porta expressed profound admiration for Sarpi:

> In Venice we encountered Reverend Master Paul, of Venice, then Provincial of the Order of Servants, and now its most worthy Procurator. He was devoted to these same studies. Not only are we not ashamed to acknowledge that we have learned something from him, but we are proud of it, for, of all those whom we have happened to meet, we have never known anyone more learned and subtler than him, who was born to encyclopedic knowledge. He is the splendor and ornament not only of the city of Venice, or of Italy, but of the entire world.[6]

The scientific and philosophical 'thoughts' from his Roman period — to which other scattered, undated annotations on Apollonius' *Conics*, Euclidean and Archimedean geometry, Archimedean and pseudo-Archimedean statics, and Guidobaldo Del Monte's *Mechaniche*, should probably also be linked — illuminate Sarpi's approach to the sciences. On the one hand, he grappled with the problem of the mathematization of natural philosophy; on the other, he evinced a tendency to unify sensory and perceptual phenomena within atomistic theory, articulated a cohesive view of the sensory faculty and the intellect, and challenged the Peripatetic theories of disparate faculties and the distinction between agent intellect and possible intellect, as well as the very concept of form itself, which Sarpi, in alignment with Epicurean theories, conceptualized in materialistic terms.[7]

However, it was only with his return to Venice in 1588 that Sarpi found the tranquillity he needed to resume his studies and dedicate himself to them: "[... He] returned from Rome to quietude and to his studies, [...] and all the time that

---

sive de oculo visus organo, in *De visione voce auditu*, Venice: Francesco Bolzetta, and Padua: Lorenzo Pasquato, 1600, p. 93).

[6] "Venetiis eidem studio inuigilantem cognouimus R. M. Paulum Venetum, ordinis Seruorum tunc prouincialem, nunc dignissimum procuratorem, a quo aliqua didicisse non solum fateri non erubescimus, sed gloriamur, quum eo doctiorem, subtilioremque, quotquot adhuc videre contigerit neminem cognouerimus, natum ad Encyclopediam: Non tantum Venetæ vrbis, aut Italiæ, sed orbis splendor, & ornamentum": Giovanni Battista Della Porta, *Magiæ naturalis libri viginti*, Frankfurt: André Wechel, 1591, pp. 288–289. In his second printed work, the young Galileo paid tribute to Sarpi, too: "[...] among those connoisseurs of the Mathematical sciences, was the Most Reverend Father, Master Paolo, of the Servites, theologian of the Most Serene Lordship, of whom I can without any hyperbole affirm that no one in Europe surpasses him in knowledge of these sciences" ("[...] tra i quali intendentissimi delle scienze Matematiche, vi erano il M. Reuer. Padre Maestro Paolo de i Servi Teologo della Sereniss. Signoria, del quale posso senza iperbole alcuna affermare, che niuno l'auanza in Europa di cognizione di queste scienze": *Difesa Contro alle Calunnie & Imposture di Baldessar Capra Milanese* [1607], in *OG* 2, p. 549).

[7] As Libero Sosio suggests in "Fra Paolo Sarpi e la cosmologia" (cited, footnote 3), p. CI, it was perhaps the study of the *Optics* of Alhazen and Witelo (*Opticæ Thesaurus. Alhazeni arabis libri septem, nunc primùm editi. Eiusdem liber De crepusculis & Nubium ascensionibus. Item Vitellionis Thuringopoloni libri X*, Basel: Eusebius & Nikolaus Episcopius, 1572) that prompted Sarpi to abandon Aristotle and reconsider the ancient philosophy of the Presocratics, as well as the physical-astronomical theories of the Stoics.

remained from his divine offices [...] he spent amongst books. At that time, he committed to writing some of his natural, metaphysical, and mathematical thoughts; later, upon re-reading them, he did not hold them in high esteem, and was wont to remark, 'Oh! What childish thoughts passed through my mind'. Yet, I am quite certain that learned men, upon perusing them, would not deem them childish".[8]

2.3 From Aristotle to Copernicus

An early stage in the evolution of Sarpi's astronomical views is documented in 'thought' no. 1, penned in 1578. This entry, which opens the collection — almost programmatically, one is tempted to say — amounts to a consideration of method:

> We possess certain knowledge of both the being and the cause of those things that we know precisely how to construct. Of those we ascertain merely by experience, we know the being, but not the cause. By conjecturing it, we merely seek the possible; but amongst many causes we deem possible, we cannot be certain which one is true — which, as we can observe, holds true for the descriptions of the theories of celestial objects, and would similarly befall those who examine a clock for the first time. [...][9]

Here, Sarpi seems to acquiesce to the traditional view — articulated, amongst others, by Andreas Osiander in the unsigned Preface to Copernicus' *De revolutionibus*,[10] but also, prior to him, by Simplicius,[11] and subsequently adopted by Thomas

---

[8] "[...] ritornato da Roma alla quiete et a' suoi studii, [...] e tutto il tempo che gl'avanzava da' divini offizii [...] lo spendeva ne' libri. Scrisse in quel tempo alcuni suoi pensieri naturali, metafisici e matematici, i quali dopo rivedendo, non ne faceva stima, e soleva dire: 'Oh! che puerizie mi passavano per la mente'. Et io sono ben sicuro che vedendogli gl'uomini dotti, non le stimaranno puerizie": Micanzio, *Vita del padre Paolo* (cited, footnote 4), p. 1322.

[9] "Sappiamo certo e l'esser e la causa di quelle cose di far le quali abbiamo perfetta cognizione; di quelle che solo conosciamo per isperienza sappiam l'esser, ma non la causa; conghietturandola poi, cerchiamo solamente quella che è possibile, ma tra molte cause che troviamo possibili non possiamo certificarsi qual sia la vera: il che si vede avvenir nelle descrizioni delle teorie celesti, ed averebbe a chi vedesse di prima faccia un orologio. [...]": *Pensieri naturali*, no. 1, p. 3. In what follows, my understanding of Sarpi's *Pensieri* is based on Sosio's detailed commentary.

[10] See *De revolutionibus*, p. 2.8–14 (Engl. transl., p. xx): "For it is the duty of an astronomer to compose the history of the celestial motions through expert and careful study. Then he must conceive and devise the causes of these motions or hypotheses about them. Since he cannot in any way attain to the true causes, he will adopt whatever suppositions enable the motions to be computed correctly from the principles of geometry for the future as well as for the past. [...] these hypotheses need not be true nor even probable. On the contrary, if they provide a calculus consistent with the observations, that alone is enough" ("Est enim Astronomi proprium, historiam motuum cœlestium diligenti & artificiosa obseruatione colligere. Deinde causas eorundem, seu hypotheses, cum ueras assequi nulla ratione possit, qualescunque excogitare & confingere, quibus suppositis, ijdem motus, ex Geometriæ principijs, tam in futurum, quam in præteritum recte possint calculari. [...] Neque enim necesse est, eas hypotheses esse ueras, imo ne uerisimiles quidem, sed sufficit hoc unum, si calculum obseruationibus congruentem exhibeant").

[11] Simplicius, *In De caelo*, I 2 (= p. 32.29–32 Heiberg; English translation by Ian Mueller, *On Aristotle On the Heavens 1.2–3*, London-New York: Bloomsbury Academic, 2011, p. 66.29–34,

Aquinas[12] — that the astronomer is precluded from accessing the true causes of the phenomena under investigation. The astronomer may describe them, offer hypothetical explanations, but can never ascertain their causes with certitude: knowledge of celestial phenomena is destined to remain uncertain. In this quotation, Sarpi adopts a traditional distinction, prevalent of Paduan Aristotelianism: that between certain and probable knowledge, between deduction and induction. Empirical knowledge, proceeding from effects to causes, must relinquish certainty and content itself with advancing hypotheses, as identical effects may be accounted for by recourse to diverse causes. In so doing, Sarpi articulates a position aligned with tradition. However, to a discerning reader such as himself — and to others before him, beginning with Rheticus, who had personally overseen the drafting of *De revolutionibus*, and Tiedemann Giese, Copernicus' friend and patron, and after them to Giordano Bruno, who expressed his conviction with particular incisiveness[13] — the anonymous Preface to *De revolutionibus* perhaps seemed rather distant from Copernicus' own intentions.[14] In any event, at the time of committing this 'thought'

slightly modified): "it is evident that the disagreement concerning these hypotheses does not constitute a reproach <to Aristotle>. The crux of the matter lies in determining by what hypothesis the phenomena might be preserved, and thus there is nothing surprising if different individuals sought to preserve the phenomena by recourse to different hypotheses" ("δῆλον δέ, ὅτι τὸ περὶ τὰς ὑποθέσεις ταύτας διαφέρεσθαι οὐκ ἔστιν ἔγκλημα· τὸ γὰρ προκείμενόν ἐστι, τίνος ὑποτεθέντος σωθείη ἂν τὰ φαινόμενα; οὐδὲν οὖν θαυμαστόν, εἰ ἄλλοι ἐξ ἄλλων ὑποθέσεων ἐπειράθησαν διασῶσαι τὰ φαινόμενα").

[12] Thomas Aquinas, *In libros De caelo et mundo*, II, lectio 17, 2 (= *Sancti Thomae Aquinatis Doctoris Angelici Opera omnia iussu impensaque Leonis XIII. P. M. edita*, Vol. III: *Commentaria in libros Aristotelis De caelo et mundo, De generatione et corruptione, et Meteorologicorum ad codices manuscriptos exacta*, Rome: Typographia Polyglotta, 1886, pp. 186a-187b): "However, it is not necessary that the hypotheses to which [past astronomers] had recourse be true: for, whilst such hypotheses enable us to save the phenomena, it ought not, nevertheless, be asserted that these are true, since one might perhaps account for the apparent motions to the stars by some other means, which men have not yet conceived" ("Illorum tamen suppositiones quas adinvenerunt, non est necessarium esse veras: licet enim, talibus suppositionibus factis, apparentia salvarentur, non tamen oportet dicere has suppositiones esse veras; quia forte secundum aliquem alium modum, nondum ab hominibus comprehensum, apparentia circa stellas salvantur").

[13] See *La cena de le Ceneri* [1584], edited by Giovanni Aquilecchia, in Giordano Bruno, *Œuvres complètes*, Vol. II: *Le souper des cendres*, Paris: Les Belles Lettres, 1994, pp. 127, 129; English translation by Hilary Gatti, *The Ash Wednesday Supper*, Toronto-Buffalo-London: The University of Toronto Press, 2018, p. 91: "[…] a certain preliminary letter attached to it by I know not what ignorant and presumptuous ass. This person claims that he is doing a favour to the author (but perhaps his real concern is to supply lettuces and berries for other asses so that they should not go hungry) by issuing a warning to the readers before they start reading the book and consider its statements" ("[…] certa epistola superliminare attaccata non so da chi asino ignorante e presuntuoso; il quale (come volesse iscusando faurir l'autore, o pur a fine che anco in questo libro gli altri asini, trovando ancora le sue lattuche e frutticelli, avessero occasione di non partirsene a fatto deggiuni), in questo modo le avvertisce, avanti che cominciano ad leggere il libro e considerar le sue sentenze").

[14] The name of the true author of the Preface, circulated among the astronomers of that era (Petrus Apianus, Hieronymus Schreiber, Michael Mästlin, Johannes Prätorius), would only receive public confirmation with Kepler. In the *verso* of the title page of his *Astronomia nova αἰτιολόγητος, seu*

to paper, Sarpi remained bound by a traditional stance and had not yet cultivated specific interests in astronomy, and in Copernicus in particular.

A second 'thought', no. 27, appears to lead directly to Copernicus:

> Were one situated on the Moon and to observe the Earth, through its monthly rotation one would perceive upon the face of the Earth all that appears on the face of the Moon, all its phases, and both eclipses.[15]

Here, the (optical) relativity of motion is articulated with clarity, and within an astronomical context. The phenomenon is well known; Copernicus himself alludes to it, quoting Virgil: "As we recede from the harbor, the lands and cities recede".[16] Forty years subsequent to Copernicus, Bruno would also mention it: "the Moon is no more heaven to us than we are to the Moon".[17] Both *La cena de le Ceneri* and *De l'infinito, universo e mondi* date to 1584 — thus, after Sarpi's 'thought' — and the

---

physica coelestis (1609) — referencing a passage from one of Petrus Ramus' works (*Scholarum mathematicarum libri unus et triginta*, Basel: Eusebius & Nikolaus Episcopius, 1569, p. 50) — Kepler states: "It is a most absurd tale, I admit, to demonstrate natural phenomena through false causes, but this tale is not in Copernicus" ("Fabula est absurdissima, fateor, Naturalia per falsas demonstrare causas: sed fabula haec non est in Copernico"). And, after naming Osiander, he declares: "Thus Copernicus does not mythologize [i.e. he does not merely intend to present a new mathematical model, useful for calculation but not claiming to offer an actual description of reality], but seriously presents paradoxes [i.e., he offers a new picture that openly clashes with the received opinion]: that is, he philosophizes [i.e., he performs the task of natural philosophers, whose aim is to provide a truthful account of reality], which is what you wished of an astronomer" ("Non igitur μυθολογεῖ Copernicus, sed serio παραδοξολογεῖ, hoc est, φιλοσοφεῖ: quod tu in Astronomo desiderabas": in *Johannes Kepler gesammelte Werke*, edited by Max Caspar et al., Munich: C. H. Beck, 1937–2017 [henceforth, *KGW*], Vol. 3, p. 6.37–38; English translation by William H. Donahue, *Astronomia Nova*, Santa Fe, NM: Green Lion Press, 2015, p. 4).

[15]"Se uno fusse nella Luna e guardasse la terra, mediante il moto mestruo vedrebbe in Terra tutte l'apparenze della Luna, tutte le figure e tutti due gli eclissi": *Pensieri naturali*, no. 27, p. 36. See also ibid., no. 266, p. 231: "'Can you, like him, spread out the skies, hard as a cast metal mirror?'. *Job* 37, 18. If the Earth does not move, the galaxy shows that the heavens are moving" ("'Tu forsitan cum eo fabricatus es coelos, qui solidissimi, quasi aere fusi sunt'. Iob 37, 18. Se la Terra non si muove, dimostra la galassia muoversi il cielo").

[16]Copernicus (*De revolutionibus*, I 8, p. 29.10–11) quotes Virgil, *Aeneid*, III 72: "Prouehimur portu, terræque urbesque recedunt". The comparison between the Earth in motion and a ship receding from the harbor, yet appearing stationary to those aboard — so much so that one imagines it is the lands and cities that are receding — is already found in Aristotle (*De insomniis*, 2, 460 b 22–27) and other classical authors, such as Cicero (*De divinatione*, II 120) and Lucretius (*De rerum natura*, IV 387–390). The comparison is also frequently encountered in medieval authors such as Nicole Oresme (*Le livre du ciel et du monde*, II 25), John Buridan (*Quæstiones super libris quattuor De caelo et mundo*, II 22), Nicholas Cusanus (*De docta ignorantia*, II 12), and after Copernicus Virgil's verse gained new notoriety, becoming a *locus classicus*: see, for example, Kepler, *Astronomia nova*, in *KGW* 3, p. 28.34–35; and Campanella, *Apologia pro Galileo* [1622], edited by Michel-Pierre Lerner, Paris: Les Belles Lettres, 2001, p. 112. See also *De revolutionibus*, Vol. III, pp. 107–108.

[17]"[...] non più la luna è cielo a noi, che noi alla luna": Bruno, *La cena de le Ceneri* (cited, footnote 13), p. 51; Engl. tr., p. 37. See also *De l'infinito, universo e mondi* [1584], edited by Giovanni Aquilecchia, in Giordano Bruno, *Œuvres complètes*, Vol. IV: *De l'infinito, universo e mondi / De l'infini, de l'univers et des mondes*, Paris: Les Belles Lettres, 2006[2], e.g. pp. 183, 185.

absence of references to Bruno in *Pensieri naturali* suggests that Sarpi probably became acquainted with these works at a later juncture. It would be an error, however, to overinterpret this 'thought', singling it out as a sort of "precursor" to Galilean relativity of motion, just as it would be fallacious to discern, in Sarpi's words, a direct echo of Copernicus. In this particular 'thought', Sarpi seems merely to posit an equality of nature between the Earth and the Moon. Indeed, his statement may be situated within the ambit of a philosophy of nature inspired by classical thought: after all, Plutarch himself, in the *De facie in orbe lunae*, had already posited that the Moon is a sort of "celestial Earth".[18]

---

[18] Plutarch, *De facie in orbe lunae*, 925 A-B: "you see that it is ludicrous to deny the Moon's Earth-like qualities on the grounds of its separation from the lower region, whilst simultaneously asserting it is a star, even though we observe it thousands of kilometers away from the upper circumference, as if it were submerged within a well" ("ὁρᾶτε ὅτι γελοῖόν ἐστιν, εἰ γῆν οὐ φήσομεν εἶναι τὴν σελήνην, ὅτι τῆς κάτω χώρας ἀφέστηκεν, ἄστρον δὲ φήσομεν, ὁρῶντες ἀπωσμένην τῆς ἄνω περιφορᾶς μυριάσι σταδίων τοσαύταις ὥσπερ <εἰς> βυθόν τινα καταδεδυκυῖαν"). The debate on the similarity between the Earth and the Moon is extensive, both in antiquity and in the medieval period, and involving authors such as Macrobius, Averroes, Albertus Magnus, Augustine Nipho, and Francis Patricius. In the *De motu libri X* (Florence: Bartolomeo Sermartelli, 1591) Francesco Buonamici, Galileo's professor at the University of Pisa, wrote, "Lunam terram cælestem vocant" ("They call the Moon a celestial Earth": p. 906E).

Buonamici refers to Proclus' commentary on the *Timaeus*: see, for example, *In Platonis Timaeum commentaria*, ed. Ernst Diehl, Leipzig: B. G. Teubner, 1903–1906, Vol. II, p. 48.17 (= ed. Gerd Van Riel, Oxford: Oxford University Press, 2022, Vol. III, p. 66.13; English translation by Dirk Baltzly, in Proclus, *Commentary on Plato's Timaeus*, Vol. III, Book 3, Part 1: *Proclus on the World's Body*, Cambridge: Cambridge University Press 2007, pp. 97–98): "For the Moon is an ethereal earth" ("γῆ μὲν γὰρ αἰθερία ἡ σελήνη"); Vol. III, p. 142.12–14 (= ed. Van Riel, Vol. IV, p. 183.16–18; English translation by Dirk Baltzly, in Proclus, *Commentary on Plato's Timaeus*, Vol. V, Book 4: *Proclus on Time and the Stars*, Cambridge: Cambridge University Press 2013, p. 242): "After all, it is customary to call the Moon 'earth' since it plays the same sort of role in relation to the Sun that the Earth plays in relation to fire" ("τὴν μὲν γὰρ σελήνην γῆν εἰώθασι καλεῖν, ὡς τοῦτον ἔχουσαν τὸν λόγον πρὸς ἥλιον, ὃν ἡ γῆ πρὸς τὸ πῦρ": both of these passages are accompanied by the citation of an Orphic fragment, fr. 91 Kern = fr. 155 Bernabé); Vol. III, p. 172.20–21 (= ed. Van Riel, Vol. V, p. 14.10–11; English translation by Harold Tarrant, in Proclus, *Commentary on Plato's Timaeus*, Vol. VI, Book 5: *Proclus on the Gods of Generation and the Creation of Humans*, Cambridge: Cambridge University Press 2017, p. 50–51: "For Orpheus called the Moon a 'celestial earth'" ("καὶ γὰρ οὐρανίαν [καὶ] <γῆν> τὴν σελήνην 'Ορφεὺς προσηγόρευσε", with reference to the same Orphic fragment). In fact, however, Buonamici appears to misinterpret Proclus, who, in his commentary on Plato's *Timaeus*, distinguishes Earth (the celestial body) from earth (the element). Indeed, the initial quotation from Proclus reflects the opinion of the Pythagoreans, who believed that the four elements exist in two distinct series: one below the Sun (ethereal) and one above it (celestial). Accordingly, "γῆ μὲν γὰρ αἰθερία ἡ σελήνη" does not signify that the Moon is a celestial Earth, but rather that in the ether (i.e., below the Sun), the Moon constitutes the element 'earth'. Indeed, the text proceeds: Mercury is water, Venus is air, the Sun is fire. Then, in the heavens (i.e., above the Sun), Mars is fire, Jupiter is air, Saturn is water, and the sphere of the fixed stars is earth. To corroborate the thesis that the Moon is (the element) earth, Proclus cites an Orphic fragment. However, after presenting the opinion of the Pythagoreans, Proclus refutes it: the individual elements are not exclusively associated with the celestial spheres; rather, all four elements are present within each celestial sphere. Therefore, Proclus' passage was misinterpreted, or at the very least forced, to convey a meaning different from the author's original intent.

In those selfsame years, Sarpi began to evince an interest in mechanics. And precisely in contemplating natural and violent motion — a classical distinction within Aristotelian philosophy[19] — do we discover another potential avenue to astronomy, and particularly to the conundrums posed by Copernicanism. Indeed, in 'thought' no. 100, Sarpi critiques the Aristotelian distinction between these two types of motion, equating impetus with an inherent disposition, along the lines of an approach already discernible in Buridan:

> The violent and the natural are not two distinct natures, but two modes of one and the same nature; wherefore every motion is natural with respect to the passive principle, nor would a stone ascend were it not naturally disposed to be moved. The distinction between the violent and the natural is deduced from the principle, which is either external or internal: natural being the virtue that resides within and has formed the body and furnished it with its means; and violent being the one that is introduced by an external force and swiftly dissipates, because it lacks the means to sustain its action, and encountes an opposing and stronger one from within, such that the former has the subject disposed in its own manner, and contends, repelling the latter. Initially, the latter is stronger and prevails, but it is then compelled to yield, and submits [...].[20]

The *editio princeps* of Proclus' commentary on the *Timaeus*, edited by Simon Griner (Grynaeus) and Johannes Herbster (Oporinus), was published in Basel in 1534 as a complement to Plato's *Opera omnia*, and was likely the catalyst of the debate that developed in the latter half of the sixteenth century: see Alain-Philippe Segonds, "Proclus: astronomie et philosophie," in Jean Pépin, Henri Dominique Saffrey (éd.), *Proclus lecteur et interprète des anciens. Actes du colloque international du CNRS Paris (2–4 octobre 1985)*, Paris: Éditions du CNRS, 1987, pp. 319–334, especially p. 319. Proclus' arguments were echoed by Giordano Bruno: see, for example, Bruno, *La cena de le Ceneri* (cited, footnote 13), p. 167 (Engl. transl., p. 121): "[...] when Lucian says that the Moon is another Earth, inhabited and cultivated like this one, he says it only to ridicule those philosophers who claim that there are many Earths (and in particular the Moon, whose similarity to this globe of ours becomes evident the closer she is to us). He was mistaken about that, and no less ignorant or blind than anybody else" ("se quando Luciano disse la luna essere un'altra terra cossì abitata e colta come questa, venne a dirlo per burlarsi di que' filosofi che affermorno essere molte terre (e particolarmente la luna, la cui similitudine con questo nostro globo è tanto più sensibile, quanto più è vicina a noi) lui non ebbe ragione: ma mostrò essere nella comone ignoranza, et cecità"). See also *De l'infinito, universo e mondi* (cited, footnote 17), p. 185: "the Moon (which is another Earth)" ("la luna (che è un'altra terra)"); and p. 209: "And just as of the Moon from this Earth, which is another Moon, some parts appear to be brighter, others less bright, so of the Earth from that Moon, which is another Earth, some parts appear to be different, due to the variety and difference of the areas of its surface" ("E sì come della luna da questa terra, ch'è un'altra luna, appaiono diverse parti altre più altre men luminose, cossì della terra da quella luna, ch'è un'altra terra, appaiono diverse parti per la varietà e differenza de spacii di sua superficie").

[19] See Aristotle, *Physica*, IV 8, 215 a 1–6, and VIII 4, 254 b 7–14.

[20] "Il violento e il naturale non sono due nature diverse, ma due modi di essere d'una stessa natura; imperciò che ogni moto è naturale rispetto al principio passivo, né il sasso muoverebbesi all'insù quando naturalmente ad esser mosso atto non fosse. La distinzion del violento e naturale si deduce dal principio, che è fuori o dentro, naturale sendo la virtù che è dentro ed ha formato il corpo e fattigli gli stromenti: quella violenta, che da forza esterna è introdotta, la qual tosto manca, perché non ha stromenti da continuar l'azione, ed una contraria più potente ne trova di dentro, sì che la prima ha il soggetto a modo suo disposto, e combatte, e scaccia la seconda; la seconda è più potente da principio e vince, ma poi a ceder è forzata e perde": *Pensieri naturali*, no. 100, p. 123.

In the concluding sentence, Sarpi also approaches Buridan in accepting the thesis that the heavens are not impelled by intelligences, as tradition dictated, but by a sort of *impetus* (a "virtue" or "force"): "Thus we can observe a virtue that moves circularly with violence, as in the spinning top, and that is natural in certain things, as in the stars".[21]

The critique of Aristotelian mechanics persists in 'thought' no. 130, wherein the notion of the Earth's motion emerges:

> Stillness is not an absence of straight motion, but an end: accordingly, since a thing is not at its proper place, it moves to attain it. By contrast, stillness <...> may be an absence of circular motion, because nature moves circularly everything that it moves, and does so with perpetual purpose, so that it does not cease. Therefore, were the Earth to rest at the center of the world, it would receive no motion naturally, but nature would be the principle of rest within it.[22]

Notwithstanding the conditional clause, which renders the final sentence somewhat ambiguous, and the circumscribed scope within which the entire 'thought' operates (Sarpi remains at the level of definitions), one cannot but discern a convergence with the Copernican way of reasoning. Indeed, it appears that Sarpi observes that the Earth's stillness, a cornerstone of Aristotelian cosmology, contravenes a principle of Aristotle's physics, according to which nature is a principle of motion; alternatively, Sarpi entertains the possibility of the Earth's motion and explores the tenets of Aristotelian physics to ascertain whether irrefutable arguments exist against it. Here, as elsewhere, the notably discontinuous and synthetic character of *Pensieri naturali* precludes a definitive answer.

The same interpretative difficulty is found in *Pensieri naturali*, nos. 135–136, written shortly after 1585. Indeed, 'thought' no. 136 (a direct continuation of the preceding one), reads:

> Accordingly, no motion is irregular; rather, the congeries of many regular motions appears irregular to those who do not know them all: we observe that this has already been the case in assessing the motion of the planets. Therefore, whosoever comprehends perfectly all the simple motions that concur into one which appears irregular, he would know the periods of all of them. Consequently, every motion is regular and can be predicted by anyone who apprehends all its causes. Such, however, is God alone.[23]

---

[21] "Dunque possiam vedere una virtù che muova circolarmente con violenza, come nella trottola, ed ella sia naturale in qualche cosa, come nelle stelle": ibid., p. 124.

[22] "La quiete non è privazion del moto retto, ma fine, imperciò che, essendo la cosa fuori del suo luogo, vi si muove per istarvi; bensì può esser privazion del moto circolare la <...> quiete, perché la natura muove circolarmente tutto quel che muove e con perpetua intenzione, non a ciò che si riposi. Adunque se la Terra fosse stabile in mezzo del mondo, ella non avrebbe verun moto dalla natura, ma la natura in lei principio di quiete sarebbe": ibid., no. 130, p. 148.

[23] "Laonde niun moto è irregolare, ma la congerie di molti regolari, a chi non gli sa tutti, un irregolare apparisce: vediamo esser già stato così nel giudicar il moto dei pianeti. Chi dunque conoscesse perfettamente tutti i moti semplici che concorrono a ciascuno il qual paia irregolare, li periodi di tutti conoscerebbe: dunque regolare è ogni moto, e predire si può da chi le sue cause tutte sappia. Questi peraltro è solo Iddio": ibid., no. 136, p. 152. Firm belief in the rationality of the world system derives from Aristotle's idea of the absolute perfection of the celestial world, which stands in radical contradistinction to the terrestrial realm. The embrace of Copernicanism does not spring

Sarpi reiterates that nature is perfectly rational and regular, ever consistent: disorder, irregularity and chance can only be apparent; every phenomenon is the product of an intricate web of causal connections wherein all is necessarily determined by eternal and unchanging causes, and only our ignorance may lead us to claim otherwise: "If, in some cases, the present or the past are unknown to us because we cannot refer them to all their causes, we no longer perceive them as necessary, but we declare of them: perhaps it was not, perhaps it is not — just as we say of the future: perhaps it will not be".[24]

This applies, in particular, to the motion of the planets, which served as the classical exemplar of perfection: the observed irregularities (above all, retrograde motions) — which posed such considerable challenges to the great astronomers of antiquity, such as Eudoxus, Callippus, Apollonius or Ptolemy, and which continued to puzzle Sarpi's contemporaries, from Regiomontanus to Copernicus himself, from Tycho Brahe to Kepler — can only be apparent, a consequence of our inadequate analytical faculties. It was precisely the disagreement amongst the great *auctoritates* of the past that prompted Copernicus (as we read in the opening quotation, which, I believe, encapsulates the entire spirit of the Renaissance) to formulate the heliocentric hypothesis: being in disagreement, these figures could not be authoritative, and therefore it was permissible for him to attempt to pursue alternative paths.

The regularity of nature is the fundamental premise of all ancient astronomers; what varies is the hypothesis employed to elucidate apparent irregularity. With 'thoughts' nos. 135–136, Sarpi appears to be opening the door to Copernicanism: having observed that the various hypotheses advanced to explain apparent irregularity are incompatible and insufficient, why not adopt a different perspective?

---

from this certainty (which is perfectly satisfied by the Ptolemaic system, too), but rather from the capacity to transcend sensory perception, that is to say, to reject empirical evidence by appealing to human reason. In the Second Day of the *Dialogue Concerning the Two Greatest Systems of the World*, Galileo would have his spokesperson Salviati say: "the philosopher-mathematician, when he wants to recognize in the concrete the effects which he has proved in the abstract, must deduct the material hindrances" ("il filosofo geometra vuol riconoscere in concreto gli effetti dimostrati in astratto, bisogna che difalchi gli impedimenti della materia": *OG* 7, p. 234.5–7; Engl. transl., p. 207, slightly modified). Later, whilst discussing the tower experiment, Galileo would state: "It is [...] better to put aside the appearance, on which we all agree, and to use the power of reason either to confirm its reality or to reveal its fallacy" ("Meglio è [...] che, deposta l'apparenza, nella quale tutti convenghiamo, facciamo forza co 'l discorso, o per confermar la realtà di quella, o per iscoprir la sua fallacia": *OG* 7, p. 281.6–8; Engl. transl., p. 256). And in the Third Day, he would commend Aristarchus and Copernicus for their courage in allowing reason to prevail over observational evidence: "nor can I ever sufficiently admire the outstanding acumen of those who have taken hold of [the Copernican] opinion and accepted it as true; they have through sheer force of intellect done such violence to their own senses as to prefer what reason told them over that which sensible experience plainly showed to the contrary" ("né posso a bastanza ammirare l'eminenza dell'ingegno di quelli che l'hanno ricevuta e stimata vera, ed hanno con la vivacità dell'intelletto loro fatto forza tale a i proprii sensi, che abbiano possuto antepor quello che il discorso gli dettava, a quello che le sensate esperienze gli mostravano apertissimamente in contrario": *OG* 7, p. 355.19–24; see also ibid., 30–33 [Engl. transl. p. 328]).

[24] "Se in qualche caso il presente o il passato ci sono incogniti perché non possiamo riferirgli a tutte le loro cause, non più per necessari gli abbiamo, ma diciam di loro: forse non fu, forse non è, come diciam del futuro: forse non sarà": *Pensieri naturali*, no. 135, p. 152.

2.4 Immersion in *De Revolutionibus*

It is in 'thought' no. 202, dating from Sarpi's Roman period (1585–1588), that we first encounter the name of Copernicus:

> Against Copernicus it is alleged that the parallel lines are lost from sight before they are observed to converge: which I believe to be false, because two distant lights appear as one.[25] Against the same, who considers the [Earth-Sun] distance, or semi-diameter of the annual motion, to be imperceptible in comparison with the eighth sphere, it must be stated that every minutest star is larger than it, because every such star is larger than any imperceptible quantity.[26]

In Chap. 10 of Book I of *De revolutionibus*, prior to presenting the diagram of the new world system, Copernicus addresses a potential objection to heliocentrism: if the Earth orbits the Sun, some stars, when observed from opposing points of the Earth's orbit, should exhibit parallax; however, as this is phenomenon is not observed, the Earth must be stationary at the center of the universe. The stars, however, are far more distant than was presumed in Copernicus' era, and the Earth's orbital motion would occasion virtually imperceptible shifts. Copernicus accounts for the absence of detectable parallax by contending that the immense height of the sphere of the fixed stars would, in effect, nullify the parallax caused by the Earth's annual displacement along its orbit. This implies, however, that the Sun, when positioned at a distance comparable to that of the fixed stars, would be invisible to us, whilst the fixed stars, despite their enormous distance from us, remain perfectly visible. Hence, even the smallest of the visible stars must be larger not only than the Sun, but also than a sphere possessing a diameter equivalent to the Earth-Sun

---

[25] See *De revolutionibus*, I 6, p. 25.4–7 (Engl. transl., p. 13): "Yet a line drawn from the Earth's surface [to a point in the firmament] must be distinct from the line drawn from the Earth's center [to the same point]. Nevertheless, because these lines are immense in relation to the Earth, they become like parallel lines. Because their terminus is enormously remote, they appear to be a single line. For, in comparison with their length, the space enclosed by them becomes imperceptible, as is demonstrated in the *Optics*" ("cum tamen necesse sit aliam esse lineam quæ a superficie terræ, & quæ a centro, sed propter immensitatem respectu terræ fiunt quodammodo similes parallelis, quæ præ nimia distantia termini apparent esse linea una, quando mutuum quod continent spacium ad earum longitudinem efficitur incomparabile sensu, eo modo quo demonstratur in Opticis").

[26] "Contro Copernico allegasi che le parallele prima si perdon di vista che a concorrer si veggano: il che credo esser falso, perché due lumi lontani paiono uno. Al medesimo che mette la distanza o semidiametro del moto annuo insensibile nella ottava sfera è necessario dire che ogni minima stella sia maggior di esso, perché ogni tale di ciascuna quantità insensibil è maggiore": *Pensieri naturali*, n. 202, p. 182. See *De revolutionibus*, I 10, p. 39.15–19 (Engl. transl., p. 22): "Yet none of these phenomena appears in the fixed stars. This proves their immense height, which makes even the sphere of the annual motion, or its reflection, vanish from before our eyes. For, every visible object has some measure of distance beyond which it is no longer seen, as is demonstrated in the *Optics*" ("Quod autem nihil eorum apparet in fixis, immensam illorum arguit celsitudinem, quæ faciat etiam annui motus orbem siue eius imaginem ab oculis euanescere. Quoniam omne uisibile longitudinem distantiæ habet aliquam, ultra quam non amplius spectatur, ut demonstratur in Opticis"). As for the quote in the previous note, the reference is to Euclid, *Optica*, 3, p. 4.26–27 Heiberg.

distance — that is, the radius of the Earth's orbit. If the Earth-Sun distance (the "semi-diameter of the annual motion") is negligible compared to the colossal size of the sphere of the fixed stars (the "eighth sphere"), then "every minutest star", being visible — that is, "perceptible" — must be larger than that distance, and consequently even larger than the Sun, which, if situated at the distance of the fixed stars, would be invisible. This effectively diminishes the Sun's significance — a notion that Copernicus and his predecessors found difficult to accept.[27]

This appears to be a species of *reductio ad absurdum* of the Copernican hypothesis — an *impasse* in Sarpi's gradual approach to heliocentrism, and one that seems, at least for the time being, insurmountable. So much so, in fact, that Sarpi's interest in celestial phenomena waned for the remainder of his Roman period (during which, as noted previously, Sarpi lacked the leisure and opportunity to devote himself to his studies, as he might havce wished). Astronomy does not recur in *Pensieri naturali* until Sarpi's return to Venice, between 1588 and 1589. In a marginal note appended to 'thought' no. 202, penned subsequent to its original composition, we read:

> He [Copernicus] would contend that the star is a second or third sphere, and perhaps the entire synod around a Sun.[28]

The content of this note is akin to 'thought' no. 518, composed around 1592 (the year of Galileo's arrival in Padua, a few years after Sarpi's return from Rome), and likely coeval with it:

> [...] as the semidiameter of the annual motion is imperceptible in the eighth sphere, it must be smaller than any minutest star; but in the darkness it could be that the star we observe is not the Sun itself, but the synod, which is as expansive as the Sun's rays, just as from a distance we do not perceive the candle, but the sphere of the rays that surround it.[29]

Copernicus, in essence, addresses the objection raised in 'thought' no. 202 by positing that, in fact, we do not discern the bodies of stars, but the much larger spherical

[27] According to the Copernican hypothesis, his detractors would contend, the Sun would be an ordinary star, devoid of special importance, contrary to the assertions of a long-standing tradition. This tradition, whilst retaining the Earth at the center of the cosmos, ascribed to the Sun the role of the purest source of light, warmth, and life. However, this very same tradition is invoked by Copernicus himself, who posits that the Sun is entitled to occupy the central position owing to its beauty and importance: the Sun must reside at the center because it is the lamp of the world (see *De revolutionibus*, I 10, p. 38, which cites Hermes Trismegistus and Sophocles). Galileo would likewise appeal to this perspective (in his letter to Piero Dini on 23 March 1615, in *OG* 5, pp. 297–305), recalling the "metaphysics of light" of Ficinian inspiration, characteristic of Florentine Neoplatonism, with explicit references also to the *Psalms* (e.g., 18, 6 = 19, 4 KJV: "In sole posuit tabernaculum suum").

[28] "Egli [Copernico] direbbe che la stella è una seconda o terza sfera, e forse tutto il sinodo attorno un sole": *Pensieri naturali*, no. 202 (*marginal note*), p. 182. Copernicus, Sarpi suggests, might be referring to the fact that what we observe of a star is not merely its physical body, but rather its body enveloped by the various spheres formed by its rays (synod).

[29] "[...] essendo il semidiametro del moto annuo insensibile nell'ottava sfera, bisogna che sia minor d'ogni minima stella; ma al bruno potrebb'esser che la stella che si vede non fosse il Sole ma il sinodo, il qual è tanto grande quanto i raggi di quel sole giungono, sì come noi di lontano non vediam la candela ma la sfera de' raggi che la circondano": ibid., no. 518, p. 379.

portion of space illuminated by their rays. In so doing, he effectively justifies the claim made in *De revolutionibus*, Book I, Chap. 10, thus preserving the plausibility of the Copernican hypothesis without fundamentally altering the scale of the universe or drastically undermining the Sun's significance.

Assuming that each star is enveloped by a system ("synod") of elementary spheres and a large sphere of vapors,[30] what we perceive is not the star itself, but the extent of space illuminated by its rays.[31] If this premise holds, the acceptance of the Copernican hypothesis, while necessitating a profound restructuring of the cosmos, does not wholly revolutionize it, thereby maintaining an acceptable scale for the universe and, crucially, avoiding deprecation of the Sun's traditionally attributed role.

As one peruses *Pensieri naturali*, an increasingly profound comprehension of *De revolutionibus* becomes evident, alongside a growing conviction in the Copernican viewpoint. However, it is often necessary to interpret between the lines. This is exemplified, for instance, in 'thought' no. 384, wherein Copernicus' name appears for the second time:

> According to Aristotle, straight motion cannot reverse in a continuous manner: it is necessarily required to introduce rest.[32] Nonetheless, if a straight motion is caused by two circular ones (as in Copernicus' demonstration), which produces the trepidation, then necessarily the two straight motions of going and returning are continuous.[33]

---

[30] In the preceding 'thought' (no. 517), Sarpi ponders whether a sphere of vapors, akin to that which causes twilight on Earth, might exist around the Moon and the other celestial bodies: "Around the Moon there is perhaps a sphere of vapors similar to that which causes our twilights; their mixture is either not a body but only space, because it would prevent us from seeing with its depth supplementing the surface, or else it is merely a body made of solar light. This twilight sphere may still be around all the stars: in this case, whenever it happens, we may answer the question, what is that we see and call sky, and how can it shed light?" […]" ("Intorno alla Luna vi è forse una sfera di vapori simile alla nostra crepuscolina, e l'intreccio fra queste o non è corpo ma solo spazio, perché impedirebbe la vista colla sua profondità supplente alla superficie, ovvero è solo corpo di lume solare. Questa crepuscolina sfera può esser ancora intorno a tutte le stelle, con che, quando sia, vien sciolto il quesito che cosa è quello che vediamo e chiamiam cielo, e com'è capace di lume […]": *Pensieri naturali*, no. 517, p. 379). The existence of a sphere of vapors around the Earth, Moon and celestial bodies is proposed by Alhazen: see *Opticæ Thesaurus* (cited, footnote 7), pp. 283–288.

[31] As Libero Sosio suggests (*Pensieri naturali*, commentary on no. 202, p. 184), the idea that enables Sarpi to overcome the objection to the Copernican hypothesis probably stems from his reading of Nicholas of Cusa, *De docta ignorantia* (1440); see Nikolaus von Kues, *Werke*, Vol. I, edited by Paul Wilpert, Berlin: Walter de Gruyter, 1967, pp. 1–100: p. 64.

[32] This is referred to as the *punctum reflectionis*, or *quies media*; see also the discussion in Galileo's *De motu*, within the section "in which it is shown, contrary to Aristotle and the received view, that in the point of reflection there is no rest" ("in quo contra Aristotelem et communem sententiam ostenditur, in puncto reflexionis non dari quietem": *OG* 1, pp. 323–328).

[33] "Vuol Aristotile che un moto retto non possa continuo ritornare, ma sia necessario intrometter quiete: non di meno se un retto vien causato da due circolari, come nella dimostrazion di Copernico, che fa la trepidazione, necessariamente i due moti retti d'andar e tornare sono continui": *Pensieri naturali*, no. 384, p. 290.

Here, Sarpi grapples with a problem of mechanics: that of the interval of rest which, according to Aristotle, distinguishes the (violent) motion of projection and the (natural) motion of fall (*quies media*).[34] He does so by contrasting Aristotle's thesis with an example drawn from *De revolutionibus* — no longer, however, from Book I, which is cosmological in nature and accessible even to non-astronomers, but from Book III, Chap. 4 ("Quomodo motus reciprocus siue librationis ex circularibus constet", that is, "How reciprocal motion, or libration, consists of circular motions"), in which Copernicus addresses a technical issue, namely, the precession of the equinoxes.

A similar indication of the meticulous attention with which Sarpi engages with Copernicus' text is found in 'thought' no. 390, which presents a critique of the Platonic theory of the tenfold proportion of the elements,[35] and in which Sarpi demonstrates his acquaintance with astronomical distances, according to both Ptolemy and Copernicus. Likewise, in 'thought' no. 395, Sarpi evinces skepticism regarding the existence of celestial orbs and motive intelligences.[36] 'Thought' no. 395 hold particular significance from the Copernican perspective: for if the denial of solid orbs presented no minor difficulty for those who upheld the diurnal rotation of the celestial sphere around the Earth (since it was necessary to explain how so many stars could avoid falling and maintain their relative positions, despite not being anchored in a solid orb), this difficulty vanishes if one posits the rotation of the Earth, which adequately accounts for the apparent motion of the stars around it.

Finally, in 'thought' no. 467, Sarpi considers Cusanus' infinite world, whose center is everywhere and whose circumference is nowhere:[37]

> Were the place infinite, the Earth would not be in the middle, but would be where it is; for now there is no force that could move it from its site, nor would there be any reason for it

---

[34] See Aristotle, *Physica*, VIII 8, 263 a 1–3: "[…] that which reverses its direction along a straight line is required to cease moving. It is, therefore, not possible for a continuous motion along a straight line to be eternal" ("ἀνάγκη ἄρα στῆναι τὸ ἀνακάμπτον ἐπὶ τῆς εὐθείας. οὐκ ἄρα ἐνδέχεται συνεχῆ κίνησιν εἶναι ἐπὶ τῆς εὐθείας ἀΐδιον").

[35] See Plato, *Timaeus*, 32 b 3–c 4.

[36] The latter almost assumes the guise of a methodological directive: "Why is it that, even if we have not seen Constantinople, we still credit those who inform us of its existence, whereas those who tell us, for example, that there are celestial orbs, or that they are animated by intelligence, we do not credit? For the former is uttered by someone who (we are certain) might be able to know it, whereas the latter by someone who (we submit) knows no more than we do; just as one, lacking ears, would credit accounts of sounds from others whom he perceives to possess ears" ("Perché mai, se ben veduto non abbiam Costantinopoli, crediamo non di meno a chi ne dice che v'è, ed a chi ci dica, per essempio, che vi sono gl'orbi celesti, ovvero che son mossi da intelligenza, non crediamo? Perché il primo dicesi da uno cui siam certi poterlo sapere, ma il secondo da un si dice, che ne sa quanto noi; come, se uno fosse privo d'orecchie, crederebbe de' suoni quello che da altri, quali vede aver orecchie, detto gli venisse": *Pensieri naturali*, no. 395, p. 300).

[37] See Cusanus, *De docta ignorantia* (cited, footnote 31), p. 63: "Therefore, the machine of the world will be as if it had a center in any place, and a circumference in none, since God, who is anywhere and nowhere, is both the circumference and the center" ("Unde erit machina mundi quasi habens vndique centrum, et nullibi circumferentiam, quoniam circumferentia et centrum deus est qui est vndique et nullibi").

> to proceed more to one place than to another. Rather, it would be the center, because in the infinite *centrum ubique* [the center is anywhere].[38]

Here, however, Sarpi sappears to have in mind also (and perhaps primarily) Copernicus. Consider *De revolutionibus*, Book I, Chap. 9:

> Accordingly, since nothing prevents the Earth from moving,[39] I suggest that we should now consider also whether several motions suit it, so that it can be regarded as one of the planets. For, it is not the center of all revolutions. This is indicated by the planets' apparent non-uniform motion and their varying distances from the Earth. These phenomena cannot be explained by circles concentric with the Earth. Therefore, since there are many centers, it will not be by accident that the further question arises whether the center of the universe is identical with the center of terrestrial gravity or with some other point. For my part I believe that gravity is nothing but a certain natural desire, which the divine providence of the Creator of all things has implanted in parts, to gather as a unity and a whole by combining in the form of a globe. This tendency is present, we may suppose, also in the Sun, the Moon, and the other brilliant planets, so that through its operation they remain in that spherical shape which they display. Nevertheless, they swing round their circuits in various ways.[40]

---

[38] "Se il luogo fosse infinito, la Terra non sarebbe in mezzo, ma sarebbe dov'è; perché ora non v'ha forza che dal suo sito possa muoverla, né ragion v'ha perché ella più in un luogo che in un altro andasse. Bensì centro ella sarebbe perché nel infinito *centrum ubique*": *Pensieri naturali*, no. 467, p. 351. This 'thought' dates from 1588, and Sarpi perhaps already had in mind Bruno's Italian dialogues, which were published 4 years prior. See, in fact, *De la causa, principio et uno* [1584], edited by Giovanni Aquilecchia, in Giordano Bruno, *Œuvres complètes*, Vol. III: *De la causa, principio et uno / De la cause, du principe et de l'un*, Paris: Les Belles Lettres, 2016^2, p. 277: "If the point does not differ from the body, the center from the circumference, the finite from the infinite, the maximum from the minimum, surely we can state that the universe is all center, or that the center of the universe is everywhere; and that the circumference is in no part, however different it is from the center; or else that the circumference is everywhere, but the center is not to be found, as it differs from it" ("Se il punto non differisce dal corpo, il centro da la circonferenza, il finito da l'infinito, il massimo dal minimo, sicuramente possiamo affirmare che l'universo è tutto centro, o che il centro de l'universo è per tutto; e che la circunferenza non è in parte alcuna, per quanto è differente dal centro; o pur che la circonferenza è per tutto, ma il centro non si trova in quanto che è differente da quella"); and again, in *De l'infinito, universo e mondi* (cited, footnote 17), p. 339: "for in the universe there is no middle point, nor circumference: but (if you will) the middle point is in everything, and any point may be taken to be part of some circumference, with regard to some other middle point, or center" ("perché nell'universo non è mezzo né circonferenza: ma (se vuoi) in tutto è mezzo, et in ogni punto si può prendere parte di qualche circonferenza, a rispetto di qualche altro mezzo o centro").

[39] At the end of the preceding chapter, Copernicus merely wrote: "You see, then, that all these arguments make it more likely that the Earth moves than it is at rest. This is especially true of the daily rotation, as particularly appropriate to the Earth" ("Vides ergo quod ex his omnibus probabilior sit mobilitas terræ, quam eius quies, præsertim in cotidiana reuolutione, tanquam terræ maxime propria": *De revolutionibus*, I 8, pp. 31.26–32.2; Engl. transl., p. 17): that is, it is an exclusive characteristic of the Earth.

[40] *De revolutionibus*, I 9, p. 32.5–18; Engl. transl., pp. 17–18 ("Cum igitur nihil prohibeat mobilitatem terræ, uidendum nunc arbitror, an etiam plures illi motus conueniant, ut possit una errantium syderum existimari. Quod enim omnium reuolutionum centrum non sit, motus errantium inæqualis apparens, & uariabiles eorum a terra distantiæ declarant, quæ in homocentro terræ circulo non possunt intelligi. Pluribus ergo existentibus centris, de centro quoque mundi non temere quis dubitabit, an uidelicet fuerit istud grauitatis terrenæ, an aliud. Equidem existimo, grauitatem non aliud esse, quam appetentiam quandam naturalem partibus inditam a diuina prouidentia opificis uniuersorum, ut in unitatem integritatemque suam sese conferant in formam globi coëuntes. Quam affec-

Once the architectural structure of the Aristotelian cosmos, wherein space presents privileged directions and, in some measure, exerts physical action upon bodies, is dissolved, and once the notion of an overall undifferentiated space is conceded, there remains no rationale for Earth to occupy one specific location over another. If, according to Aristotle, the center of the world coincides with the center of the *gravitas* of bodies, that is, with the center of the Earth — which, for this reason, necessarily occupies (and is at rest at) the center of the world — Copernicus unequivocally separated the Earth's center of the world's center. In so doing, he multiplied the centers of attraction, as each celestial body ultimately constitutes a distinct gravitational system. Sarpi is not so much concerned with arguing that, in an infinite universe, the Earth would not move, but rather that we cannot maintain that the Earth occupies a central position: this is the pivotal step that signals his full embrace of Copernicanism.

2.5 Full Acceptance of the Copernican Hypothesis

Acceptance of the Earth's motion — not as a purely mathematical hypothesis introduced to "save" the phenomena, but as a physical theory, intended to describe reality as it truly is — involved several difficulties. Advocates of the Aristotelian theory identified these as grounds for refuting the geo-kinetic hypothesis. Most of these objections were mechanical and dynamical in nature. In 'thought' no. 540, Sarpi examines one such difficulty:

> Heavy bodies on Earth would not disintegrate, both because, given the vast size of the Earth, the tangent and the circle coincide, and because the virtue that holds them together is greater — since a heavy body falling downward would descend more than fifteen miles per minute; and finally, because our heavy bodies are not at the extremity, but at the midpoint of their circular motion.[41]

tionem credibile est etiam Soli, Lunæ, cæterisque errantium fulgoribus inesse, ut eius efficacia in ea qua se repræsentant rotunditate permaneant, quæ nihilominus multis modis suos efficiunt circuitus").

[41] "Li gravi in Terra non si dissiparebbono, sì perché in tanta grandezza della Terra la contingente e la circolare son la stessa, come perché la virtù di star unito è maggiore a motivo che un grave all'in giù verrebbe più di quindici miglia al minuto, finalmente perché i gravi nostri nell'estremo non sono ma nel mezzo della circolazione": *Pensieri naturali*, no. 540, p. 401. Possibly, Sarpi's source is Giovanni Battista Benedetti, who, in *Diversarum speculationum Mathematicarum, & Physicarum Liber* (Turin: Heirs of Nicola Bevilacqua, 1585), contributed to the discovery of the modern concept of inertia. He postulated the tendency of all bodies to persist in their state of motion along the tangent to their trajectory: "[...] The second reason is the fact that every heavy body, either by nature or by the force of motion, naturally follows a straight path, as we can clearly see by throwing stones with a slingshot, and twirling the arm" ("[...] Secunda causa est, quia quoduis graue corpus, aut per naturam, aut per vim motum, rectitudinem itineris naturaliter appetat, quod clarè cognoscere possumus, proijciendo lapides funda, & circumducentes brachium": p. 287); see also ibid., p. 159: "any part of the body that moves by itself, because of an impulse impressed within it by any external motive force, has a natural inclination to follow a straight line, not a curved one" ("quælibet pars corporea, quæ à se mouetur, impetu eidem à qualibet extrinseca virtute mouente impresso, habet naturalem inclinationem ad rectum iter, non autem curuum") and,

What would happen, then, if the Earth were animated by a *vertigine diurna* (daily rotation), as Galileo would say? The assumptions necessary to address this question are laid out in the preceding 'thought':

> Everything that is moved circularly, if it is detached from the source of force, moves along the tangent, and such circular motion may cease; however, if the force that holds it whole is stronger, [the body] does not disintegrate; the opposite occurs if the force is weaker.[42]

In other words: a body moved in a circular path will tend to continue in a straight line along the tangent if it is not held to its circular course by a sufficiently strong force. In *De revolutionibus* (Book I, Chaps. 7–8), Copernicus addresses a related objection — attributed to Ptolemy[43] — that, if the Earth rotates swiftly on its axis, it would break apart. Copernicus replies that such disintegration could result only from violent motion, whereas the Earth's rotation is natural, and therefore cannot result in disintegration.[44]

---

similarly, p. 161. However, the concept is also found in Plutarch, *De facie in orbe lunae*, 923 C-D: "Yet the Moon is saved from falling by its own motion and the rapidity of its revolution, just as bullets placed in a sling do not fall because they are made to revolve in a circle" ("καίτοι τῇ μὲν σελήνῃ βοήθεια πρὸς τὸ μὴ πεσεῖν ἡ κίνησις αὐτὴ καὶ τὸ ῥοιζῶδες τῆς περιαγωγῆς, ὥσπερ ὅσα ταῖς σφενδόναις ἐντεθέντα τῆς καταφορᾶς κώλυσιν ἴσχει τὴν κύκλῳ περιδίνησιν").

42 "Ogni cosa che sia mossa circolarmente, se si discontinua da dov'è la forza, muovesi per la contingente, ed esso moto circolare ha virtù di discontinuare, ma se la forza dello star unito è maggiore non dissipa, bensì ne [*lege*: se] è minore": ibid., no. 539, p. 400.

43 See Ptolemy, *Almagest*, I 7, Vol. 1, p. 25.9–14 Heiberg; English translation by Gerald J. Toomer, *Ptolemy's Almagest*, Princeton: Princeton University Press, 1998, p. 45.

44 See *De revolutionibus*, I 7, p. 27.12–24 (Engl. transl., p. 15): "Therefore, remarks Ptolemy of Alexandria, if the Earth were to move, at least in a daily rotation, […] a motion would have to be exceedingly violent and its speed unsurpassable to carry the entire circumference of the Earth around in 24 hours. But things which undergo an abrupt rotation seem utterly unlikely to stick together, and seem more likely, if they have been produced by combination, to fly apart unless they are held together by some bond. The Earth would long ago have burst asunder, he says, and dropped out of the heavens (which is quite ridiculous); and, what is more, living creatures and any other loose weights would by no means remain unshaken. Nor would objects falling in a straight line descend perpendicularly to their appointed place, which would meantime have been withdrawn by so rapid a movement. Moreover, clouds and anything else floating in the air would be seen drifting always westward" ("Si igitur, inquit Ptolemæus Alexandrinus, terra uolueretur, saltem reuolutione cotidiana […] concitatissimum esse motum oporteret, ac celeritatem eius insuperabilem, quæ in XXIIII. horis totum terræ transmitteret ambitum. Quæ uero repentina uertigine concitantur, uidentur ad collectionem prorsus inepta, magisque unita dispergi, nisi cohærentia aliqua firmitate contineantur: & iamdudum, inquit, dissipata terra <extra> cælum ipsum (quod admodum ridiculum est) excidisset, & eo magis animantia atque alia quæcunque soluta onera haudquaquam inconcussa manerent. Sed neque cadentia in directum subirent ad destinatum sibi locum, & ad perpendiculum, tanta interim pernicitate subductum. Nubes quoque & quæque alia in aëre pendentia semper in occasum ferri uideremus"). In I 8, p. 28.1–9 (Engl. transl., p. 15), he continues: "Yet if anyone believes that the Earth rotates, surely he will hold that its motion is natural, not violent. But what is in accordance with nature produces effects contrary to those resulting from violence, since things to which force or violence is applied must disintegrate and cannot long endure. By contrast, that which is brought into existence by nature is well-ordered and preserved in its best state. Ptolemy has no cause, then, to fear that the Earth and everything Earthly will be disrupted by a rotation created through nature's handiwork, which is quite different from what art or human intelligence can accomplish" ("Verum si quispiam uolui terram opinetur, dicet utique motum esse naturalem, non uiolentum. Quæ uero secundum naturam Sunt, contrarios operantur effectus his

In 'thought' no. 540, however, Sarpi advances beyond Copernicus. While Copernicus accepts the Aristotelian distinction between natural and violent motion, Sarpi offers three arguments to demonstrate that the Earth would neither disintegrate, nor would buildings collapse, nor would objects and people be flung into space.[45] First, he argues, the Earth is so vast that the tangent and circumference effectively coincide. Secondly, the cohesive force exceeds the extrusive force: indeed, a heavy object falls faster than a body moving along the tangent.[46] Thirdly, heavy bodies, although situated on the Earth's surface, are not located at the utmost periphery, but rather in an intermediate zone. For if the spheres of earth and water are within the terrestrial surface, then above them lies those of air and of fire. Despite the limitations of Sarpi's reasoning, his effort to reinforce Copernicus' position by supplementing philosophical reflections with physical analysis is noteworthy.

If, however, the Earth's axial rotation is not sufficient to cause it to disintegrate, it is nevertheless capable of producing a significant mechanical effect: the tides. Over the centuries, the phenomenon of the *flusso e reflusso* (ebb and flow) of the sea has elicited various explanations. Both Ptolemy and Kepler, for instance, attributed the tides to the influence of the Moon — though their theories, perceived as astrological in nature, were rejected by both Aristotelian physicists and Galileo.[47] In 1595, Sarpi offered a different perspective. Likely influenced by Cesalpino — who linked the tides to terrestrial motion — explained them as a consequence of the

quæ secundum uiolentiam. Quibus enim uis uel impetus infertur, dissolui necesse est, & diu subsistere nequeunt: quæ uero a natura fiunt, recte se habent, & conseruantur in optima sua compositione. Frustra ergo timet Ptolemæus, ne terra dissipetur, & terrestria omnia in reuolutione facta per efficaciam naturæ, quæ longe alia est quam artis, uel quæ assequi possit humano ingenio").

[45] In Galileo's formulation, many years later: "Now there remains the objection based upon the experience of seeing that the speed of whirling has a property of extruding and discarding material adhering to the revolving frame. For that reason it has appeared to many, including Ptolemy, that if the Earth turned upon itself with great speed, rocks and animals would necessarily be thrown toward the stars, and buildings could not be attached to their foundations with cement so strong that they too would not suffer similar ruin" ("Resta ora l'instanza fondata su 'l veder per esperienza come una vertigine veloce ha facultà di estrudere e dissipare le materie aderenti alla machina che va in volta; per lo che pareva a molti, ed anco a Tolomeo, che quando la Terra si rigirasse in sé stessa con tanta velocità, i sassi e gli animali dovessero esser scagliati verso le stelle, e che le fabbriche non potessero con sì tenace calcina esser attaccate a i fondamenti, che esse ancora non patissero un tale eccidio": *Dialogo sopra i due massimi sistemi del mondo*, in *OG* 7, p. 214; Engl. transl., p. 188).

[46] See also *Pensieri naturali*, no. 149, p. 158. In the *Dialogo sopra i due massimi sistemi del mondo*, Galileo would express a very similar idea: "the stone would not be separated from the Earth, because its motion away in the beginning would be so very minute that its inclination toward the center of the Earth would be a thousand times stronger" ("la pietra non si separerà dalla Terra, poiché il suo allontanarsene su 'l principio sarebbe tanto e tanto minimo, che ben mille volte più vien ad esser l'inclinazione che ha il sasso di muoversi verso il centro della Terra [...]": *Dialogo sopra i due massimi sistemi del mondo*, in *OG* 7, p. 221.31–34; Engl. transl., p. 195). The entire discussion, in which Galileo takes up and improves upon the Copernican argument, can be found on pp. 214.30–244.15 (Engl. transl., pp. 188–217).

[47] See Ptolemy, *Tetrabiblos*, II 13, 8, lin. 1049–1052 (p. 157 Hübner); Kepler, *Astronomia nova*, in *KGW* 3, p. 26.3–32 (Engl. transl., pp. 25–26). See also Galileo, *Dialogo sopra i due massimi sistemi del mondo*, in *OG* 7, pp. 442.11–449.34 (Engl. transl., pp. 416–423).

Earth's motions as described by Copernicus.[48] It must be noted, however, that the motion posited by Cesalpino differs from the one assigned to the Earth by Copernicus: Cesalpino's model proposes that motion is transmitted from one sphere to the next, beginning with with the outermost and reaching, in diminished form, the Earth at the center of the universe.[49]

Sarpi expounds his own theory of tides in 'thoughts' nos. 568–571. 'Thought' no. 568 lays the groundwork: with respect to the size of the Earth's orbit (the "orb of the annual motion"), the Earth's diameter, according to Sarpi, measures 6′.[50]

[48] By 1595, Galileo had already been in Padua for 3 years. His *Discorso del flusso e reflusso del mare* (later revisited in the Fourth Day of the *Dialogo sopra i due massimi sistemi del mondo*) is dated 8 January 1616. It was penned in the form of a letter to Cardinal Alessandro Orsini, who had been elevated to the cardinalate just over 2 weeks prior. The *Discorso* marks Galileo's final attempt to avert the condemnation of *De revolutionibus* (which a Decree of the Congregation of the Index suspended, *donec corrigatur*, on 5 March 1616), and it follows the so-called "Copernican letters" of 1613–1615.

[49] See Andrea Cesalpino, *Peripateticarum Quæstionum Libri Quinque*, Venice: Giunta, 1571, III, *quaestio V* ("Maris fluxum & refluxum ex motu terræ non Lunæ fieri," ff. 59*v*-61*r*): "Therefore, neither the Moon, nor the winds, nor anything similar can impart such motion to water. There remains, then, but one explanation: that of being moved by the movement of the container, which we have deemed accidental" ("Nec igitur luna, nec venti, nec aliquid huiusmodi aquam dimouere tali motu queant. Relinquitur autem vnus modus tantum, vt scilicet moueatur ad motum continentis, quem per accidens esse diximus": f. 60*r*-*v* C); "From what has been said, it is therefore clear that I have resolved all doubts concerning the ebb and flow of the sea, assuming a certain, however small, movement of the Earth" ("Ex dictis igitur patet solui omnes dubitationes, quæ circa maris fluxum, ac refluxum contingunt, supposita quadam terræ motione parua tamen": f. 61*r*). In the revised and expanded second edition of the work (1593), the passages appear in *quaestio VI* (ff. 70*r*-71*v*), f. 71*r* A and f. 71*v* E, respectively.

[50] It is challenging to ascertain precisely how Sarpi arrived at this particular value for the Earth's diameter. He likely began with *De revolutionibus*, IV 19, p. 300.7 (Engl. transl., p. 207), wherein Copernicus, discussing the Sun's distance (measured in Earth's radii), correctly attributes the value 1210 to Ptolemy (see *Almagest*, V 15, Vol. 1, p. 425.19–20 Heiberg; Engl. transl., p. 257); "But", Copernicus adds, "after Ptolemy other astronomers found that the foregoing conclusions did not agree well enough with the phenomena, and reported other findings about these topics. [...] Hence they believe it follows that, at apogee, the distance of the Sun from the Earth is 1146 parts" ("Alij uero post Ptolemæum, quoniam inuenerunt haud satis congruere hæc apparentijs, alia quædam de his prodiderunt. [...] e quibus sequi putant apogæi Solis a terra distantiam esse partium M.C.XLVI": ibid., p. 300.14–15, 23–24; Copernicus is probably referring to al-Battāni, who in *Opus astronomicum*, I 30, p. 61 Nallino, gives the value 1146; but see also Regiomontanus, *Epytoma in Almagestum*, V 21, p. 143 Schmeidler). If, then, the radius of the Earth's orbit is equal to 1146 parts, the length of the orbit (which, for Copernicus and Sarpi, as well as Galileo, is circular), assuming $\pi = 22/7$ (a common value, at the time: already Archimedes, in Κύκλου Μέτρησις [*De quadratura circuli*], prop. 3, showed that $223/71 < \pi < 22/7$; the approximation of π by 22/7, moreover, is implicit in the calculations leading to the values given in *Pensieri naturali*, no. 669, p. 509), equates to approximately 7203.43 Earth radii. Expressed in minutes of arc, the Earth radius is thus equal to 2′ 59″; the diameter, to 5′ 59″. This, however, does not definitely prove that Sarpi commenced with the value of 1146 for the Earth-Sun distance, since any value close to this figure yields the same result, when approximated to the nearest minute of arc: in fact, starting from the value established by Ptolemy (i.e. 1210: see *supra*), one arrives at a measurement of the Earth's diameter equal to 5′ 40″; commencing instead from the value provided by Copernicus (1179: see *De revolutionibus*, IV 19, p. 301.14; Engl. transl., p. 207), the result is 5′ 49″. As is apparent from Sarpi's own words — "The diameter of the Earth of the annual motion subtends 6 minutes, and possibly less than that [...]" ("Il diametro della Terra del moto annuo sottende minuti 6, e non so che meno [...]": *Pensieri*

Given that the center of the Earth ("the center of annual motion") traverses a distance of approximately 360°/365 per day — that is, just under 1° or 60′ — the Earth advances daily along its orbit by a distance equivalent to ten times its diameter.[51] Since the Earth completes one full rotation around its axis in 24 h, a point on the equator moves, during the 12 h of night, a distance equivalent to one Earth diameter on the plane of the ecliptic. To this must be added five Earth diameters due to the Earth's motion. During the 12 h of daylight, by contrast, one Earth diameter — representing the distance covered by the same in the opposite direction due to diurnal rotation — must be subtracted from the five diameters of orbital motion. As Sarpi explains:

> in the upper part [i.e., during the night] the diurnal motion in 30 minutes advances by 1/5, and equally retards in the other 30 [minutes] in the lower [i.e., during the day]. [...] Hence every point on the surface at times moves quickly, at times at a mean speed, at times slowly, as in the case of the Moon.[52]

The sea waters are consequently subject to significant acceleration and deceleration every day, stemming from the interplay of the Earth's diurnal and annual motions. It is for this reason, Sarpi suggests, that they ebb and flow — much like the water within a bowl, which responds to the motion of its container and the properties inherent to the fluid itself:

> Every water carried in a basin, at first remains behind, and rises at the end, because it has not yet received motion well; and when [the basin] stops, it keeps moving by the motion it received and rises at the front. Seas are waters in basins, so that with the Earth's annual motion they produce the same effect — at times fast, at times at mean velocity, at times slow during the daily rotation — that is seen by moving the basin in various directions. When the seas are so vast that they cover <a quarter of the entire sphere>, with parts of them in the fast-moving section and others in the section of mean velocity, the difference will be more pronounced. And the more so if they cover half of the sphere, with some parts in the fast section and others in the slow.[53]

---

naturali, no. 568, p. 423) — he was likely not intent on an exact calculation, but rather on the overall reasoning, and thus proceeded with approximations.

[51] In fact, the Earth's orbit around the Sun is about 924,375,700 km, and the Earth's average diameter is about 12745.59 km; imagining — as Sarpi and Galileo believed — the orbit to be circular, the Earth's diameter is approximately 17.87″ (around 0.005°), which is slightly more than $1/10^6$ of the orbit.

[52] "[...] nella parte superiore il moto diurno in 30 minuti promove 1/5, e negl'altri 30 nell'inferiore altretanto ritarda. [...] Onde ogni punto sulla superficie ora è veloce, ora mediocre, ora tardo, come nella Luna": *Pensieri naturali*, no. 568, p. 423.

[53] "Ogni acqua in un catino portata, nel principio del portarla resta indietro, e si alza dall'ultimo, perché ricevuto non ha ancora bene il moto; e quando si ferma segue a muoversi per il moto ricevuto e dalla parte avanti si alza. I mari sono acque ne' catini, onde col moto della Terra annuo fanno quel effetto che, ora sendo veloce, ora tardo e mediocre per il diurno, vedesi nel muovere il catino diversamente. E quando i mari sieno tanto grandi che tengano una <quarta [*lege*: quarta parte della sfera]>, sì che parte sia nel veloce, parte nel mediocre, vi sarà maggior differenza, e più ancora se la metà della sfera tengano, sì che parte nel veloce, parte nel tardo sia": ibid., no. 569, p. 424.

This reasoning is strikingly similar to Galileo's later explanations in his *Discourse on the ebb and flow of the sea* (1616) and in the Fourth Day of the *Dialogue Concerning the Two Greatest Systems of the World* (1632),[54] though Galileo's account is admittedly more refined.[55]

---

[54] From the Fourth Day of the *Dialogo sopra i due massimi sistemi del mondo*: "We have already said that there are two motions attributed to the terrestrial globe; the first is annual, made by its center along the circumference of its orbit about the ecliptic in the order of the signs of the Zodiac (that is, from west to east), and the other is made by the globe itself revolving around its own center in twenty-four hours (likewise from west to east) around an axis which is somewhat tilted, and not parallel to that of its annual revolution. From the composition of these two motions, each of them in itself uniform, I say that there results an uneven motion in the parts of the Earth" ("Due aviamo detto essere i moti attribuiti al globo terrestre: il primo, annuo, fatto dal suo centro per la circonferenza dell'orbe magno sotto l'ecclittica secondo l'ordine de' segni, cioè da occidente verso oriente; l'altro, fatto dall'istesso globo, rivolgendosi intorno al proprio centro in ventiquattr'ore, e questo parimente da occidente verso oriente, benché circa un asse alquanto inclinato e non equidistante a quello della conversione annua. Dalla composizione di questi due movimenti, ciascheduno per sé stesso uniforme, dico resultare un moto difforme nelle parti della Terra": *OG* 7, pp. 452.12–20; Engl. transl., p. 426). "Since this contrariety exists in the motion of the parts of the terrestrial surface when it is turning around its own center, it must happen that in coupling the diurnal motion with the annual, there results an absolute motion of the parts of the surface which is at one time very much accelerated and at another retarded by the same amount" ("Stante dunque tal contrarietà di moti nelle parti della superficie terrestre, mentre che ella si rigira intorno al proprio centro, è forza che, nell'accoppiar questo moto diurno con l'altro annuo, risulti un moto assoluto per le parti di essa superficie terrestre ora accelerato assai ed ora altrettanto ritardato": ibid., pp. 453.4–8; Engl. transl., p. 427). "From this we conclude that just as it is true that the motion of the whole globe and of each of its parts would be equable and uniform if it were moved with a single motion, whether this happened to be the annual or the diurnal, so is it necessary that upon these two motions being mixed together there results in the parts of the globe this uneven motion, now accelerated and now retarded by the additions and subtractions of the diurnal rotation upon the annual revolution. Now, if it is true (as is indeed, as proved by experience) that the acceleration and retardation of motion of a vessel makes the contained water run back and forth along its length, and rise and fall at its extremities, then who will make any trouble about granting that such an effect may — or rather, must — take place in the ocean waters? For their basins are subjected to just such alterations; especially those which extend from west to east, in which direction the movement of these basins is made. Now this is the most fundamental and effective cause of the tides, without which they would not take place" ("Concludiamo per tanto, che sì come è vero che il moto di tutto il globo e di ciascuna delle sue parti sarebbe equabile ed uniforme quando elle si movessero d'un moto solo, o fusse il semplice annuo o fusse il solo diurno, così è necessario che, mescolandosi tali due moti insieme, ne risultino per le parti di esso globo movimenti difformi, ora accelerati ed ora ritardati, mediante gli additamenti o suttrazioni della conversion diurna alla circolazione annua. Onde se è vero (come è verissimo, e l'esperienza ne dimostra) che l'accelerazione e ritardamento del moto del vaso faccia correre e ricorrere nella sua lunghezza, alzarsi ed abbassarsi nelle sue estremità, l'acqua da esso contenuta, chi vorrà por difficultà nel concedere che tale effetto possa, anzi pur debba di necessità, accadere all'acque marine, contenute dentro a i vasi loro, soggetti a cotali alterazioni, e massime in quelli che per lunghezza si distendono da ponente verso levante, che è il verso per il quale si fa il movimento di essi vasi? Or questa sia la potissima e primaria causa del flusso e reflusso, senza la quale nulla seguirebbe di tale effetto": ibid., pp. 453.23–454.2; Engl. transl., pp. 427–248). Similarly in the *Discorso del flusso e reflusso*, in *OG* 5, pp. 381.13–383.26.

[55] Galileo modulates the effects of the "most fundamental cause" (the very same identified by Sarpi) by recalling several "specific accidents" that affect the phenomenon of tides to varying degrees: see *Dialogo sopra i due massimi sistemi del mondo*, in *OG* 7, pp. 454–456; Engl. transl.,

In 'thought' no. 570, Sarpi identifies four factors affecting tidal intensity of the tides. First, the size of the basin: tides are not observed in small lakes and seas. Second, the Earth-Sun distance: variations in this distance throughout the year result in proportional changes in tidal strength. Third, basin shape and orientation: these characteristics modulate the intensity of tides. Fourth, seasonal changes: tidal characteristics vary from season to season.[56]

Sarpi's theory culminates in 'thought' no. 571, where he outlines the specific mechanics of high and low tides. During acceleration, high tide occurs as the water resists the motion due to inertia, then returns to equilibrium; during deceleration, low tide occurs. As Sarpi explains:

> Accordingly, the motion of the sea will result from the composition of two elements: first, the water lagging behind, and rising, when the motion turns from slow to fast, and of its natural tendency to reach equilibrium; and second, catching up with motion, when this turns from fast to slow, and therefore rising, by its natural tendency to go back to equilibrium.[57]

While I cannot delve deeper here, it is evident from reading Sarpi's *Pensieri naturali* that his knowledge of *De revolutionibus* was both profound and exhaustive. His understanding was not confined to the cosmomological and introductory Book I; rather, he delved into the intricate, astronomical specifics found in the later Books. This is particularly clear in in 'thoughts' nos. 637–643, 653–659 and 665–667. Each of these is intimately linked to *De revolutionibus* — be it through data, calculations, or various considerations. Furthermore, Copernicus is explicitly named in four 'thoughts': nos. 639, 640, 643 and 655.

I conclude this brief account of Sarpi's growing acceptance of the Copernican hypothesis, as shown in his *Pensieri naturali* (despite their fragmented nature), by

---

pp. 428–431). To these he adds what he calls "secondary causes" (ibid., pp. 457.9–462.11; Engl. transl., pp. 431–436), in describing which he suggests a possible explanation for an entire tidal cycle extending for 12 h. This is in contrast to the "primary cause" alone — the combination of the Earth's annual revolution around the Sun and its daily rotation around its own axis — which can only account for a cycle extending for 24 h (ibid., pp. 458.21–459.6; Engl. transl., pp. 432–433).

[56] For the first and third factors, see Galileo, *Dialogo sopra i due massimi sistemi del mondo*, in *OG* 7, pp. 457.11–458.20 and p. 459.7–32 (Engl. transl., pp. 431–432 and p. 433), respectively. The second factor may perhaps be found in Galileo's explanation of the monthly cycle of the tides: see ibid., pp. 477.17–479.13; Engl. transl., pp. 452–454 (Galileo, however, considers the distance of the Sun from the Earth-Moon system, which varies according to the Moon's position in its orbit around the Earth, affecting the primary cause of tides with monthly cyclicity). Finally, the fourth factor may perhaps be found in Galileo's explanation of the annual tidal cycle, where the tilt of the Earth's axis of rotation with respect to the plane of the ecliptic comes into play: see ibid., pp. 482.19–484.32; Engl. transl., pp. 457–460.

[57] "Il moto del mare sarà dunque un moto composto di due cose: prima, del restar l'acqua indietro, ed alzarsi, quando il moto passa dal tardo al veloce, e del suo naturale di voler tornare all'equilibrio; l'altra di seguir il moto quando al tardo dal veloce passa, e perciò alzarsi col suo naturale di tornare all'equilibrio": *Pensieri naturali*, no. 571, p. 426; see Galileo, *Dialogo sopra i due massimi sistemi del mondo*, in *OG* 7, p. 454.8–21 (Engl. transl., p. 428).

focusing on 'thought' no. 653. After the introduction of the telescope,[58] Sarpi offers several considerations and poses a few questions:

> That all the epicycles of the superior planets are the same size as the annual orb [i.e. the Earth's orbit] and equal to one another. That all their motions are equal to the motion of the annual orb relative to the planet itself. That the differences in latitudes are proportional to the Earth's approaching to and receding from [a planet] during its motion along the annual orb.
>
> The motions of the largest orbits are slower: why is the last one very fast? The motions are all towards the rear: why is the motion of the latter in advance?
>
> No motion along a higher orbit causes motion along a lower orbit: all motions are caused by the *primum mobile*.
>
> Venus has the Sun at the center of its orbit, as evidenced by its phases: why not all the other [planets]? Their motions cannot be known, except at their commutation.[59]
>
> The diversity in the magnitudes of the higher apogees and perigees is in that proportion. The magnitudes vary according to the annual motion. The retrogradations of Saturn are minimal and take a long time, those of Mars are large and take a very short time.
>
> Why do the motions of a planet along its orbit and epicycle must be equal to half of the motion of the Sun? Why, in the superior planets, the motion of the epicycle is equal to the motion of the Sun relative to the planet?[60]

As is the case in almost all other instances, this 'thought' draws attention to a few critical problems within Ptoleamic astronomy. Sarpi's various assertions lack any justifications, and questions are left unanswered — as if to imply that what is stated is beyond dispute, and what is inquired either possesses a widely acknowledged

[58] We may date this 'thought' thanks to the reference to the phases of Venus, first observed in the fall of 1610 (as evidenced by the anagram sent by Galileo to Giuliano de' Medici, in Prague, on 11 December 1610: *OG* 10, no. 435, p. 483) and communicated by Galileo himself to Sarpi, after a protracted silence, on 12 February 1611 (see *OG* 11, no. 476, pp. 48–49).

[59] In medieval astronomy and astrology, the expression *in consequentia* (εἰς τὰ ἑπόμενα, namely, "towards the rear") refers to motions that follow the natural order of the Zodiac signs (from Aries to Piscis, on the ecliptic), that is, eastward motions, also called direct motions. The converse, *in praecedentia* (or *in antecedentia*: εἰς τὰ προηγούμενα, namely, "in advance"), refers to motions against the order of the signs, namely, westward motion, or retrograde motions. By *commutatio* ("change", "alteration", or "commutation") medieval astronomers referred to a change in motion or reversal of direction (*commutatio motus*): either from direct to retrograde motion, or from retrograde to direct motion; in this context, *commutatio* describes the stationary points, that is, the points at which planets appear to halt before altering the direction of their motion. See Toomer, *Ptolemy's Almagest* (cited, footnote 43), p. 20.

[60] "Che tutti li epicicli de' superiori sono uguali all'orbe annuo e tra loro. Che tutti li moti loro sono uguali al moto dell'orbe annuo alla stella. Che le differenze delle latitudini sono a proporzion dell'accostarsi e discostarsi della Terra per il moto dell'orbe annuo. I moti de' più gran giri son più tardi: perché l'ultimo è velocissimo? I moti tutti in consequenza: perché quello in precedenza? Niun giro superiore muove l'inferiore, mossi venendo tutti dal primo. ♀ ha per centro il ☉ per le sue φάσεις: perché non tutti gli altri? Non si posson conoscere i lor moti, se non per quello della comutazione. La diversità delle grandezze de' superiori apogei e perigei è a quella proporzione. Le grandezze secondo il moto annuo si variano. Li regressi di ♄ minimi sono e lunghissimi, que' di ♂ grandi e brevissimi. Perché i moti del pianeta nel orbe suo e nell'epiciclo hanno da essere al mezzo moto del Sole uguali? Perché il moto del epiciclo è uguale al moto del Sole al pianeta ne' superiori?": *Pensieri naturali*, no. 653, pp. 494–495.

answer, or, more likely, none at all: Ptolemaic astronomy, it seems, proves inadequate in accounting for these issues.

Sarpi commences by noting the equivalent duration of motion on the epicycle of the superior planets: this duration corresponds precisely to a complete revolution of the Earth around the Sun (a sidereal year), as the planets' motion on the epicycle is merely a consequence of the Earth's orbital motion. The subsequent reference to the variation of latitudes pertains to the concluding book of *De revolutionibus*, wherein Copernicus scrutinizes the theory of planetary latitudes, announcing at the outset of Book VI that "this variation is geared to the motion and revolutions of the Earth's grand circle".[61]

Sarpi proceeds to observe that as the distance of the spheres from the system's center increases, their rotational velocity diminishes. He then poses a direct question: why, therefore, does the sphere of the fixed stars rotate with such celerity? Moreover, all these motions progress from west to east (*in consequentia*): why does the last one proceed in the opposite direction, from east to west (*in praecedentia*)? Whilst the former question remains unanswerable, the solution to the latter becomes evident when adopting a Copernican viewpoint: specifically, the outermost sphere is immobile, as it is the Earth that is in motion (upon its own axis). Hence the subsequent assertion: contrary to Aristotle's declaration, motion is not transmitted from one sphere to another, commencing with the outermost sphere (the *primum mobile*); instead, we perceive motion owing to the Earth's rotation.

Just like the Moon, Venus — as Galileo ascertained in the fall of 1610 — also appears crescent-shaped (*falcata*). Following the telescopic discoveries outlined a few months prior in *Sidereus nuncius*, this represented a further blow to the received view, and another piece of evidence supporting the Copernican hypothesis: "Now, we are made certain", Galileo wrote to Sarpi in early 1611, "that Venus revolves around the Sun, and not beneath it (as Ptolemy believed): were that the case, it would never appear as anything other than less than half a circle; nor would it be above it (as Aristotle thought): for, were it above the Sun, it would never be observed crescent-shaped, but always significantly more than half, and almost invariably perfectly round".[62] If this holds true for Venus, Sarpi rhetorically ponders, why should it not be the case for the other planets as well?

The subsequent remark concerning the *commutatio* is explicitly grounded in the Copernican hypothesis: since our vantage point (the Earth) is itself in motion, our understanding of planetary motion is solely derived from the changes in their

[61] "[...] and this variation is geared to the motion and revolutions of the Earth's grand circle" ("Quæ [...] uarietas ad motum ac reuolutiones orbis magni terræ reguletur": *De revolutionibus*, VI 1, p. 440.5–6; Engl. transl., p. 307).

[62] "Or eccoci fatti certi che Venere si volge intorno al sole, e non sotto (come credette Tolommeo), dove mai non si mostrerebbe se non minore di mezzo cerchio; né meno sopra (come piacque ad Aristotele), perchè se fusse superiore al sole, non si vedrebbe mai falcata, ma sempre più di mezza assaissimo, e quasi sempre perfettamente rotonda": Galileo to Sarpi, February 12, 1611, in *OG* 11, no. 476, p. 48.93–98.

apparent movement, that is, the variation in the angle each planet forms with the Sun, as observed from the Earth.[63]

In the ensuing statement, Sarpi remarks that the disparities noted in the proportions between the apogees and perigees of the superior planets are dependent on the combined orbital motion of both the Earth and the individual planets. Due to the projection of their image onto the background of the fixed stars, from the Earth's perspective — with the Earth itself orbiting the Sun — the retrograde (apparent) motion of the superior planets varies according to their distance from the Earth. Consequently, this motion will have a shorter path but a longer duration for Saturn (the outermost planet), and a longer path but a shorter duration in the case of Mars (which, among the superior planets, is the closest to the Earth).[64]

The 'thought' concludes with two queries, which, once more, we can address by adopting the Copernican perspective. Initially, Sarpi observed that, for the superior planets, the duration of their motion along the epicycle is equal to the period of the

[63] Even the terminology is Copernican: as Rheticus notes, "What [...] the ancients termed the 'planet argument', my master [Copernicus] calls the 'motion of commutation' of the planet, since we calculate, by means of it, the phenomena that are the product of the Earth's motion along the grand circle [...]" ("Quod [...] Veteres argumentum planetae dixerunt, hoc D. Praeceptor motum commutationis planetae vocat, quia per eum apparentias ratione motus terrae in orbe magno contingentes ratiocinamur [...]": *Narratio prima* [1540], edited by Henri Hugonnard-Roche and Jean-Pierre Verdet, with the collaboration of Michel-Pierre Lerner and Alain-Philippe Segonds, Warsaw: Ossolineum, 1982, p. 74.141–144). Much of Book V of *De revolutionibus* is devoted to *commutatio*; at the opening of Chap. 1, for example, we read: "Two entirely different motions in longitude appear in them. One is caused by the Earth's aforementioned motion, and the other is each one's own proper motion. I have decided without any impropriety to call the first one a motion of commutation, since it is this which makes all of them appear to have stations, progressions, and retrogradations. These phenomena appear not because the planet, which always moves forward with its own motion, is erratic in this way, but because a sort of commutation is produced by the Earth's motion according to the difference in size of their orbits" ("Bini longitudinis motus plurimum differentes apparent in ipsis. Vnus est propter motum terræ quem diximus. Alter cuiusque proprius. Primum non iniuria motum commutationis dicere placuit, cum ipse sit qui in omnibus illis stationes, progressiones, & regressus facit apparere, non quod planeta sic distrahatur, qui motu suo semper procedit, sed quod per modum commutationis sic appareat, quam efficit motus terræ pro differentia & magnitudine illorum orbium": V 1, pp. 328.23–329.5; Engl. transl., p. 227). See also *De revolutionibus*, IV 2, p. 251.25–26 (Engl. transl., p. 176), and footnote 4 *ad loc.* (Vol. III, pp. 305–306). In *Epitome astronomiae Copernicanae* (1618–1621), Kepler draws attention to the ambiguity of the term chosen by Copernicus: "Since *parallaxis*, in Greek, is equivalent to what, in Latin, is denoted by *commutatio*, the ambiguity must be avoided: both are angles, but the former refers to the angle of the anomaly [or motion] of commutation, whereas the latter to commutation itself" ("Cumque Parallaxis Graecè sit idem, quod latinè commutatio, cauenda est ambiguitas; vtrumque est angulus, ille Anomaliae Commutationis, hic Commutationis ipsius": *KGW* 7, p. 426.1–4; see also pp. 425.13–426.8). See also Emmanuel Poulle, "La théorie épicyclique selon Ptolémée au Moyen Age," in *Avant, Avec, Après Copernic. La représentation de l'Univers et ses conséquences épistémologiques*, Paris: Blanchard, 1975, pp. 51–60, especially pp. 59–60; and Noel M. Swerdlow, Otto Neugebauer, *Mathematical Astronomy in Copernicus's De revolutionibus*, New York: Springer, 1984, Part I, p. 301.

[64] The duration and amplitude of retrogradations for the superior planets, from the inside out, are approximately: Mars, 72 days and 16° (occurring every 780 days); Jupiter, 121 days and 10° (occurring every 399 days); Saturn, 138 days and 7° (occurring every 378 days).

Earth's revolution around the Sun. Now he inquires: why must such motions be equal to the mean motion of the Sun (that is, the Earth's orbital motion around the Sun)?

The list provided by Sarpi, though merely illustrative (it comprises almost exclusively geometric-kinematic, rather than physical, considerations), unmistakably draws attention to the superior rational order, and thus the greater plausibility, of the Copernican system. According to this system, all anomalies in planetary motions — which Ptolemy and his predecessors had to account for by introducing homocentric spheres, eccentrics and epicycles, without, however, ever managing to explain their number and size, or the planets' velocities along them, except by resorting to *ad hoc* hypotheses, introduced simply to "save the phenomena" — can be explained as (apparent) consequences of the Earth's motion. Indeed, although Eudoxus, Callippus and Ptolemy certainly succeeded in producing a model that accounted for the observations,

> they could not elicit or deduce the most important thing, namely, the structure of the universe and the true symmetry of its parts. On the contrary, their experience was just like someone taking from various places hands, feet, a head, and other pieces, very well depicted, it may be, but not for the representation of a single person; since these fragments would not belong to one another at all, a monster rather than a man would be put together from them. Hence in the process of demonstration, or *methodos*, as it is called, those who employed eccentrics are found either to have omitted something essential, or to have admitted something extraneous and wholly irrelevant. This would not have happened to them, had they followed sound principles. For if the hypotheses assumed by them were not false, everything which follows from their hypotheses would be confirmed beyond any doubt.[65]

[65] "Rem quoque præcipuam, hoc est mundi formam, ac partium eius certam symmetriam non potuerunt inuenire, uel ex illis colligere. Sed accidit eis perinde, ac si quis e diuersis locis, manus, pedes, caput, aliaque membra, optime quidem, sed non unius corporis comparatione, depicta sumeret, nullatenus inuicem sibi respondentibus, ut monstrum potius quam homo ex illis componeretur. Itaque in processu demonstrationis, quam μέθοδος uocant, uel præterijsse aliquid necessariorum, uel alienum quid, & ad rem minime pertinens, admisisse inueniuntur. Id quod illis minime accidisset, si certa principia sequuti essent. Nam si assumptæ illorum hypotheses non essent fallaces, omnia quæ ex illis sequuntur, uerificarentur proculdubio": *De revolutionibus*, p. 7.17–28 (from the dedicatory letter to Pope Paul III; Engl. transl., p. 4). With these words, Copernicus clearly contradicts what Andreas Osiander asserted in the Preface to the work (see ibid., p. 2.8–14: see above, footnote 10). Indeed, Copernicus maintains that in astronomy, where effects must follow from causes, nothing true can ever result from something false, save by means of argumentative stratagems that have no place in a rigorously demonstrative science. See Kepler, *Mysterium cosmographicum* (1596), in *KGW* 1, p. 15.5–9 (English translation by Alistair M. Duncan, *The Secret of the Universe*, New York: Abaris Books, 1981, p. 75): "[…] I have never been able to agree with those who rely on the model of an accidental demonstration, which infers a true conclusion from false premises by the logic of the syllogism. Relying, as I say, on this method they argued that it was possible for the hypotheses of Copernicus to be false and yet for the true phenomena to follow from them as if from authentic postulates" ("[…] nunquam assentiri potui illis, qui freti exemplo accidentariae demonstrationis, quae ex falsis praemissis necessitate Syllogistica verum aliquid infert. Qui, inquam, hoc exemplo freti contendebant, fieri posse, vt falsae sint, quae Copernico placent hypotheses, et tamen ex illis vera φαινόμενα tanquam ex genuinis principijs sequantur"). See also Kepler, *Apologia pro Tychone contra Ursum* [1600], in Nicholas Jardine, Alain-Philippe Segonds, *La Guerre des astronomes. La querelle au sujet de l'origine du système géo-héliocen-*

Crucially, the assertions and questions presented in 'thought' no. 653 bear witness to Sarpi's complete and considered adherence to Copernicanism, an understanding achieved through several years of meticulous study of *De revolutionibus*.

2.6 After *Sidereus Nuncius*

In the ensuing years — owing to dialogues with fellow intellectuals in Padua during the late sixteenth and early seventeenth centuries, particularly with Galileo, and to the telescopic discoveries of 1609–1610 (in which Sarpi played a not insignificant role)[66] — Sarpi's embrace of Copernicanism deepened and evolved through

trique à la fin du XVIe siècle, Vol. II/2, Paris: Les Belles Lettres, 2008, pp. 265.23–267.14 (= *KGW* 20.1, p. 26.5–42); Kepler's in-depth discussion concerning what constitutes an astronomical hypothesis occupies the entirety of Chap. 1 of the work: see ibid., pp. 250–276 (= *KGW* 20.1, pp. 18–30).

[66] Micanzio, *Vita del padre Paolo* (cited, footnote 4), pp. 1391–1392: "of various inventions of instruments, he avoided even letting it be known that he was the author; [...] of the spyglass, known in Italy as Galileo's [but] invented in Holland, he understood its functioning when, after one was presented at the Most Serene Republic with a request of one thousand gold coins, it was given to father Sarpi, so that he could test it and ascertain its potential uses, and give his opinion of it. And since he was not permitted to open it and examine its interior, he surmised what it might be, and discussed this with Galileo, who found father Sarpi had hit the mark" ("di varie invenzioni d'istromenti ha fuggito anco il lasciar sapere ch'esso ne fosse auttore; [...] dell'occhiale, detto in Italia del Galileo, trovato in Olanda, fu da lui penetrato l'artifizio, quando, presentatone uno alla serenissima signoria con manda di milla zecchini, fu al padre [Sarpi] dato carico di far le prove a che potesse servire e dirne il suo giudizio; e perché non gl'era lecito aprirlo e vedere, imaginò ciò che esser potesse, e lo conferì col signor Galileo, che trovò il padre aver dato nel segno"). Then, Micanzio proceeds to combine Sarpi's interests, intelligence and considerable practical attitude with his discreet character: "But it is worth noting that many instruments devised by mathematicians and astronomers, which they described with such solemnity that making them is difficult and using them much more so, he made with his own hands and gave models to artisans; by reducing those instruments to such ease and simplicity, he seemed to have the Earth and the heavens in his head" ("Ma è cosa notabile che tanti istromenti dagl'auttori delle matematiche et astronomiche inventati, ma descritti con tante solennità, che 'l fargli è difficile e l'usargli molto più, egli facendogli anco di sua mano e dando i modelli ad artefici, gli riduceva a tanta facilità e semplicità, che pareva avere nella sua testa la terra et i cieli": ibid.). In *Sidereus nuncius* (1610), Galileo does not mention Sarpi's name, nor does he hint at the role he played: this lack of recognition in the discovery of the telescope and subsequent astronomical observations, as well as Galileo's departure from Padua and his rapprochement with the Jesuits and the Roman Curia, are some of the causes of the drastic cooling of relations between Galileo and his Venetian friends after 1610, including above all Paolo Sarpi (and Giovan Francesco Sagredo): see Gaetano Cozzi, "Galileo Galilei, Paolo Sarpi e la società veneziana", in Id., *Paolo Sarpi tra Venezia e l'Europa*, Turin: Einaudi, 1979, pp. 135–234, especially pp. 179–183. On Sarpi's role, see Mario Biagioli, "Did Galileo Copy the Telescope? A 'New' Letter by Paolo Sarpi," in Albert Van Helden, Sven Dupré, Rob van Gent, Huib Zuidervaart (eds.), *The Origins of the Telescope*, Amesterdam: KNAW, 2010, pp. 203–230. See also Eileen Reeves, *Galileo's Glassworks: The Telescope and the Mirror*, Cambridge MA-London: Harvard University Press, 2008; Massimo Bucciantini, Michele Camerota, Franco Giudice, *Il telescopio di Galileo. Una storia europea*, Turin: Einaudi, 2012 (English translation by

engagement with his interlocutors. However, it never attained a public dimension, unlike that of Galileo.

It was not until 1616, following the Congregation of the Index's condemnation of *De revolutionibus*,[67] that Sarpi felt compelled to issue a public statement. A decade earlier, on 28 January 1606 — at a period of considerable tension between the Republic of Venice and Rome, shortly after the commencement of the dispute that would culminate in the Interdict (in which Sarpi himself would play a pivotal part)[68] — he had been appointed *consultore in iure* of the Republic, a position he would hold for 17 years.[69] The Decree of the Congregation of the Index condemning Copernicus' work had reached the Venetian Holy Office in early May, and Sarpi had been requested for a "consultation" regarding its publication and dissemination within the Serenissima's territory. After briefly mentioning the other books proscribed in the decree, Sarpi turned his attention to *De revolutionibus* and its author:

> Nicolaus Copernicus was a catholic priest, a public lecturer at the University of Rome, and very close to the holy memory of Pope Paul III, both during his cardinalate and subsequently, upon his elevation to the papacy. His book was published just under a century ago: it was seen and read across Europe, and its author was widely regarded as the most learned in the profession of astronomy the world has ever known, so much so that the calendar reform, championed by Pope Gregory XIII, is founded upon his theory. For these reasons, the suspension of the book cannot be published without drawing attention to the novelty of suspending an old text, seen globally, and uncensured by the Council of Trent, or in Rome.[70]

Catherine Bolton, *Galileo's Telescope: A European Story*, Cambridge, MA: Harvard University Press, 2015); and Eileen Reeves, *Evening News: Optics, Astronomy, and Journalism in Early Modern Europe*, Philadelphia: University of Pennsylvania Press, 2014.

[67] See *OG* 19, pp. 322–323; English translation, "Decree of the Index (5 March 1616)", in *The Galileo Affair: A Documentary History*, edited and translated by Maurice A. Finocchiaro, Berkeley-Los Angeles-London: University of California Press, 1989, pp. 148–150.

[68] In December 1605, a bitter dispute flared up following the promulgation, in Venice, of laws that curtailed ecclesiastical property rights and the indictment of two clergymen for common crimes. This conflict, which resonated across Europe, officially persisted until April 1607. It pitted two formidable powers against each other: the Venetian State and the Church, confronting each other with no room for compromise. The former was resolute in asserting the supremacy of its laws over all citizens, lay and religious alike, whilst the latter was equally determined to secure recognition as a distinct entity, transcendent of national borders and civil governance, answerable solely to the pope in Rome.

[69] As Sarpi himself would effectively write: "The remit of law consultant is to declare what is lawful in the matter, case, or affair presented to him. In disputes, a sound knowledge of the laws is all that is required to pronounce upon what is legitimate, for cases invariably present themselves in universal terms: they do not vary, and what is once rightly deliberated shall prove eternally serviceable" ("L'ufficio del consultor in iure è rispondere quel che è di raggione nel fatto o caso over negozio che gli vien proposto. Per rispondere *de iure* nelle scole non si vi ricerca altro che una buona cognizione delle legi perché li casi si propongono in termini universali che non ricevono varietà e quel che una volta è ben discuso serve per sempre": "Carico di Consultor in iure della Republica. 15 settembre 1618", in Paolo Sarpi, *Opere*, edited by Gaetano and Luisa Cozzi, Milan-Naples: Ricciardi, 1969, pp. 464–467: p. 464).

[70] "Nicolò Copernico fu un prete catolico, publico lettor nello Studio di Roma, e molto familiare della santa memoria di papa Paulo III mentre era cardinale, et anco doppo creato papa; et il suo libro è stato stampato già poco meno di 100 anni, veduto e letto da tutta Europa con stima che

Whilst maintaining a degree of formal detachment, necessitated by his official capacity, Sarpi did not refrain from recalling the significance of Copernicus as a "renowned astronomer",[71] a catholic, and a confidant of Pope Paul III (to whom *De revolutionibus* was dedicated). Nor did he omit to highlight the awkwardness of a decree concerning a book printed almost a century prior, widely circulated and held in high esteem by many — so much so, Sarpi notes, with a hint of sarcasm, as to form the basis of the calendar reform associated with another pope, Gregory XIII.[72] Sarpi then continued:

> […] with all this, considering on the other hand that this kind of doctrine in no way damages the power of princes nor favors it, and the temporal authority can derive no benefit whatsoever from it, nor does this affect the art of printing in this State, […] moreover, as there are very few individuals who are professional astronomers, there is no need ever to worry that a scandal might arise, I should believe that granting the prohibition and the suspension […] could not cause any harm to public life.[73]

In conclusion, then, Sarpi endorsed the publication of the decree on Venetian territory, a stance he could scarcely have avoided. His statements represent formal assertions of principle: indeed, he was obliged to uphold the agreement between the Republic and Rome, ensuring that acceding to the Holy Office's request would not be misinterpreted as yielding to an external authority. However, alongside the justifications for not opposing the request — given that it in no way infringes upon "the power of princes" nor can the pope's temporal authority benefit from it — Sarpi added an unsolicited comment: after all, he wrote, considering the minuscule number of professional astronomers, the sole individuals capable of actually

quell'auttore sia stato il più dotto nella professione di astronomia che il mondo abbia mai avuto, anzi che sopra la dottrina di quello è fondata la correzzione dell'anno, fatta da papa Gregorio XIII. Per queste cause la suspensione del libro non è per riuscire senza che sia admirata questa nova introduzzione di suspendere un libro vecchio, veduto da tutto il mondo, e per il passato non censurato né al concilio di Trento né in Roma": Paolo Sarpi, "Sopra un decreto della Congregazione in Roma in stampa presentato per l'Illustrissimo Signor Conte Del Zaffo" (7 May 1616), in Sarpi, *Opere* (cited, footnote 69), pp. 603–605: p. 604. Sarpi's full text is also available in *De revolutionibus*, Vol. III, pp. 632–635 (text and French translation), 677–679 (notes).

[71] "[c]elebre astronomo": ibid., p. 603.

[72] As the members of the Roman College knew only too well, the calendar reform served as a key differentiator between the Catholic and the Reformed nations. The latter had, indeed, declined to adopt it precisely because of its papal promotion. For this very reason, in 1543 — not many years after Luther's notable stance with his 95 Theses, in 1517 — Andreas Osiander chose not to affix his signature to the Preface of *De revolutionibus*, which was dedicated to Pope Paul III. The Council of Trent, commencing shortly thereafter, in 1545, and concluding in 1563, formally endorsed the Church's resolute counter-reaction, thereby reasserting its authority.

[73] "[…] con tutto ciò, considerando dall'altra parte che questa sorte di dottrina non tocca in alcun modo la potestà de principi né la favorisce, e l'auttorità temporale non può riceverne beneficio alcuno, né meno questo tocca l'arte della stampa di questo Stato, […] essendo anco pochissime le persone ch'attendono alla professione d'astronomia, non si può manco temer che possi nascer scandolo, pertanto stimerei che il conceder la proibizione e suspensione […] non possi esser di alcun pregiudicio publico": Sarpi, "Sopra un decreto della Congregazione in Roma in stampa" (cited, footnote 70), p. 605.

comprehending a highly technical text such as Copernicus', there should be no apprehension that the publication of the decree might provoke any scandal.

Contrary to what has been suggested — albeit authoritatively[74] — these words do not betray a belittlement of the importance of the new science, and of the Copernican theory in particular. As can be inferred even from a cursory reading of *Pensieri naturali*, Sarpi's adherence to Copernicanism is meticulously considered, profoundly convinced and solidly substantiated: it is difficult to imagine that he regarded the Copernican theory as a hypothesis in the weak sense of the term, as Osiander did, rather than in the strong sense, as Copernicus (and indeed Kepler and Galileo alongside him) intended.[75] In his consultation, Sarpi approached the issue from the perspective of the Republic of Venice, for which the censorship of *De revolutionibus* posed no difficulty: Copernicus' work had very few readers, it held no bearing on political authority, and, not being printed in Venice, its condemnation caused no detriment to Venetian printers.

Sarpi would not emulate Galileo. Galileo's natural philosophy emerged from the interplay of reason and experience: its purview was a sphere destined to progressively expand without impediment, yet strictly within its own domain, which was utterly distinct from that of the absolute and the divine. The new science neither required nor accommodated the divine: it did not seek it, it did not conflict with it, and it did not judge it — nor did it wish to be judged by those unfamiliar with it. In

[74] According to Gaetano Cozzi, the progressively widening chasm that, from 1610 onwards, estranged Galileo from his erstwhile Venetian friends, was brought about "[by] the reaffirmation of a divergence on the scientific level, [by] Galileo's embarking upon a path and towards objectives that his friends neither shared nor comprehended [...]. They failed to understand, in other words, Galileo's confident and combative embrace of Copernicanism, to the extent of centring his entire scholarly endeavour upon it. For his Venetian friends, whose intellectual fervour revolved around the problems of mechanics, into which Galileo had introduced them, thus opening up perspectives they deemed revolutionary, Copernicanism remained merely a mathematical hypothesis — compelling, but in need of adjustments and further verification, to be approached with caution, dispassion, and scientific rigour. To embark on a struggle for it within a Church they perceived as impervious to all spiritual and cultural needs seemed absurd, futile, even ridiculous" ("il ribadirsi di un distacco sul piano scientifico, l'avviarsi di Galileo per una strada e verso mete che gli amici non condividevano e non comprendevano [...]. Essi non capivano, in altre parole, quel volgersi sicuro e battagliero di Galileo verso il copernicanesimo, fino ad incentrarvi tutto il suo impegno di studioso. Per gli amici veneziani, la cui passione era fissa verso i problemi di meccanica cui Galileo li aveva introdotti, spalancando prospettive che ritenevano rivoluzionarie, il copernicanesimo restava un'ipotesi matematica. Convincente, ma bisognosa di ritocchi e di verifiche, da trattare con cautela e con freddezza e rigore scientifici: ingaggiar una lotta per esso, in seno a una Chiesa che ritenevano chiusa ad ogni esigenza spirituale e culturale, sembrava assurdo, inutile, fin ridicolo": Cozzi, "Galileo Galilei, Paolo Sarpi and la società veneziana" [cited, footnote 66], pp. 218–219).

[75] For Osiander, 'hypothesis' is to be understood in a weak sense: as a conjecture advanced out of sheer computational necessity, a means of "saving the appearances" in the absence of a superior alternative: see the text cited in footnote 10, from the Preface to *De revolutionibus* ("De hypothesibus huius operis"). For Copernicus, by contrast, 'hypothesis' is to be interpreted in a strong sense: as the very bedrock of reasoning, the foundational principle from which the order of the cosmos is to be derived; see, for example, ibid., I 11, p. 44.7–8 (Engl. transl., p. 25): "We shall use this movement as a principle and hypothesis for the demonstration of the other [movements]" ("Quo tanquam principio & hypothesi utemur in demonstrationibus aliorum").

short, it referred to a different book. The book of the scientist is wordly, finite and quantifiable; the book of God speaks of the supernatural and may be accessed through alternative means — but the two remain distinct. The philosophy underpinning this novel perspective deviated not only from the Aristotelian tradition but also from Galileo's own earlier endorsement of Copernicanism. In the words of Eugenio Garin:

> Once the barriers of the sky were down and the hierarchical and finalistic conceptions abandoned, the physical world unified and the logico-cognitive processes clarified, Galileo's Copernicanism took on new dimensions. Once knowledge of both the sky and the Earth became "natural", i.e. scientific, the ultimate relationships change, and so do the boundaries of the realm of man. Where the heavens, the stars, and space come together; when a unified science sets out to identify the laws of every field of possible experience, the ultimate problems arise again in a different form.[76]

After 1610, the Copernican hypothesis, previously defended solely in theory, could now be fully embraced: the "great things" announced on the title page of *Sidereus nuncius* proved decisive for the affirmation of the heliocentric theory, but the path to its definitive acceptance as an actual description of the structure of the universe was far longer and more arduous than perhaps Galileo himself had anticipated. Old and new evidence had to be supplemented with a novel understanding of philosophy as an autonomous human endeavour, to be assessed solely by its own standards. In the years leading up to the publication of the *Dialogue* and those following the trial of 1633, Galileo adopted an open stance and ultimately secured for science the right to err. Unlike Sarpi (and their conteporary Descartes), he was no longer willing to wear a mask.[77]

---

[76] "Cadute le barriere del cielo, cadute le concezioni gerarchiche e finalistiche, unificato il mondo fisico, chiariti i processi logico-conoscitivi, il copernicanesimo galileiano assume nuove dimensioni. Fattasi «naturale», ossia scientifica, la conoscenza così del cielo come della terra, i rapporti ultimi cambiano; e cambiano i confini del regno dell'uomo. Nel punto in cui i cieli, le stelle, gli spazi, vi rientrano; nel momento in cui una scienza unificata si dispone a individuare le leggi di ogni campo di possibile esperienza, i problemi ultimi si ripropongono diversamente": Eugenio Garin, *Scienza e vita civile nel Rinascimento italiano*, Rome-Bari: Laterza, 1965, p. 163.

[77] "I am of such a nature that, like a chameleon, I take on the character of those I converse with; yet, while I unwillingly absorb the temper of the secretive and gloomy, I readily and gladly adopt that of the cheerful and open. I am forced to wear a mask: no one in Italy can do without one" ("Ego eius ingenii sum, ut velut chamaeleon, a conversationibus mores sumam; verum quos ab occultis et tristibus haurio, invitus incordio, hilares et apertos, sponte et libens recipio. Personam coactus fero, licet in Italia nemo sine ea esse possit": Sarpi to Jacques Gillot, 12 May 1609, in Paolo Sarpi, *Lettere ai Gallicani*, edited by Boris Ulianich, Wiesbaden: Steiner, 1961, no. III*, pp. 131–133: p. 133); "Quod ad Iesuitas attinet, bene quidam eorum dixit: Iesuita est omnis homo; chamaeleontem habes" ("As far as Jesuits are concerned, one of them said it well: every man is a Jesuit, you have a chameleon before you": Sarpi to Jacques Leschassier, 3 August 1610, ibid., no. XLI*, pp. 87–89: p. 87). A few years later, Descartes would note: "Just as actors, warned not to let their shame appear on their faces, put on a mask, so I, about to step onto this theater of the world — where until now I have been merely a spectator — enter the stage wearing a mask" ("Ut comoedi, moniti ne in fronte appareat pudor, personam induunt: sic ego, hoc mundi theatrum conscensurus, in quo hactenus spectator exstiti, larvatus prodeo": *Cogitationes privatæ*, in *Œuvres de Descartes*, edited by Charles Adam and Paul Tannery, Vol. X, Paris: Librairie Philosophique J. Vrin, 1966,

Acknowledgements This is the last work I had the good fortune to discuss at length with John Heilbron. We talked about it during our last walks in Padua, in September 2023. As always, he was encouraging, and offered many suggestions. We also drafted another paper and sketched the contents of a joint book, both on Galileo. For all this, and much more, I cannot thank him enough.

My gratitude to Gerald J. Toomer for his assistance and help, and to Concetta Luna for several comments on an earlier version of this work.

Stefano Gattei is Associate Professor of history and philosophy of science at the University of Trento, Italy. He works on contemporary issues in the philosophy of science, as well as the history of ideas and the history of cosmology in early modern times, with special reference to Galileo and Kepler. His most recent publications include *On the Life of Galileo* (Princeton University Press, 2019), a critical edition and annotated translation of Regiomontanus' 1464 Padua oration (*Studia graeco-arabica*, 14/1, 2024), and *Feyerabend in Dialogue* (Springer, 2024, co-edited with Roberta Corvi). He is currently working on a book on Campanella and Galileo, and on a critical edition of Kepler's *Strena seu De nive sexangula* (1611).

---

pp. 213–256: p. 213). See also Giodano Bruno: "[…] studious Dissimulation, to whom Jupiter grants permission to occasionally appear in the heavens — not as a goddess but, at times, as a handmaid of Prudence and, at times, as a shield of Truth" ("[…] la studiosa Dissimulazione: a cui Giove fa lecito che talvolta si presente in cielo, e non già come Dea, ma come tal volta ancella della Prudenza, e scudo della Veritade": *Spaccio de la bestia trionfante* [1584], edited by Giovanni Aquilecchia, in Giordano Bruno, *Œuvres complètes*, Vol. V/1–2: *Expulsion de la bête triomphante*, Paris: Les Belles Lettres, 1999: Vol. V/2, p. 299).

Chapter 3
Wise Guy

Eileen Reeves

Abstract This essay has as its focus Galileo's Letter to the Grand Duchess Christina (1615), purportedly a sober meditation on the relevance of recent astronomical discoveries to a Scriptural exegesis. It argues that the letter is instead a pastiche of three celebrated orations of the Second Sophistic, "The Philosopher" "How the Philosopher Should Speak," and "The Sophist," composed between 355 and 360 CE by the pagan natural philosopher, rhetorician and statesman Themistius. The latter figures here not as the familiar paraphrast of Aristotle's works, but rather as a speaker sufficiently persuasive and ambiguous to have served both Christian and non-Christian emperors in an epoch of enormous confessional conflict. More generally, Galileo's familiarity with the Second Sophistic is part of a broad cultural movement of particular importance in an age of absolutism.

Keywords Galileo · Christina · Themistius · Second sophistic · Orations

> For by commanding the sun to stop 'in the midst of heaven' [Joshua] may not have meant it to freeze at noon, as most commentators had assumed, but rather, for the benefit of later *sophisticates,* he had alluded to the sun's true place in the universe.[1]

"The things that we know are a fraction of those that we don't." So said Galileo in his *Letter to the Grand Duchess Christina* of 1615, paraphrasing a remark made by the fourth century C.E. philosopher and rhetorician Themistius, itself a paraphrase of Aristotle's *On the Soul.*[2] What we know of that letter, an ill-advised attempt to

[1] John Heilbron, *The Sun in the Church. Cathedrals as Solar Observatories* (Cambridge, MA: Harvard University Press, 1999) 17. Emphasis mine, and in memory of one of the wisest guys.

[2] Galileo Galilei, "Lettera a Madama Cristina," in Galilei, *Opere*, edited by Antonio Favaro, 20 vols. (Florence: G. Barbèra, 1890–1909) 5: 320; this and all subsequent translations mine unless otherwise acknowledged. See also Galileo Galilei, *Scienza e Religione. Scritti Copernicani,* edited

E. Reeves (✉)
Princeton University, Princeton, NJ, USA
e-mail: ereeves@princeton.edu

S. Gattei (ed.), *The Incomparable Historian*, Archimedes 75,
https://doi.org/10.1007/978-3-032-25678-2_3

resolve the apparent conflict between heliocentrism and Scripture, has increased a little of late: in 2018, Salvatore Ricciardo rediscovered Galileo's holograph *Letter to Benedetto Castelli,* a bold prototype of the same argument, in the archives of the Royal Society, where it had languished undisturbed for several centuries through a clerical error, the transcription of "1613" as "1618."[3]

This essay returns to the letter to Christina, making two interrelated claims. It argues that the letter scales up from the pallid paraphrase of Themistius to an elaborate pastiche of three of that philosopher's most celebrated orations, "The Philosopher," "How the Philosopher Should Speak," and "The Sophist," and that this showy performance is an early modern reprise of the cultural movement of the Second Sophistic, with particular implications for the widowed Grand Duchess's authority in and beyond the Medici Court. While Galileo's early and generally critical reaction to Themistius' youthful work as Peripatetic paraphrasist is familiar within the context of his studies of motion around 1589–1592,[4] I am arguing that his subsequent engagement with this figure as a sophistic orator is more substantial, strategic, and favorable. That these claims are not already part of what we know about the letter is due less to Galileo's subtlety than to a different sort of clerical error. The anti-Copernican measures of the Decree of the Congregation of the Index in 1616 and Galileo's trial, condemnation, and abjuration in 1633 sharply limited the circulation of the original Italian text in Catholic regions, precisely when readers turned to the cultural upstarts of the Second Sophistic, including the late bloomer Themistius.[5]

3.1 So Long a Day

The origins of the letter lay in a lunch at the Medici palace in mid-December 1613; the Benedictine Benedetto Castelli, Galileo's former student and a newly appointed professor of *mathematica* at Pisa, provided his mentor with a post-prandial account. When Castelli described observing the Medici stars—the four largest of Jupiter's moons—the other guests encouraged him, but an antagonist soon emerged. Though

---

by Massimo Bucciantini and Michele Camerota (Rome: Donzelli, 2009) 49 n 33. Themistius, *On Aristotle On the Soul,* translated by Robert B. Todd (London and New York: Bloomsbury, 2013) Book 3, Chap. 3. 87,29; 110. Themistius' remark, made in the context of *On the Soul* 3. 427b 1–3, was a commonplace: see for example Tommaso Garzanti, *Piazza Universale* (Venice: Gio. Battista Somasco, 1586) 742.

[3] Michele Camerota, Franco Giudice, and Salvatore Ricciardo, "The Reappearance of Galileo's Original Letter to Benedetto Castelli," *Notes and Records of the Royal Society* **73** (2018): 11–28.

[4] Paolo Galluzzi, *Momento. Studi Galileiani* (Rome: Edizioni dell'Ateneo & Bizzarri, 1979) 98–100, 125–127, 172 n. 73, 319 n. 23; Michele Camerota and Mario Helbing, "Galileo and Pisan Aristotelianism: Galileo's "De Motu Antiquiora" and the *Quaestiones de Motu Elementorum* of the Pisan Professors," *Early Science and Medicine* **5.4** (2000): 319–365.

[5] For a recent English translation and discussion of the events of 1615–1616, see *The Trial of Galileo. 1612-1633,* edited by Thomas Mayer (Toronto: University of Toronto Press, 2012) 1–105.

Cosimo Boscaglia, professor of philosophy at Pisa, conceded the existence of the satellites, he objected that however plausible Galileo's telescopic discoveries, the notion of a mobile earth defied belief because of its evident conflict with Scripture.

After the slow and solemn meal, Castelli was asked to continue the conversation in the Grand Ducal chambers. There he found the Grand Duke Cosimo II with his consort Maria Maddalena, his mother Christina, his kinsmen Don Antonio de' Medici and Duke Paolo Giordano II Orsini, and Boscaglia. Invited to explain the compatibility of a mobile earth and Scripture, Castelli described his performance as that of a lofty theologian and of a knight-in-arms defending the faith. He enlisted aid from Don Antonio and Duke Orsini, winning over most of his interlocutors. Christina's resistance was designed to prolong the discussion; the "battle," Castelli wrote, lasted "two full hours." Boscaglia fell silent.[6]

Castelli's previous letters offer more context. Arriving in Pisa in early November 1613 to begin teaching, he had been warned by Archbishop Francesco Bonciani not to discuss the Copernican world system in public lessons. He had promptly agreed to teach nothing of the sort, adding that in 24 years Galileo had likewise avoided such material. His inaugural lecture went very well and was attended by numerous students and colleagues, Castelli wrote; by early December, he had twenty-five or thirty well-born private pupils, their evening strolls giving rise to envy and malice.[7]

His first meal at the Medici court was a dress rehearsal for the upcoming discussion of Scripture and astronomy. The Medici asked him about his public and private lessons, Christina nodded approvingly at the mention of Euclid, and the philosophers praised the late instrument-maker and professor Antonio Santucci. When Castelli observed that his predecessor's "transition from shoemaker to lecturer in *mathematica* had been a great achievement, and that he understood the first six books normally read [of Euclid's *Elements*] and Sacrobosco's *Sphere*," the philosophers conceded that he had been a confusing, unmethodical teacher. The Benedictine, noting that he had heard as much from students, reassured the Grand Dukes of his own pedagogical prowess.[8]

A professor of logic, Canon Andrea Bellavita, had defended the hypothesis of a mobile earth the evening before in a private academy, Castelli added, and intended to argue *against* that position later that day; the Grand Duchess smiled at him as Bellavita announced this plan, and he himself said nothing.[9] That silence was evidently typical neither of Castelli nor of his rival Boscaglia. The latter was known for introducing young students to Latin literature and Greek philosophy, for his oratorical abilities, poetic compositions, and lordly manner. Neither his treatises on natural philosophy nor his paraphrases of Aristotle's *On the Heavens* and *Physics* was

---

[6] Galilei, *Opere* 11: 605–606.

[7] Galilei, *Opere* 11: 590, 594, 596, 601, 605.

[8] Galilei, *Opere* 11: 604–605. See also Galilei, *Opere* 11: 332, and Giorgio Strano, "The Heavens at the Medici Court: Antonio Santucci's Cosmological Models," in Richard Dunn, Silke Ackermann, and Giorgio Strano, ed., *Heaven and Earth United: Instruments in Astrological Contexts* (Leiden: Brill, 2018): 180–209.

[9] Galilei, *Opere* 11: 604–605.

published, but well after his death, Boscaglia was remembered as a loquacious, voracious fixture at the Grand Dukes' banquets.[10]

Galileo's *Letter to Castelli* confirms that the Scriptural passage under scrutiny involved a miracle wrought by Moses' successor, the spy, spymaster and holy warrior Joshua. After Joshua had waged a series of battles to lead the Israelites into Canaan, the ruler of the royal city of Gibeon capitulated to him in advance of any conflict. Five other kings in the region decided to punish the Gibeonites for their surrender, and Joshua was summoned to the besieged city. In a crescendo of violence, the Israelites slaughtered the allied Canaanites, pursued the survivors, and saw many struck down by a hailstorm. Realizing that a longer day would allow him to annihilate the remaining Canaanites, Joshua said,

> Move not, O sun, toward [Gibeon], nor thou, O moon, toward the valley of Ajalon. And the sun and the moon stood still, till the people revenged themselves of their enemies. Is not this written in the book of the just? So the sun stood still in the midst of heaven, and hasted not to go down the space of one day. There was not before nor after so long a day, the Lord obeying the voice of a man, and fighting for Israel.[11]

The incident was especially well known to the Medici, as Antonio Tempesta had dedicated etchings of this and other Biblical battles to the Grand Duke in 1613. It is likely, too, that one of the several "stories of Joshua" in tapestries owned by Don Antonio was devoted to this prominent event.[12] Given Joshua's command, Christina asked, how could the passage be read as anything other than an allusion to a mobile sun and a stationary earth? Though Castelli's letter offered no explicit mention of *Joshua* 10, his self-presentation as someone at once engaged in battle and capable of speaking of sacred matters, summoned by the powerful and resisted by other elites, and acutely conscious of the passage of time, re-enacts a pantomime of the Biblical hero. Galileo had already discussed the passage in Castelli's presence; as he reminded his former student, some if not all of the explanation offered to the Grand Duchess was his own.[13]

The *Letter to Castelli* anticipates the basic premises of the *Letter to the Grand Duchess Christina.* After gesturing to the smouldering envy that accompanies displays of prowess, Galileo observed that Scripture can never lie, but that its effort to speak in the simplest terms to the unschooled meant that it could not always be taken literally, as in those instances where God is depicted with hands, feet, and eyes, and as prone to anger, remorse, hatred, and forgetfulness. Commentators might err by placing excessive emphasis on the reality of such features, as such insistence would engender contradictions, heretical belief, and blasphemy. Galileo further contrasted the polyvocal aspect of Scripture, dictated by the Holy Spirit, but

---

[10] Paganino Gaudenzi, *De Progressu philosophiae apud romanos* (Pisa: Amadori Massa and Lorenzo Laudi, 1643): 377; Giovanni Targioni Tozzetti, *Notizie degli ingrandimenti delle scienze fisiche accaduti in Toscana,* 3 volumes (Florence: Giuseppe Bouchard, 1780) 1: 88–90.

[11] Douay-Rheims *Bible*, *Joshua* 10: 12–14.

[12] Pier Filippo Covoni, *Don Antonio de' Medici al Casino di San Marco* (Florence: Tipografia cooperativa, 1892) 223.

[13] Galilei, *Opere* 5: 385.

adjusted in rhetorical terms to diverse audiences, with the inflexible decrees of Nature, likewise bound by divine laws, but indifferent to the metric of human perception and comprehension.[14]

The gap between Scripture and Nature suggested the inadvisability of using the one domain to interpret the other. He could readily believe, Galileo wrote, that the single goal of Scripture was to persuade men of those precepts necessary to their salvation and yet beyond the reach of human discourse; no other field of study but rather the Holy Spirit would render such notions plausible.

> But I do not see any need to believe that the same God who endowed us with senses, speech, and minds, wanted to subordinate the use of those faculties in order to offer us other means of reaching that same knowledge. And this is especially true in those disciplines that have a minimal presence and ambiguous conclusions in Scripture: such is exactly the case of astronomy, which has such a small part that neither the planets nor even their names appear.[15]

Despite his insistence on the mutual irrelevance of Scripture and natural philosophy, Galileo returned to the passage from *Joshua.* Like Castelli, he depicted such discussions as a battle, and especially deplored his rivals' immediate turn to "an inevitable, overwhelming weapon, the sight of which terrifies the most skilful and accomplished warrior," rather than the expected reliance on empirical and logical arguments.[16] As he saw it, this weapon—the pre-emptive assertion of Scriptural authority over that of any other discipline—was meant to mask the fundamental weakness of a position that had no other evidence to offer, and to circumvent further discussion from adversaries they knew to be stronger. Boscaglia's initial objection to the notion of a mobile earth, the chill the remark imposed on the conversation, and his silence after Castelli's explanation conform to this pattern.

Galileo's agenda was to show that the passage was incompatible with the Ptolemaic world system, and that the celestial configuration described involved a centrally located sun and a mobile earth, neatly complementing his own recent investigations of the solar body. Noting that in the Ptolemaic cosmos the whirling of the outermost shell of the heavens meant that the sun and stars travelled from east to west, completing a circuit in 24 hours and bringing the appearance of day and night, he observed that the sun did not keep pace with its stellar backdrop, but fell a bit behind each day. Over the course of a year, the sun thus slowly drifted from west to east through the zodiac. This retrograde movement was the true motion of the solar body; it defined the seasons and the year itself, just as the rapid and contrary circuit of the stars produced day and night. In such an arrangement, Joshua's request that the sun stop its own eastward movement would mean that it would be swept along with the stars; this outcome would shorten rather than lengthen the day.[17]

---

[14] Galilei, *Opere* 5: 281–283.

[15] Galilei, *Opere* 5: 284.

[16] Galilei, *Opere* 5: 285.

[17] Galilei, *Opere* 5: 286–287.

Galileo then elaborated the easy adaptation of the Copernican model to the miracle at Gibeon. He reminded Castelli that his recent examination of the sunspots—a series of letters addressed to the Augsburg Humanist Marcus Welser, contesting the Ptolemaic views of the Jesuit Christoph Scheiner—had revealed that the centrally located sun rotated on its own axis, completing a full circuit over the course of a lunar month. Supposing that this movement ceased by divine decree, all planetary motions would also be halted until the sun, like the hub of a great wheel, resumed its rotation. The earth, too, would stop its own axial rotation, suspending the movement that produced sunset until the Canaanites had been defeated.[18]

The *Letter to Castelli* predates that addressed to Christina by some 18 months; one copy was sent from Florence in February 1615 by the Dominican preacher Niccolò Lorini to the Prefect of the Congregation in Rome with certain passages underlined.[19] Like Galileo, Lorini was closely involved with the Medici court, and their clash was not new. In the fall of 1612, the priest had assured Galileo by letter that rumors notwithstanding, he had "not spoken of philosophical matters against anyone," that he had never dreamed of saying such things, that his utterance, designed simply to show that he "wasn't a dead body," was that "*Ipernico*, or whatever his name is, seems opposed to Sacred Scripture." The mistake implied that Lorini had only read but never heard the name "Copernicus" and that he had mistaken an abbreviation for the first two letters as an *I*.[20]

While Galileo called Lorini a "stumbling sermonizer," he might have known that the man was not as uninformed as *Ipernico* suggested. In 1615, as the preacher alerted the authorities to the *Letter to Castelli,* he published a book of sermons endorsing the compatibility of natural philosophy and Scripture, but also stating that an abundance of knowledge would imperil faith.[21] In 1617, a year after the Decree of the Congregation of the Index, in a book of sermons addressed to Maria Magdalena de' Medici, he denounced the "empty notion of Nicolas Copernicus that the earth moves."[22] Lorini's emphasis on paradoxes involving the conversion of the rising sun into the setting one, or the metamorphosis of East into West, and his commentaries on the uselessness of the *canne vuote* or hollow reeds of *Matthew* 11: 8 when so many were sceptical of Galileo's *cannocchiale* or telescope suggest envious attention to the astronomer's claims.[23] And the preacher's countless startling comparisons—Saint Stephen's joy in his martyrdom for Christ as "a snowball fight between friends," salvation as the best of pharmaceutical "secrets," the Holy Spirit's role in Mary's conception of Christ as a smelting process, and God's contribution as delicate needlework on her womb—typify those commentators whose emphasis on

[18] Galilei, *Opere* 5: 287–288.

[19] Galilei, *Opere* 5: 264.

[20] Galilei, *Opere* 11: 427, 461.

[21] Niccolò Lorini, *Preparazione e frutto al santissimo natale di Cristo divisa in XX prediche* (Florence: Gio Donato, Bernardino Giunti, e Compagni, 1615) 6, 370.

[22] Niccolò Lorini, *Elogii delle piu principali s. donne del sagro calendario e martirologio romano* (Florence: Zanobi Pignoni, 1617) 17.

[23] Lorini, *Preparazione,* 77–82, 142, 327, 365, 384–385.

a deity with hands, feet, eyes, and unseemly emotions would produce, in Galileo's view, contradiction, heretical belief, and blasphemy.[24]

3.2 Court Portraits

Galileo returned to these arguments in his *Letter to the Grand Duchess Christina* in the spring or summer of 1615. The bold sketches of various unnamed individuals and types distinguish this letter from its antecedent; because Galileo dispensed entirely with references to the meal that provoked it, a slightly larger, transhistorical community emerges here. Among this letter's other elaborations are its frequent references to writers associated with the early church, and its shadowy vignettes of the main protagonists in this conflict. Ecclesiastical authorities of the mid-fourth to mid-fifth century CE are especially prominent. In addition to his references to Jerome (342–420 CE), translator of the Bible, and to Augustine (354–430 CE), whose *On the Literal Interpretation of* Genesis he often cited, Galileo likely intended to include Ambrose (339–397 CE), as the familiar Latin hymn with which his text closes, "Most Holy God of the Heavens," was sometimes attributed to that Church Father.[25]

Galileo portrayed himself as one whose celestial discoveries prompted envy and scepticism in various schools and sects, as amused by uninformed reactions to such findings, but fastidious about his reputation in matters of piety among the populace, as ready to submit to religious authorities, as familiar with malicious gossip about him, but willing to forgive it, and above all as committed to pursuing an argument in orderly fashion until he has resolved it.[26] As he has originally planned to address the letter to a churchman, possibly Castelli, Christina had little in the way of personal features; Galileo described her in the opening sentence as one familiar with his recent discoveries, and midway through the letter, as one who remembered Antonio Santucci's very late and unexpected conversion to heliocentrism.[27] The last three references to Christina are entirely generic directives, and there is no indication that it was she who had encouraged Galileo's student, endorsed the study of Euclid, smiled at Bellavita's facile attitude to cosmology, posed the crucial question during the meal of December 1613, or prolonged Castelli's subsequent discussion.

Nor does Galileo depict his rivals with individuating detail; they are simply envious, hypocritical, ignorant of natural philosophy, unfamiliar with hermeneutical principles of the Church Fathers, quick to invoke ecclesiastical authority over other disciplines, and unaware of the intellectual and cultural consequences of this

---

[24] Lorini, *Preparazione,* 106–107, 110; *Elogii,* 273, 290.

[25] See for instance Bartolommeo Gavanti's statement that "Caeli Deus Sanctissime" was attributed to Ambrose in certain breviaries in his *Commentaria in rubricas missalis et breuiari Romani* (Rome, Francesco Cavallo, 1628) 2: 47.

[26] Galilei, *Opere* 5: 309–311.

[27] Galilei, *Opere* 5: 265–266, 275.

approach. But his strategic citation of a well-known letter of Jerome supplements and nuances this hazy group portrait. At stake here is that Church Father's account of those who meddle in Scriptural interpretation with the same misguided ease with which they manipulate Homeric and Virgilian texts into patchwork confections entirely remote from the poets' original intention:

> The gabby old bag, the doddering elder, the smooth-talking sophist, all encroach on Scripture, tear at it, teach it before they have learned it. Others with furrowed brow weigh out big words and wax philosophical about the Bible among working women; and still others--for shame!--learn from the ladies what they then teach to the men. And as if that were not enough, they hold forth to others with a certain glibness--or better, boldness--about things they themselves don't grasp. I pass in silence over those who, like me, have come from secular literature by chance to Sacred Scripture, have soothed the popular ear with flowery phrases, and have thought that whatever they said was a divine decree. They have not deigned to know what the Prophets and Apostles believed, but adapt their unsuitable evidence to their private purposes, as if it were a superb rather than positively perverted pedagogical practice to distort those intentions and to twist a resistant Scripture to their will.[28]

This passage was familiar to many readers, as it figured in the prefaces of various editions of the Vulgate, had been translated to Italian, was popular with polemicists, including Roberto Bellarmino, and enlivened the occasional sermon.[29] While the characters so irritating to Jerome correspond to early modern narrative and comedic stereotypes, they also have a disconcerting relevance to the protagonists of this Galileian episode. Christina, particularly in the silent role assigned her in Galileo's letter, is far too restrained to bear much resemblance to the "gabby old bag," but as a middle-aged widow not invulnerable to garden-variety misogyny and ageism. Father Lorini, two decades her senior, and distinguished by the reference to "Ipernicus or whatever his name is," could be mistaken for Jerome's doddering elder. The vain and voluble Boscaglia resembled the slick sophist, and the glib Canon Bellavita, one who held forth among working women. The latter's role as supervisor in the Grand Duchess's ongoing project of rehabilitating former prostitutes would have enhanced the original passage.[30]

Jerome's reference to those who had turned from secular literature to Scripture, managing this transition with showy speech, and offering their insights as divine

[28] Galilei, *Opere* 5: 323, citing Jerome's Letter to Paulinus of Nola, Epistle 53.7

[29] See for example Kaspar Schatzgeyer, "Preface," in *Omnia Opera* (Ingolstadt: Alexander Weissenhorn, 1543) second unnumbered page; *Epistole di S. Girolamo dottore della chiesa,* translated by Giovanfrancesco Zeffi (Venice: Giunti, 1562) fol. 202 r-v; Roberto Bellarmino, *Disputationes de controversiis Christianae Fidei, adversvs huius temporis hæreticos,* 3 volumes (Ingolstadt: David Sartorius, 1587) 1: 184; *Biblia Sacra Vulgatae Editionis Sixti Quinti* (Venice: Heirs of Damiano Zenaro, 1605) thirteenth unnumbered page; Marc'Antonio Sanseverino, *Quaresimale predicato ne' più famosi pulpiti d'Italia* (Naples: Luc'Antonio di Fusco, 1664) 113.

[30] On Boscaglia, see especially Philostratus's depiction of Protagoras in *Lives of Sophists;* for Bellavita's activities, see *Capitoli, constituzioni, et regole per le Conuertite di Pisa e loro Conuento fondato, e dotato dalla sereniss. madama Crestina di Loreno, gran duchessa di Toscana* (Pisa: Giovanni Fontani, 1615) 4. For an unflattering comparison of Bellavita and Boscaglia to Galileo, see Angelo Fabroni, *Historia academiae Pisanae* (Pisa: Gaetano Magnaini, 1792) 2: 373–374.

decrees, was a serviceable description of the career path of Francesco Bonciani, Archbishop of Pisa, who had devoted his youth and wealth to literature, oratory, and diplomacy, and who had entered the priesthood in midlife.[31] In 1613, he had warned Castelli against mentioning heliocentrism in lectures; by 1616, after the anti-Copernican measures of the Congregation of the Index, he would promulgate decrees emphasizing the profession of faith expected of clergy, canon and civil lawyers, philosophers, physicians, public and private professors of mathematics, literature, music, medicine and surgery, rectors and schoolmasters, booksellers, and typesetters. These texts, adopting the militaristic idiom of the Counter-Reformation, noted that bishops might take up weapons, and that Inquisitors were permitted armed escorts in defending the Church against presumed heretics.[32]

The caricatural roles of those in the Grand Duchess' orbit provided a cautionary tale for Christina herself, while the allegorical personifications of Nature and Scripture in Galileo's letter constituted two rhetorical extremes best avoided by the ruler: the first was inflexible in her decrees and indifferent to her subjects, and the second overly concerned with a low and popular appeal. The ideal was the figure of Theology herself, presented to Christina by Galileo's rivals as "Queen of the Sciences," an imperial discipline to which other, lesser fields must inevitably yield.[33] This was a polite fiction: Philosophy held this title, and until a fire in 1596, anyone in the Cathedral of Pisa would have seen in the famous marble statues beneath the pulpit that discipline appearing in such a role among the seven liberal arts. Galileo elaborated the basis on which Theology's authority was maintained, dismissing the possibility that its pre-eminence derived from the fact that it, rather than the works of Archimedes, Ptolemy, Boethius, and Galen, offered the best of geometrical, astronomical, musical, and medical precepts. Her standing rested instead on her elevated, unique subject matter: nowhere else, except through her revelations, does one discover salvation. Theology's concern with the picayune details of inferior sciences, irrelevant to this goal, only imperilled her status.[34]

This last observation could be interpreted as yet another warning to Christina: a detached and remote perspective, rather than meddling supervision, better suited both the allegorical Queen of the Sciences and the Grand Duchess. What complicates Galileo's remark is its elaboration; extending theologians' authority to disciplines in which they had no training, he wrote, would be as if "an absolute prince, knowing that he might freely command and be obeyed, wanted, without being either a physician or an architect, medicine and construction to be practiced according to his views, despite the heavy toll on all who were ailing, and the imminent collapse of buildings."[35] While the Medici promoted themselves as architects and physicians

[31] Antonio Felice Mattei, "Franciscus IV Boncianus," in *Ecclesiae Pisanae Historia,* 2 volumes (Lucca: Leonardo Venturi, 1772) 2: 219–225.

[32] *Synodus dioecesana Pisana* (Pisa: Giovanni Fontani, 1616) 7–8, 11, 59.

[33] Galilei, *Opere* 5: 324.

[34] Galilei, *Opere* 5: 324–325.

[35] Galilei, *Opere* 5: 325.

of the state, the family included those who adopted these roles in literal rather than metaphorical fashion, Don Giovanni and Don Antonio de' Medici.

These two men had been strategically displaced from the epicenter of power; as illegitimate offspring of the Grand Dukes Cosimo I and Francesco I, they benefitted from the family's wealth and authority, bore the title "prince," but could under no condition act as absolute rulers. As a military engineer, Don Giovanni designed buildings and stage sets for festivals, and such efforts had resulted in a conflict with Galileo in 1608: the floating structure this prince proposed for a Medici wedding celebration seemed to him in danger of submersion. Don Giovanni was the dedicatee of a treatise contesting Galileo's telescopic observations and gesturing to references to heavenly bodies in Scripture in early 1611 and of another objecting to his account of buoyancy in late 1612; he would also be the recipient of news of the anti-Copernican ruling of March 1616.[36]

Don Giovanni's nephew Don Antonio, by contrast, was a warm supporter of Galileo; Castelli described his generosity "as truly that of a prince."[37] Galileo was far from convinced, however, by the alchemical practices to which this patron was devoted, and such doubts were acute from January to June 1614, when Prince Francesco de' Medici, the son of Christina and the late Grand Duke Ferdinando, like Galileo himself, was seriously ill. In early May Castelli wrote that the Prince's life was in the hands of a Paracelsian healer in Don Antonio's orbit, and he died soon thereafter. Christina, having lost her husband and one son, understood the limited control that even Galileo's notional absolute ruler commanded over the physical realities of the built environment and the body. But in his description of the public prayers, alms and pious works offered by the Medici in the weeks before the prince's death, Castelli observed to Galileo that "the Grand Duchess, in particular, does violence to God, so to speak," as if in her despair she had substituted for an inexorable, inflexible, and largely inscrutable Nature some more approachable entity. Such, in fact, would be that Old Testament deity endowed with a body, forgetful of the past, unaware of the future, and given to bouts of anger and repentance.[38]

3.3 Second Sophistic

At first glance, Galileo's letter to Christina lacks stylistic features typical of his other works: there is little linguistic exuberance, and less still of the exaggerated, rather caustic courtesy with which he approached rivals. This difference is doubtless due to the gravity of the subject matter, and the exalted status, age, and gender of his eventual addressee. But the fact that the letter so often yields more complicated

[36] Galilei, *Opere* 4: 32 n. 3, 65; Massimo Bucciantini, "Reazioni alla condanna di Copernico," *Galilaeana* **1** (2004): 3–19.

[37] Galilei, *Opere* 11: 606.

[38] For references to the ill health of both Galileo and Francesco de' Medici in this period, see Galilei, *Opere* 12: 23, 30, 39, 41, 43, 49–50, 51, 54–57, 58, 61, 66, 70, 79–80.

interpretations just beyond its apparent simplicity, even as it maintains deniability, narrows the gap between this text and Galileo's more familiar stylistic and rhetorical tendencies. Much of what he denounces here is duplicity, rather than the usual observational ineptitude, mathematical incompetence, inconsistency, poor logic, or unthinking allegiance to Aristotelianism. While there is an obvious distinction between the subterfuges censured in others—professional envy masked by simulated piety, deliberate misreading of Scripture, feigned intellectual expertise, a strategic recourse to ecclesiastical authority—and Galileo's own artifice, we can readily appropriate Jerome's "smooth-talking sophist" to describe this aspect of his performance. More generally, we might categorize the swift surprise of his resolution of *Joshua* 10, after the text's measured movement and its steady resistance to Scriptural interpretation, as a sophist's manoeuvre.

Though the *Letter to the Grand Duchess Christina* follows Jerome in using "sophist" as a byword for speakers whose flashy arguments lack intellectual rigor, I am proposing instead that the entire text is in sympathy with the Second Sophistic as a cultural movement. Before turning to the Themistian orations on which Galileo drew, consider this sketch of the movement. Around 230 CE its influential exponent Lucius Flavius Philostratus composed a series of biographies describing the professional activities and personal conflicts of Greek-speaking orators living under imperial Roman rule and ranging from the late first century CE through his own day.[39] The cultural phenomenon he constructed and defended in his *Lives of the Sophists*, known as the "Second Sophistic," was heir to the original Sophists, intellectuals who had flourished as charismatic speakers in the fifth century BCE, whose popular appeal devolved over time into a conflict with more substantive forms of philosophical discourse, whose best-known representative was Gorgias of Leontini, and whose characteristic move was to "make the weaker argument the stronger."[40] Philostratus' history was continued, in a darker world and with less flair, by the Sophist Eunapius (346–414 CE) in his *Lives of the Philosophers.*

Broadly, the Second Sophistic involved a Hellenic, often nostalgic orientation in a world where economic and political power had shifted to Rome, a commitment to competitive, public, and stylish performances sometimes involving role-playing, repurposed literary commonplaces, and tests or ambiguous asides aimed at a knowing audience. Frequently occupying a court-based role as interpreter, ambassador, or political adviser, these sophists insisted upon their place within a genealogy of prominent teachers and successful students, and emphasized their familiarity with the visual arts and their strong identification with cities in the Eastern reaches of the Empire. The movement's greatest early representative was Dio Chrysostom (40–120

---

[39] Philostratus, *Lives of the Sophists,* and Eunapius, *Lives of the Philosophers,* translated and introduced by Wilmer Wright (Cambridge, MA: Harvard University Press, 1921).

[40] For references to this trait, see Aristophanes, *Clouds* 112, 883–884; Plato, *Apology* 18b8–9; Aristotle, *Rhetoric* 1402a24–28; Cicero, *Brutus* 29–31.

CE); Themistius (317–388 CE), much influenced by Dio, was the outstanding figure of the late Second Sophistic.[41]

The early modern interest in the Second Sophistic is unsurprising, particularly in the court culture of absolutist states, in antiquarian circles, in Mannerist writers, and in the competitive world of preachers. Its appeal was especially strong in Italy, where cities such as Venice, Florence, Rome, and Naples, like Athens, Corinth, Smyrna, Alexandria, and Constantinople under the Roman Empire, sought to maintain their bygone cultural prestige in the face of diminished economic, political, and military power. Galileo's acquaintance with the movement likely dates to 1590, when he was in close contact with Jacopo Mazzoni, whose resistance to Copernicus would result in 1597 in the astronomer's first public endorsement of the heliocentric cosmos.[42] In defending Dante's work against departures from literary and ethical norms, Mazzoni traced the devolution of the Sophists' relationship with philosophers, drew frequently on Philostratus, and used the label "Second Sophistic" for the first time in the vernacular. He argued that legitimate poetry offered a persuasive and absorbing vision, that its goal was both aesthetic and ethical, and that to meet expectations of a credible, edifying, and beguiling alternative to this world, such poetry, like Sophistic, incorporated the real *and* the false in logical and organic fashion.[43] Mazzoni portrayed himself as a Sophist, cited Chrysostom's orations, and compared the intellectual value of contradiction in the discovery of truth, that combination of resistance and obliquity, to the sun's movement: not only did it run against the course of the stars, it was also at a slight angle to it. And the terms he used to defend Dante's reliance on images from the physical world to convey abstractions anticipate Galileo's emphasis on Scripture's invocation of an embodied God to appeal to an unschooled audience.[44]

Galileo encountered the Second Sophistic again in the Paduan *Accademia de' Ricovrati*, whose members were intellectuals devoted above all to performance, argumentation, and reworking literary commonplaces to their own ends. Galileo's friend Canon Giovanni Belloni offered a commentary on that institution's emblem in June 1600, discussing the devolution of the term "sophist," and his feigned alarm at "wicked, libidinous princes, garrulous clergymen, or professors of fallacious

---

[41] See Graham Anderson, *The Second Sophistic: A Cultural Phenomenon in the Roman Empire* (London and New York: Routledge, 1993); Kendra Eshleman, *The Social World of Intellectuals in the Roman Empire: Sophists, Philosophers, and Christians* (Cambridge, UK: Cambridge University Press, 2012); *The Oxford Handbook of the Second Sophistic,* edited by Daniel S. Richter and William A. Johnson (Oxford, UK: Oxford University Press, 2017). On the Second Sophistic in the sixteenth century, see the foundational work of Eric MacPhail, *The Sophistic Renaissance* (Geneva: Droz, 2011) and more lately Teodoro Katinis, *Sperone Speroni and the Debate over Sophistry in the Italian Renaissance* (Leiden: Brill, 2018).

[42] Galilei, *Opere* 2: 193–202; 10: 44–45; see John Heilbron, *Galileo* 110–112.

[43] Heilbron emphasizes a similar outlook in Galileo's interest in "the realistic treatment of the marvelous" in literary and, as needed, in philosophical matters; see *Galileo,* 20–21.

[44] Mazzoni's statements about the Sophists occur mostly in his Proem, where the numbers correlate with sections rather than pages; see Iacopo Mazzoni, *Della difesa della commedia di Dante* (Cesena: Bartolommeo Raverii, 1587) 4, 11, 16, 40, 45–48, 52–61, 73–74; see also pages 412–413.

doctrines" is a typical example of Sophistic ambiguity, a passage designed to escape the notice of the cruel ruler, and more importantly, to test the audience.[45] Given that two founding members of the Academy—Galileo and his colleague Cesare Cremonini—would suffer for their close association with "fallacious doctrines," Belloni's moment of jocularity has a prophetic edge. Belloni's brother Camillo, another professor of philosophy at Padua, would denounce Cremonini in 1604, provoking an investigation of Galileo as well. If there was a test here, the audience failed it, or failed to take it seriously.[46]

Galileo was also closely involved with two other humanists who were especially interested in Themistius' orations. Those works, delivered between 350 and 385 CE, are traditionally divided into the categories of imperial and private, though the earliest printed versions of the original and in Latin translation featured an admixture of the two. The imperial sort, addressed to a series of Christian and non-Christian rulers, combines praise (often unearned) and advice (often unheeded), with strategic adjustments to each new regime. As we might expect in an orator who survived dramatic shifts in policy over the course of 35 years, Themistius generally advocated a moderate approach to the belligerent and the vanquished, and on occasion endorsed religious pluralism as a corollary to a similar diversity of opinion in natural philosophy. The private orations, less concerned with historical occasions than with cultural issues, are often ironic or hyperbolic in tone, and might feature a test of sorts for the elite audience.[47] In both cases Themistius insisted, unsurprisingly, on the relevance of philosophy to the state.

Girolamo Mercuriale, a professor of medicine whose friendship with Galileo dates to the 1580s, appears more concerned with the imperial addresses, as brief references emerge in 1577 in his treatise on plague, and in 1601 in an updated edition of his best-selling treatise on exercise.[48] The most crucial of these allusions was the earliest, where in tacit acknowledgment of the criticism generated by his approach to the recent plague, particularly from his friend Girolamo Donzellini, a physician and translator of eight orations, Mercuriale noted on the first page of his

---

[45] Giovanni Belloni, *Discorso intorno all'antro delle ninfe* (Padua: Lorenzo Pasquati, 1601) fol. 50–53; on the *sermo coloratus,* veiled speech, a species of innuendo, see "A Glossary of Rhetorical Terms" in Philostratus, *Lives of the Sophists,* 570.

[46] Antonino Poppi, *Cremonini, Galilei e gli inquisitori del santo a Padova* (Padua: Centro studi antoniani, 1993).

[47] See *Politics, Philosophy, and Empire. Select Orations of Themistius,* translated with an introduction by Peter Heather and David Moncur (Liverpool: Liverpool University Press, 2001); *The Private Orations of Themistius,* translated, annotated, and introduced by Robert Penella (Berkeley: University of California Press, 2000); John Vanderspoel, *Themistius and the Imperial Court* (Ann Arbor: The University of Michigan Press, 1995); Themistius, *Discorsi,* translated and introduced by Riccardo Maisano (Turin: UTET, 1995).

[48] Girolamo Mercuriale, *De Pestilentia* (Venice: Paolo Meietti, 1587) fol. 1, and *De Pestilentia* (Venice: Giunti, 1601) fol. 1; *De Arte Gymnastica,* fourth edition (Venice: Giunti, 1601) 78, 103, 119. The latter references are not found in the first three editions, published in 1569, 1573, and 1587.

work that "The philosopher Themistius said that disputes about both His own nature and His works were pleasing to God."[49]

Gian Vincenzo Pinelli, who would host Galileo when he arrived in Padua in 1592, had several manuscripts with both sorts of orations, the most valuable having been acquired from the Benedictine monastery Santa Giustina in that city.[50] In early 1582 the English astronomer Henry Savile sent Pinelli the so-called "Twelfth Oration," purportedly a Latin translation made by the Hungarian diplomat, former Catholic, and natural philosopher Andreas Dudith, of a vanished Greek original.[51] Though aspects of this imperial address strongly resembled Themistius' genuine fifth oration, which maintained that men of different faiths engaged in competitive worship of one and the same deity, Pinelli rejected the attribution, as would others, but the text would nonetheless be published in 1605. By 1613, it was retrofitted with an ersatz version of the original Greek, and addressed to a different emperor; the oration or excerpts occasionally appeared thereafter, often as a matter of philological or antiquarian interest rather than as an endorsement of religious freedom."[52] Between 1599 and 1601, Pinelli asked the Scotsman Thomas Seget to undertake Latin translations of the Greek orations he possessed in manuscript form. It is not clear that the translations were completed, much less that they survive, but it is noteworthy that Seget was involved with the *Accademia de' Ricovrati* from February 1600, that he and Galileo were devoted to each other from 1599 through 1610 and perhaps beyond, and that the orations under scrutiny included the three which the astronomer would use in composing the letter to Christina. Even in his final illness in 1601, Pinelli remained interested in other manuscripts of the orations, seeking more information on unpublished Themistian material recently transferred from

---

[49] The assertion condenses arguments in the fifth oration, and ascribes to Themistius an observation that the latter had attributed to Heraclitus. See Heather and Moncur, *Politics, Philosophy, and Empire,* 168–170; Themistius, *Discorsi,* 280–283; Vanderspoel, *Themistius and the Imperial Court,* 148–154. See also Girolamo Mercuriale, *On Pestilence: A Renaissance Treatise on Plague,* translated and introduced by Craig Martin (Philadelphia: University of Pennsylvania Press, 2022) 1–20, 28; Samuel Cohn, *Cultures of Plague: Medical Thinking at the End of the Renaissance* (Oxford: Oxford University Press, 2010) 161–207, especially 162–167, 177–180, 200.

[50] Ciro Giacomelli, "Libri e biblioteche greche a Padova," in *Tra Oriente e Occidente. Dotti bizantini e studenti greci nel rinascimento padovano*, edited by Nicolò Zorzi and Ciro Giacomelli (Padua: Padua University Press, 2022) 25–42, especially 26–27, 32, 34–35, 39.

[51] Robert Goulding, "Who Wrote the Twelfth Oration of Themistius?" *Journal of the Warburg and Courtaud Institutes* **63** (2000): 1–20 and Gábor Almási, "The Riddle of Themistius' Twelfth Oration and the Question of Religious Tolerance in the Sixteenth Century," *Central Europe* **2.2** (2004): 83–108.

[52] Themistius, *Orationes sedecim*, translated by Dénis Petau (La Flèche: Jacques Rezé, 1613) 403–417. See also [Themistius], *Thesaurus Principium,* translated by Georg Rem (Frankfurt: Johannes Bringerus, 1614) 225–230; *Deductio das ist Nothwendige Ausführung* (N.P.: 1620) 1: 228; Michael Wendland, *Dissertatio catholica de pio magistratus officio* (Gera: Martin Spiess, 1614) 335–348; Théophile Raynaud, *Theologia naturalis* (Lyon: Claude Landry, 1622) 414; Janus Gruter, *Florilegi magni* (Strasbourg: Lazar Zetzner, 1624) 2: 589–590; Henry Stubbe, *An Essay in defense of the good old cause* (London: n.p., 1659) 72–74.

Salamanca to Brussels.[53] By 1609 Pinelli's Themistian manuscripts were in the possession of Cardinal Federico Borromeo.[54]

3.4 How the Sophist Should Speak

Galileo's initial turn to Themistius' orations was likely prompted by his adversaries' attention to this figure in 1611–1612 during the controversy over floating bodies. Not only did Vincenzo de Grazia and Giorgio Coresio draw on the paraphrases of *On the Heavens* and *Physics* in this debate, but the latter, a physician and professor of Greek at Pisa, opened his criticism of Galileo's treatise on hydrostatics with an allusion to the healthy competition generated by discord in natural philosophy, and the insulting suggestion that his adversary had merely intended to provoke discussion. Galileo would have none of these or other "sophisms;" he gestured—in a single, ferocious sentence—to the Themistian notion of the evolution of natural philosophy, to the freedom to resist or reinterpret the words of canonical authors in that domain, and to his own contribution to the miniscule fund of knowledge in the abyss of the sciences.[55]

Galileo is far bolder in his appropriation of Themistius' "The Philosopher," "How the Philosopher Should Speak," and "The Sophist," in his letters to Castelli and Christina. These private orations were likely composed and delivered in Constantinople between 355 and 360 CE, when Themistius' early work as an Aristotelian paraphrasist, begun in 347, was coming to an end.[56] Galileo perhaps had access to the physician Donzellini's Latin translations of "The Philosopher" and "The Sophist," published in 1559 along with six others.[57] He might also have seen a collection of seven Imperial Orations—six genuine addresses, and the factitious oration on religious freedom—presented in Latin by the Augsburg poet and jurisconsult Georg Rem and published in 1605. If so, Rem's paratext was of relevance to him: it included a message from Marcus Welser expressing eagerness for the publi-

[53] Stefano Gattei, "The Wandering Scot. Thomas Seget's *album amicorum,*" *Nuncius* **28** (2013): 345–463, especially 354, 435–436; Gilbert Tournoy, "Isaac Casaubon and Greek Scholarship in the Low Countries," *Erudition and the Republic of Letters* **4** (2019): 330–351, especially 345–348.

[54] Marcia Grendler, "A Greek Collection in Padua: The Library of Gian Vincenzo Pinelli (1535-1601)," *Renaissance Quarterly* **33.3** (1980): 386–416; Giuseppe Pascale, "Ricerche sulla tradizione manoscritta delle orazioni di Temistio," *Aevum* **84.2** (2010): 361–402, on 364–367, 371.

[55] Galilei, *Opere* 4: 23, 27, 31, 65, 203, 209, 213, 407–409, 467, 653. For other allusions to equivocal positions and terminology, novelty, performative oratory, and to Themistius himself, see *Opere* 4: 77, 177, 248, 342, 364, 366, 424, 425, 433, 718.

[56] Henry J. Blumenthal, "Themistius, the last Peripatetic Commentator on Aristotle?" in *Aristotle Transformed,* edited by Richard Sorabji (Ithaca, NY: Cornell University Press, 1990) 113–123; Vanderspoel, *Themistius and the Imperial Court,* 37; *Private Orations,* 4, 15–16, 19–20, 30, 121–122.

[57] Themistius, *Orationes octo elegantissimae ac eruditione varia refertissimae,* translated to Latin by Girolamo Donzellini (Basel: Pietro Perna, 1559).

cation, a reference to other orations which Rem had seen in 1585 in Pinelli's library, and a description of the concocted oration as one not mentioned in the many allusions to Themistius in Cardinal Baronius' writings. At the very least, such remarks implied that these works were crucial for prominent members of the Catholic world; Galileo might also have noted Rem's suggestion that the puzzling silence about Themistius in Eunapius's *Lives of the Philosophers* was motivated by envy or religious bigotry.[58]

It is also possible that Galileo drew upon Denis Pétau's bilingual edition of 1613, which included the first published Latin version of "How the Philosopher Should Speak." But as a significant portion of that oration was unavailable to Pétau, Galileo perhaps consulted the original text, or in the case of his letter to Christina, relied on Pierre Pantin's Latin translations of six orations alongside the Greek, posthumously published in 1614.[59] The editions of Rem, Pétau, and Pantin all had oblique connections to Galileo's contemporaneous solar observations. Though Rem offered no details about how he had access to Dudith's Latin version of the so-called "Twelfth Oration," the likely conduit was the astronomer Johannes Praetorius in Altdorf, whose study of the sunspots was mentioned by Galileo's rival Christoph Scheiner in 1612.[60] Pétau's notice to the reader emphasized the intermediary role of the addressee of both Scheiner's and Galileo's treatises on sunspots, Marcus Welser, in the textual emendation of two orations.[61] Pantin's edition was addressed to the Dutch statesman Cornelius van der Myle, the dedicatee of a treatise on sunspots published in February 1612, which Federico Cesi, founder of the first scientific society, the Academy of the Lynxes, sent to Galileo that fall.[62]

Consider "The Philosopher." Themistius' aim was to establish himself as a serious philosopher and to suggest that his detractors fell short of the Platonic notion of this role; the genuine practitioner was well-born, and thus erudite, motivated neither by envy nor by material gain, intolerant of deception, and uninterested in gossip and meddling. While Galileo's self-presentation, especially in the letter to Christina, conforms to this bland paragon of probity, intelligence and discretion, the depiction of the enemy is more consequential. Beyond the vignettes of a grammar-master, a babbler, and an Aristotelian pedant, Themistius suggested that these squabbles were

---

[58] Themistius, *Orationes sex augustales … adjecimus & septimam Themistij orationem ad Valent. Imp. pro libertate relligionis, Latinè*, translated to Latin by Georg Rem (Amberg: Johann Schönfelt, 1605); these references occur on the unnumbered pages in the prefatory material and in Rem's "Themistii Euphrada Vita," likewise unnumbered, at the end of the volume.

[59] Themistius, *Orationes sedecim;* Themistius, *Orationes aliquot non editae,* translated by Pierre Pantin (Leiden: Jan Paedts, 1614).

[60] Almási, "The Riddle," 89 n 37, 101; Christoph Scheiner, *Accuratior Disquisitio,* in Galilei, *Opere* 5: 62.

[61] Petau, "Ad lectorem candidum," in Themistius, *Orationes 16;* Riccardo Maisano, "La critica filologica di Petau e Hardouin e l'edizione parigina del 1684 delle orazioni di Temistio," *Archivum historicum societatis iesu* **43** (1974):267–300, on 284, 286.

[62] *De maculis in sole animadversis … Batavi dissertatiuncula* (Leiden: Plantin Raphelengius, 1612); Galilei, *Opere* 11: 404, 409–410.

not mere irritants, but also had deleterious societal effects. Drawing on personal experience, but exaggerating for comic effect, he noted that

> if for the love of learning, someone sitting on a couch in his room and with nothing above and beyond that to concern him, consumes the texts of Aristotle and Theophrastus, [the false philosophers] hear of it immediately, and having roused everyone in the city, call in the king's eyes to learn if this killer, this parricide, this poisoner still sees the light of day [literally, still considers the sun]. And the greatest charge against him, says the accused, is that he was dismissive of [Aristotle's] *Categories,* and that he argued with some foreigner about what was in the subject [of a proposition]. And if he were to touch on syllogisms, or write something about Nature, this very act would be seditious, and they would shout that the despot had to be swept from the city.[63]

This was a serio-comic sketch of Galileo's situation. His method of observing sunspots, credited to Castelli, had taken place in a dark room, and he had rejected several Aristotelian texts and his follower Theophrastus' *On Weather Signs*, all of which argued for spots very distant from the sun. To reduce their conflict over buoyancy to the blandest terms, his Greek antagonist Coresio had argued with him over Aristotelian categories and propositions, and had noted his implicit departure from Theophrastus's views. While Galileo's conjectures and discoveries were not denounced as seditious, nor he as a tyrant, a number of preachers in Florence were quick to condemn his findings in their sermons, and high-ranking functionaries of the Medici court were often asked about his work, as if to comment upon its potential for civic disruption.[64]

Consider, too, Themistius' explanation of the public's general inability to recognize philosophical pretense. Incompetent and uncredentialed artisans, architects, physicians, musicians, and dancers are promptly identified through the evidence of the senses, but fraudulent philosophers remain undetected. Their exotic, overbearing style,

> stringing together certain obsolete little words, darkened by their departure from everyday usage, deafens us. In this way they bring to mind the harsh 'inquisitor' [*ruspator*] and 'defense' [*defensaculum*] and that sort of terror, so that listeners are shocked, shivering with fright, and all but speechless, as they say happens to men when they find that a wolf has been watching them.[65]

In his Latin translation of the Greek words notable for their obsolescence and obscurity, Pétau used *ruspator* and *defensaculum* rather than *inquisitor* and *propugnaculum,* the more obvious and usual terms, possibly because the latter pair often

[63] Themistius, *Orationes octo elegantissimae,* 34–35; Themistius, *Orationes sedecim,* 26; see also *Private orations,* 76–77; *Discorsi,* 714–715.

[64] Galileo Galilei and Christoph Scheiner, *On Sunspots,* translated and introduced by Eileen Reeves and Albert Van Helden (Chicago: The University of Chicago Press, 2010) 10–11, 80–82, 104–105, 126–127; Galilei, *Opere* 4: 204, 219, 242, 261, 263; Luigi Guerrini, *Galileo e la polemica anticopernicana a Firenze* (Florence: Edizioni Polistampa) 2009.

[65] Themistius, *Orationes octo elegantissimae,* 29; Themistius, *Orationes sedecim,* 21. See also see also *Private orations,* 73 and n. 26; *Discorsi,* 710–711 and n. 42.

occurred together in documents concerning the Holy Office.[66] Donzellini, writing in 1559, had omitted the troublesome terms altogether, and for good reason: he had already been denounced twice by the Inquisition for heresy, and would be brought to trial again in 1560, 1577, 1578, and 1587. Repeatedly sanctioned for his interest in works on the Index, but also recognized as a gifted physician, in the final encounter he was put to death by drowning in Venice's lagoon.[67] The conclusion of the passage, that one who sees that a wolf has been observing him will lose the power of speech, was still current, if not wholly credited, in early modernity, as was the corollary that the wolf sighted first by a man will be weakened and unable to flee.[68]

In sum, if we read the passage as suggesting that pseudo-philosophers maintain their authority through the use of intimidating language, that they reduce resistance with terms such as "inquisitor" and "defense," and that they, like wolves, are a menace to the flock that is the Church, we encounter a scenario familiar to Galileo. As he noted in his letter to Christina, his rivals implied that they had multiple arguments against Copernicanism at their disposal, but in fact depended upon a single line of defense. He described this powerful instrument only as "an inevitable and tremendous weapon, the mere sight of which terrifies their adversary," but the shift in emphasis from the oral domain of the dispute to an uncanny visual effect resulting in silence, or rather a certain circumspection, is precisely that of the precedent in Themistius' oration. Crucially, Galileo's adaptation of the passage naturalized the otherwise extraneous detail about wolves. In insisting that his opponents, despite their weapon, seemed to him "the first to take fright, and sensing that they are unable to resist their enemy, try not to let him approach," Galileo acknowledged the potential reversibility of this power dynamic. His membership in the Academy of the Lynxes would have enhanced the suggestion: that animal, endowed with acute sight, was described in Latin and all Romance vernaculars as a type of wolf.[69]

Even where Galileo departs from his Themistian model, traces of the original remain, and account for the odd syntax and lexicon of certain passages. In criticizing his enemies for gossiping, Themistius cited an Aesopic tale to the effect that everyone wears two shoulder bags, one on his chest and the other on his back, and thus views only the vices of his neighbors, and none of his own; he then insisted

---

[66] The most notable instance was in *Immensa aeterni Dei*, the papal bull of 1588, where the phrase *firmissimum Catholicae fidei propugnaculum*, "the most robust defense of Catholic faith," occurred in the provisions regarding the Congregation of the Inquisition; see *Bullarum, diplomatum et privilegiorum sanctorum Romanorum*, 25 volumes (Torino: Franco e Dalmazzo, 1857–1885) 8: 986–987.

[67] Alessandra Celati, "Heresy, Medicine and Paracelsianism in Sixteenth-Century Italy: the Case of Girolamo Donzellini (1513–1587)," *Gesnerus* **71.1** (2014): 5–37; Hannah Marcus, *Forbidden Knowledge: Medicine, Science and Censorship in Early Modern Italy* (Chicago: The University of Chicago Press, 2020) 43–45, 47.

[68] For a list of classical and Patristic references to the belief connecting mutism or aphasia to being seen first by a wolf, see Daniel Ogden, *The Werewolf in the Ancient World* (Oxford, UK: Oxford University Press, 2021) 27 n 40.

[69] On the relationship of the *lupus cervarius* to the lynx, see the Lyncean Johannes Faber, *Animalia mexicana descriptio* (Rome: Iacopo Mascardi, 1628) 519–530.

ironically that he much preferred the sight of his own flaws, and nothing but the merits of others.[70] Galileo further modified the fable, gesturing to the practice of *infamiae nota,* where Roman censors recorded a citizen's alleged infamy, and stating that he was motivated to defend himself because of the "false notes with which [anti-Copernican philosophers] so unjustly desired to load onto [his] back." This odd image also recalls the association of infamy with *notae* elsewhere in Donzellini's translation, and with weight elsewhere in Pétau's text.[71]

"How the Philosopher Should Speak" alludes directly to the astronomical issues with which Galileo's letter is concerned, and shares several features with "The Philosopher:" both present the occasion of the oration as a banquet, rivals as professional combatants, and the goal of the discipline as a civic contribution and a matter of public health. Though Themistius begins by deploring the fact that a few men, former teachers of philosophers, police its practice and argue against its wider diffusion, he soon reveals that he has been accused of attempting some innovation in the discipline and of proposing new gods. These self-appointed censors view him as a sophist, as he does them.

Themistius' strategy here anticipates Galileo's approach in the letter, for just as the orator shows that his antagonists introduced a spurious novelty in deviating from the Platonic definition of a philosopher, so the astronomer implies that his critics have innovated in their reckless disregard for the Patristic norms of Scriptural interpretation. With a sophist's readiness to argue for *and* against any position, Themistius also depicts innovation as a necessary feature of all human endeavors, offering a quick history of philosophy itself, and showing its evolution through the changes brought by Thales, Anaximander, Anaxagoras, Socrates, Plato, and Aristotle. His emphasis falls more upon the presentation, preservation, and branching of the discipline than on its own particular and internal arguments, and he claims his place as the latest contributor by insisting on philosophy's relevance to a broad public.

Themistius then ventriloquizes the roles of Philosophy, now personified as a reclusive and imperious Lady, and of the abandoned city's leading citizens, the latter being drawn from the small and elite audience attending his oration. In their effort to coax Philosophy from her retreat, the city fathers have divided the discipline into divine matters and human affairs. In theory, only the latter category is relevant to them, but their way of bracketing the former is crucial for Galileo's letters to Castelli and Christina:

> We have no quarrel with the first of these [two] divisions, nor do we ask you to busy yourself explaining to us why the celestial bodies are not all borne along together with one motion through the heavens, and why instead there are those few that struggle to move against the majority in the opposite direction. We are asking you neither why the whole of

[70] Themistius, *Orationes octo elegantissimae,* 50; Themistius, *Orationes sedecim,* 39; see also *Private orations,* 85; *Discorsi,* 728–729.

[71] Galilei, *Opere* 5: 313; Themistius, *Orationes octo elegantissimae,* 86; Themistius, *Orationes sedecim*, 105. Galileo also used the verb *addossare* in the debate over buoyancy; see *Opere* 4: 128.

> heaven seems to progress in one direction rather than in the other, nor even what it was that was revealed to you beyond it.[72]

The elaborate list of things of no interest to the city's representatives—thunder, lightning, rain, winds, snow, hail, rainbows, and the salty sea—are no sooner articulated than dismissed. "Finally," they assure Philosophy, "do not teach us about the earth itself, how it is inclined downward and yet remains aloft. Instead, let all these things be sequestered within your academy, in the service of your disciples, since we are perhaps already too slow and too old for such study, were we to pursue that higher wisdom."[73]

Clearly the matters set aside here are those relevant to Galileo in interpreting *Joshua* 10, but this is not to say that his letters on this Scriptural passage are an indictment of Themistius' views. It is rather that Galileo endorses that philosopher's template of progressive innovation by proposing his own novelties—both his celebrated astronomical discoveries and his sense that these "divine matters" were no threat to the social fabric—as the latest in an ongoing process of the discipline's development. A resplendent passage midway through the letter reads like an exalted panegyric to the philosophical matters the city's representatives have curiously delineated, particularly because his invocation of the "Book of the Heavens" allowed Galileo to conflate rather than to oppose the divine and the celestial. Arguing that censuring Copernicus' work would be futile, given the abundance of new telescopic data, Galileo suggests that all astronomical study, and even looking at the heavens, would need to be forbidden, and at enormous cost.

> Let no one believe that reading the sublime concepts inscribed in the [Book of the Heavens] is limited to the mere sight of the splendor of the Sun and the stars, and their appearance and departure, for that is all that animals and common man perceive. Rather, here there are mysteries so deep and ideas so exalted that they have not been fully fathomed by the vigils, the labors, and the study of hundreds upon hundreds of the most acute minds during the endless investigations of thousands of years.[74]

The oration ends, as does this letter, with a focus upon the sun. Themistius shifts from a meditation on Apollo to a list of things men might learn, provided they are sufficiently alert, from the daily and seasonal signals generated by the solar body itself. Galileo follows a discussion of the sun's hub-like position with the first stanza of a hymn from vespers, "Most Holy God of the Heavens," which recalls that on the fourth day the Creator "fashion[ed] the flaming wheel of the sun / and establish[ed] the path of the moon / and the track of the wandering stars." His emphasis on the

[72] Themistius, *Orationes aliquot non editae*, 74; see also *Private orations,* 159; *Discorsi,* 880–881 and n. 82.

[73] Themistius, *Orationes aliquot non editae*, 74–75; see also *Private orations,* 159; *Discorsi,* 880–881.

[74] Galilei, *Opere* 5: 329.

literal meaning of these words, moreover, contradicts the conventional, unconvincing insistence on the figurative sense.[75]

Galileo also draws on "The Sophist," composed around 360 CE. Abrupt in its ending, it was typically paired in early modernity with a fragment devoted to the question of a discipline that governed all others.[76] As in previous orations, Themistius deputizes his audience as a jury, declares that he had been accused of sophistry, arrogance, and innovation by his enemies, and compares the latter to savage but hapless warriors. Theirs is a cyclical illness, he explains, and it is envy; this scourge routinely emerges among the less gifted in tandem with the flowering of philosophy. The careers of Pythagoras, Socrates, Plato, Aristotle, and now his own have all undergone attacks from rivalrous inferiors. Galileo sympathized with such arguments; he and his supporters often referred to envious antagonists in precisely these years.[77]

"The Sophist," unlike its precedents, does not draw on the motif of philosophy as a feast, but converts that figure of speech to a botched banquet, and something shifting from metaphorical to literal referents. Themistius tells his audience that it would be stale, even rancid stuff if he reminded them that the term "sophist" was once a reverential title, and asks if he has dished up arrogance to students otherwise moderate in their tastes, but later turns to minute accounts of how much oil and wheat he had distributed among his disciples. When he emphasizes how little he resembles Plato's degraded sophist—one who trafficks in knowledge, whose rhetoric offers a sham reality, and whose quarry is the rich, idle student—food becomes the emblem of this lifestyle. As Themistius sees it, "whoever claims philosophy, dragging that name about, gulping it down while he weighs up brined ham and haggles with household managers while divvying up the wine and victuals, is a fraud in philosopher's cloak;" we might regard Galileo's absolute silence about the meal that occasioned his letter to Christina as a sober version of this argument.[78]

Other biographical particulars in this oration—Themistius' gestures to his pedagogical success and to his performance at court, the casual depiction of his paraphrases of Aristotelian texts as a youthful exercise, or the reference to his triumphal reception in the dominant, strategically unnamed city of Rome in 357 CE—would have been pertinent for Galileo as well, but find little or no echo in the letter to Christina. What the Themistian and Galileian works share, however, is an explicit investigation of the claims made for a governing discipline. This inquiry, already broached in "The Philosopher," emerges as the conclusion to "The Sophist;" Donzellini, unlike Pétau, makes no reference to the suture of two different texts here. At issue is the reason a particular discipline—presumably philosophy—is said

[75] Themistius, *Orationes aliquot non editae,* 80; see also *Private orations,* 163; *Discorsi,* 888–889; Galilei, *Opere* 5: 348; Giovanni Battista Possevino, *Hinni sacri del breuiario romano* (Venice: Andrea Baba, 1613) 135–137.

[76] Themistius, *Orationes octo elegantissimae,* 169–171; Themistius, *Orationes sedecim*, 129–131.

[77] Miles Chappell, "Cigoli, Galileo, and *Invidia,*" *Art Bulletin* **57.1** (1975): 91–98.

[78] Themistius, *Orationes octo elegantissimae,* 153; Themistius, *Orationes sedecim*, 115; see also *Private orations,* 119; *Discorsi,* 798–799.

to occupy this role. Themistius cautions his audience that this superior sort of knowledge "neither gets involved with angles, nor pines away in the company of numbers," but maintains its regal status and exercises its authority. The basis of its power was its ability to discern which arts were healthy for a well-governed city and should be allowed, and which imperil it and must be banished. But in acknowledging the complication of this scheme—that most objects or fields could be put to evil as well as good uses—Themistius restores some measure of autonomous self-regulation to individual trades and disciplines.[79]

In discussing his rivals' insistence that theology occupied the role of governing discipline, Galileo's letter to Christina both endorses and revises Themistius' argument. His assertion that theology could not claim to be Queen of the Sciences on the basis of content mirrored this fragment, and his statement that she "does not descend from her royal throne to the lowest and most humble speculations of those inferior disciplines" is a sanitized version of Themistius' droll reminder that philosophy neither dallies with angles nor daydreams over numbers. The suggestion that theology's sublime status derives from her lofty subject matter and from the revelation on which salvation depends allows Galileo to isolate that discipline rather than making it, like Themistius' philosophy, central to a city's wellbeing. But the most interesting index of his engagement with Themistius is his improvement upon two details of this fragment. In describing philosophy's detached supervision of other and lower disciplines, Donzellini had written "it rules over them as does a certain *architectum*," a designer or director of activity, and both translators had emphasized Philosophy's familiar civic function as a vigilant monitor of public health.[80] As if to acknowledge the inadequacy of these figurative, rather bloodless roles, as well as to capitalize on the skepticism concerning Don Giovanni as military architect and Don Antonio as alchemical healer, Galileo converted the original argument and its pallid elaborations to literal terms. Themistius opens his notional trial by pledging to abandon metaphors; Galileo closes his with a similar gesture.

3.5 Costume Drama

Beyond the echoes of the orations in the letters to Castelli and the Grand Duchess, Galileo would have found additional Themistian features appealing: both were devoted to gardening, especially viticulture, appreciated irony, enjoyed performing, but disliked, or claimed to dislike, extemporaneous speaking. Rather than develop these commonalities, however, by way of conclusion I want to argue for Galileo's identification with the sophist.

[79] Themistius, *Orationes octo elegantissimae,* 169–170; Themistius, *Orationes sedecim*, 129; see also *Private orations,* 235–236.

[80] Themistius, *Orationes octo elegantissimae,* 169–170; Themistius, *Orationes sedecim*, 130.

We might regard his deployment of Themistius' dictum from his paraphrase of Aristotle's *On the Soul*—"the things that we know are a fraction of those that we don't"—as both a typical test of his audience, and a suggestion of the otherwise unacknowledged role of this thinker. The fact that his antagonist Boscaglia had, in texts that fell far short of their Themistian antecedents, written paraphrases of Aristotelian works might have added to appeal of Galileo's daring appropriation of "The Philosopher," "How the Philosopher Should Speak," and "The Sophist." The inadequacies of the letter to Christina—its distance from epistolary models—disappear, moreover, if we regard it instead as an oration, a paraphrase or rather pastiche of Themistius' extant orations. In this respect, it would be not unlike the concoction "In Support of Religious Freedom," though it is improbable Galileo intended such an outcome. His letter concludes nonetheless with a distracting suggestion about its own genre: readers would have known that during vespers the segue to the hymn—in this case, "Most Holy God of the Heavens"—was "Let my prayer ascend, Lord," *Dirigatur, domine, oratio mea.*[81]

One might argue that Galileo in the guise of Themistius was a quick costume drama, a cameo appearance for the occasion. But before his trial and abjuration, Galileo nonetheless described his intellectual tendencies in writing the *Dialogue concerning the Two Chief World Systems* in terms drawn from the sophists' playbook. In discussing the tides and the sunspots, in particular, he wrote, he had presented hypotheses he regarded as weak and inconclusive as stronger and more robust than their counterarguments, and had given in "to the natural pleasure everyone takes in his own subtleties, and in showing himself as cleverer than the common man in finding, even for false propositions, ingenious and apparent reasons for their likelihood."[82] Ever the wise guy.

Eileen Reeves, Professor Emerita of Comparative Literature, Princeton University. Her research interests concern the relationship of early modern science, literature, and art, especially in the ambit of Galileo Galilei. Her publications include *Painting the Heavens* (Princeton, 1997), *Galileo's Glassworks* (Cambridge MA, 2008), *Evening News* (Philadelphia, 2014), and with Albert Van Helden, *On Sunspots* (Chicago, 2010).

[81] See for example *Breuiarium romanum, ex decreto Sacrosancti Concilij Tridentini restitutum* (Paris: Jacques Kerver, 1574) fol. 43.

[82] Galilei, *Opere* 19: 296.

Chapter 4
"In fibris corporis planetae": Johannes Kepler and the Fabric of the Earth

Luca Ciancio

Abstract The 'invention' of *strata lapidea* seems to have been the result of a process of recognition, within the body of the earth, of naturally generated and orderly structures that could be read geometrically. During the early seventeenth century an important and still largely unacknowledged contribution to this process was made by Johannes Kepler. Notably, in some crucial chapters of his *Astronomia nova* (1609) and in the *Epitome astronomiae Copernicanae* (1618–1621), Kepler speculated on the existence of filaments, fibres and shells in the body of the sun and the planets and described their possible structures and arrangements in a series of illustrations. He also extensively discussed their functions as structures endowed with a motive virtus that played a central role in his dynamic physical astronomy.

Keywords Johannes Kepler · Fibres of the earth · Magnetic philosophy · Theory of the earth · Post-Aristotelian meteorology

4.1 Introduction

In his *Istoria universale* (1697), the astronomer and antiquarian Francesco Bianchini (1662–1727) included extensive excerpts from an excavation report carried out at the foot of Vesuvius in 1689 which contained an accurate description of "several strips of earth that seemed to have been arranged in an order as if they were floors

---

Since the beginning of my career I have regarded John Heilbron as a model of style, rigor and originality. I would like to thank him for his kindness and for having been a constant stimulus to improve the quality of my work. I would also like to express my gratitude to Renato G. Mazzolini and Stefano Gattei for discussing with me various aspects of this research.

L. Ciancio (✉)
Department of Cultures and Civilizations, University of Verona, Verona, Italy
e-mail: luca.ciancio@univr.it

S. Gattei (ed.), *The Incomparable Historian*, Archimedes 75,
https://doi.org/10.1007/978-3-032-25678-2_4

or pavements, placed horizontally one on top of the other."[1] Assuming a uniform rate and manner of sedimentation, Bianchini suggested that it might be possible to date these superimposed "floors or pavements" on the basis of the chronology of Vesuvius' eruptions and archaeological finds from the Roman era that had been discovered in them.[2] The scope and consequences of this new method of studying the past can hardly be overestimated. Here, perhaps for the first time in a classic work of chronology, Bianchini, the "incomparable Monsignor", was applying a stratigraphic method to the study of the past cross-referencing evidence found in 'monumental' antiquities—civil and natural—together with traditional literary sources.[3]

Over the course of the eighteenth century the extension of this approach to the main areas of historical erudition and natural history had the effect of revolutionizing historiography, antiquarianism and geology, but this transformation itself required two fundamental acquisitions: one was the recognition of the presence within the terrestrial crust of rock layers, arranged on top of each other in an orderly sequence and which could be distinguished on the basis of their contents; the other was the emergence of an interpretation of these layers not as being the result of an original act of creation or of a single diluvial catastrophe, but of sedimentary processes that took place over historical time and that were still going on. It is now becoming increasingly clear that Enlightenment natural historians such as Vallisneri, Buffon and Hutton re-established the idea from Antiquity that the terrestrial crust had, in addition to a natural genesis, a history whose timescale was vastly longer than that accepted by sixteenth and seventeenth century scriptural chronologists.[4]

[1] "[…] alcune striscie di terra, che parevano disposte con ordine, quasi fossero suoli, ò pavimenti, collocati orizzontalmente l'uno sull'altro." Francesco Bianchini, *La istoria universale provata con monumenti, e figurata con simboli de gli antichi* (Roma: Stamperia Antonio De' Rossi, 1697), 246–251, 246. Here Bianchini repeatedly uses the term *strata lapidea* referring to Borelli's *Historia et meteorologia* (see n. 6).

[2] This seems to be the first printed mention of the finds which led, in 1738, to the epoch-making discovery of the buried city of Herculaneum. See Marcello Venuti, *Descrizione delle prime scoperte dell'antica città d'Ercolano* (Roma: Stamperia del Bernabò e Lazzarini, 1748), 51–54.

[3] John L. Heilbron, *The Incomparable Monsignor. Francesco Bianchini's world of science, history and intrigue* (Oxford: Oxford University Press, 2022), 46–47. On Bianchini's geological views Nicoletta Morello, "Tra Diluvio e vulcani. Le concezioni geologiche di Francesco Bianchini e del suo tempo," in Luca Ciancio and Gian Paolo Romagnani, eds., *Unità del sapere, molteplicità dei saperi. Francesco Bianchini (1662–1726) tra natura, storia e religione* (Verona: Quiedit, 2010), 185–206.

[4] Paolo Rossi, *The Dark Abyss of Time: The History of the Earth & The History of Nations from Hooke to Vico* (Chicago: University of Chicago Press, 1984); François Ellenberger, *Histoire de la géologie. 1 Des Anciens à la première moitié du XVII siècle* (Paris: Technique et Documentation—Lavoisier, 1988); Rhoda Rappaport, *When Geologists Were Historians, 1665–1750* (Ithaca and London: Cornell University Press, 1997); Martin Rudwick, *Bursting the Limits of Time* (Chicago and London: University of Chicago Press, 2005). Ivano Dal Prete urges us to re-think the history of deep time in *On the Edge of Eternity. The Antiquity of the Earth in Medieval and Early Modern Europe* (Oxford: Oxford University Press, 2022).The traditional view, according to which the concept of "deep history" is a specifically modern-age acquisition, is more clearly formulated in

Nonetheless, the notion of stratum as a structural unit of the crust had already been formulated in the 1660s. It was deliberately and appropriately used by the physician and anatomist Nicolaus Steno (Niels Stensen, 1638–1687) in the work *Canis Carchariae dissectum caput* (1667). Two years later, Steno provided a sophisticated etiological explanation of the spatial arrangement of the strata and their content in his *Prodromus*, a work considered to be foundational for the modern earth sciences.[5] The definitive acceptance of the notion of stratum and of the mechanical-extensional approach connected to it is documented in the *Historia et meteorologia incendij aetnaei* (1670) by Giovanni Alfonso Borelli (1608–1679), a natural philosopher who participated along with Steno in the circle of the Galilean Academy of Cimento.[6]

However, the origin of the concept of stratum appears to be more complex than the scholarship has hitherto supposed.[7] The 'invention' of *strata lapidea* seems to have been the result of a process of recognition, within the body of the earth, of naturally generated and orderly structures that could be read geometrically. This process took place over a long span of time, between the early sixteenth and mid seventeenth centuries, in conjunction with a more general revival of interest in the sublunary world stimulated by new findings in meteorology, geography, chemistry and natural history.[8] There is a general consensus that a turning point in these events

---

Martin J.S. Rudwick, *Earth's Deep History. How It Was Discovered and Why It Matters* (Chicago and London: The University of Chicago Press, 2014), 79-97.

[5] Nicolaus Steno, *Elementorum myologiae specimen [...] cui accedunt Canis Carchariae dissectum caput* et *Dissectus piscis ex Canum Genere* (Florentiae: Ex Typographia sub signo Stellae, 1667), 91, 103; Nicolaus Steno, *De solido intro solidum naturaliter contento dissertationis prodromus* (Florentia: Ex Typographia sub signo Stellae, 1669), 26–29. An English translation in Troels Kardel and Paul Maquet, *Nicolaus Steno: Biography and Original Papers of a 17th Century Scientist* (Berlin: Springer, 2018[2]), 699, 784.

[6] Giovanni Alfonso Borelli, *Historia et meteorologia incendij Aetnaei anni 1669* (Regio Iulio: in Officina Dominici Ferri, 1670), 4.

[7] *Stratum* is the past participle of the Latin verb *sterno* (to spread, scatter, strew; but also to extend, level, pave, cover). It implies a temporal process of deposition. See Egidio Forcellini and Jacopo Facciolati, *A New and Copious Lexicon of the Latin Language* (Boston: Wilkins, Carter and Co., 1845), 846–847. Also Egidio Forcellini, *Lexicon totius latinitatis lexicon* (Lipsiae: in Libraria Hahniana, 1835), V, 179 (*sterno*), 189 (*stratum*); P.G.W. Glare, ed., *Oxford Latin Dictionary* (Oxford: Oxford University Press, 2012), II, 2014 (*stratum*).

[8] An early understanding of the sedimentary origin of strata can be attributed to Leonardo da Vinci. See Jean Paul Richter, ed., *The Notebooks of Leonardo da Vinci* (New York: Dover Publications, 1970), vol. 2, 182 (935). See now Domenico Laurenza, "Leonardo's Theory of the Earth: Unexplored Issues in Geology from the Codex Leicester," in Fabio Frosini and Alessandro Nova, eds., *Leonardo da Vinci on Nature: Knowledge and Representation* (Venice: Marsilio, 2015), 257–267. Useful insights on the structure of the Earth also came from architectural theory, especially the Vitruvian tradition: Daniele Barbaro, *I dieci libri dell'architettura di M. Vitruvio* (Vinegia: Francesco Marcolini, 1556), 30.52, 49.63. See also n. 96.

was the '(archi)-tectonic' and dynamic view of the earth proposed by Descartes in Part III of his *Principia philosophiae* (1644), a view he illustrated in a series of influential explanatory diagrams.[9] However, during the early seventeenth century an important and still largely unacknowledged contribution was made by Johannes Kepler. Notably, in some crucial chapters in his *Astronomia nova* (1609) and in the *Epitome astronomiae copernicanae* (1618–1621), Kepler speculated on the existence of filaments, fibres and shells in the body of the sun and the planets and described their possible structures and arrangements in a series of illustrations; he also extensively discussed their functions as structures endowed with a motive *virtus* that played a central role in his dynamic physical astronomy.

Kepler's thinking in this regard was quite detailed and persistent despite its ambiguities, changes of mind and the problems he left unresolved. As we shall see, according to his hypothesis, the fibres (or filaments) are physical structures that constitute the body of the sun and of the planets, and whose characteristics and behavior explained fundamental celestial dynamics. Thus, fibres and filaments are the very heart of the new causal astronomy. I shall not go into the details of the mathematical discussions developed by Kepler and for that I refer readers to Bruce Stephenson's classic *Kepler's Physical Astronomy*.[10] The main aim of my paper is to attempt to improve our understanding of the geometric shape of these hypothetical structures and their arrangement in planetary bodies. To do this I will first draw on the texts and visual representations chosen by Kepler. The task will not be easy due to ambiguities in the texts which, at least to some extent, reflect waverings and uncertainties in Kepler's own thinking. In order, therefore, to shed more light on what they meant, I shall also look at the sources that inspired this theory and examine one significant episode of its reception. My second aim, no less arduous from the historical-interpretative standpoint, is to try to understand the relationship between Kepler's astronomical hypothesis of the fibres and those of his writings devoted to subterranean meteorology in which he sought to imagine the internal structures of the earth and where he described its main phenomena as indications that it had a soul. While this may not lead to a fully satisfactory and coherent understanding of what Kepler's fibres consisted of, it will however allow us to explore the

[9] Renatus Cartesius, *Principia philosophiae* (Amstelodami: Apud Ludovicum Elzevirium, 1644). Jacques Roger, "La théorie de la terre au XVII siècle," *Revue d'Histoire des Sciences* **26** (1973), 23–48; Ellenberger, *Histoire de la géologie. 1*, 216–224; Bernard Joly, *Descartes et la chimie* (Paris: Vrin, 2011), 89–132 et Annexe 2; Stephen Gaukroger, *Descartes' System of Natural Philosophy* (Cambridge: Cambridge University Press, 2002), 317–322.

[10] In Kepler's writings, *filamentum* is generally equivalent to *fibra*, but it might also refer to the sub-component of a *fibra*. Bruce Stephenson, *Kepler's Physical Astronomy* (Princeton: Princeton University Press, 1994), 67–75, 110–118, 120–137, 146–178, 196–197. See also James R. Voelkel, *The Composition of Kepler's Astronomia Nova* (Princeton and Oxford: Princeton University Press, 2001); Rhonda Martens, *Kepler's Philosophy and the New Astronomy* (Princeton: Princeton University Press, 2000). From the perspective of the history of Earth sciences see Vincent Deparis and Hilaire Legros, *Voyage à l'intérieur de la Terre. De la géographie antique à la géophysique moderne. Une histoire des idées* (Paris: CNRS Editions, 2002), 83–90; Kerry V. Magruder, "Global Visions and the Establishment of Theories of the Earth," *Centaurus* **48** (2006), 234–257, 236–238.

relationship between the vitalistic *physiologia* he consistently espoused and the emergence of a mechanical-structural vision of the planet.[11]

4.2 Filaments, Fibres and Shells in the *Astronomia nova*

Kepler discussed these structures in four crucial sections of the *Astronomia nova*: chapters 34 and 36 (Part III), chap. 57 (Part IV), chap. 63 (Part V).[12] In Chap. 34 he argued that the sun's motive power on the planets was the product of a *species immateriata*[13] which it emitted in the form of a spherical surface that radiates outward in straight lines at an infinite velocity in the manner of light, itself moved by the rotational motion around its own axis. For the purpose of explanation Kepler introduced a double analogy between the sun and the magnet based on the work by William Gilbert.[14] The sun's motive power, just like the attractive power of the magnet, flows out spherically creating an *orbis virtutis* that grows weaker with increas-

[11] Patrick J. Boner, "Kepler's Living Cosmology: Bridging the Celestial and Terrestrial Realms," *Centaurus* 48 (2006), 32–39; Id., "Life in the Liquid Fields: Kepler, Tycho and Gilbert on the Nature of the Heavens and Earth," *History of Science* 46 (2008), 275–297; Id., *Kepler's Cosmological Synthesis. Astrology, Mechanism and the Soul* (Leiden-Boston: Brill, 2013). On the various non- and anti-Aristotelian currents of thought that paved the way for the rise of mechanical philosophy see Daniel Garber, "Remarks on the Pre-history of the Mechanical Philosophy," in Daniel Garber and Sophie Roux, eds., *The Mechanization of Natural Philosophy* (Dordrecht: Springer, 2013), 3–26.

[12] Johannes Kepler, *Astronomia nova*, in Johannes Kepler, *Gesammelte Werke*, Herausgegeben von Max Caspar et al. (Munchen: C.H. Beck'sche Buchhandlung, 1993–2009), 21 vols., III (1990), 4–424, hereafter KGW. Johannes Kepler, *New Astronomy*, New Revised Edition, translated by William H. Donahue (Santa Fe, NM: Green Lion Press, 2015). Also useful is the new Italian translation by Franco Mancinelli: Giovanni Keplero, *Nuova Astronomia* (Atri: Hatria Edizioni, 2021).

[13] Carla Rita Palmerino, "Bodies in Water like Planets in the Skies: Uses and Abuses of Analogical Reasoning in the Study of Planetary Motion," in Massimo Bucciantini et al., eds., *Mechanics and Cosmology in the Medieval and Early Modern Period* (Florence: Olschki, 2007), 145–168; Boner, "Life in the Liquid Fields", cit.; Miguel Angel Granada, "A quo movetur planetae? Kepler et la question de l'agent du mouvement planétaire après la disparition des orbes solides," *Galilaeana* **7** (2010), 111–141. On the problem whether *species immateriata* should be translated with "enmattered" instead of "immaterial" (as I am inclined to believe), see Sheila J. Rabin, "Was Kepler's species immateriata substantial?" *Journal for the History of Astronomy* **36,** (2005), 49–56, and Granada, "A quo movetur planetae?", cit., note 86.

[14] In 1603 the axial rotation was still attributed to an animal faculty, not to the fibres: "Cuilibet facultati magneticae adjungo facultatem animalem, convertendi corpus suum circa axem corporis [...]." KGW XIV, 355. For a history of magnetism and the origin of Kepler's magnetic cosmology see now Christoph Sander, *Magnes. Der Magnetstein und der Magnetismus in den Wissenschften in den Frühen Neuzeit* (Leiden-Boston: Brill, 2019), 323–338. On Gilbert's natural philosophy see J.A. Bennett, "Cosmology and the Magnetical Philosophy, 1640-1680," *Journal for the History of Astronomy* **12** (1981), 165–177; Gad Freudenthal, "*Theory of Matter and Cosmology in William Gilbert's* De magnete," *Isis* **74** (1983), 22–37; Stephen Pumfrey, "Magnetical philosophy and astronomy, 1600-1650," in René Taton and Curtis Wilson, eds., *The General History of Astronomy* (Cambridge: Cambridge University Press, 1989), 45–53, Fritz Krafft, "The New Celestial Physics

ing distance. In the case of the sun, this is a power that lays hold (*prensare*) of celestial objects and moves them around its body, not one of attraction.[15] Secondly, given the fact that "if a little strip of iron is placed in a middle position between the heads of the magnet at the side", it arranges itself parallel to the "straight fibres [...] extended throughout its length", one might therefore suppose that in the sun there is also "a directing force, and consequently that it has circular fibers all set up in the same direction, which [is] indicated by the zodiac circle."[16] Hence, these are circular fibres, not straight ones as in magnets, arranged circularly (*circumporrectas*) in the same direction of the zodiac, that is, of the equator. Because of their non-attractive motive function filaments and fibres should then be defined as quasi-magnetic.[17] Since the fibres move in an easterly direction following the axial rotation, the *species immateriata* that flows out from them appears to be able to move the planets with a velocity that depends on three variables: distance that makes the virtue denser or more rarefied; the volume of each planet presenting a different surface on which the solar *virtus* can act; the opposing resistance of each planetary body to the solar *virtus*. This resistance or tendency to a state of rest (*inclinatio ad quietem*) also varies with the amount of matter (*copia materiae*) or density of each planet.[18]

In Chap. 34 Kepler also adopts a second important suggestion from Chap. 1, book 6 of Gilbert's work in which the author convincingly demonstrates that the earth is a gigantic magnet. Based on this, Kepler theorized (*plausibile est*) that, like the sun, the earth also contains magnetic fibres whose *species immateriata* flowing out into the surrounding space is the cause—more precisely the concause—of the

of Johannes Kepler," in Sabetai Unguru, ed., *Physics, Cosmology and Astronomy, 1300–1700* (Dordrecht: Kluwer, 1991), 185–227, 195–198.

[15] "Solis corpus est circulariter magneticum et convertitur in suo spacio, transferens orbem virtutis suae, quae non est attractoria sed promotoria." KGW XV, 172.93–95.

[16] Kepler, *New Astronomy*, 289. "Ut vero magnes non omni parte trahit, sed filamenta (ut ita dicam) seu fibras (motoriae virtutis sedem) rectas habet per longum extensas, ita ut ferri lingulam, si medio loco inter capita magnetis a latere consistat, non attrahat, sed tantummodo parallelon suis fibris dirigat: ita credibile est, in Sole non esse ullam vim Planetarum attractoriam, ut in magnete (accederent enim ad Solem tantisper, donec cum ipso conjungerentur penitus.) sed tantum directoriam, ideoque fibras habere circulares in eam plagam circumporrectas, quae monstratur a circulo zodiaco." (KGW III, 246.4–12). In *De magnete*, however I can find no reference to filaments and fibres as the seat of the magnetic force. The term *fibra* in used once with regard to the living body. To designate the underground structures Gilbert employs *vena* (vein) and *viscera terrae* (bowels of the earth). William Gilbert, *De magnete* (Londini: Excudebat Petrus Short, 1600), 9, 34 (*fibras*), 40. However, the term *magneticis fibrarum* appears in his posthumous *De mundo* (Amstelodami: Apud Ludovicum Elzevirium, 1651), 126.4.

[17] On the heuristic value of this hypothesis see K.'s letter to Maestlin, 5.3.1605, n. 335: "Sol [...] est magneticus vel quasj (similitudinem enim volo, non pertinaciter rem ipsam)". KGW, XV 172. And yet Kepler makes the point, "The example of the magnet [...] is a very pretty one, and entirely suited to the subject; indeed, it is little short of being the very truth." Kepler, *New Astronomy*, 289; KGW III, 246.19–20.

[18] Stephenson, *Kepler's*, 143–144. If the density of ether is nothing or negligible, any possible irregular density of celestial bodies, in other words, a non-compact or partially hollow internal structure, would have significant effects on the dynamics of planetary motion. See also n. 43.

moon's motion.[19] But in this case, the passage that describes its characteristics seems to contradict the one dealing with the arrangement of the fibres in the solar body. Describing the arrangement of the fibres in the body of the earth Kepler wrote: "And because of this very thing, that it has magnetic fibres intersecting the line of its motion at right angles, those fibres therefore lie in various circles about the poles of the earth parallel to its motion."[20] The phrase "magnetic fibres intersecting the line of its motion at right angles" would lead us to interpret the entire statement as an anticipation of the existence of rectilinear fibres perpendicular to the ecliptic plane later discussed in Chap. 63. These fibres, Kepler proposed, arranged themselves circularly around the terrestrial axis creating in this way a series of circles parallel to the ecliptic. In any event, the need to explain lunar motion confirms the presence in the earth of circular fibres arranged parallel to the equator, as Kepler repeated in the conclusion: "I know that the earth's filaments and its motion indicate the equator [...]."[21]

That the solar fibres are circular and "ordered according to zodiacal longitude" is confidently repeated by Kepler in Chap. 36 where he again speculates on the intensity of the sun's motive action and its weakening.[22] Here he explains that "the rays from all the filaments of the solar body (standing up from one hemisphere of the body) flow down and converge on the individual points of the path of any of the planets as well as on those poles that are above the poles of the sun's body, and the body of the planet is transported according to the measure of the density of this entire *species* compounded of all the filaments."[23] Since the motive action is only fully exerted when the planets are exposed to the action of the fibres that are arranged in the same direction, that is, of the hemisphere that is oriented towards them, the planets do not deviate from the ecliptic because it is only in that position that the *species immateriata* emitted by the fibres acts in concert. To provide a geometric demonstration of this thesis Kepler introduced the first image of solar fibres (Fig. 4.1). This is a hybrid image: not a pure geometric diagram but also a visual representation of their physical arrangement on the body of the planet. The circular fibres are shown (in profile) as lines parallel to the segment ζκ with the planet in ζ rotating in the intermediate region of the zodiac between the poles, that is, on the

---

[19] Kepler solved the problem of a possible reciprocal interference between the sun's motive force and that of the Earth by considering the enormous difference between the amount of matter of the sun and of the Earth and the terrestrial fibres' limited range of action (*orbis virtutis*). For a detailed discussion see Chap. 37.

[20] Kepler, *New Astronomy*, 289. "[...] et ob id ipsum, quia fibras habet magneticas, lineam motionibus suae rectis angulis intersecantes [-em], ideo illae fibrae variis circulis motioni parallelis polos telluris circumsistant [...]." KGW III, 246.

[21] *Ibidem*.

[22] "Nam et filamenta magnetica solis supra ponebantur in longum tantummodo porrigi, non etiam versus polos aut aliorsum." KGW 251.1–2.

[23] Kepler, *New Astronomy*, 294. "[...] tamen ab omnibus Solaris corporis filamentis (ab uno hemisphaerio corporis stantibus) radii defluunt et confluunt tam ad puncta singula itineris alicujus Planetae, quam ad ipsos polos polis corporis Solis imminentes; et Planetae corpus vehitur ad modulum densitatis, hujus integrae speciei, ex filamentis omnibus compositae." KGW III, 251.16–20.

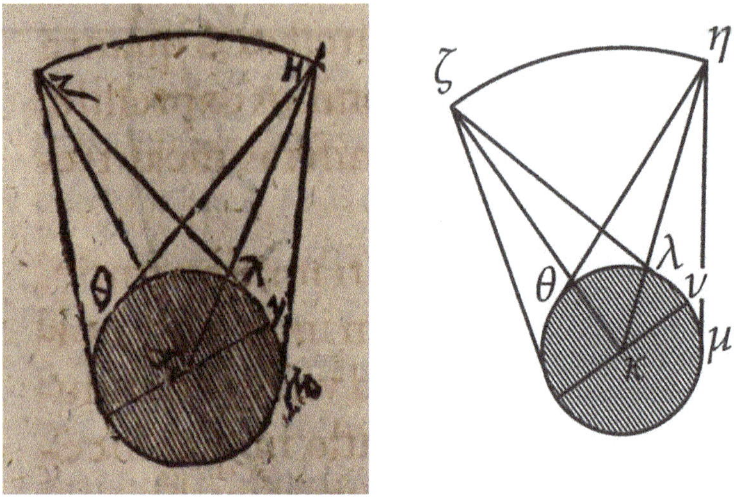

Fig. 4.1 J. Kepler, *Astronomia nova*, 1609, 181. (KGW III, 251). Bayerische Staatsbibliothek München, urn:nbn:de:bvb: v2_bsb11196682_00223

plane of the ecliptic. If the planet were at η, it would also be exposed to the rotation of fibres λμ which are moving in the opposite direction and which are concentric to the pole ν and this would counteract on the pulling force. Therefore, the planet will neither go backward, nor will go to the poles.

Later in Chap. 57, Kepler used the hypothesis of the planets' quasi-magnetic nature to assign a probable natural cause for the libration of Mars on its elliptical orbit and also to avoid resorting to an explanation based on animal power or the planet's having a mind. Here again, in order to illustrate his geometric demonstration, Kepler introduced another hybrid image (Fig. 4.2) which shows the fibres arranged parallel to the central axis of the planet DA. As an aid for the imagination the illustration also suggests a resemblance between the magnetic "directive" property of the planet and the steering action performed by an oar on the movement of a boat placed in a circularly rotating stream of water in which the stream represents the solar *virtus*.

The arrangement of the filaments parallel to a central magnetic axis—"its globe consists of an infinite number of physical lines, as it were, parallel to DA"[24]—with one head seeking out the sun, the other fleeing it, allows the planet, attached to the centre B, to oscillate longitudinally. In this case the image shows the planet from the top and so the fibres that compose it are rectilinear, arranged horizontally and parallel to the plane of the ecliptic. Calculating the forces based on the law of the lever, Kepler showed that the changing position and distance from the sun in the

[24] Kepler, *New Astronomy*, 415. "[…] globus constat ex infinitis quasi lineis Physicis ipsi DA parallelis." KGW III, 352.32. "Sit ergo loco totius corporis, omniumque ejus filamentorum medius axis DA ad speculandum propositus." KG WIII, 352.35–36.

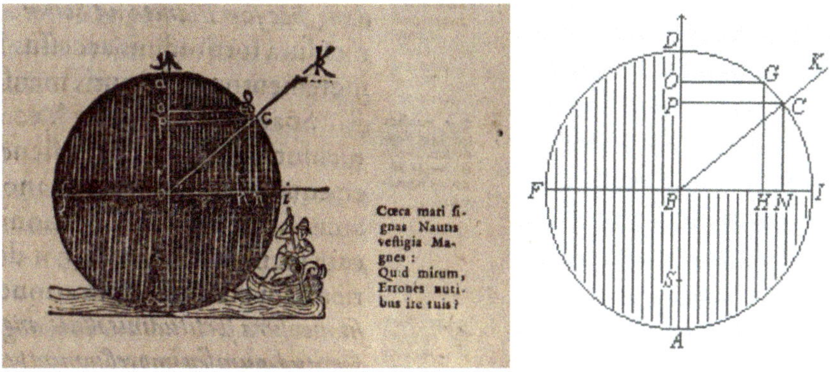

Fig. 4.2 J. Kepler, *Astronomia nova*, 1609, 274. (KGW III, 353). Bayerische Staatsbibliothek München, urn:nbn:de:bvb: bsb11196682_00316

movement of translation along the orbit, caused a slight natural libration of the body of Mars with respect to the sun without the intervention of any external mind.[25] This confirms that celestial dynamics is the result of the combined action of the sun's motive force and of the planet's own force; in this way, and only in this, the magnetic polarity of the planets cooperate in producing their trajectories around the sun and the anomalies observed by astronomers.[26]

In Chap. 63, the magnetic theory of libration provides the inspiration for a similar physical hypothesis regarding the variation in Mars' latitude (the planet's angular distance from the ecliptic). The development of this hypothesis had important implications for the internal structure of the earth and orientation of its fibres. Here again Kepler postulated that inside the planet "in place of the whole body and all its filaments", there is a magnetic axis or diameter always directed towards those fixed stars under which lay the limits of the planet's latitudinal motion. This axis, remaining parallel to itself at all parts of the orbit, had a virtue similar to the directive faculty of an oar, able to deflect the planet from the path of the solar image moving it from north to south of the solar ecliptic and back.[27] This was evidently a magnetic

[25] KGW III, 352–354. The mathematical demonstration is discussed in Stephenson 1994, pp. 110–117. See also K's letter to Maestlin 12/22.12.1616, n. 750: "Dico enim librarj planetam in linea recta versus Solem appetentia magnetica, quae in diversis fibrarum inclinationibus sit diversae fortitudinis; [...] Hypothesis est physica, quia habet exempla physica magnetis, est physica hoc est naturalis, quia vera, et educta ex ipsa natura interna corporum planetae et Solis, est physica, quia modos omnes tenet motionum naturalium." (KGW XVII, 202). In the last section of the chapter Kepler discusses the problem of the reciprocal magnetic occultations between the planets and with the solar force (KGW III, 363.23–38).

[26] KGW III, 255.20–30. "[...] virtutem nudam et solitariam in ipso Planetae corpore residentem, non esse sufficientem transportando de loco in locum suo corpori, quod pedibus alis et pinnis caret, quibus in aura aetheria nitatur." KGW III, 256.17–20.

[27] Kepler, *New Astronomy*, 460; KGW III, 389–390. For a comprehensive explanation see Stephenson 1994, 130–133 and Fig. 27.

axis different from that used in Chap. 57 to account for libration: it too was made up of parallel rectilinear fibres, but inclined with respect to the ecliptic plane. Since these axes must have been the effect of existing structures in the body of the planet and these two powers "differ in time and position", there was no choice to admit the presence of distinct structures within the planet. With the exception of rotational movements around their own axis, Kepler concluded, "there is [...] nothing to impede their residing in one and the same planetary body as a whole [...]. Thus if the planets are moved like the moon, which does not rotate, [...] nothing prevents our saying that the two are interwoven, as the weft is interwoven with the warp."[28]

Kepler would later return to and clarify the hypothesis of interwoven fibres in the *Epitome*. In the case of the earth, which rotates around the sun and also around its own axis, this solution was, therefore, inapplicable. For this reason Kepler put forward the hypothesis that there also exists "some interior globe within this globe of the earth, to which the diurnal motion of the earth's exterior does not penetrate, but which is held in place by a very strong inclination towards certain sidereal positions, so as not to follow the revolving exterior of the body."[29] The idea of there being a spherical nucleus, compact and stationary, implied that the interior globe was physically detached from the outer crust which, thus, had the form of a separate shell.[30] In this case, the circular fibres present in the crust and parallel to the equator, superimposed themselves over the rectilinear fibres of the nucleus (oriented along the axis) rather than intertwining with them. It is worth noting that decades later Edmund Halley also assumed that earth's crust and core were detached. To explain the slow changes observed in the variation of the compass (but with no reference to fibres) Halley proposed his "extravagant" hypothesis of a hollow earth, made up of an outer shell that is one of three nested shells, at the center of which there is a solid sphere.[31]

[28] Kepler, *New Astronomy*, 462. "[...] nihil impedit asserere, intextas esse mutuo virtutes utrasque, ut subtegmina sunt intexta staminibus." KGW III, 392.33–34.

[29] Kepler, *New Astronomy*, 463. "An igitur et in hoc Telluris globo sit interior aliquis globus, ad quem diurnus Telluris exterioris motus non penetret, sed qui fortissima inclinatione ad certa Fixarum loca retineatur, quo minus exterius corpus revolutum sequatur?" KGW III, 393.19–22.

[30] In his *Epitome* K. would use the term *exterior cortex*. See KGW VII, 92.17. Underneath this, one also had to imagine a space occupied by a fluid that allowed it to rotate around the core.

[31] According to Halley's hypothesis, the internal sphere has a slightly slower axial rotation velocity than that of the crust. Edmund Halley, "An Account of the cause of the Change of the Variation of the Magnetical Needle, with an Hypothesis of the Structure of the Internal parts of the Earth," *The Philosophical Transactions of the Royal Society* (1692), 563–578, 568. See Peter W. Sinnema, "We have adventured to make the Earth Hollow": Edmund Halley's Extravagant hypothesis," *Perspectives on Science* **22**/4 (2014), 423–448.

4.3 Filaments and Fibres in the *Epitome*

Kepler's *Epitome astronomiae copernicanae,* the first textbook of heliocentric astronomy and a work that acquired a wide popularity in Europe, was in fact a detailed and systematic account of his own astronomy in which celestial phenomena found a convincing explanation through the mathematical study of their physical causes.[32] The nature and function of the magnetic fibres existing in the sun and planets is discussed in Book I, Part V (1618), in Book IV, Part II and Part III (1620) and in Book V, Part I (1621). In these sections Kepler largely reiterated the hypotheses he put forward in *Astronomia nova* but with a more detailed treatment and accompanied by new illustrations. For a better understanding of the nature of the fibres it is useful to first look at what Kepler wrote in Book I, Part V, where his aim is to demonstrate that the constant inclination of the earth's axis of rotation is due to a faculty "nested" in the parallel filaments arranged around the axis, everywhere with equal weight.[33] The proof (*proba*) of Kepler's assertion consists in the fact that within the terrestrial globe there exists a faculty that directs magnets and all magnetic bodies according to the direction (*tractum*) of the axis. Thus, "since it is clear that the terrestrial globe has a rectilinear form that follows the orientation of the axis" it is not absurd to attribute to this form the constant direction of the terrestrial axis.[34] For this reason the fibres parallel to the axis are by no means a mathematical fiction, but rather real structures that constitute the body of the earth, "manifest" and therefore somehow visible.

As regards to the rotatory motion around its axis, according to Kepler the impetus given by God at the beginning of the world was later plausibly transferred to the corporeal faculty of the fibres, "fibres parallel to the axis, but which surround it circularly, like rectilinear threads are intertwined with circular ones in a woof."[35] It is an unmistakable indication of the spatial arrangement of the fibres that determine

[32] Stephenson, *Kepler's*, 139. Johannes Kepler, *Epitome astronomiae copernicanae*, in KGW VII, 4–530. Johannes Kepler, *Epitome of Copernican Astronomy & Harmonies of the World*, translated by Charles Glenn Wallis (Amherst: Prometeur Books, 1995). See also the French translation: Jean Kepler, *Abrégé d'astronomie copernicienne*, traduit par Jean Peyrout (Paris: Librairie Blanchard, 1988).

[33] "Vel interna et positiva facultas naturalis in fibris rectilineis axi parallelis, tuendi sese in suo sito primaevo. Nam volutionis facultas nidulatur in his ipsis totius globi filamentis, circa axem circulariter, aequali undique pondere, circumjectis, quibus innixa corpus torquet." KGW VII, 87.27–31.

[34] "Equidem inest globo Telluris, secundum tractum axis, et fibrarum axi parallelarum, facultas in se dirigendi magnetes et magnetica omnia, magnetibus vero mutua inest facultas, hanc plagam Telluris quaerendi. Cum igitur manifestum sit, Globum telluris informatum esse forma corporea rectilinea secundum Tractum axis primi motus, non absurde eidem formae tribuitur etiam haec constantia directionis axis in plagam semper eandem. KGW VII, 87.40–88.5. This is an "interna figuratio rectilinea." KGW VII, 329.8, 43.

[35] "Verisimile est, hanc ipsam primae rotationis continuatam speciem in terra, transformatam esse, seu coaluisse in talem facultatem corpoream; et sic in fibras terrae, dispositas secundum ductum motus ipsius, inolevisse; sic ut fibrae omnes intelligantur rectilineae ut prius dictum, et axi parallelae, sed circulariter axem circumstantes, veluti subtemina recta staminibus circularibus intexta."

the stability of the axis of daily rotation, but are in relationship with other fibres that are concentric with the equator. As Kepler explains, it is, in fact, very likely that the motor faculty exerted by the earth on the moon depends instead on the existence of circular motor fibres.[36] So, in what manner are the two types of fibre arranged in relation with each other? Perhaps, Kepler replies, like the fibres in the human stomach, as suggested by the anatomists: "The physicians have an example of this intertwining of the fibres in the substance of the stomach – not double as in our case, but clearly triple – in which similarly three orders of fibres intertwine reciprocally and distribute three faculties of the stomach, attractive, retentive and expulsive."[37] I will return later to this suggestion.

In order to facilitate comprehension among a public of non-specialists and students, Kepler included two xylographic images, rather crude but which scholars today cannot fail to take into careful consideration. Instead of images mixing geometry and reality such as those employed in the *Astronomia nova*, here he chose two more realistic, perspectival images both showing inner sections of the globe.[38] The first of these (Fig. 4.3a) shows two hemispheres obtained from a section of the globe along a north-south axis. In the second (Fig. 4.3b), the earth is shown as a sort of magnetic top cut in two halves along the terrestrial equator. The first image shows the rectilinear fibres parallel to the axis of rotation, the second shows the circular fibres concentric and parallel to the equator.[39]

In the case of the rectilinear fibres parallel to the axis (Fig. 4.3a), they occupy the entire body of the planet and surround the axis arranging themselves around it. As for the circular fibres (Fig. 4.3b), in view of what Kepler said in the *Astronomia nova* they must occupy not only the interior, but also appear near the surface in order to be able to emit their quasi-magnetic force. This is confirmed by the fact that again in Fig. 4.3b, while the hachures on the edge of both hemispheres are there to convey a sense of three-dimensionality, the parallel lines at the equator drawn on the surface

---

KGW VII, 89.41–90.4. Kepler estimated that "vicies centena milia" diurnal rotations had occurred and that therefore the Earth was approximately 5.479 years old. KWG VII, 89.32.

[36] "Quod si prius causam directionis axis satis probabiliter contulimus in formam corpoream, secundum fibras rectas, axi parallelas: iam multo probabilior fiet etiam altera globi informatio, secundum fibras circulares, indeque dependens facultas motoria: cum etiam hoc praedictum sit, globum eundem, secundum illarum tractum habere rationem quiescentis, secundum harum circumductum, rationes mobilis." KGW VII, 90.23–28.

[37] "Exemplum hujus fibrarum implexionis non geminae tantum ut hic, sed plane triplicis, habent Medici in substantia ventriculi, qui consimiliter inter tres illos fibrarum ordines, implexos mutuo, tres ventriculi facultates distribuunt, attractricem, retentricem, expultricem." KGW VII, 90.29–32. K. attributed the same faculties to the Earth in KGW XV, 163.80.

[38] Isabelle Pantin, "Kepler's Epitome: New Images of an Innovative Book," in Sachiko Kusukawa and Ian Maclean, eds., *Transmitting Knowledge. Words, Images, and Instruments in Early modern Europe* (Oxford: Oxford University Press, 2006), 217–237, 233–237.

[39] The editors of Kepler's *Gesammelte Werke* felt it necessary to correct the original image on page 121 and replace it with an illustration showing the fibres parallel to the diameter on the equatorial plane and not concentric to the equator. KGW, VII, 90. The same image is reproduced in Johannes Kepler, *Kurze Darstellung der copernicanischer Astronomie* (Würburg: Königshausen et Neumann, 2010), 113.

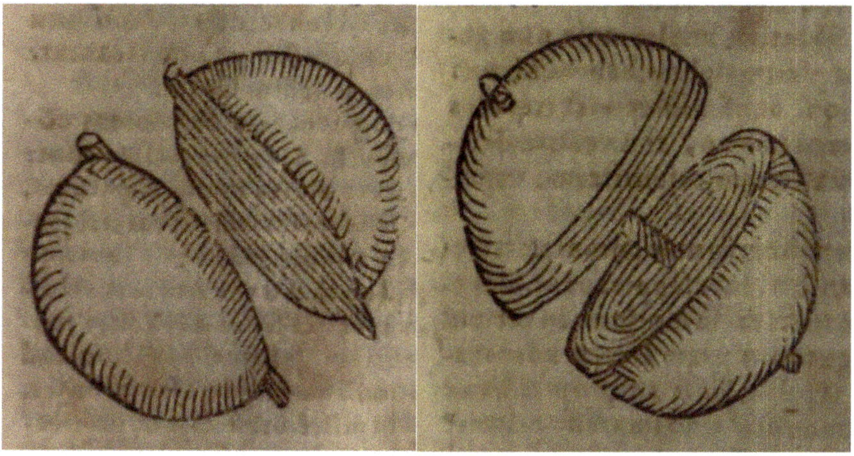

Fig. 4.3 (**a**) J. Kepler, *Epitome*, 1618, 117. (**b**) J. Kepler, *Epitome*, 1618, 121. Bayerische Staatsbibliothek München, urn:nbn:de:bvb: v2_bsb10060988_00149; urn:nbn:de:bvb: v2_bsb10060988_00153

of the upper hemisphere are unmistakably fibres. Thus, these circular fibres are arranged both concentrically to the axis, and also superposed from the equator to the poles. Now, as the fibres are literally filaments, threadlike structures adjacent to each other, both horizontally and vertically, it does not seem possible to interpret these two illustrations as precursors to a stratigraphy. The straight and circular lines traced inside the hemispheres have the sole purpose of indicating the orientation of the fibres and do not represent beds or layers viewed in profile.

Kepler provides more detail of how the fibres are interwoven in Book IV, part III which is devoted to the irregularities of planetary motion, reiterating the explanation of the probable cause of Mars' libration given in Chap. 57 of *Astronomia nova*. Indeed, he asked, "what prevents one and the same globe from having twofold rectilinear threads stretching out from the whole body, so that from some it receives a libration in altitude, and by the others is rowed backward and forward?"[40] He accompanied this statement with a rather crude and scarcely legible image (Fig. 4.4), but which leaves no doubt of the fact that the fibres are intertwined in the manner of a woven fabric.

The illustration is accompanied by a text in which Kepler drew an analogy between the variously intersecting movements of water in a river: "Thus on the surface of rivers a threefold movement of the parts is discerned – each movement proceeding in its own direction: the first is the flow of water, the second that of the waves, which that flow casts crosswise to the banks in a continuous series, the third comes from the wind. For if a contrary wind blows obliquely, it roughens the surface of the waters, and starts another series of smaller waves moving in their own

[40] Kepler, *Epitome,* 109; KGW VII, 346.14–16. Figure 4, printed at page 605 of the original edition (1620), was expurgated from KGW VII, 345. The editors of the American and of the French translations (see n. 32) both opted for redrawing it, although they did not agree on which texture it was supposed to represent.

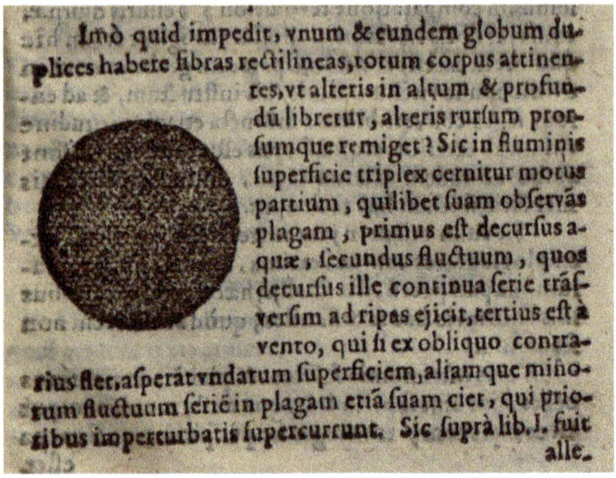
Imò quid impedit, vnum & eundem globum du-
plices habere fibras rectilineas, totum corpus attinen-
tes, vt alteris in altum & profun-
dū libretur, alteris rurſum pror-
ſumque remiget? Sic in fluminis
ſuperficie triplex cernitur motus
partium, quilibet ſuam obſeruās
plagam, primus eſt decurſus a-
quæ, ſecundus fluctuum, quos
decurſus ille continua ſerie trāſ-
verſim ad ripas ejicit, tertius eſt à
vento, qui ſi ex obliquo contra-
rius flet, aſperat vndatum ſuperficiem, aliamque mino-
rum fluctuum ſeriē in plagam etiā ſuam ciet, qui prio-
ribus imperturbatis ſupercurrunt. Sic ſuprà lib. I. fuit
alle-

Fig. 4.4 J. Kepler, *Epitome*, 1620, 605. Bayerische Staatsbibliothek München, urn:nbn:de:bvb:v2_bsb11301955_00193

direction […].”[41] The analogy, which is immediately followed by a reference to the previously mentioned triple interconnection of the fibres in the stomach, served to illustrate the idea that an intertwining of physical structures oriented in different directions would not prevent the simultaneous transmission of divergent and autonomous driving forces.

Nonetheless, in consideration of planetary phenomena like the retrogression of the limits and progression of the absides, Kepler wondered whether “the threads productive of latitude [will] remain inwoven with the threads of longitude.”[42] In his response, which was negative and the same as he gave in *Astronomia nova* regarding lunar libration, he evoked an ancient image drawn from biology: “[…] here too we can seek within the outer crust a separate globe – like the yolk within the white of egg – supplied with its own threads, and able to revolve according to the same laws, and with the strength of its forces different from that of the outer crust, if need be; so that both of them […] can be deflected by the same outside cause, with different measures of speed […].”[43] Also for this hypothesis, the anatomy of the stomach provided further possible confirmation. In fact, as Kepler observed, the three

[41] “Sic in fluminis superficie triplex cernitur motus partium, quilibet suam observans plagam, primus est decursus aquae, secundum fluctuum, quos decursus ille continua serie transversim ad ripas ejicit, tertio est a vento, qui si ex obliquo contrarius flet, asperat undarum superficiem, aliamque minorum fluctuum seriem in plagam etiam suam ciet, qui prioribus impertubatis supercurrent.” KGW VII, 346.16–21. Curiously, our Fig. 4 is not reproduced in KGW.

[42] “Si Limites retrocedunt, Apsides progrediuntur; non haerebunt intextae, fibrae latitudinis administrae, fibris longitudinis, quas utrasque eidem globo dedisti?” KGW VII, 346.30–32.

[43] *Ibidem*, 110; “Possumus igitur etiam hic quaerere in exterior crusta, separatum globum velut in ovi albumine vitellum, fibris suis instructum, et ad earum leges convertibilem, distincta etiam fortitudine virium ab exterior crusta, si opus est […].” KGW VII, 346.35–39. The same hypothesis had been suggested to explain the irregularity in attitude: *Ibidem*, 97; KGW VII, 335.45–336.2. A plausible source is Merula, *Cosmographia generalis*, 26. The structural analogy between an egg and the Earth was later taken up and illustrated in Thomas Burnet, *The Sacred Theory of the Earth* (London: R. Norton, 1691), 62–63.

superimposed tunics that constitute the stomach may become active one independent of the others. Thus, the terrestrial fibres responsible for the impulse exerted on the moon and the conservation of the axial inclination were not arranged in the manner shown in Figs. 4.3a, b but had to be thought of as separate—the ones constituting the external crust, the others the nucleus—and were able to react differently to the same external cause. Kepler concluded ironically that if ancient astronomy had situated solid and invisible crystalline spheres in the celestial region that was so transparent as to appear empty, then no one should be offended by the proposal to suppose that there was something on the inside of the planets since these were "visible and palpable bodies."[44]

4.4 Fibres in Anatomy and the Natural History of the Earth

Thus, in an attempt to summarize Kepler's ambiguous hypotheses, in the bodies of the sun, Mars, the moon and probably in the other celestial bodies there would seem to be three types of fibres: the first, which are rectilinear and arranged parallel around the axis of rotation, guarantee the constant axial orientation of the planet during its long ecliptic movement and account for the variation of latitude with respect to the ecliptic plane. Next, there are rectilinear fibres parallel to the ecliptic plane that can cause the planet's libration and influence the longitude during the course of its revolving motion. Finally, in the sun—but also in the earth and, in the light of Galileo's telescopic discoveries, also in Jupiter—there are circular fibres parallel to the equator whose diameters decrease as we move from the surface inward and from the equator to the poles, drivers of the entire *machina mundi* including the movement of the satellites. In the planets without satellites, the rectlinear fibres intertwine almost at right angles in the manner of a woven fabric, either biological or artificial, in such a way as to occupy the entire mass. In the case of the earth, however, it would seem necessary to envisage, for the circular fibres of the crust, an arrangement separate from those in the inner nucleus which Kepler thought was composed of rectilinear fibres parallel to the axis.

While the orientative and motive functions of these structures appear to have been sufficiently explored and, above all, amenable to mathematical theorization in accordance with observed data, their physical characteristics and their actual material composition raise many questions. Are these the same structures that sometimes emerge from beneath the surface of the earth's crust? If so, to what depth do they reach? Is there a relationship between the fibres and mountain ranges, between their

[44] "Non indignabitur igitur, nos in globis, qui sunt corpora cospicua et palpabilia, simile qui fabricari." KGW VII, 347.3–4. The planetary motions "are not furnished physically by those two [imaginary] lines but by the inclination of the real threads (*per inclinationem realium fibrarum*) of the planet's body." KGW VII, 364.34. The irregularities of the moon confirm the existence of a globe within its globe, "ut nucleus intra corticem", having a different rotation from it. KGW VII, 350.25–26.

internal composition and their orientation? Does the magnetic force they contain depend on some particular chemical-mineral composition of theirs? Kepler says nothing about this in his astronomical works, and above all, he never draws any explicit correspondence between the quasi-magnetic fibres and the rocky structures studied by physical geography and natural history. In order to try to get a better understanding of his thinking we have no choice but to explore the sources behind this hypothesis.

The most obvious of these, which Kepler himself repeatedly mentioned, were from the medical literature, especially anatomy.[45] It is worth remembering that in the pursuit of his studies of optics Kepler had acquired a deep knowledge of the anatomy of the eye and it was in the *Ad Vitellionem paralipomena* (1604) that the term "fibre" appeared for the first time in his printed works.[46] The works Kepler read in this field included the *Universa medicina* by Jean Fernel (1497–1558), the anatomical work of Felix Platter (1536–1614) and especially the *Anatomia praghensis* by his friend Johannes Jessenius (1566–1621), a Polish physician and pupil of Hieronymus Fabricius ab Aquapendente.[47] Jessenius represented a link with the scientific legacy of the Paduan school of anatomy, a tradition that had accumulated a detailed knowledge of the architecture of the living body. Already in the early sixteenth century—a period also characterized by great advances in textile technology—leading members of that school had suggested that the fibre was "the main operative building block" of the organic body and of its faculties.[48] Andreas Vesalius, in particular, had devoted several pages to fibres in the *De humani corporis fabrica* in which he sought to elucidate the ideas formulated by Galen in his *De naturalibus facoltatibus* on the fundamental structures of the living body.

Vesalius had clearly distinguished between the structure (and function) of veins from that of common membranes. While membranes are simple bodies, veins consist of a tunic whose special characteristic is that it is made up of fibres oriented in

[45] KGW VII, 90.29–32. See also *Ibidem*, 346.21–24. Essential in this regard is Jonathan Regier, "Kepler's Theory of Force and His Medical Sources," *Early Science and Medicine* **19** (2014), 1–27.

[46] KGW II, 149.14, 161.18. In his *Paralipomena* Kepler stressed how the anatomists appeared to him to depict and describe "the globe of the eye" "as if the overall form of the eye […] were to represent the image of an onion." KGW II, 147.24–27. This comparison may have suggested an analogous representation for the globe of the earth. For the use of the same analogy in Cabeo's *Philosophia magnetica* (1629) see n. 56.

[47] Jean Fernel, *Universa medicina* (Lutetiae Parisiorum: Apud Andream Wechelum, 1567), 3.6, 17.51; Johannes Jessenius, *Anatomiae, Pragae Anno MDC* (Witebergae: Excudebat Laurentius Seuberlich, 1601), 38v. On Jessenius see Tawrin Baker, "Why all this Jelly? Jacopo Zabarella and Hieronymus Fabricius ab Aquapendente on the Usefulness of the Vitreous Humor," in Peter Distelzweig et al., eds., *Early Modern Medicine and Natural Philosophy* (Dordrecht: Springer, 2016), 59–88, 78–82. Another interlocutor of Kepler on this subject was Johann Georg Bernegger. See KGW XVI, 166–167.

[48] On how different fibre concepts evolved over the long term, see Tobias Cheung, "Omnis Fibra ex Fibra: Fibre Economies in Bonnet's and Diderot's Models of Organic Order," in Tobias Cheung, ed., *Transitions and Borders between Animals, Humans and Machines 1600–1800* (Boston-Leiden: Brill, 2010), 66–104, 70–74. See also Mirko Grmek, "La notion de fibre vivante chez les médecins de l'école iatrophysique," *Clio Medica* **5** (1970), 297–318.

different directions. Since, as he argued, there are three types of motion, Nature has created three types of fibres—straight, oblique and transversal—each endowed with a particular function: "Straight fibres transmit attraction, oblique fibres retention and transversal fibres expulsion."[49] The illustration of the three types of fibres (Fig. 4.5a) was followed by a second image (Fig. 4.5b) which was needed in order to clarify the fibrous fabric of the tunic that, for example, constituted the vena cava. The straight fibres (D), Vesalius wrote, are arranged according to the "longitude" of the organ. The oblique ones (B-C), are circular and intersect in the form of the letter "x"; in this way they are able to retain everything contained in the organ's cavity. Transversal fibres (A) are positioned exactly according to the "latitude" of the body, surrounding it in the manner of the oblique fibres but forming concentric rings.[50]

As for the relationship of spatial connection among the fibres, in his commentary of the first illustration, Vesalius defined it as a triple interweaving (*implexus*) between the different types, exactly as Kepler seems to have understood this in the *Epitome*.

The notion of fibre also plays an important role in the work of Georg Agricola (1494–1555), the German humanist physician whose ideas in the field of metallurgy and mineralogy were known to Kepler. In Book III of the *De re metallica* (1556) Agricola developed a detailed typology of two subterranean structures, veins and fibres, of particular interest from the point of view of mining.[51] Vein was the term more commonly used in ancient natural history—in particular by Pliny and Seneca—to indicate both subterranean channels where minerals are deposited and where waters flow, and their contents. On the basis of traditional observations made by German prospectors and miners, Agricola distinguished a variety of veins based on width, density and direction; veins are separated by *commissurae* (seams or joints of rocks) formed of undifferentiated rock. *Fibrae* (Fig. 4.6a) are simply the

[49] "Ac rectae quidem fibrae attractioni conducunt, obliquae autem retentioni, transversae expulsioni." Andrea Vesalius, *De humani corporis fabrica* (Basileae: Johannes Oporinus, 1543), 358; Claudius Galenus, *De naturalibus facultatibus libri tres, a Thoma Linacro Anglo interprete* (Parisiis: Ex officina Christiani Wecheli, 1537), 61–62. On Galen's reliance on limited mechanical explanations within a non-mechanistic framework see Domenico Bertoloni Meli, *Mechanism. A Visual, Lexical, and Conceptual History* (Pittsburg: Pittsburg University Press, 2019), 12–16, 14, 38–47 (on Vesalius' structural approach).

[50] "Obliquae autem, quae secundum obliquum organum obducuntur. Atque hae circulares sunt: sed quum in duplici sint differentia, invicem contrariantur, ac se mutuo x literae, aut potius duorum invicem implexorum circulorum modo decussatim intersecant [...]." Vesalius, *De humani corporis fabrica*, 358. For a comparison between visual strategies in Renaissance anatomy and astronomy see Martin Kemp, "Temples of the Body and Temples of the Cosmos: Vision and Visualization in the Vesalian and Copernican Revolution," in Brian S. Baigrie, ed., *Picturing Knowledge* (Toronto: University of Toronto Press, 1996), 40–85, 75–82.

[51] Georgius Agricola, *De re metallica Libri XII* (Basileae: Apud Hieron. Frobenium et Nicolaum Episcopium, 1556); English trans. by Herbert Clark Hoover and Lou Henry Hoover (New York: Dover Publications, 1950). Hiro Hirai, *Le concept de semence dans les theories de la matière* (Turnhout: Brepols, 2005), 111–134. Bernardo Cesi, *Mineralogia sive Naturalis Philosophiae Thesauri* (Lugduni: Sumptibus Iacobi et Petri Prost, 1636), 127.

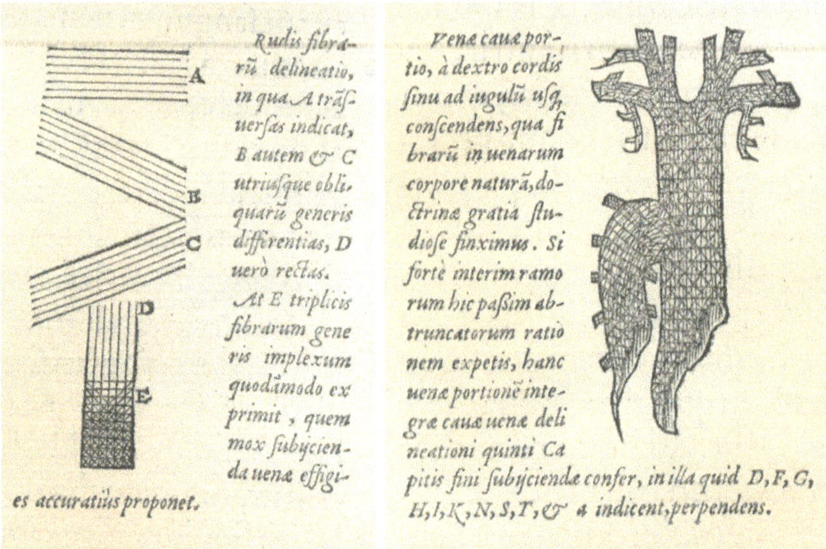

Fig. 4.5 (**a**) and (**b**) A. Vesalius, *De humani corporis fabrica*, 1543, 358. Bayerische Staatsbibliothek München, urn:nbn:de:bvb: v2_bsb11200121_00374

thinnest veins; they have the same characteristics and they connect through various directions with the larger ones.[52]

Agricola pointed out that these structures had another characteristic: their geographical orientation, easily observable both on flatland and in the body of mountains. If, as Fig. 4.6a shows, fibres and veins intertwine at extremely variable angles, the veins in Fig. 4.6b appear to be arranged along recognizable geographic-astronomical directions. In particular, Agricola stated, the deep veins are characterized by the fact that they stretch from east to west, while others are arranged from south to north.[53] It is significant that Agricola follows his description of the orientation of veins with instructions on how to build a special instrument using the principle of the magnetized needle that can help mineral prospectors to identify the direction of outcroppings.

That mineralogical literature was possibly one of the sources inspiring the hypothesis of the arrangement of fibres both in the direction of the earth's axis and parallel to the equator is confirmed by Jesuit Niccolò Cabeo's (1586–1650) interpretation of Kepler's theses in his anti-Gilbertian work *Philosophia magnetica* (1629), which he later repeated in his influential *Commentaria* on Aristotle's

[52] "Fibrae porro, quae sunt venulae, in socias, in dilatata, in incumbentes, transversa autem venam secat: oblique dissidens eam quasi decussat: socia cum ipsa iungitur: dilatata venae dilatatae instar per eandem penetrat: sed fibra dilatata aeque ac profunda solet esse socia." Agricola, *De re metallica*, 50.

[53] Agricola, *De re metallica*, 36–38.

Fig. 4.6 (**a**) and (**b**) G. Agricola, *De re Metallica* (1556), 51, 35. Bayerische Staatsbibliothek München, urn:nbn:de:bvb: v2_bsb10199044_00063; urn:nbn:de:bvb: v2_bsb10199044_00047

Meteorologica (1646).[54] In demonstration of the fact that the terrestrial globe, in all its mass and in every part, possesses a magnetic force that holds it together, ordered and oriented towards the poles, Cabeo argued that when the rivers' erosive action uncovers the internal parts of the mountains (*latera*), we can observe "the veins of the earth, arranged along the entrails of the mountains, stretching parallel to the axis of the world – I mean physically speaking and according to sense – from one pole to the other; cutting the horizontal plane of the horizon at an angle referring to the height of the pole."[55] This arrangement made it possible to compare the Earth's

[54] Niccolò Cabeo, *Philosophia magnetica in qua magnetis natura penitus explicatur* (Prostant Coloniae: Apud Ioannem Kinckium, 1629), 9–12, 57–60. "Terrae venae, ut late dixi in mea philosophia Magnesica, et observavi et video observatum a Keplero et aliis, excurrunt a polo ad polum; et aliquando etiam ab ortu in occasum [...]." Niccolò Cabeo, *Quatuor libros in Meteorologicum Aristotelis commentaria, et quaestiones* (Romae: Typis Haeredum Francisci Corbelletti, 1646), II, 250. On Cabeo see Ugo Baldini, *Legem impone subactis. Studi su filosofia e scienza dei Gesuiti in Italia (1540–1632)*, (Roma: Bulzoni, 1992), 95–98; Craig Martin, *Renaissance Meteorology. Pomponazzi to Descartes* (Baltimore: The Johns Hopkins University Press, 2011), 106–124.

[55] "[...] observatum, inquam, est venas illas terrae, quae excurrunt per montium viscera, excurrere parallelas ad axem Mundi, loquor physice, et ad sensum, a polo ad polum, et incidere ad planum horizontale in angulo referente altitudinem poli". Cabeo, *Philosophia magnetica*, 58.

internal structure with that of an onion. The earth, in fact, "is oriented according to the order of the poles and, so to speak, joined together the same way an onion is joined together; so, at the centre runs the axis and around it are arranged like leaves, first the smaller (parts) and then the increasingly larger and larger ones. In the same way, when we cut an onion we can discern the leaves joined together which arrange themselves around the pole or the axis."[56]

The image Cabeo used to visualize the structure of mountains (Fig. 4.7) enables us to understand that by "veins of the earth" "quibus Mons conflatur" he was referring to what we call strata. It also had the dual purpose of demonstrating the continuity of the strata on the opposite sides of a valley oriented from south to north, and especially of emphasizing its inclination. His careful exploration of the mountains of Lombardy and the Apennines had convinced him that the earth's veins—structures different from metallic veins—invariably ("nunquam errare vidi") emerged out of the soil at an inclination of approximately 45° to the plane of the horizon "ut facit axis polaris".

Further investigations, extended to other regions of the globe, might, he claimed, confirm the hypothesis that the inclination of the strata corresponded exactly to the respective heights of the poles. For the time being, Cabeo remarked, this hypothesis had found confirmation in the work of Kepler: "This I had observed and written when I discovered that the same thing had been observed and published in Germany by the erudite Kepler; I was delighted and until it is not observed elsewhere by

Fig. 4.7 N. Cabeo, *Philosophia magnetica*, 1629, 58. Bayerische Staatsbibliothek München, urn:nbn:de:bvb:v2_bsb10860874_00084

[56] "[…] terram universam dirigi per ordinem ad polos, et quasi coagmentari ea ratione, qua coagmentatur caepe; ita ut per medium excurrat axis, et circa axem circumponatur veluti folia primum minora, deinde maiora, et maiora. Sicuti enim, si scindatur caepa, cernuntur folia illam coegmentatoria excurrere circa polum, seu circa axem." *Ibidem*, 58.

others, I will hold to this conclusion with all my strength."[57] Even in the case when the veins sometimes "ran parallel to the equator falling at right angles to the earth's axis", in his view the proposition remained true that the earth is "joined together according to the orientation of the poles, and composed like crusts that arrange themselves around the axis, indicating a force that orients them in the direction of the poles."[58]

4.5 Towards a New Physics of the Earth

In keeping with the approach he adopted in his major astronomical works, when Kepler was writing about quasi-magnetic fibres, it made perfect sense for him to concentrate on their geometric characteristics and to ingeniously use their heuristic value to explain phenomena like the driving power exerted by the sun on the planets, lunar librations and the different latitude of orbits. It is, however, significant that neither in these works, nor in his correspondence, nor even in those works where he dealt with topics like meteorology and physical geography, did Kepler ever discuss the question of the specific material that composed the fibres. This separation of contexts of discourse, which was largely the result of his focused interest and main area of expertise in astronomy, may be explained if we imagine that Kepler attached a special epistemic importance to the physical-astronomic hypothesis of the fibres.[59] If this perspective is adopted, Kepler's physical astronomy would seem incapable of interacting with disciplines like physical geography or the natural history of the earth. Nevertheless, there is one biographical circumstance that justifies the appropriateness of exploring the possibility of such an interaction. In 1621 Kepler evoked his youthful passion for cosmography prompted by his reading of the *Exotericae Exercitationes* by Julius Caesar Scaliger (1484–1558). The fact that Kepler had developed opinions "on the elements, on the nature of fire, on the origins of springs, on the ebb and flow of the sea, on the shape of the continents" and that he had only abandoned the idea of publishing them after the appearance of *Mysterium Cosmographicum* (1597) would appear to justify our looking for a correlation between the new concepts of the structure of the earth introduced by causal

[57] "Hoc observaveram ego, et scripseram, quando inveni ab erudito Keplero hoc idem in Germania observatum, et vulgatum, ex quo vehemnter gavisus sum, et utinam ab alijs etiam alibi observeretur, sicuti enixe opto." *Ibidem*, 59.

[58] "[...] in quo casu nihilominus vera manet propositio terram esse coagmentatam, et quasi compactam per ordinem ad polos, [...]. Cum igitur terra sit hoc modo coagmentata per ordinem ad polos, et composita quasi quibusdam crustis excurrentibus circa axem, signum est vim habere se dirigendi per ordinem ad polos". *Ibidem*, 59. KGW VII, 87.40–88.5.

[59] On Kepler's epistemology see Nicholas Jardine, *The Birth of History and Philosophy of Science* (Cambridge: Cambridge University Press, 1984); Martens, *Kepler's Philosophy*, cit.; Nicholas Jardine, "Kepler as Transgressor and Amalgamator of Disciplines," *Journal for the History of Astronomy* **40** (2009), 375–380.

astronomy and the ideas coming out of sublunary physics, especially hypotheses concerning subterranean physics.[60]

This is obviously not the place for a comprehensive discussion of Kepler's meteorology and its relationship with the ideas of Aristotle, Tycho and Scaliger or with astrology and celestial physics in general.[61] However, fragments of a post-Aristotelian physics of the earth are scattered throughout his entire work, above all in little explored sections of his main astronomical works and in some "meteorological" writings devoted to the internal structure of the earth. While I shall make no prior attempt to identify in them any theoretical consistency that Kepler himself never managed to achieve, they can be gathered and reordered as evidence of his efforts to lay the groundwork for a new idea of our planet, an aim Kepler seems to have consciously pursued.[62]

It is in Book I of the *Epitome* devoted to spherical astronomy that Kepler provides a number of clues about his idea of the earth's structure. After describing and explaining the circular and continuous arrangement of the waters, a fact long known to sailors through experience and observation and confirmed by Magellan's circumnavigation, Kepler illustrated the regular distribution and circulation of the waters on the continents and around them by drawing on ideas from scholars of hydraulics.[63] The most important theoretical novelty arose in the context of his demonstration that the earth was round. Kepler explained the terraqueous globe's natural tendency to a spherical shape by referring to the new doctrine of gravity which was illustrated at greater length in his *Introduction* to *Astronomia nova*: "We see the body of the earth and of the water endowed with a corporeal force that unites and

[60] KGW VIII, 15.5–11. "Kepler had originally planned to expand his research to the terrestrial regions and to present his findings in a series of 'cosmographical essays'. However, as he explains, this plan failed because […] the exploration of the sublunary world yielded unsatisfactory results." See Kuni Sakamoto, *Julius Caesar Scaliger Renaissance Reformer of Aristotelianism. A Study of His* Exotericae Exercitationes (Leiden-Boston: Brill, 2016), 1–2.

[61] Kepler himself provided a detailed list of his pronouncements on sublunary nature in his *Harmonices mundi*, Book IV, cap. VII. KGW VI, 264. Sheila J. Rabin, "Kepler's Attitude Toward Pico and the Anti-Astrology Polemic," *Renaissance Quarterly* **50**/3 (1997), 750–770; Miguel Angel Granada, "Kepler v. Roeslin on the Interpretation of Kepler's *Nova*: (1) 1604-1606," *Journal for the History of Astronomy*, **36** (2005), 299–319; Boner, "Kepler's Living Cosmology, cit., 33–37; Massimo Bucciantini, *Galileo e Keplero. Filosofia, cosmologia e teologia nell'Età della Controriforma* (Turin: Einaudi, 2007); Boner, "Life in the Liquid Fields", cit.; Miguel Angel Granada, "Novelties in the Heavens Between 1572 and 1604 and Kepler's Unified View of Nature," *Journal for the History of Astronomy* **40** (2009), 393–402, Adam J. Mosley, "Heaven and Earth in the Late-Sixteenth Century: Tycho and Kepler on the Sub- and Supra Lunary," in Miguel Angel Granada and Edouard Mehl, eds., *Nouveau ciel, nouvelle terre* (Paris: Les Belles Lettres, 2009), 143–156; Miguel Angel Granada, "After the Nova of 1604," *Journal for the History of Astronomy* **62** (2011), 353–390; Boner, *Kepler's Cosmological Synthesis,* cit.

[62] "Cum autem celeberrimis nonnullis Philosophiae et Medicinae Professoribus visus sim novam philosophiam condere, eamque verissimam; omni cura fovenda et educanda est tenella plantula, ut sunt nova omnia, ut radices agat in animis philosophantium;" KGW VI, 264.26–30.

[63] KGW VII, 32–33.

attracts any kind of body, a force which is usually called gravity."[64] To explain the arrangement of the waters in relation to the solid parts and to account for their cohesion, Kepler raised the question of the possible endogenous structure of the globe, a question which is almost entirely absent from cosmographic manuals like those by Gemma Frisius and Paulus Van Merle:[65] "Since, therefore, this entire globe of the earth is everywhere surrounded by waters, a fluid material that does not confine itself inside walls, it is not impossible that the inside of the Earth is accessible from all sides by gigantic conduits; for that reason the earth could be similar to a perforated vessel, made up of continents that are like the pieces of a mosaic, full of water on the inside.[66] This is a crucial passage, but its conciseness raises a series of difficult problems of interpretation.

With *olla pertusa* Kepler may have had in mind a globular, perforated jar or pot, mentioned by *scriptores rei rusticae* such as Cato and by Pliny in his *Naturalis Historia*.[67] Such containers were widely employed for agricultural purposes like transplanting plants, sowing, layering, or transporting trees over long distances.[68] He may also have had information about the large use of perforated pots (*ollis perforatis*) (Fig. 4.8) in melting procedures as illustrated in Book XII of Georg Agricola's *De re metallica*.[69]

If therefore the terrestrial globe resembles a terracotta recipient made up of pieces that are joined together, that is to say, the continents, these pieces should contain the seas in their surface cavities, alongside the dry lands. In fact, if we consider the already known extent of the oceans, the openings envisaged in the recipient

---

[64] "Videmus corporis Terrae et Aquae inesse vim corpoream, vniendi sibi corpora quaecunquae, atrahendique, quam vim vulgò gravitatem dicunt." KGW VII, 34. According to the *Astronomia Nova*, "Gravitas est affectio corporea, mutua inter cognata corpora, ad unitionem seu conjunctionem (quo rerum ordine est facultas Magnetica) ut multo magis terra trahat lapidem, quam lapis petit Terram." KGW III, 25.21–23.

[65] Petrus Apianus and Gemma Frisius, *Cosmographia, sive Descriptio universi Orbis* (Antverpiae: apud Ioan. Bellerum, 1584); Paullo Merula, *Cosmographia generalis libri tres* (Amstelodami: Officina Plantiniana Raphelengis, 1605). The latter is a work by the Dutch jurist, historian and geographer Paul van Merle (1558–1607), whose vast classical erudition and extensive knowledge made this an extraordinary encyclopedia of geo-cosmography available for contemporaries. Both works are mentioned in KGW XV, 388.

[66] "Cùm ergò totus hic globus Telluris vndique circumfluant aquis, materiâ fluidâ, seque ipsam ad latera non terminante; nec sit improbabile, Terram etiam interius tubis ingentibus esse commeabilem undique, adeòque Terra forte sit instar ollae pertusae, constans Continentibus ceu testis, intus aquâ refertis: [...]." KGW VII, 34.

[67] M. Porci Catoni, *Quae extant, ab Ausonio Popma Frisio collecta et restituta* (Lugduni Batavorum: Ex Officina Plantiniana, 1590), 41 (par. 52); Gaius Plinius Secundus, *Historiae mundi libri XXXVII* (Coloniae Allobrogorum: Apud Petri et Ian Chovet, 1615), 347 (*ollas perforatas*).

[68] Wilhelmina F. Jashemsky et al., eds., *Gardens of the Roman Empire* (Cambridge: Cambridge University Press, 2018), 552.

[69] "The sulphurous mixture [...] may be heated in similar pots, but with perforated bottoms." Agricola, *De re metallica* (English trans. n. 47) 580. In a letter of 1608, the Paracelsian alchemist and physician Joachim Tanckius (1557–1609) suggested to K. to compare the functioning of the *ventriculum* with an *olla*. KGW XVI, 211–212.

Fig. 4.8 G. Agricola, *De re Metallica* (1556), 467. Bayerische Staatsbibliothek München, urn:nbn:de:bvb: b v2_bsb11193455_00483

model should not correspond to the seas but to the "gigantic conduits" that bring water to the depths. These, in turn, can only be located at the bottom of the ocean basins. The waters, thus, circulate both above and around the continents (rivers, oceans), inside them (conduits on the crust and, as we shall see, caverns in mountains and springs) and probably also underneath them. So far, it is not clear why there should be a connection between surface circulation and deep circulation, but Kepler deals with the problem later. Finally, what substances are to be found in the core and what phenomena take place there is not made clear. In any case, as he repeats in Book I Part V, "the earth is spherical, full inside and narrow."[70] This would appear to exclude the hypothesis that the central core is the site of something akin to the great abyss mentioned in the Bible, or Tartarus described in Plato's *Phaedo* and in other texts that were well known to Kepler.[71] What is certain thus far is that

[70] "[…] terra enim globosa est et intus plena et angusta, speciemque quandam cordis habet; caelum amplissimu et excavatum, per quod hanc facultatem diffundi oportet circulariter; cum e contrario possit anima aliqua in Terrae centro poni radicata, […] indeque velut è corde humano per speciem sui continuata excurrere in omnes globi suis partes;" KGW VII, 83.34–39.

[71] KGW VII, 92. Plato's *Phaedo* is mentioned in *Harmonice mundi*, KGW VI, 269. 21. On the geographical myth in *Phaedo* [111c-113c] see Otto Baensch, "Die Schilderung der Unterwelt in Platons Phaidon," *Archiv für Geschichte der Philosophie* **16**, (1903), 189–203; John S. Morrison, "The Shape of the Earth in Plato's Phaedo", *Phronesis* **4**, (1959), 101–119. Further speculations on the subterranean regions were put forward in the Pseudo-platonic dialogue *Axiochus* edited by the humanist Rodolphus Agricola (1444–1485): Plato, *Opera a Marsilio Ficino traducta, adjectis ad ejus vitae et operum enarrationem Axiocho ab Rodulpho Agricola; et Alcyone ab Augustino Datho*

Kepler envisioned a discontinuity between the crust structured in the manner of a mosaic and a region underneath it which is not empty, a discontinuity that might be in line with the hypothesis he put forward in the *Astronomia nova*, that the quasi-magnetic fibres constitute separate shells.

Further important suggestions may be found in the *Epitome* and in the *Harmonice mundi*, but in order to understand how Kepler conceived the structure and dynamics of the underground we first need to look at an important passage in *De stella nova* (1606) in which Kepler stressed the indispensable role of heat and moisture. Here Kepler had speculated that in the body of the earth there exist twin faculties (*gemina facultas*): "[…] one that attracts marine waters into the secret places of concoction, the other that expels the concocted vapors like a kind of sweat […]."[72] According to the (pseudo-aristotelian) Book IV of the *Meteorologica*, concoction is a maturing process "initiated by the thing's own heat, even though external aids may contribute to it." In some cases the result of this process can be "a thing's nature, in the sense of its form and essence." In others it is "the realization of some latent form, as when moisture takes on a certain quality and quantity when cooked or boiled or rotted or otherwise heated."[73] In overall agreement with this view, Kepler regarded this process as part and parcel of a natural capacity pervading the sublunar and supralunar spheres to continually regenerate the decaying matter through the establishment of new forms.[74]

In Book I Part V of the *Epitome*, these and other processes of transformation are considered to be certain signs of the presence of a soul in the "heart" of the earth.[75] The first trace (*vestigium*) of a soul was the existence of a "perpetual and sensible"

---

traslatis (Paris: Iodoco Badio, 1522); Rodophus Agricola, *Nonnulla opuscula* (Basileae: Andream Gratandum, 1518), 6–7.

[72] "In Telluris globo gemina est facultas (quod meteora attinet) altera attrahendi aquas marinas in occultas concotionis sedes, altera expellendi vapores concoctos per quendam quasi sudorem;" KGW I, 317.13–15. See Boner, "Life in the Liquid Fields," cit., 283–284. In *Astronomia nova* he mentions the simple property of the waters of rivers to flow down towards the centre of the earth: "Sic in flumine simplex aquae proprietatis est ad centrum terrae descendere." KGW III, 255.11.

[73] Aristotle, *Meteorologica*, edited by H.D.P. Lee (Cambridge Mass.: Harvard University Press, 1978), 297–301, 299. In an autobiographical sketch dated 1597 Kepler recorded his close agreement with the content of this section of the book: "In libro quarto Meteororum inhaesit praecipue disputando". KGW XIX, 329. Malcom Wilson, *Structure and Method in Aristotle's* Meteorologica. *A More Disorderly Nature* (Cambridge: Cambridge University Press, 2013), 274. For a contemporary, non-Aristotelian view of concoction see Francis Bacon, *Sylva Silvarum* (London: John Haviland for William Lee, 1639), 179.

[74] On Kepler's combination of a Platonic concept of form with the material and efficient causes see Norma E. Emerton, *The Scientific Reinterpretation of Form* (Ithaca and London: Cornell University Press, 1983), 40–41, 168–169.

[75] Kepler had already alluded to this in *Astronomia. Pars optica* (1604), stressing the existence of a diffuse subterranean heat which was obvious from the Aethnean fires and from thermal waters. KGW II, 34–36. K. often discussed the role of subterranean heat in analogy with the inner heat of the human body theorized by Fernel. See for example KGW XVI, 91–92. The presence in the Earth of an "architectonica naturalis facultas" is discussed in *De stella nova* (1606). KGW I, 267–270.

subterranean heat diffused in the body of the earth.[76] In the second place, specific effects of the soul had to be considered processes such as the generation of metals, the exudation (transpiration) of internal humidity which was the origin of the rivers, perpetual humid and arid exhalations that are the cause of various meteors. Thirdly, indications of the earth having a soul were the properties of hot and inflammable substances extracted from its viscera like sulphur and marcasite. These substances, in fact, can be converted into light which is kindred to the soul. As Kepler explained, all of this "may be observed only in the outer crust of the earth: what surprising things are stored in its vast internal bosom of one thousand seven hundred miles?"[77] In this crucial passage, Kepler continued to leave unanswered the question of the nature and dynamics of the core, while repeating the hypothesis of a discontinuity between outer crust and nucleus. Of particular importance is the fact that the passage contains a precise measure of the dimensions of the "vast internal bosom", a figure whose derivation is unclear but from which, in turn, it was possible to estimate the quantity of matter in the nucleus.[78]

In his *Harmonice mundi*, Kepler again took up the discussion of the various phenomena "heating and agitating the body of the earth", here as part of a more comprehensive discussion of the relationships between sublunary physics and astral influences.[79] Such phenomena, he confirmed, can be explained as the effect of a soul "diffused through the whole body of the earth", but "rooted in some particular part of it exactly as the human soul is in the heart" and emerging from it "as if from a hearth, fount or inner sanctum."[80] This passage could well be interpreted as a precise

[76] The very essence of the sublunary soul "seems plainly to be a sort of flame (*flamma quaedam*) [...]", of a kind without which "a sensible subterranean heat would never persist actively in bare matter. " Johannes Kepler, *The Harmony of the World*, trans. E.J. Ziton, A.M. Duncan, J.V. Field, (Philadephia: The American Philosophical Society, 1997), 366; KGW VI, 271.9–11.

[77] "Et haec spectantur tantummodo in exteriori cortice terrae: quanto mirabiliora censes condi spaciosissimo ejus sino interiore, per mille septingenta milliaria?" KGW VII, 92.17–19.

[78] It also indicates a significant departure from the hypothesis of a high density magnetic nucleus suggested in Gilbert's *De magnete*, Book I, cap. XVII: "But the strong foundation of the Globe, its great mass, is that terrene body, far surpassing in quantity the whole aggregate of fluids and waters whether in combination with earth or fire [...]; and this mass makes up most of the globe, constituting nearly its whole interior framework [...]. But the true earth-matter we hold to be a solid body homogeneous with the globe, firmly coherent [...]". William Gilbert, *De Magnete* (Londini: Excudebat Petrus Short, 1600) 40, 42. Jacqueline Biro, *On Earth as in Heaven. Cosmography and the Shape of the Earth from Copernicus to Descartes* (Saarbrucken: VDM Verlag Dr. Muller, 2009), 65–66.

[79] Kepler believed that the earth soul "sets the weather in motion (*movet tempestates*) at the appearance of aspects [...]." Kepler, *The Harmony of the World*, 322–323, 362–363; Kepler dealt with the relationship between astral influences, the Earth's soul and meteorology in *De fundamentis astrologiae certioribus* (1602) and he returned to it again in *Apologia pro opera Harmonices mundi* (1622) in answer to criticism by Robert Fludd. See KGW IV, 12–35 e KGW VI, 409–420. J.V. Field, "A Lutheran Astrologer: Johannes Kepler," *Archive for History of Exact Sciences* **31**/4, (1984), 189–272.

[80] "[...] ex quo ceu foco, fonte, vel penetrali, per speciem suam exit in circumfusum Terris Oceanum et superfusum utrisque Aerem." Kepler, *The Harmony of the World*, 322; KGW VI, 237.11–13.

indication that the earth's soul is in a central location, while the double reference to the human heart and to the hearth confirm the idea of a direct relationship between the activity of the soul and the generation of subterranean heat. When "the wandering stars are in a harmonic configuration with their rays at the earth", he explains, the earth soul perceives the harmonic proportions and "has much power of heating and agitating the body of the earth and the underground workshops in its mountainous regions." Therefore, the exhalations of vapor and clouds producing the great majority of meteorological phenomena take place principally in those areas and imply the existence of water deposits in the caverns of the mountains.[81]

In the *Harmonice*'s epilogue devoted to *De natura sublunari*, Kepler provided more details regarding the *fabrica* that sustains the planet's many vital functions. The interior of the earth's crust must be traversed by channels through which the water of the oceans flows down to the depths "just as other animate beings consume food and drinks" in keeping with what he had said earlier about the processes of concoction. In addition, in the crust wide caverns are to be found in which the exhalations are collected and where minerals are generated.[82] Besides the function of digestion, these channels are necessary to account for the regular absorption of the waters which flow down from the mountains to the seas, given that "the sea never overflows from the pouring in of so many rivers."[83] It would therefore appear to be confirmed that there is a communication between the waters that surround the globe and the more internal waters, a conspicuous flow made possible by a network of conduits which from the ocean basins traverse the outer crust and reach the depths of the terrestrial body.[84]

This model was the culmination of a gradual and irreversible break with the physics of *De caelo* (Book II, XIV) which held that the earth is a compact spherical body since "every portion of earth has weight until it reaches the centre, and the jostling of parts greater and smaller would bring about not a waved surface, but

---

[81] *Ibidem* 323; KGW VI, 237.22–30.

[82] Kepler's relationship with alchemy and mineralogy is examined in Miguel Angel Granada, "After the Nova of 1604: Roeslin and Kepler's Discussion on the Significance of the Celestial Novelties (1607-1613)," *Journal for the History of Astronomy* **42** (2011), 353–390; Jonathan Regier, "Stars, Crystals and Courts: Johannes Kepler and Anselmus Boëtius de Boodt," in Patrick Boner, ed., *Kepler's New Star (1604)* (Leiden-Boston: Brill, 2020), 107–128; Miguel Angel Granada, "Johannes Kepler. The Sun as the Heart of the World," *Journal for the History of Astronomy* **53**/2 (2022), 133–140.

[83] "Oportet igitur, ut sicut Animantia reliqua cibum potumque hauriunt, sic etiam Tellus, canalibus certis, aliquid materiae trahat: ex qua illa tam multiplicia excoquat, quia ex nihilo nihil fit; trahit autem sorbetque aquam marina; quae causa est cur aeternis tot fluminum infusionibus mare nunquam redundet." *Ibidem* 364; KGW VI, 269.7–11.

[84] This could be one of the sources of inspiration for Athanasius Kircher's hypothesis of the *hydrophilacia* suggested in his *Mundus Subterraneus* (1665). See Nicoletta Morello, "Nel corpo della Terra. Il geocosmo di Athanasius Kircher," in Eugenio Lo Sardo, ed., *Athanasius Kircher. Il museo del mondo* (Roma: De Luca, 2001), pp. 178–196; Luca Ciancio, "*Immoderatus fervor ad intra coërcendus*. Reactions to Athanasius Kircher's central fire in Jesuit Science and Imagination," *Nuncius*, **33**/3, 464–504.

rather compression and convergence of part and part until the centre is reached."[85] But it also represented a departure from the doctrine of the *Meteorologica* (Book I, IX-XIV and Book II, I-IV) according to which the circulation of waters only concerned the outermost crust and is associated with a slow and perennial migration of the seas over the surface of the globe. Kepler's distancing from Aristotle was also evident in his interpretation of the traditional doctrine of the two exhalations, humid and dry. According to Kepler—who reiterated the theory of the humanist Rudolph Agricola (and "indeed of all farmers")—their origin and circulation must take place "not solely on the surface of the earth, but also inside it in subterranean caverns, and in the passages of the mountains."[86] And it was not primarily due to the heating produced by the sun's rays, but by the activity of the terrestrial soul. Also in this case, if all materials of the weather arise from the depth of the earth, its most interior body is far from being a mass of compact rock.

Other dynamics occurring under the surface complete the picture of an earth in continuous transformation in all of its parts. Certain subterranean regions are modified following what are veritable diseases of the earth that result from a malfunctioning of the abovementioned digestive faculty: "Thus I was right to advise in my book *On the New Star*, page 173, that all the faculties of digestion, the attractive, the retentive and the expulsive should be looked for in it, seeing that those diseases are states of them."[87] Finally, there is the possibility that, like in other animals, the earth's soul governs a sort of respiration that can bring water and air from the surface to the depths and from the depths back to the surface, perfectly consonant with the movement of the tides caused by the magnetic attraction of the moon. This hypothesis might find a sympathetic audience among the philosophers, Kepler remarked, "especially if there were in addition some indication of parts in the depths of the earth which are flexible, which would take the role of lungs or gills."[88]

---

[85] Aristotle, *On the Heavens*, in Jonathan Barnes, ed., *The Complete Works of Aristotle. The Revised Oxford Translation* (Princeton: Princeton University Press, 1995), 2 vols., I, 488 (297 a, 8–11). "So what is heaviest and coldest, that is, earth and water, separates off at the centre or round the centre [...]." Aristotle, *Meteorologica*, cit., 19 (340 b, 19–21). According to the Pseudo-aristotelian *De mundo*, "Next to it [the aerial element] the whole earth has been formed, firmly fixed in the lower position, at the midmost centre of the Universe, closely compacted, immovable and unshakable." Aristotle, *On the Universe*, in Jonathan Barnes, ed., *The Complete Works of Aristotle*, 628 (392 b, 32–34).

[86] "[...] intus etiam in cavernis subterraneis, in meatibus montium". *Ibidem*, 363; KGW VI, 268.21–22. This may be a reference to Rodolphus Agricola, *Elucubrationes aliquot lectu dignissimae* (Coloniae: Per Ioannem Gymnicum, 1539), 294–295.

[87] Kepler, *The Harmony of the World*, 365; KGW VI, 270.13. The text contains indirect references to K.'s correspondence with David Fabricius on various earthly phenomena such as earthquakes. *Ibidem* 366; KGW VI, 270.22–29.

[88] *Ibidem*, 366; KGW VI, 270–27–32. This hypothesis was criticised by Libert Froidmont (1587–1653), one of the leading meteorologists of the first half of the seventeenth century. Significantly, Froidmont interpreted Kepler's theories as closely related to those of Georg Agricola who in his view had stressed too much the role of subterranean fires: Libert Froidmont, *Meteorologicorum libri sex* (Antverpiae: Ex officina Plantiniana, 1627), 158, 224. Martin, *Renaissance Meteorology*, cit., 100–104.

Out of this brief account of Kepler's subterranean meteorology, whose foundations were in a magnetic conception of gravity and in the idea of a centrally located earth soul that creates new forms, we can detect the emergence of a new terrestrial physics, one that was certainly fragmentary but whose general features were not incompatible with the products of physical-astronomical thinking. In working out his views, Kepler drew on the theoretical resources developed by natural philosophers like Gilbert, physicians and anatomists like Fernel, Jessen and Vesalius, naturalists like Georg Agricola as well as humanists, travelers and geographers. This theoretical repertoire, much of which taken from the works of the *novatores*, not only allowed him to cast off a good deal of the cumbersome burden of Aristotelian physics, but also to set the conditions needed to imagine the possibility of bringing the new sublunary physics into harmony with the quasi-magnetic structures he introduced in his astronomical works.

4.6 Conclusion

In his *Praefatio* to the *Tabulae Rudolphinae*, to illustrate the historical relationship between astrology and astronomy, Kepler used the metaphor of the concentrically arranged growth rings of plants visible in the section of the trunk: "[...] just as the years reveal themselves in the fibres of trees, so in the entire composition of the most divine art [astronomy] there appear to be some clues as to its origins [...]."[89] For Kepler, thus, the temporal meaning of the structure of vegetable fibres was clear. He was unable to say the same for the fibres and filaments he hypothesized for the body of the earth. While he had considered the possibility that the terrestrial globe had not appeared fully created from the beginning of time, Kepler could not identify in the planet's structures any "clues as to its origin".[90] Indeed, the problem of the planet's origin had very little to do with his main aim which was to construct a heliocentric physical astronomy that would be an alternative to any explanation of celestial motion in terms of motive intelligences inside each planet. With this aim in mind, the traditional chronology and patient recording of celestial events were enough to perform the calculations that were required.[91] In essence, Kepler was an astronomer and chronologist like Monsignor Francesco Bianchini but, unlike

[89] "Verumtamen, ut in arborum fibris anni, sic in tota divinissimae artis compositione lineamenta quaedam apparent ortus hujus; [...]" KGW X, 36.

[90] "Equidem Terra, cùm primùm à Deo creata, aut et ipsa fluida quaed am et mollis massa fuit, posteriusque induruit; et tunc eadem vera sunt etiam de terrâ, quae de aquis sunt dicta; aut creata est in sua soliditate et duritie primaevâ; et tunc ne sic quidem verisimile est, aliam illi figuram inditam, quàm acceptura fuit à seipsa, si mollis initio fuisse." KGW VII, 34–35.

[91] On Kepler's approach to chronology see Anthony Grafton, "Humanism and Science in Rudolphine Prague: Kepler in Context," in Id., *Defenders of the Text. The Tradition of Scholarship in an Age of Science, 1450–1800* (Cambridge Mass.: Harvard University Press, 1991), 178–203; Anthony Grafton, "Chronology, Controversy, and Community in the Republic of Letters," in Id., *Worlds made by Words. Scholarship and Community in the Modern West* (Cambridge Mass.:

Bianchini, he did not possess the archeologist's particular sensitivity for the historicity attached to any material record.

Considered together, Kepler's ideas on subterranean physics never coalesced into a coherent theory of the earth. The array of texts and conjectures I have attempted to understand and correlate resembles more an ongoing worksite, a variety of exploratory hypothesis not sufficiently coordinated and corroborated. Having said that, Kepler's suggestions were certainly read and received, as may be seen in the case of Cabeo, and they contributed to spreading a new type of approach to the mysterious depths of the earth which had hitherto been the almost exclusive domain of Paracelsian chemists. Notwithstanding his theoretical waverings and the interpretative doubts he raises, Kepler's contribution to the construction of a new physics of the earth, "a tender little plant to be cherished and reared with every care, as must all its innovations",[92] remains an important one.

In his account of the likely causes of the planetary motion we can identify at least two significant contributions on the pathway that led to a post-Aristotelian understanding of the natural world. First, Kepler formulated a physical concept of the fibre as a basic structural unit of the terrestrial body and he successfully showed that upon this concept it was possible to build a physical theory which explained the movement of the planets by the sun's motive force and which agreed with the data observable at the time.[93] Interwoven and superimposed fibres, while being very different from Nicolaus Steno's strata, were material structures endowed with specific extensional characteristics. The introduction of this notion began the process of substituting qualitative concepts like "vein" or "viscera" belonging to vitalistic *physiologia*.[94] While it is clear that the metaphors used by Kepler in the field of

---

Harvard University Press, 2009), 114–136; Daniel Rosenberg and Anthony Grafton, *Cartographies of Time* (New York: Princeton Architectural Press, 2010), 81–86.

[92] Kepler, *The Harmony of the World*, 358. KGW VI, 264.29–30. See also n. 62.

[93] On the reception of Kepler's hypothesis, it should be noted that information from the mining sector was used by some to refute it. In 1631, on the basis of a close comparison with Georg Agricola's observations, Jean Baptiste Morin concluded that the direction of the fibres did not correspond to that imagined by Kepler and therefore the planetary structure he had suggested was false and the fibres fictitious. See Ioanne Baptista Morino, *Famosi et antiqui problematis de telluris motu, vel quiete* (Parisiis: Apud Authorem iuxta Pontem novum, 1631), 42–43, 47–48. Even more mocking was the opinion expressed by Libert Froidmond: "Interrogent deinde metallarios Germaniae, et Leodienses carbonarios, an viscera terrae effodiendo, nervos, aut musculos, aut ullas omnino fibras deprehenderint, quibus animalia motum solem exequi. Negabunt, imo effuse ridebunt, scio. Et nos faciamus." Libert Froidmond, *Ant-Aristarcus sive Orbis-terrae immobilis* (Antverpiae: Ex Officina Plantiniana Balthasaris Moreti, 1631), 79. Although the fibres hypothesis could be used in an anti-Copernican vein, it received some attention from Jesuit astronomers like Athanasius Kircher and Giambattista Riccioli: Athanasius Kircher, *Magnes sive de arte magnetica libri tre* (Romae: Ex Typographia Ludovici Grignani, 1641), 484–485; Giambattista Riccioli, *Almagestum novum* (Bononia: Ex typographia Haeredis Victorij Benatij, 1651), 275–278. For Gassendi's more nuanced reaction see Kuni Sakamoto, "The German Hercules's Heir: Pierre Gassendi's Reception of Keplerian Ideas," *Journal of the History of Ideas* **70**/1 (2009), 69–91, 74.

[94] To grasp the novelty of this it is worth comparing Kepler's structural vision with the traditional anatomo-physiological perspective adopted by Jean-Baptiste Morin in his *Nova mundi sublunaris*

subterranean meteorology such as shell, yolk, *olla pertusa*, certainly cannot be equated with a consistent vision of the terrestrial *fabrica*, they undeniably introduced important notions of a mechanical-structural character.[95] Secondly, the corpus of illustrations of the terrestrial body and fibres elaborated by Kepler was a significant innovation in the iconography of the new physics. These images indicate the spread of a new conception of how natural structures were spatially constituted, a conception which could draw its visual and theoretical resources from the domains of anatomy, mineralogy, architecture and technology.[96] New concepts, metaphors and images, supported the attempt—corroborated by mathematics and observation—to apply physical-geometric coordinates to the arrangement of subterranean structures so that hypotheses could be formulated that restored the unity between the terrestrial and celestial worlds.

Kepler's theories left many unanswered questions and unexplored issues. Firstly, he never correlated the fibres' function as structures containing and distributing the magnetic force with any specific physico-chemical composition. Secondly, he never attempted to understand these structures as the product of generative processes taking place on a cosmological scale over the course of time. Fibres, minerals and zoomorphic stones existing within rocks remained for him the product of self-explanatory seminal forces. Finally, Kepler failed entirely to recognize pervasive geomorphological processes like sedimentation and erosion. While the earth appeared to be composed of structural components arranged architectonically in the manner of fibres in a tree trunk, there was as yet no one able to interpret the origin of this architecture as being the result of purely physical processes. It was only with the emergence of a new mechanical and corpuscular philosophy in the central decades of the seventeenth century that all the conceptual tools became available for a dynamic and historical interpretation of the earth's structure.[97]

---

anatomia (Parisiis: Apud Nicolaum Du Fossé, 1619), 116–144. On ancient sources see Liba Taub, "Physiological Analogies and Metaphors in Explanations of the Earth and the Cosmos," in Manfred Horstmanshoff, Helen King, Claus Zittel, eds., *Blood, Sweat and Tears. The Changing Concepts of Physiology in Early Modern Europe* (Leiden-Boston: Brill, 2012), 41–63.

[95] Kepler's famous 1605 declaration to Herwart von Hohenburg is worth mentioning here: "Scopus meus hic est, ut Caelestem machinam dicam non esse instar divinj animalis, sed instar horologij [...], ut in qua pene omnis motuum varietas ab una simplicissima vi magnetica corporalj, utj in horologio motus omnes a simplicissimo pondere. Et docet hanc rationem physicam sub numeros et geometriam vocare [...]." KGW XV, 146.57–62.

[96] On the transformation of European 'spatialities' see Denis Cosgrove, "Ptolemy and Vitruvius: Spatial Representations in the Sixteenth-century Texts and Commentaries," in Antoine Picon and Alessandra Ponte, eds., *Architecture and the Sciences. Exchanging Metaphors* (Princeton: Princeton Architectural Press, 1999), 1–29; Martin Kemp, *Seen/Unseen. Art, Science, and Intuition form Leonardo to Hubble Telescope* (Oxford: Oxford University Press, 2006), 35–40. On Kepler as a reader of Daniele Barbaro see Nicholas Jardine, "The Place of Astronomy in Early-modern Culture," *Journal for the History of Astronomy* **29**/1 (1998), 49–62, 53–54.

[97] A specific contribution to realize the sedimentary nature of strata came from the Galilean 'science of waters': hydraulics, hydrostatics, origin of springs. One of the first images of a stratigraphical sequence appeared in Bernardino Ramazzini, *De fontium mutinensium admiranda scaturigine*.

Luca Ciancio, Associate Professor of History of Science, University of Verona, Italy. After devoting his early career to the history of geology and natural history during the Enlightenment, in recent years he has focused his attention to the emergence, in the early modern era, of a new meteorology of the underground regions.

Tractatus physico-hydrostaticus (Mutinae: Typis Haeredum Suliani Impressorum Ducalium, 1691), 20.

Chapter 5
A Jesuit Journey from Heaven to Earth

Ivana Gambaro

Abstract As early as 1628, Nicolas-Claude Fabri de Peiresc recognized the importance of developing an extensive network of collaborators to collect astronomical observations and improve geographical knowledge. He also understood the need to involve the leaders of the Society of Jesus in this project. Later, Athanasius Kircher SJ took up the challenge and developed an ambitious research program—the *Consilium Geographicum*—based on the extraordinary network of Jesuit colleges spread throughout the world; this program was subsequently continued by Giovan Battista Riccioli SJ. By reconstructing the path that led Riccioli from celestial observations to the determinations of longitude and latitude presented in his *Geographia et hydrographia reformatae*, this study analyzes data related to lunar eclipses, corroborated by archival material, to illuminate the concrete research practices employed by Riccioli—practices that at times diverged from the Aristotelian tradition, thereby contradicting the methodological claims present in his printed works.

Keywords Athanasius Kircher · Giovan Battista Riccioli · Seventeenth century astronomy · Seventeenth century geography

---

The subjects dealt with in this contribution were partly addressed by the author at the *Tenth Conference of the European Society for the History of Science* (ESHS) held in Brussels (Belgium), from 7 to 10 September 2022, within the Symposium *Dissemination and Exchange of Scientific Knowledge in Early Modern Europe: Interactions among Scientific Communities and Religious Milieus*.

I. Gambaro (✉)
Department of Antiquities, Philosophy and History, University of Genoa, Genoa, Italy
e-mail: ivana.gambaro@unige.it

S. Gattei (ed.), *The Incomparable Historian*, Archimedes 75,
https://doi.org/10.1007/978-3-032-25678-2_5

5.1 A French *Savant* and a German Polymath

On 4 March 1628, in a Europe troubled by the Thirty Years' War, from the peaceful city of Aix-en-Provence, a French *savant*, a Prince of the Republic of Letters, outlined his ambitious project to his correspondents: the brothers Pierre and Jacques Dupuy. Pierre was a lawyer in the French Parliament, while Jacques was the prior of Saint Sauveur-lès Bray:

> I must convince the Cardinal [Francesco Barberini, nephew of the pope Urban VIII], via the Generals of the Jesuits and other religious orders versed in mathematics in various places in the world, to instruct them to make observations [of the next solar eclipse] wherever it appears, and to prepare for this end a rulebook for the order that they must follow along with the necessary instruments [...] in order to establish the true longitudes of the most notable places where they are, at least for those in Europe, and especially those who live in the Azores and the Canary Islands, Constantinople, Jerusalem, Alexandria and other places which are of note, and famous in Antiquity.[1, 2]

Nicolas-Claude Fabri, Seigneur de Peiresc (1580–1637), was a *savant*, lawyer, astronomer, scholar and was also known for his tireless letter writing. Like Marin Mersenne (1588–1648), he operated at the center of a dense network of correspondents, embodying a kind of *éminence grise* within the Republic of Letters. Although he did not publish extensively during his lifetime, Peiresc, from his home in Aix, utilized this network and his rich collection of books to support other scholars by offering gifts or loans of materials, financial assistance, or simply advice. As early as the 1620s, he had conceived the idea of long-distance collaborations, recognizing the potential benefits for astronomy and geography. His aim was to enhance geographical knowledge through systematic observation of astronomical events and thus, with more accurate cartography, to ensure safer navigation and,

---

[1] English translation from Alexandra Bamji, Geert H. Janssen, Mary Laven, *The Ashgate research companion to the Counter-Reformation* (Farnham UK: Ashgate, 2013) 327. Original in French: "*Et par ce que dans deux ans ou environ il y doibt avoir une ecclypse de soleil, il faut que je mette en teste au cardinal* [Francesco Barberini, neveu du pape Urban VIII] *de faire commander, par les generaulx des Jesuistes, et aultres ordres qui ont des religieux versez en mathematique en divers lieux du monde, de leur commander d'en faire des observations partout où elle apparoistrà, et de faire dresser exprez pour cet effect ung petit formulaire de l'ordre qu'il y fault tenir et des instruments qui y sont necessaires à peu prez, pour y faire ce qui se pourrà, afin de regler les vrayes longitudes des plus celebres lieux de leur residance, au moings pour ceux de l'Europe, et notamment pour ceux qui resident aux Azores et Canaries, et pour Constantinople, Hierusalem, Alexandrie et aultres lieux plus notables et plus celebres dans l'antiquité*". On Peiresc see Peter N. Miller, *Peiresc's Europe. Learning and Virtue in the Seventeenth Century* (New Haven: Yale University Press, 2000) and Peter N. Miller, *Peiresc's Orient. Antiquarianism as Cultural History in the Seventeenth Century* (Farnham, UK: Ashgate, 2012).

[2] Letter of Peiresc to the brothers Du Puy, n. xcix, Aix, 4 mars 1628 in *Lettres de Peiresc aux frères Dupuy* publiées par Philippe Tamizey de Larroque, tome 1er (décembre 1617-décembre 1628), Paris: Imprimerie Nationale, 1888, 545–555, quotation on p. 549. See also Paula Findlen (ed.). *Empires of Knowledge: Scientific Networks in the Early Modern World* (London, New York: Routledge, Taylor & Francis Group, 2019) 80–86.

hypothetically, to develop an effective method for calculating longitude. He began by requesting his correspondents to collect simultaneous observations of celestial events expected in various locations, and soon he realized that the vast community of missionary scholars could accomplish this task more efficiently.

In October 1632, he encountered Father Athanasius Kircher (1602–1680) in Aix. Kircher had fled to Avignon from Würzburg, which had fallen to the Protestant armies of Gustav II Adolf of Sweden the previous October. The German Jesuit captivated Peiresc with his encyclopedic knowledge and extensive linguistic skills, which included, according to him, the ability to decode hieroglyphics. The French *savant* developed a collaboration with Kircher that lasted several years and, thanks to his contacts in Rome, he succeeded in facilitating Kircher's transfer to the Eternal City.[3]

In 1635, Peiresc asked Kircher to observe a total lunar eclipse on 28 August in Rome, where the Jesuit had finally moved in 1633. The purpose was to deduce the distance between various locations by comparing observational data from different places. Kircher, assisted by Melchior Inchofer (1584–1648), made the requested observations and sent the data to Peiresc, who also received data from two other observers in Rome and other locations in Italy and in the eastern Mediterranean area. Through these observations, Peiresc came to change the estimated size of the eastern Mediterranean by approximately 1000 km, and pleased with these initial achievements, he wrote from Aix to Gabriel Naudé (1600–1653):[4]

> I am deeply obliged to you for the great care you took in ensuring that yesterday's eclipse was observed by Argoli in Padua, Camillo Glorioso in Naples and P. Incofer in Rome [and the observations will be compared] with those of Germany, Holland, Paris and ours both by Monsieur Gassend in Digne […] and by the R.P. Théophile Minuti of the Minimes […] and others. […] This will offer an excellent opportunity to correct the distances between all the places where something similar has been observed.[5]

[3] Several studies on Kircher have been published, here I mention only a few: Daniel Stolzenberg, *Egyptian Oedipus: Athanasius Kircher and the Secrets of Antiquity* (Chicago: University of Chicago Press, 2013); John E. Fletcher, Elizabeth Fletcher (ed.), *A Study of the Life and Works of Athanasius Kircher, 'Germanus Incredibilis'* (Amsterdam: Brill Publications, 2011); Paula Findlen (ed.), *Athanasius Kircher: The Last Man Who Knew Everything* (New York: Routledge, 2004).

[4] Findlen, ref. note 2, 84.

[5] "*Vous m'avez infiniment obligé de tant de soing que vous avez prins pour faire observer l'eclipse d'hier par l'Argoli de Padoue, le Camillo Glorioso de Naples et le P. Incofer à Rome* [et les observations seront comparées] *avec celles de l'Allemagne, de Hollande, de Paris et les nostres tant de Monsieur Gassend à Digne* […] *que de R.P. Theophile Minuti des Minimes* […] *et des aultres.* […] *Il y aura beau moyen de regler les distances de tous les lieux où l'on aura observé quelque chose de pareil* [Transl. by the author]". Letter of Peiresc to G. Naudé, Aix, 29 aoùt 1635, in *Lettres de Peiresc* publiées par Philippe Tamizey de Larroque, ref. note 2. See also Françoise Charles-Daubert, "Libertinage et érudition. Peiresc relais de l'Europe savante" in Anne Reinhold, *Peiresc ou la passion de connaître*, Colloque de Carpentras, Novembre 1987 (Paris: J. Vrin, 1990) 41–60.

The following year Peiresc mobilized the Superior General Muzio Vitelleschi (1563–1645), who, in the first months of 1637, informed him that he had "given the necessary order to all his missionaries to make observations of the next eclipses" aided by a small booklet compiled by Pierre Gassendi (1592–1655) containing instructions for remote observers.[6] However, in June 1637 Peiresc died before his network could develop around the authoritative center of Aix.

In the late 1630s the center of the network moved to Rome where the figure of Kircher played a particularly important role. Influenced by Peiresc and stimulated by Gassendi and Mersenne, the German Father structured and expanded the Jesuit network of information exchange that radiated from the *Collegio Romano* where he held a highly prestigious role. In 1639 he began developing a project, the *Consilium Geographicum* (Geographical Plan),[7] to improve longitude measurements of places where Jesuit Colleges were located, so much so that some years later Gassendi could speculate that Kircher had picked up the baton from the illustrious Peiresc and continued his work: "Peireskius and Kircherus were an Example to all men, by so much Expense, Care and Labour, advancing the common good [...] in these days of ours there are few Kirchers, and never a Peireskius".[8]

5.2 A Symbolic Olive Tree

By the mid-seventeenth century, Rome held significant influence both intellectually and religiously and was home to the headquarters of the Society of Jesus, the most highly centralized religious order of the early modern era. An intriguing visual synthesis of the "Jesuit empire" around 1646 is captured in the *Horoscopium Catholicum Societatis Iesus*, also known as the Ignatian Tree (Fig. 5.1), a schematic map which symbolically portrays an olive tree rooted in Rome, with St. Ignatius of Loyola depicted at its base in prayer, extending its branches to all locations where Jesuit missions were present.

The map identifies 466 locations across 39 provinces spanning five continents, reached by numerous ships depicted behind St. Ignatius, which connect Rome with the farthest reaches of the globe. The branches, adorned with leaves bearing the names of the provinces, serve as a link between the Jesuit missions and Rome, the root from which everything sprang, while the sundials placed in each province indicate the local time in relation to that of Rome. This visual representation provides a comprehensive view of the extensive Jesuit mission network, showcasing the spread of their influence worldwide. Additionally, Psalm 113 ("*From East to West,*

[6] Findlen, ref. note 2, 83–87, quotation on p. 86.

[7] The *Consilium* was a comprehensive academic work that combined cartography, natural history, anthropology, human culture, and history. See Marcelo Aranda, "Deciphering the Ignatian Tree. The Catholic Horoscope of the Society of Jesus", in Findlen, ref. note 2, 106–125.

[8] Pierre Gassendi, *The Mirrour of True Nobility and Gentility, Being the Life of the Renowned Nicolaus Claudius Fabricius, Lord of Peiresk ... Written by the learned Petrus Gassendus, Englished by W. Rand* (London: Humphrey Moseley, 1657), quotation on p. 287 from the *Praise of Peireskius* at the end of Book VI.

Fig. 5.1 *Horoscopium Catholicum Societatis Iesus*. (Kircher, ref. note 9)

praiseworthy is the name of the Lord"), translated into 34 different languages spoken in the provinces where Jesuit were present, underscores the universal nature of their missionary endeavors.

The *Horoscopium*, included by Kircher in his *Ars magna lucis et umbrae* (1646), was chosen by the Jesuit polymath as a powerful and spectacular communication tool to celebrate the extension of the 'Jesuit empire' and the universality of the Catholic Church. Here, the propagandistic intent was deliberately privileged over the objective of visual clarity that could have been achieved by simply indicating the Jesuit missions on a map.[9] Recently, the Ignatian Tree has been analyzed in depth by several scholars.[10] I mention it here to illustrate the network that Kircher relied on to realize his ambitious project: the Geographical Plan (*Consilium Geographicum*), which aimed to reform geographical knowledge in a similar manner to what had occurred half a century earlier when the calendar reform solved the problem of Easter calculations.[11]

The calculation of longitude at sea has always been regarded as a crucial problem for sailors and, after King Philip II of Spain, followed by Philip III in 1598, offered generous rewards for a simple method to determine longitude at sea, several attempts were made by renowned scholars, including Galileo. Certainly, observations associated with a lunar eclipse would have been useful, but as eclipses were infrequent, the method was not suitable to practical use during navigation. Consequently, other techniques had to be developed.

The plan, which aimed to better define the longitudes of places, did not at this stage rely on eclipse observations, but on observations of magnetic deviations, and sought to achieve its goal by collecting an impressive amount of data and by specifying the methods, instruments, locations, and times according to which the observations were to be conducted. A more general plan would have included the collection of observations on tides and winds, on waters and rivers, on animals and plants, and finally on the customs and laws of human communities on the five continents. Seeking to map the geography of the world more accurately, Kircher and his Jesuit collaborators were devising a method both to document nature and to celebrate the ever-expanding Christian empire of Faith.

The worldwide extension of missionary activity offered in principle the opportunity to identify the longitudes and latitudes of all places that hosted Jesuit communities. *Confrères* who travelled to remote places were trained to collect observational data on astronomical events, magnetic deviations etc.

Each Jesuit center thus became an element of the Earth's surface matrix, with data reaching Rome from each location by various routes, as was the case with the

[9] Athanasius Kircher SJ, *Ars magna lucis et umbrae* (Romæ: sumptibus Hermanni Scheus, 1646). The *Horoscopium* is inserted between p. 552 and p. 553.

[10] Aranda's chapter in *Empires of Knowledge* (ref. note 2) analyses the Ignatian Tree in depth and illustrates the contributions of various scholars.

[11] Agustín Udías SJ, "Athanasius Kircher and Terrestrial Magnetism: The Magnetic Map", *Journal of Jesuit Studies*, 7 (2020) 166–184.

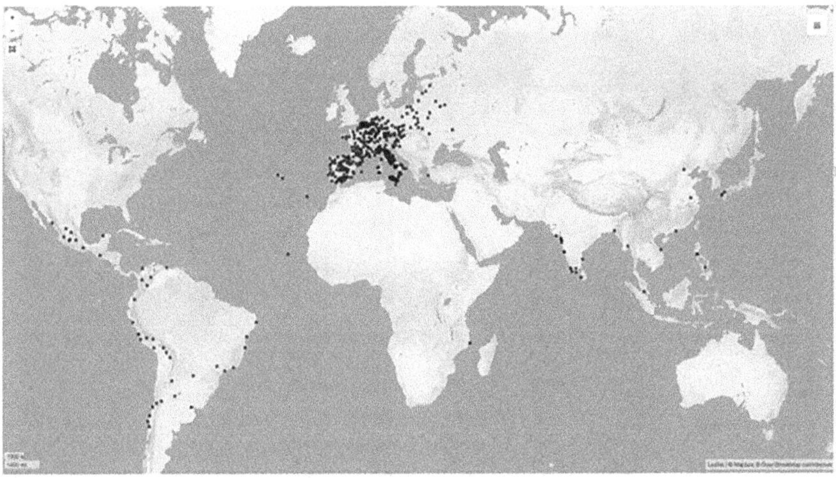

Fig. 5.2 A map describing the Jesuit world by mid-seventeenth century (Aranda, ref. note 2, 112)

missions in China, where the Jesuits used to send identical letters to Rome both eastward and westward, often in multiple copies (Fig. 5.2).[12]

Given the serious problem of determining longitude at sea, the concept was to measure it by observing the deviation of a compass needle from the North as defined by the North Star. It appears that Christopher Borri SJ developed a method to address this issue and applied it during his return journey to Europe from a mission in Asia. However, the explanation of the method, the documents recording the measurements, and the instrument, *Naugnomon* or nautical gnomon, were all lost.[13]

Strongly desired by General Vitelleschi, Kircher's plan succeeded in collecting a large amount of data. The vast majority came from Europe, other observations were made in Aleppo, Alexandria, the Cape Verde Islands, Constantinople, Guinea and some came from Asia: Goa, Canton, Macao. In total 42 Jesuit observers and 21 other scholars contributed to the collection of data, which were included in the book *Magnes sive de arte magnetica* published in 1641 (Fig. 5.3).[14]

In addition to valuable data on magnetic declinations, the book extolled the marvelous and mysterious qualities of magnets and reported a large number of curiosities and oddities, capturing the interest of many readers. It was so successful that a

[12] Leaving for China, the Swiss-German Jesuit Johann Schreck (1576–1630) recommended to his European correspondents: "Write to me [multiple copies of a letter] as much by East as by West, and they will certainly be delivered." Findlen, ref. foonote 2, 72.

[13] Angelo Mercati, "Notizie sul gesuita Cristoforo Borri e su sue 'inventioni' da carte finora sconosciute di Pietro della Valle il Pellegrino", *Pontificia academia scientiarum, Acta*, t.15, 1951, 25–46.

[14] Athanasius Kircher SJ, *Magnes sive de arte magnetica*, (Romae: ex Typographia Ludovici Grignani, 1641[1]) 439–455.

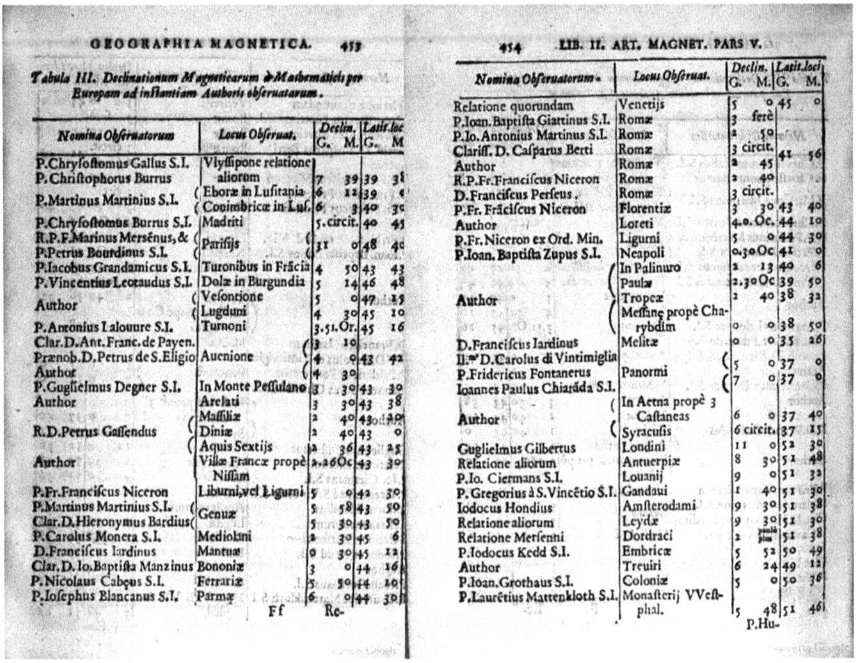

GEOGRAPHIA MAGNETICA. 453

Tabula III. Declinationum Magneticarum à Mathematicis per Europam ad instantiam Authoris obseruatarum.

Nomina Obseruatorum	Locus Obseruat.	Declin. G.	Declin. M.	Latit. loc. G.	Latit. loc. M.
P.Chrysostomus Gallus S.I. / P.Christophorus Burrus	Vlyssipone relatione aliorum	7	39	39	38
P.Martinus Martinius S.I.	Eboræ in Lusitania	6	12	39	[illegible]
	Conimbricæ in Lus.	6	3	40	30
P.Chrysostomus Burrus S.I.	Madriti	5. circit.		40	45
R.P.F.Marinus Mersēnus, & / P.Petrus Bourdinus S.I.	Parisijs	11	0	48	40
P.Iacobus Grandamicus S.I.	Turonibus in Frācia	4	50	43	43
P.Vincentius Leotaudus S.I.	Dolæ in Burgundia	5	14	46	48
Author	Vesontione	5	0	47	15
	Lugduni	4	30	45	10
P.Antonius Lalouure S.I.	Turnoni	3.51.Or.		45	16
Clar.D.Ant.Franc.de Payen.	Aucnione	3	10		
Prænob.D.Petrus de S.Eligio		4	0	43	42
Author		4	30		
P.Guglielmus Degner S.I.	In Monte Pessulano	3	30	43	30
Author	Arelati	3	30	43	38
R.D.Petrus Gassendus	Massiliæ	2	40	43	20
	Diniæ	2	40	43	0
	Aquis Sextijs	2	36	43	25
Author	Villæ Francæ propè Nissam	2.26Oc		43	30
P.Fr.Franciscus Niceron	Liburni, vel Ligurni	5	0	42	30
P.Martinus Martinius S.I.	Genuæ	5	58	43	50
Clar.D.Hieronymus Bardius		5	30	43	50
P.Carolus Monera S.I.	Mediolani	2	30	45	6
D.Franciscus Iardinus	Mantuæ	0	30	45	12
Clar.D.Io.Baptista Manzinus	Bononiæ	3	0	44	16
P.Nicolaus Cabęus S.I.	Ferrariæ	5	50	44	10
P.Iosephus Blancanus S.I.	Parmæ	6	0	44	30

Ff Re-

454 LIB. II. ART. MAGNET. PARS V.

Nomina Obseruatorum.	Locus Obseruat.	Declin. G.	Declin. M.	Latit. loci G.	Latit. loci M.
Relatione quorundam	Venetijs	5	0	45	0
P.Ioan.Baptista Giattinus S.I.	Romæ	3	ferè		
P.Io.Antonius Martinus S.I.	Romæ	2	50		
Clariss. D. Casparus Berti	Romæ	3	circit.	41	56
Author	Romæ	2	45		
R.P.Fr.Franciscus Niceron	Romæ	2	40		
D.Franciscus Perseus.	Romæ	3	circit.		
P.Fr. Frācifcus Niceron	Florentiæ	3	30	43	40
Author	Loreti	4.0.Oc.		44	10
P.Fr.Niceron ex Ord. Min.	Ligurni	5	0	44	30
P.Ioan.Baptista Zupus S.I.	Neapoli	0.30Oc		41	0
Author	In Palinuro	2	13	40	6
	Paulæ	2.30Oc		39	50
	Tropeæ	2	40	38	32
	Messanę propè Charybdim	0	0	38	50
D.Franciscus Iardinus	Melitæ	0	0	35	26
Ill.mus D.Carolus di Vintimiglia	Panormi	5	0	37	0
P.Fridericus Fontanerus		7	0	37	0
Ioannes Paulus Chiarāda S.I.					
Author	In Aetna propè 3 Castaneas	6	0	37	40
	Syracusis	6 circit.		37	15
Guglielmus Gilbertus	Londini	11	0	52	30
Relatione aliorum	Antuerpiæ	8	30	51	48
P.Io. Ciermans S.I.	Louanij	9	0	51	32
P. Gregorius à S.Vincētio S.I.	Gandaui	1	40	51	30
Iodocus Hondius	Amsterodami	9	30	51	38
Relatione aliorum	Leydæ	9	30	51	30
Relatione Mersenni	Dordraci	2	paulò plus	51	38
P.Iodocus Kedd S.I.	Embricæ	5	52	50	49
Author	Treuiri	6	24	49	12
P.Ioan.Grothaus S.I.	Coloniæ	5	0	50	36
P.Laurētius Mattenkloth S.I.	Monasterij VVestphal.	5	48	51	46

P.Hu-

Fig. 5.3 Magnetic declinations observed in Europe (Kircher, ref. note 14, 453–454)

second edition was published in 1643, followed by a third edition in 1654. However, both the book and its author faced criticism from prominent scholars of the time.[15]

Over time the complexity of the measurements and the discovery that the deviations seemed to follow no law, as they decreased and increased in different places without following a pattern, or in the same place they showed different values over the years, convinced Kircher that he had to abandon this method if he wanted to solve the longitude problem.[16] Finally, in 1654 he declared that his *Consilium Geographicum* (Geographical Plan), with the enormous number of observations collected on magnetic deviations, had been stolen from him by one of the visitors who often crowded his Museum in Rome.[17] As a result, the long-awaited detailed map with the magnetic declinations of 518 locations never saw the light of day, and he

[15] Gassendi, Mersenne and even Descartes were curious but also rather doubtful, if not skeptical, about the many extraordinary phenomena, natural magic and occult forces treated by Kircher in the book. See Thomas L. Hankins, Robert J. Silverman, *Instruments and the Imagination* (Princeton, NJ: Princeton University Press, 1995), 14–36. Torricelli and Huygens were also critical. See Findlen, ref. note 3, especially the introduction by Paula Findlen and chapter 10 by Michael J. Gorman.

[16] Kircher, ref. note 14, 429; Athanasius Kircher SJ, *Magnes sive de arte magnetica* (Romae: Sumptibus Blasij Deuersia & Zanobij Masotti Bibliopolarum, 1654) 314 and 340.

[17] *Ibid.*, 294.

was thus relieved of the obligation to present the inconclusive outcome of this titanic effort that had absorbed so much of the Society's resources.

Later Charles François Milliet Dechales SJ, Royal professor of Hydrography in Marseille, showed in his *Cursus Mathematicus* that this method is fallacious, since magnetic declination varies at different locations along the same meridian.[18]

5.3 From a Geographical Plan to a Reform of the Geographia

As is well known, during his long and productive life Kircher developed a global network of several hundred correspondents from European and Asian locations; of these, almost a third were members of the Society of Jesus, active in Jesuit colleges or missions. Among others, he maintained a close correspondence with Giovanni Battista Riccioli SJ (1598–1671), who was teaching and conducting research at the College of Santa Lucia in Bologna. An astronomer, natural philosopher, geographer and theologian, Riccioli taught in Parma, Mantua and Bologna, where he conducted an impressive number of astronomical observations and a series of experiments on falling bodies from the Torre degli Asinelli.[19] Correspondence with Kircher shows that Riccioli was particularly interested in the *Consilium Geographicum* and in a project dedicated to solving the longitude problem:

> As for the *Consilium Geographicum* on which Your Reverence is currently working, I wish to know whether it will be published before other works. For I have some information on this matter that I wish to share with Your Reverence, and I do not believe it to be of little

---

[18] Deschales concluded that this method was based on "*a vain conjecture, as magnetic declination has no certain rule*" ("vana fuit coniectura, cum Magnetis declinatio certa careat regula" [Transl. by the author]). Claude François Milliet Dechales SJ, *Cursus seu mundus mathematicus,* 3 voll (Lugduni: ex officina Anissoniana, 1674) quotation from vol. 2, 333.

[19] Since the end of the last century and more recently Riccioli has attracted the attention of several scholars who have deeply studied his major works. Below are just a few relevent studies: Ivana Gambaro, *Astronomia e tecniche di ricerca nelle lettere di G. B. Riccioli ad A. Kircher*, Quaderni del Centro di Studio sulla Storia della Tecnica del CNR, n.15 (Genova: Università degli Studi di Genova, 1989); John L. Heilbron, *The Sun in the Church. Cathedrals as Solar Observatories* (Boston: Harvard University Press, 1999); Maria Teresa Borgato (ed.), *Giambattista Riccioli e il merito scientifico dei gesuiti nell'età barocca* (Firenze: Leo S. Olschki, 2002); Christopher M. Graney, *Setting Aside All Authority: Giovanni Battista Riccioli and the Science against Copernicus in the Age of Galileo* (Notre Dame: University of Notre Dame Press, 2015); Alfredo Dinis SJ (auth), Alvaro Balsas SJ, Batista Ricardo Barroso (eds), *A Jesuit against Galileo? The Strange Case of Giovanni Battista Riccioli* (Braga: Axioma Publicações da Faculdade de Filosofia, 2017); Flavia Marcacci, *Cieli in contraddizione: Giovanni Battista Riccioli e il terzo sistema del mondo* (Perugia: Aguaplano, 2018).

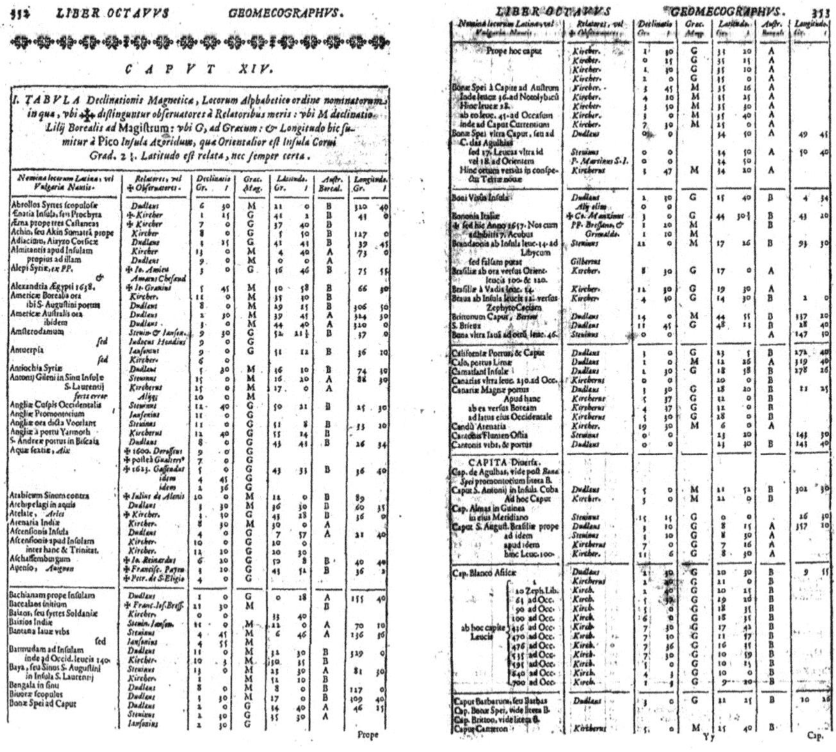

Fig. 5.4 Tables with magnetic declination, latitude and longitude (Riccioli, ref. note 21, 352–353)

> importance. All that remains […] is to write to as many people as possible, urging them to observe the eclipse this coming October, in order to correct geographical knowledge.[20]

Both exchanged numerous reports, sharing a deep interest in this subject and in astronomical, physical and magnetic measurements and observations. Kircher stated in 1654 that he had been compelled to abandon the Geographical Plan because of theft. As a result, Riccioli carried on the endeavor and, in 1661, published *Geographiae et hydrographiae reformatae*, incorporating the observations he had received from Kircher and other correspondents. The book covered many geographic topics and was considered one of the best studies on the subject at the time (Fig. 5.4).[21]

---

[20] *"Quoad Consilium Geographicum quod R.V. molitur, vellem scire an proditurum ait in lucem ante alia opera. Habeo enim non pauca quae de hoc negotio monitum optem R.V., nec puto paenitenda. Reliquum est ut […] eo fine Geographiam corrigendi scribamus ad quamplures pro eclipsi octobris futuri observanda"* [Transl. by the author]. Letter of Riccioli to Kircher, Bologna 5 July 1642 in Gambaro, ref. note 19, 52.

[21] Giovanni Battista Riccioli SJ, *Geographiæ ae hydrographiae reformatae libri duodecim* […], (Bononiæ: ex typographia haeredis Victorij Benatij, 1661).

Riccioli corrected several errors made by earlier geographers on many geographical subjects, and included an impressive amount of geographical data in the book, in particular measurements of magnetic deviations taken at various locations. The information solicited and collected by Kircher throughout the international Jesuit community, together with the data collected by Riccioli himself, made it possible to create tables with the magnetic declination, latitude and longitude of the places where the observations had been made, as well as the names of the observers who had provided them.

Riccioli meticulously organized into tables all these data, extracted also from the most well-known and qualified works available at the time on magnetism, astronomy, geography, navigation, cartography by William Gilbert, Simon Stevin, Adriaan Metius, Georges Founier SJ, Athanasius Kircher SJ, Jan Janssonius, Pierre Hérigone, and Robert Dudley. He also made use of the travel diaries of more than 20 navigators and of the communications received from scholars such as Gassendi, Mersenne, etc., and from many *confrères* who had travelled extensively in Europe, Asia and America. The locations listed in the tables reached the remarkable number of 420, with longitudes calculated from the island of Pico in the Azores archipelago.[22] However, since magnetic declination not only failed to adhere to a precise rule, but also appeared to vary over time even within a single location, the usefulness of these charts for navigators could, at best, be only temporary.

5.4 A Unique Lunar Eclipse

In addition to exchanging observations of magnetic declination, Riccioli and Kircher shared measurements of polar altitudes, reports of eclipses, and documented any other significant astronomical events over the years.

Aware of how problematic the magnetic declination method could prove to be, Riccioli focused on collecting observations of lunar eclipses at various locations, not only for their astronomical interest *per se*, but also to increase geographical knowledge by establishing the distances, in terms of degrees of longitude, between the locations where the observations were made. Most of the data is collected in the pages of his major works: *Almagestum novum* (1651),[23] *Geographiae et hydrographiae reformatae* (1661), and *Astronomiae reformatae* (1665).[24]

---

[22] Riccioli, ref. note 21, 352–358.

[23] Giovanni Battista Riccioli SJ, *Almagestum nouum astronomiam veterem nouamque* [...] *in tres tomos distributam* (Bononiæ: Ex typographia haeredis Victorij Benatij, 1651), the only tome published was the first, divided into two parts: *pars prior* and *pars posterior*.

[24] Giovanni Battista Riccioli SJ, *Astronomiae reformatae* [...] (Bononiæ: Ex typographia haeredis Victorij Benatij, 1665).

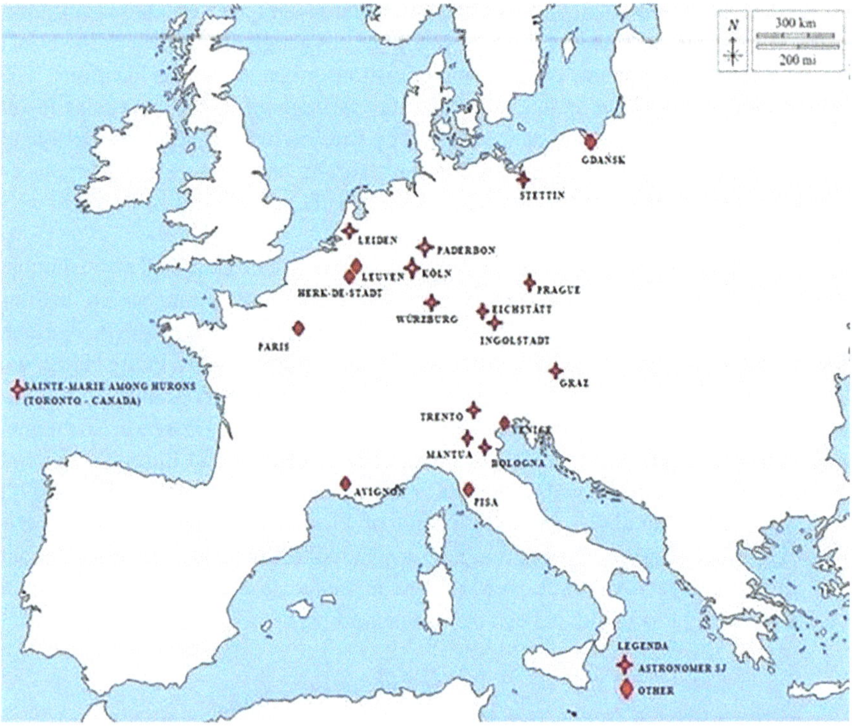

Fig. 5.5 Map illustrating the 19 European locations where the observations of the lunar eclipse of 14 April 1642 were made by various observers and collected by Riccioli (the 20th one was made at Sainte-Marie Among Hurons, Toronto, Canada)

As for the data on the eclipses it is worth recalling that, 10 years before the publication of the *Geographiae*, in the *Almagestum* Riccioli referred observations of the lunar eclipse occurred on 14 April 1642 in 20 different locations. For the first time, a report so rich in information from competent observers on a lunar eclipse appeared in print.[25] Fig. 5.5 shows the locations from which observations were received. Many observations were accurate, considering, for instance, the time of the *initium* of the eclipse in Paris and in Ingolstadt, the time difference is 40′ which implies a distance between the two cities of 10° in terms of longitude. The true distance is 9° 4′ 23″. However, other data were conflicting.[26]

In addition to its role in the geographical project pursued by Riccioli, this lunar eclipse is also interesting for other reasons, as it may shed light on the ways in which the Jesuit elaborated his experimental and observational data.

In his letters to Kircher (5 July 1642 and 6 April 1647) he referred different time intervals, measured with a pendulum, for the eclipse of 14 April 1642 he had

[25] Riccioli, ref. note 23, *pars prior*, 381–383.

[26] Riccioli, ref. note 21, 370.

observed in Bologna.[27] Data that also differed in comparison with those published in the *Almagestum novum*.[28] The following table collects the content of the two letters.

Letter 5 July 1642		Letter 6 April 1647	
Initium penumbrae	12 h 48′ 32″		
Initium verae ecl.	12 h 54′ 43″	Initium legit. Ecl.	12 h 51′ 5″
Finis totius ecl.	16 h 32′ 30″	Finis totius ecl.	16 h 36′ 19″
Duratio ecl.	3 h 37′ 47″		3 h 45′ 14″

It is difficult to understand what was the reason for these different times given for the beginning, end and duration of the same eclipse. Regarding these observations of Riccioli and his *confrères*, it can be speculated that the differences in the times of *initium* may be due to later interpretations of some data held by the Bolognese observers, while the differences in the duration of the eclipse itself could be attributed to refinements in the measurements conducted by the Jesuit on the pendulum and aimed at defining the length of the pendulum beating the second.[29]

---

[27] Gambaro, ref. note 19. The beginning of the day was assumed at noon, the eclipse therefore occurred in the early hours of 15 April 1642, according to our current conventions.

[28] However, in the *Almagestum novum* for the same eclipse Riccioli indicates different data taken by different methods:

Almagestum pars prior p. 381—first observation	*Almagestum pars prior* p. 381—second observation		
initium verae ecl.	12 h 47′ 27″	initium (estimated)	12 h 49′ 6″
finis visibilis	16 h 38′	finis (estimated)	16 h 39′ 30″
duratio ecl.	3 h 50′ 33″	duratio ecl.	3 h 50′ 24″

These two durations, both different from the previous ones, are approximated to 3 h 50′ by Riccioli in *Geographiae et hydrographiae reformatae* (ref. note 21, 370). On the same page he comments, with regret, on the many conflicting data received concerning this eclipse and notes that, as far as the duration is concerned, the average value of the data received, which range between 3 h 38′ and 4 h 2′, corresponds exactly to the duration of 3h 50′ that he himself claimed, in the *Geographiae,* to have observed in Bologna.

[29] In order to know the "*true duration*" of the eclipse we can consider, on the one hand, the data of Theodor von Oppolzer's *Canon* for the eclipse in question, the No. 4400, relating to Bologna. Transforming them into sidereal time intervals (as Riccioli operated), the eclipse duration obtained is 3 h 40′ 35″, close to the first value of the letter of July 5, 1642 (Theodor von Oppolzer, *Canon der Finsternisse*, Wien: aus der kaiserlich-koniglichen Hof- und Staatsdruckerei, 1887); private communication from Prof. A. Righini. On the other hand, with the NASA program: https://eclipse.gsfc.nasa.gov/JLEX/JLEX-EU.html for the duration we obtain 3 h 39′ 00″. According to this NASA program, in Bologna, at the Santa Lucia Observatory, in the city center, the "*partial eclipse*" began at about 00 h 47 m 30 s, and ended at 04 h 6 m 30 s (local time); private communication from Prof. V. Zanini. It should be noted that what the NASA program calls *penumbra* is not the same *penumbra* Riccioli refers to. In fact, the instructions of this program specify that "the start and end times of penumbral phases are of academic interest only, since these events are never visible to the eye". What Riccioli calls *penumbra* is probably the phase in which he perceived the Moon reddened, but in fact, the Moon had already been immersed in the *penumbra* for some time.

5.5 The Search for the Pendulum Beating the Second

The pendulum, and its law were studied by all the leading figures of seventeenth-century science. As it has been pointed out, research on the pendulum beating the second has played a major role for, among other things, "establishing the collision laws, the conservation laws, the value of the acceleration due to gravity g, and ascertaining the variation in g from equatorial to polar regions and hence discovering the oblate shape of the earth. […] Also tied up with the scientific revolution of the mid-seventeenth century was a horological revolution, and the pendulum played a central role in this as well".[30]

As it has been pointed out by Peter Dear, Riccioli's discussion of the behavior of pendulums and falling bodies in his *Almagestum novum* proceeds from a large number of "experiences, not by way of likely conjectures, but according to infallible physico-mathematical science, to certain conclusions".[31]

In fact, between 1642 and 1645 Riccioli devoted much effort to establishing the length of the pendulum *(perpendiculum)* beating the second (Fig. 5.6). With the help of many *confrères* at the observatory of Santa Lucia in Bologna, he performed repeated measurements by counting the oscillations of a pendulum, first from noon on 2 April to noon on 3 April 1642, then from the passage of the star Spica on the meridian of Bologna and the passage of the star Arcturus on the same meridian during three nights in May and June 1645. Each result obtained was the outcome of averaging the great number of measurements made by many collaborators.

In the table we can compare the results obtained by Riccioli in 1642 and in 1645. As for the symbols used, L and L′ are the lengths of the pendulums (1 Roman foot corresponds to 12 ounces), W_s and W_s' are the weights of the spheres, W_c and W_c' are the weights of the chains, T and T′ are the periods of the pendulums. The semi-periods are given in thirds (1/60 of a second) and quarters (1/60 of a third).

Consequently, it appears meaningless to draw a comparison between Riccioli's data and what we can now estimate to have been the "*true duration*" of the eclipse.

[30] Michael R. Matthews, *Time for Science Education. How Teaching the History and Philosophy of Pendulum Motion Can Contribute to Science Literacy* (Berlin: Springer Science, 2000), xix. See also at pp. 381–418, the impressive list of references, most of them regarding the history of the pendulum.

[31] "*Ideoque non ex verisimilibis coniecturis, sed ex infallibili scientia Physico-Mathematica ab experimentis sequentibus ad conclusiones certas procedemus*." (Riccioli, ref. note 23, *pars post*, 383). As for the term "physico-mathematics", a neologism that became very common in the Jesuit mathematical community and in Riccioli's scientific texts, see Peter Dear, *The Literary Structure of Scientific Argument* (Philadelphia: University of Pennsylvania Press, 1991, 146–161: English translation p. 152).

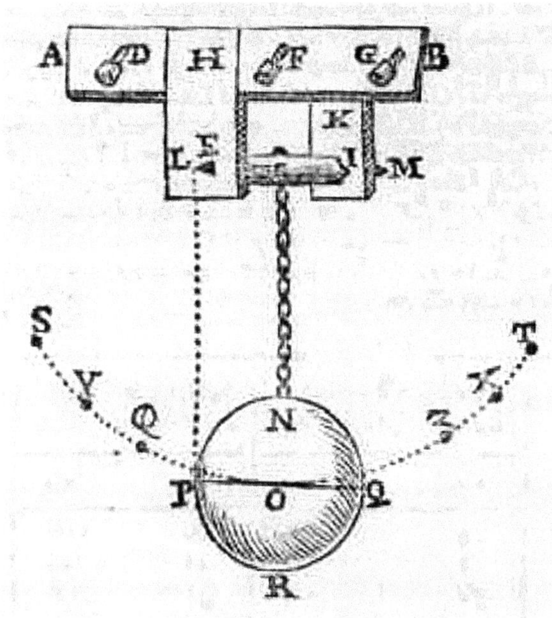

Fig. 5.6 Riccioli's pendulum (Riccioli, ref. note 23, *pars prior*, 84)

2–3 April 1642	L = 3 Roman feet and 4 ounces	W_s = 8 ounces	W_c = 3 ounces 3/5	$T/2 = 59'''4''''$
19, 28 May 1645 and 2 June 1645	L′ = 3 Roman feet and 2 ounces 67/100	W_s' = 20 ounces 1/2	W_c' = 3 ounces 2/5	$T'/2 = 59'''36''''$

Using the more reliable values obtained in May–June 1645 Riccioli recalculated the length of the pendulum beating the second[32]:

L″ = 3 Roman feet 3 ounces and 27/100

5.6 How Long Did That Eclipse Last?

If we assume that Riccioli, based on the May–June 1645 measurements, had recalculated the semi-period of the pendulum, of known length, used to observe the lunar eclipse of 14 April 1642, we can understand how he reported different data for the same eclipse in his letter to Kircher, dated 6 April 1647.

This hypothesis might shed light on how Riccioli actually handled his experimental and observational data. In fact, to determine the *true* semi-period of the

[32] A pendulum beats the second when each half-oscillation lasts one second (half of the period). However, Riccioli made an error in calculating the length of the pendulum beating the second: based on his data the correct value should be 3 Roman feet, 3 ounces, and 19/100, whereas he instead reported 3 Roman feet, 3 ounces, and 27/100 (Riccioli, ref. note 23, *pars prior*, 87–88).

pendulum employed in the measurements conducted on 2 and 3 April 1642, he certainly had to use Baliani-Galileo's law according to which the square of the period is proportional to the length of the pendulum. However, the two pendulums (of 1642 and of 1645) not only had different lengths, but were also connected to spheres of significantly different weights: in 1642 $W_s = 8$ ounces, in 1645 $W_s' = 20$ ounces ½. Consequently, the use of this law by Riccioli implies the assumption that the period does not depend on the weight of the connected sphere, as, on the contrary, an Aristotelian would have argued.

One might ask why Riccioli, in his pendulum measurements conducted in May–June 1645, did not use a sphere of the same weight as the one employed during the measurements of 2 and 3 April 1642 (and consequently during the eclipse of 14 April 1642). We can only speculate that he believed that the weight of the sphere could not affect the period of the pendulum in any way and therefore the weight had no influence on the measurements.

In other words, to determine the *true* duration of the lunar eclipse of 14 April 1642, 5 years later he had no choice but to use Baliani-Galileo's law to recalculate its duration, even though the weights of the spheres connected to the pendulums were very different.

By retracing Riccioli's steps, it is possible to determine the *true*, or at least more reliable, semi-period of the pendulum examined on 2 and 3 April 1642, which undoubtedly coincided with the one Riccioli used for the eclipse of 14 April 1642:

1″ 0‴ 33⁗ instead of 59‴ 4⁗

Finally, by recalculating the duration of the lunar eclipse of 14 April 1642 with these more reliable data on the second-beating pendulum—and adding 2 minutes to the duration, as Riccioli deemed necessary, introducing a correction of 1 minute at both the beginning and the end of the eclipse[33]—we obtain 3 h 45′ 14″ which is exactly the duration indicated in the letter Riccioli wrote to Kircher on 6 April 1647. Indeed, if Riccioli followed this path, drawing on his best knowledge derived from the measurements conducted in May–June 1945 concerning the appropriate length to assign to a seconds-beating pendulum, he could have provided Kircher in 1647 — as he in fact did— with more reliable time intervals for the eclipse of 14 April 1642.[34]

[33] See *Problema Vii Principium ac Finem legitimæ Eclipsis, aut Totalis obscurationis exacte discernere*, (Riccioli, ref. note 23, *pars prior*, 353–354).

[34] This bizarre story of the times of the eclipse observed by Riccioli changing over the years is truly unusual. See, for example, the letters in which Gassendi provided his correspondents over the years with, of course, the same time intervals for this 1642 eclipse observed in Paris: Gassendi to Hevelius, Paris *VII Kalendas Apriles* 1644 or 26 March 1644 in Pierre Gassendi, *Petri Gassendi ... Epistolae, Tomus sextus*, 182–184, part of Pierre Gassendi, Samuel Sorbière et al. (eds.), *Petri Gassendi ... Opera omnia in sex tomos diuisa ...* (Lugduni: sumptibus Laurentij Anisson, & Ioan. Bapt. Deuenet, 1658); and Gassendi to Wendelen, Paris *postridie Kalendas Apriles* 1644, or 2 April 1644, *ibid.*, 184–185. Similarly, the same data were indicated to Gassendi over the years by his correspondents when they informed him of their observations of the same eclipse: Godfried Wendelen (1580–1667) from Brussels (Wendelen to Gassendi, 11 July 1643, *ibid.*, 453–454), and Johannes Hevelius (1611–1687) from Gdansk (Hevelius to Gassendi, 23 June 1644, *ibid.*, 471–472). The time intervals reported in these letters are also given in the *Almagestum novum* (ref. note 23, *pars prior*, 381–383) for observations made in Paris, Gdansk and Herk-de-Stad.

However, he could not openly clarify his research path, being forced to follow the Aristotelian tradition as dictated by the higher authorities of the Society of Jesus. A confirmation comes from the chapter of the *Almagestum novum* dedicated to the fall of bodies, where he clearly argues that heavier bodies fall to the ground faster than lighter ones, in perfect agreement with Aristotelian physics.[35] Moreover, a few pages earlier, he had claimed that the experimental evidence contradicted the thesis held by Baliani:

> Of two pendulums equal in all respects except weight, the heavier one remains in motion for a longer time and completes more oscillations within the same time interval [...], [and one may conclude that] proposition n. 5 [...] is against Baliani, who maintains that, if two pendulums have equal length, their half-oscillations last the same amount of time [or semi-period], yet I cannot but trust what I observe with my own eyes.[36]

5.7 From Heaven to Earth

This eclipse retained its interest over the years. Book VIII *Geomecographus*, in *Geographiae et hydrographiae reformatae* published in 1661, deals with the method for measuring the longitudes of places based on the observations of eclipses, a method already analyzed in *Almagestum novum*,[37] and devotes several pages to the "*Catalogus Eclipsium Lunae in diversis Locis Terrae Marisque Obseruatarum*" where the lunar eclipse of 14 April 1642 stands out, once more, as the one with the largest number of observations collected (Fig. 5.7).

Nevertheless, Riccioli admitted that all these data presented a large number of discrepancies, a "multitudo dissidiorum". So much so that Kircher himself, "who had hoped to be able to complete the *Consilium Geographicum*, after noticing such great discrepancies among the data concerning the eclipses [observed in different places], no longer hoped to gain anything from this investigation, as evidenced by the letters he sent" to Riccioli.[38] However, Riccioli remained firmly convinced, in

---

[35] With regard to falling bodies we read: "Of two bodies of equal size, the heavier descends more rapidly, by natural motion, from the same starting point to the same endpoint than the lighter one" ("*Corporum duorum æqualis molis illud quod gravius est, naturali motu citiùs descendit ex eodem termino ad eundem terminum, quam illud quod est leuius*" [Transl. by the author], Riccioli, ref. note 23, *pars prior*, 89).

[36] In fact, Riccioli wrote "*Duorum perpendiculorum in omnibus æqualium, præterquàm in grauitate; illud quod grauius est diutiùs in motu perseuerat, et intra æquale tempus plures numero vibrationes peragit*", and concluded that "*Propositio 5 [...] est contra Balianum, qui si altitudo perpendicolorum sit æqualis, vibrationes eorum æquidiuturnas putat. Sed ego non possum oculis meis non credere*" [Transl. by the author]. (Riccioli, ref. note 23, *pars prior*, 85–86).

[37] In the *pars posterior* of the *Almagestum novum* Riccioli had already discussed the problem of longitude at length, mentioning various methods. Riccioli, ref. note 23, 608–612, in particular through lunar eclipses on pp. 608–609.

[38] "*P. Athanasius Kircher, qui ex Eclipsibus sperabat se Consilium Geographicum absolutorum, inspecta Ecliupsium tanta dissonantia, desperauit de hoc negotio, ut ex literis ad me datis constat*". Riccioli, ref. note 21, quotations on p. 370 and p. 376. See the *Catalogus Eclipsium Lunae* on pages 364–370.

368 *LIBER OCTAVVS* *GEOMECOGRAPHVS.*

Obseruatores Hora post meridiem.	*Loca obseruatiou.*	*Initium.* H.	'	*Medium.* H.	'	*Finis.* H.	'
Albertus Linemannus	*Regiomonte Prussiæ*	7	6	8	37	9	57
Io. Heuelius	*Dantisci*	7	1 ¼	8	26 ¼	9	51
Erycius Puteanus	*Louanij*	5	59	7	26	8	53
Gotefridus Vendelinus	*Hercij*	6	8	7	29	8	50
Langrenus	*Bruxellis*			7	24		
Petrus Gassendus	*Parisijs*					8	50
Io. Phocyllides	*Frenequera*					9	4 ½

XLVI. Anno 1642. Apr. 14. factas Lunaris Eclip. obseruationes per literas impetraui ab amicis, quas adhuc asseruo.

Obseruatores Hora post meridiem.	*Loca obseruat.*	*Principium.* H.	'	''	*Immersio.* H.	'	''	*Emersio.* H.	'	''	*Finis.* H.	'	''
Nos cum P. Franc. Maria Grimaldo	*Bononiæ*	12	48	6	13	52	10	15	36	22	16	40	30
PP. Middendorp. & Sturmius S. I.	*Coloniæ*	12	26	30	13	41	40						
Io. Heuelius	*Dantisci*	13	14	6	14	22	0	15	53	0	16	56	0
P. Andreas Arzet Soc. I.	*Eystadij*	12	44	0	13	53	0	15	33	0			
P. Baltassar Conradus Soc. I.	*Glacij*				14	26	0	15	48	0			
P. Henricus Marcellus Soc. I.	*Herbipoli*	12	32	54	13	37	54	15	17	36	16	31	44
Gotefridus Vendelinus	*Hercij*	12	3	0	13	19	0	14	43	0	15	42	0
P. Iacobus Viua Soc. I.	*Ingolstadij*	12	46	0	13	50	0	15	28	0	16	30	0
P. Franc. Ioseph. Bressan. Soc. I.	*Kebeci*	7	36	0	8	45	0	10	20	0			
Innominatus Kirchero	*Leydæ*	12	0	20	13	14	56				16	2	30
Gerardus Gutiscouius	*Louanij*	12	0	0									
P. Vuilelmus Vueilhamer S. I.	*Mantuæ*	12	38	27	13	45	7	15	5	0	16	25	0
PP. Mordesoun, & Lansitch S. I.	*Paderbonæ*				13	41	0	15	21	0	16	12	8
P. Georgius Furnier S. I.	*Parisijs*	12	20	0	13	11	0	14	47	0	15	51	0
vnà cum Bullialdo, & Gassendo								sed Gassend.			15	48	0
P. Vincent. Reiner. Oliuetanus	*Pisis*	12	33	0	13	49	18	15	21	23			
P. Theodorus Moretus S. I.	*Pragæ*				14	10	0	15	28	0			
D. Marcus Marci	*Pragæ*				14	9	54	15	28	27			
Laurentius Bustadius	*Stetini*				13	56	0	15	54	48			
Illustriss. D. Iulius Iustinian.	*Venetijs*	12	56	9				15	38	41			
Vide infra numero 60. quæ de his obseruationibus dicam.													

XLVII. Anno 1642. Octobr. 7. Lunaris Eclipsis factas obseruationes item per literas obtinui, quas conseruo, & sunt infrascriptæ, cum nostra.

Obseruatores Hora post meridiem.	*Loca obseruation.*	*Principium.* H.	'	''	*Immersio.* H.	'	''	*Emersio.* H.	'	''	*Finis.* H.	'	''
Nos cum P. Fr. M. Grimaldo	*Bononiæ*	14	55	50	15	56	0	17	36	40	18	44	0
P. Theodorus Moretus S. I.	*Bresnitij Bohem.*	14	57	0									
Domin. Francisc. Zupus	*Catacij*				16	17	0						
Antonius Garratus	*Ilerdæ*	14	9	30	15	11	0	16	50	10			
D. Vincentius Mutus	*Maioricæ*	14	21	8	15	24	16	17	2	20	17	59	8
D. Curzolanus	*Mediolani*	14	44	0	15	47	0						
P. Franciscus Niceronus	*Romæ*	15	2	0	16	5	30						

XLVIII. Anno 1643. Septemb. 27. præter nostram Lunaris deliquij obseruationem reliquas eo die factas ab amicis meis Vranicis obtinui, & in Litteris asseruo.

Obseruatores Hora post meridiem.	*Loca obseruation.*	*Principium.* H.	'	''	*Medium.* H.	'	''	*Finis.* H.	'	''
Gutiscouius, & Vendelinus	*Antuerpiæ*				6	52	0	8	5	30
Nos, & P. Frãc. Maria Grimaldus	*Bononiæ*	6	18	15	7	30	52	8	43	30
P. Theodorus Moretus Soc. I.	*Bresnitij in Bohemia*							8	14	0
Ioannes Heuelius	*Dantisci*							8	59	0
D. Vincentius Mutus	*Maiorica*							8	5	14
P. Io. Bapt. Zupus S. I.	*Neapoli*							8	59	40
D. Vincentius Reinerus	*Pisis*							8	13	44
Ismaël Bullialdus	*Parisijs*							7	59	36
Petrus Gassendus	*Succij prope Paris.*							8	1	38
Antonius Garratus	*Syluanelli*							8	0	0
Illustriss. D. Iulius Iustinian.	*Venetijs*							8	44	0

XLIX.

Fig. 5.7 *Catalogus Eclipsium Lunae in diversis Locis Terrae Marisque Obseruatarum* (Riccioli, ref. note 21, 368)

principle, of the usefulness of lunar eclipse data in advancing knowledge of the longitude of places, and in his *Geographiae et hydrographiae reformatae* he devoted many pages to tables where the observations were well organized and compared with each other.

In Book VIII *Geomecographus* the author included tables in which, for each location considered in the left-hand column, other locations were listed in the right-hand columns where for each eclipse (identified by its catalogue number), and for each of its phases, the observed times were compared with those recorded at the locations in the left-hand column. In the end, he discussed in detail the problems related to determining longitudes for European and extra European places.[39]

Finally, in Book IX *Mecoplaticus* Riccioli listed all known locations with their common and Latin names, latitudes and longitudes calculated from the meridian of the Palma Island in the Canary Islands (Fig. 5.8).[40] To get an idea of the precision achieved by the Jesuit, let us compare his data, for example for the city of Brussels, with the corresponding actual longitude and latitude:

According to Riccioli	Lat. 50° 48′; Long 9° 23′ 42″
According to the actual data	Lat. 50° 51′ 18″ N; Long 4° 21′ 9″ E

Riccioli listed more than 2700 locations in his tables, and, in general, the most incorrect longitudes he gave do not deviate by more than 7 or 8 degrees from currently accepted values.[41] François de Dainville (1909–1971), a well-known geographer, wrote in 1940:

> To the Italian Jesuit belongs in fact the honour of having first undertaken the great work whose necessity had already been recognized by the far-sighted minds of Peiresc, Fournier, Blaeu[42] and Varenius[43]: to amass all sea and land itineraries and the greatest number of astronomical observations, to study them, sift them and extract from them the tables of longitudes essential to the reform of geography and the correction of maps [...]. If Riccioli had the necessary resources and the collaboration of a skilled engraver to accompany, as he

[39] Riccioli, ref. note 21, 364–397, where the determination of the longitude of locations is the subject of an in-depth analysis.

[40] Riccioli, ref. note 21, 402–425.

[41] Riccioli gave his data with reference to the longitude of Palma. For Brussels he gave 27° 18′, to which the actual longitude of Palma, 17° 54′ 18″, has to be subtracted giving for Brussels 9° 23′ 42″. As is well known, it was Gian Domenico Cassini (1625–1712)—astronomer at the court of Louis XIV, after nearly 20 years of intense scientific exchange with Riccioli in Bologna—who implemented a significantly more effective method for determining longitude, a technique originally proposed by Galileo. This method made use of the Ephemerides of Jupiter's satellites (1668), which provided precise temporal references based on the satellites' occultations—celestial events that occur far more frequently than lunar eclipses. See Ivana Gambaro, "Between Heaven and Earth. The Scientific Missions of an Astronomer at the Court of the Sun King", in *Proceedings of the International Conference 'Giovanni Domenico Cassini, European astronomer of the seventeenth century on the fourth centenary of his birth'*, Bologna 18–20 June 2025, in press; and available at https://www.memsait.it/videomemorie/volume-4-2025/VIDEOMEM_4.2025.26.mp4

[42] Willem Blaeu (1571–1638), Dutch cartographer.

[43] Bernhardus Varenius (1622–1650), German geographer.

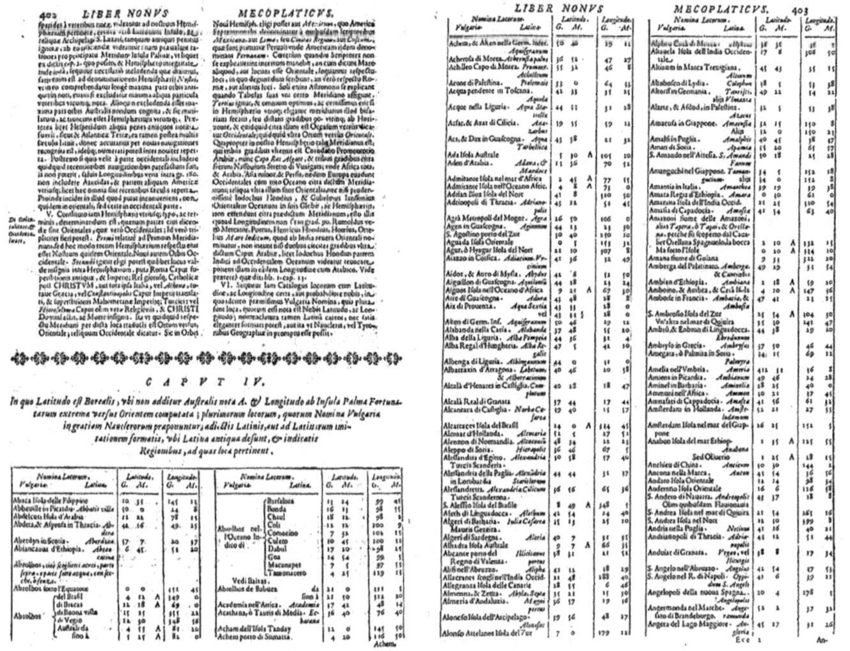

Fig. 5.8 *Catalogus Locorum cum latitudine ac longitudine certa* (Riccioli, ref. note 21, 402–403)

wished, his 'Geographia reformata' with a collection of maps based on his *Table de longitudes*, it is likely that Delisle's revolution would have been brought forward by thirty or forty years.[44]

Finally, establishing the longitude of places, especially ports, as accurately as possible could be of great importance for determining longitude during navigation, which had always been considered a crucial problem for sailors.

In fact, to overcome this challenge, Riccioli also illustrated the horological method, which involved measuring time using a timepiece, such as a sandglass or a

[44] "*Au Jésuite italien revient en effet l'honneur d'avoir entrepris le premier le grand œuvre dont les esprits très clairvoyants de Peiresc, Fournier, Blaeu, Varenius avaient déjà saisi la nécessité: amasser tous les itinéraires de mer et de terre, le plus grand nombre d'observations astronomiques, les étudier, les passer au crible et en extraire les tables de longitudes indispensables à la réforme de la géographie et à la correction des cartes* […]. *Si Riccioli avait eu les ressources nécessaires et la collaboration d'un habile graveur pour accompagner, comme il le souhaitait, sa 'Geographia reformata' d'une collection de cartes dressées d'après sa Table de longitudes, il est à croire que la révolution opérée par Delisle aurait été avancée de trente ou quarante années.* [Transl. by the author]". François de Dainville, *La géographie des humanistes* (Paris: Beauchesne, 1969^2, 445–447). Guillaume Delisle was a French cartographer known for the accurate maps of Europe and the newly explored America he drew between 1700 and 1725.

pendulum clock, set to the time of the port of departure, and comparing it with the local time measured on the ship with the height of the sun. In principle, this method would easily have made it possible to determine the value of longitude at sea, since a time interval of 1 hour corresponds to a difference of 15 degrees in longitude; however, in practice countless problems arose aboard a ship.

Riccioli, well aware of the difficulties already taken into account by other scholars, pointed out all the related problems (rocking of the ship, humidity, unfavorable weather conditions such as extreme heat or freezing temperatures, etc.) that prevented the devices available at the time from providing reliable time measurements when used aboard a ship under way.[45] As a result, around 1661, the issue still presented itself as an open challenge that would not be solved until 100 years later by John Harrison's marine chronometer.

5.8 Conclusion

The worldwide network of observers developed by Kircher yielded very interesting results, thanks to his many collaborators and also to the tireless data collection efforts of Riccioli.

So much so that, in the writer's opinion, Riccioli's most significant scientific contribution is not his beloved geoheliocentric system, so dear to him yet largely disregarded in the second half of the seventeenth century except in Catholic circles. Rather, it is his systematic and continuous collection of data, which characterized not only his research in the field of astronomy, as represented by the *Almagestum novum* (1651) and the *Astronomiae reformatae* (1665), but also extended to the field of geography, where he provided the contemporary geographers with an impressive amount of accurate information gathered in the *Geographiae et hydrographiae reformatae* (1661).

Fortunately, the collection of geographical data was a subject less likely to encounter problems with Roman censorship, unlike what had happened with a preliminary version of the *Almagestum novum*.[46] In fact, the latitude and longitude of places could have very useful practical consequences and, in principle, did not involve the need to reconcile, with difficulty, the doctrines of the Faith or the Aristotelian-Thomistic tradition with a world of recalcitrant phenomena.

---

[45] Riccioli, ref. note 21, 335–337. For sea trials of Huygens's cycloidal marine clocks see Joella G. Yoder, *Unrolling Time. Christiaan Huygens and the Mathematization of Nature* (Cambridge: Cambridge University Press, 1988). See also J. A. Bennett, "The Longitude and the New Science", *Vistas Astr.* 28 (1), 1985, 219–225.

[46] Gambaro, ref. note 19, 39–42; for Roman censorship as encountered by Riccioli see Ivana Gambaro, "Giovan Battista Riccioli, infaticabile astronomo, e la censura romana", *Giornale di astronomia*, 42 (3), 2016, 46–51; Dinis, ref. note 19, chap. 2; Ivana Gambaro, "Geo-heliocentric Models and the Society of Jesus: From Clavius's Resistance to Dechales's Mathesis Regia". *Annals of Science* 78 (3), 2021, 265–294, at p. 275.

However, even in areas of research limited to purely observational activities, when the topic touched on sensitive issues that might indicate a clear departure of the author from the Aristotelian tradition, Riccioli could not ignore the constraints coming from the *Revisores Generales* of the Society of Jesus and more generally from the cultural policy of the Order to which he belonged. Consequently, likely guided by prudence, he chose to omit some information about his research procedures, as in the case of the law of Baliani-Galileo used to determine the length of the pendulum beating the second. Elsewhere, when difficulties could arise, he was compelled, as both theologian and astronomer, to formulate his views or present his interpretations within the limits imposed both by his sincere adherence to tradition and by the internal censorship of the Society. Therefore, he wisely opted for probabilistic interpretations and for the method of multiple explanations in order to 'save the appearances'.

How, then, could a Jesuit scholar reach the truth?

On the controversial matter of cosmological models in *Almagestum novum* Riccioli had already concluded: "It is fair [...] that the Copernican hypothesis is dismissed, not only out of great reverence for ecclesiastical decrees and because it is incompatible with Sacred Scripture, but also on the basis of a deeper analysis of both hypotheses and of the reasons deduced from them, considered with the greatest care".[47]

Some years later, in 1665, even more cautiously Riccioli presented his concluding answer in *Astronomiae reformatae*: "In any controversy in which opposing reasons are equivalent, the only party to be chosen is the one favored by the Higher Authority".[48]

Acknowledgements Long ago, the late Claudio Costantini and Carlo Maccagni encouraged me to transcribe and study Riccioli's letters, which are preserved in the Archives of the Pontifical Gregorian University (APUG). This event sparked my interest in, and research on, Jesuit mathematicians and astronomers. I have fond memories of discussions with both of them. I would also like to express my deepest gratitude to the late John Heilbron for his valuable suggestions on various seventeenth-century astronomical and physical subjects. My thanks go to Alberto Righini and Valeria Zanini for providing useful information on astronomical topics. I presented an early version of this paper at the tenth ESHS conference held in Brussels (September 2022) and benefited greatly from the discussion. I am also very grateful to the editors for their careful reading and editing of my text.

[47] "*Aequum est [...] Copernicanam hypothesym non ex nimia vel mera reuerentia erga Ecclesistuca decreta, Sacrasque literas reprobari, sed etiam ex profundiore utriusq. hypotheseos inspectione, et rationibus inde per summam circumspectionem deductis*" Riccioli, ref. note 23, *pars post*, 479. Translation in Dinis, ref. note 19, 288.

[48] "*In quauis controuersia, in qua Rationes hinc inde pares fuerint, ea potiùs & sola pars est eligenda, pro qua stat maior Authoritas* [Transl. by the author]" Riccioli, ref. note 24, 86.

Ivana Gambaro Qualified as Associate Professor of Didactics and History of Physics, and Vice President of the Italian Society for the History of Physics and Astronomy (SISFA). Her main research interests are in the history of seventeenth century astronomy and physics, the history of twentieth century physics, and teacher education in physics and philosophy. Her more recent publications include "Geo-heliocentric Models and the Society of Jesus: From Clavius's Resistance to Dechales's Mathesis Regia", *Annals of Science* (2021); "L'Atlas céleste chinois et grec di Gustaaf Schlegel (1875)", in I. Chinnici & M. Gargano (eds.), *Cosmic Pages: Star Atlases in Italian Astronomical Observatories* (Naples, 2022); "Pascal e le dispute sull'esistenza del vuoto", *Nuova Lettera Matematica* (2023); and "G.B. Riccioli, a Jesuit Guide to seventeenth century Astronomy", *Quaderni di Storia della Fisica* (2024).

Chapter 6
How Huygens Avoided Figleaf Physics

Jed Z. Buchwald

Abstract John Heilbron persuasively characterized Laplace's work in the early nineteenth century as a "figleaf physics," one based on putative interactions between micro particles with only the weakest links to measurement. By contrast, a century and a half earlier, Christiaan Huygens developed a theory for double refraction based on the shape of a presumptive pulse front that was underpinned by careful measurements which specified the dimensions of his presumed model.

Keywords Huygens · Double refraction · Measurement

In a series of penetrating lectures that John Heilbron gave in Bologna in the late 1980s, he characterized the efforts of Pierre-Simon Laplace (1749–1827) and such of his followers as Jean-Baptiste Biot (1774–1862) to derive empirical relations on the basis of presumptive forces between particles as a "mathematical figleaf." For, he wrote, their "business faced the grave difficulty that, insofar as the merit of any proposed law of force was its fit with observation or experiment, it could always be altered a little as circumstances seemed to require. How build solidly on so shaky a basis?"[1] He would likely have extended the characterization to much of the ether modeling that followed the spread of the wave theory of light in the 1820s and 1830s. And he would have been correct, because the most concerted effort to derive a novel relation that could be tested, namely by Augustin-Louis Cauchy for dispersion, yielded only an infinite series with constants that could not be computed *a priori* absent the force law between ether and matter particles.[2]

[1] Heilbron, J.L., *Weighing Imponderables and Other Quantitative Science Around 1800*, 1993, Univeresity of California Press, pg. 150.

[2] Buchwald, J.Z., "Cauchy's theory of dispersion anticipated by Fresnel," *in A Master of Science History*, *Archimedes*, 2012, vol. 30: 399–416.

J. Z. Buchwald (✉)
Humanities and Social Sciences, California Institute of Technology, Pasadena, CA, USA
e-mail: buchwald@caltech.edu

S. Gattei (ed.), *The Incomparable Historian*, Archimedes 75,
https://doi.org/10.1007/978-3-032-25678-2_6

Heilbron distinguished the sort of quantification in such areas as electricity, magnetism, and heat that emerged principally after the middle of the eighteenth century from the kinds of measurements that previously obtained in optics and mechanics, not least because of the development of increasingly accurate instrumentation. "The ostensible purpose," Heilbron elaborated, "of quantifying their material was to control it, to order it, to increase its reach, to guess or make the future from it" (pg. 3). Matters were different in the seventeenth century when issues of control and reach were not so pressing. Yet then too not only Isaac Newton (1642–1727) but others sought to bind their claims to accurate measurement, if not directly to microphysical models. Indirect binding between model and measurement did however take place in the seventeenth century in at least one carefully-worked instance. That instance offers an intriguing contrast with the figleafing of Heilbron's French theorists.

To bring out the contrast, begin with Laplace's figleafing theory of capillarity. In Heilbron's apposite description, Laplace presumed that the microphysical forces generating the phenomenon act only over "insensible distances" with the result that "the force occurred only in integrals that ran from zero to infinity. No parameter representing the range—the thickness of an attracting layer, or a fraction of the diameter of a capillary tube, or the power of the distance law—survived the calculation. And this very fact, which would have been a blemish in a realist theory, was a virtue to Laplace. It enabled him to complete his calculations."[3] Here, then, not only did the forces between particles vanish from the calculation, but so did the parameters that govern the forces' actions. What was the result of this elaborate exercise in excision? Only this. That capillary rise varies inversely with the tube's diameter. Hardly a startling implication, which Laplace nevertheless had René Just Haüy (1743–1822) and Louis Jacques Thenard (1777–1857) test. They obtained a measurably stable proportion. Yet no measurement at this stage of the theory could evaluate the actual range of the acting forces, which had vanished from Laplace's sole result. A very different situation obtained in the last quarter of the seventeenth century for a theory designed to account for a particularly-puzzling optical phenomenon, a theory that, like Laplace's capillarity, originated in a microphysical model. But the similarity, we shall now see, stops there.[4]

Christiaan Huygens' (1619–1695) construction for the optical properties of the doubly-refracting crystal Iceland spar was based on his conviction that light consists in a spreading disturbance within a medium comprised of particles in direct contact with one another. These particles are so formed that they possess an inherent "springiness" thereby effecting the transmission of motion through them at a finite, though extremely fast rate. Such a motion, Huygens argued, will inevitably be transmitted to the particles that each one touches, a presumption that he encapsulated in what centuries later become known as Huygens' principle. If, now, the medium's particles are so arranged—randomly as it were—that no direction of propagation is preferred

[3] Heilbron, *op. cit.*, pg. 157.

[4] The following account of Huygens' procedures appeared in parts of Buchwald, "Huygens' Experimental Determination of the Optical Parameters in Iceland Spar", *Archive for History of Exact Sciences,* 2007, 61:67–81.

then spherical waves result. If instead the arrangement is no longer "regular," with a particular axis delineating the system's symmetry, then ovaloid waves result. So much for Huygens' microphysical model. Did he, like Laplace with capillarity a bit more than a century later, efface most of the ostensive parameters that govern the presumptive results (the ovaloids for Huygens, the effects of micro forces for Laplace) of his microphysical model? And did he like Laplace produce only a singularly-limited result? In what follows we'll examine Huygens' work to tease out the connections between measurement and the parameters of his model.

Huygens' results were eventually printed in the *Traité de la Lumière* of 1690. The documentary record as presented in part in the edition of his *Oeuvres Complètes* enables a reconstruction of much of the development.[5] The editors however both reordered the manuscript and omitted pages of calculations. A proper evaluation must accordingly examine the original material, which is preserved in Codex 9 of the Huygens notebooks at the Bibliotheek of the Rijksuniversiteit Leiden. Several omitted pages of the Codex contain numerical data from which a great deal can be concluded that is pertinent for understanding just how Huygens linked his ovaloid to measurement.

From June of 1671 through 1681 Huygens resided for the most part in Paris as a founding member of the Académie des Sciences and as a client of Jean-Baptiste Colbert (1619–1683), remaining there even during Louis XIV's (1638–1715) war on the Dutch Provinces in 1672. His first notes on the double refraction date from 1672 or 1673 and concern his early understanding of the subject after having read Erasmus Bartholin's (1625–1698) recent publication on the subject.[6] These notes contain inklings (perhaps as a result of contacts with Fr. Ignace Pardies (1636–1673)[7]) of what would evolve sometime during the next 5 years into Huygens' fully-fledged wave understanding of light.[8]

The most revealing documents were written from 1677 through 1679, beginning with an explicitly-dated diagram that depicts the geometry that Huygens had by then developed for Iceland spar, together with Latin wording on its several radii, sketches of stacked spheres, and bearing the heading 'EYRHKA 6 Aug. 1677. Causam mirae refractionis in Crystallo Islandica' (Fig. 6.1).[9] There follow two undated Latin paragraphs, each of which specifies various parameters. The first of the two[10] contains calculated values for crystal angles based on Huygens' use of

[5] *Oeuvres Complètes de Christiaan Huygens,* 1888–1950, Société Hollandaise des Sciences, La Haye, Martinus Nijhoff.

[6] Bartholini, E. (1669). Experimenta Crystalli Islandici Disdiaclastici Quibus mira & insolita Refraction detegitur, Hafniae, Danielis Paulli.

[7] Dijksterhuis, F. (2004). Lenses and Waves. Christiaan Huygens and the Mathematical Science of Optics in the Seventeenth Century, Dordrecht, Kluwer Academic Publishers., p. 110.

[8] The most insightful account of the seventeenth century background to wave optics remains A. E. Shapiro, (1973). "Kinematic optics: a study of the wave theory of light in the seventeenth century." *Archive for History of Exact Sciences* 11: 134–266.

[9] *Oeuvres,* vol. XIX, pp. 427–9; pg. 45 in Codex 9. All further references to the *Oevures* are to vol. XIX.

[10] *Oeuvres*, pg. 430; pg. 48 in Codex 9.

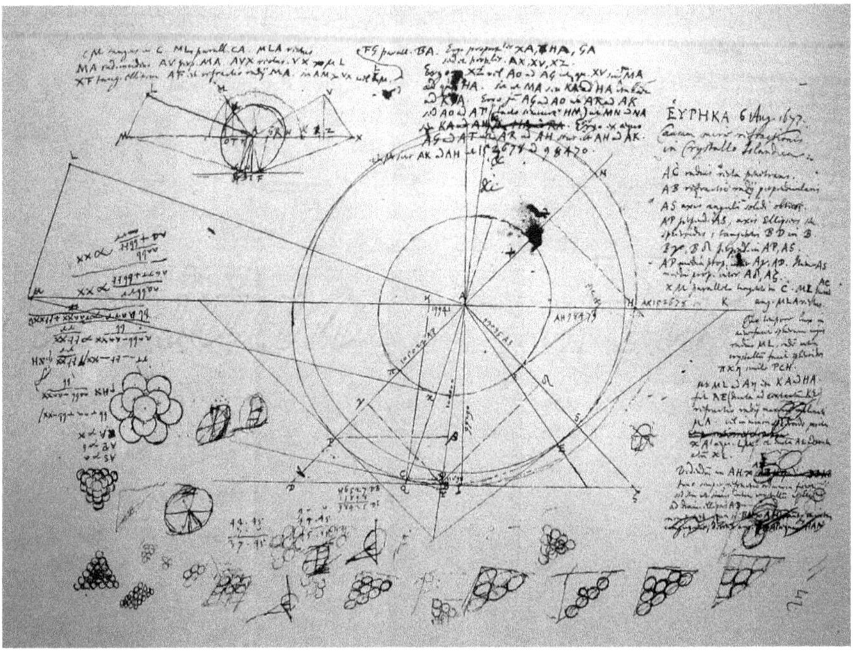

Fig. 6.1 The August 6, 1677 'EYRHKA' diagram (*Oeuvres*, insert before pg. 427; Codex 9, pg. 47)

spherical trigonometry. These values are precisely the same as the ones that can be deduced from the August 1677 diagram. The second of the two[11] is accompanied by a labeled figure that clearly shows a ray which passes through the crystal without refraction,[12] though this material dates from 2 years later (but see note 23 below). Pages 440–441 of the *Oeuvres* (extracted from another notebook) contain remarks in French (accompanied by a diagram) that Huygens titled "Observation faite le 3 Juillet 1679 qui prouve manifestement que ce n'est pas le rayon parallele aux costez du crystal qui passe sans refraction comme j'avois creu jusqu'icy". Two separately--labeled sections are next. The first is titled by Huygens "A Paris 6 Aoust 1679. EYRHKA. La confirmation de ma theorie de la lumiere et des refractions"; it examines refractions with crystals cut at various angles through the ellipsoid.[13] The second, which is undated but follows immediately in the notebook, discusses the difference between an expanding ovaloid and an expanding sphere.[14]

Although it's clear from the August 1677 'Eureka' page that Huygens had worked with Iceland spar, the extent of his experimentation at the time remains uncertain. By the summer of 1679, 2 years later, he had begun a series of careful measurements, in

[11] *Oeuvres*, pg. 430, but pg. 90 *verso* in Codex 9.

[12] *Oeuvres*, pg. 431, fig. 155 at top.

[13] *Oeuvres* pp. 441–3; Codex 9, pg. 98.

[14] *Oeuvres* pg. 443; Codex 9, pg. 99.

major part because Ole Rømer (1644–1710) had queried his remarks on the subject, possibly during a meeting on July 1 at the Paris Académie.[15] The July 3 extract noted above contains what is almost certainly Huygens' reply to Rømer, based on experiments and backed further over the next month by his work with cut crystals.

The large, central diagram on the 'EYRHKA' page from the summer of 1677 (Fig. 6.1) depicts a principal section[16] of Huygens' new surface for the extraordinary refraction of spar, together with various numbers. The smaller diagram at upper left depicts rays incident on the crystal from both left and right, while the words to the top and right provide definitions, a relation that determines the extraordinary refraction, and explanations. Just below and to the center left of the main diagram two numbers are added to produce a third. The far left of the page contains an algebraic sketch written upside down.

To understand how Huygens obtained these numbers consider first a major consequence of his structure. The basic geometry replaces the sphere of ordinary refraction with an (oblate) ellipsoid of revolution, and the consequence in question enables a refraction to be computed directly from an incident ray that lies in a principal section or in a plane normal to one. The consequence appears explicitly in Fig. 6.1 in the words just above and to the right of the small diagram at the upper left.

In Fig. 6.2 (which exaggerates the ellipticity for clarity) the line LA represents an incoming wave front whose corresponding ray is RA, drawn from a circle of radius AH; the interface is gAH, and point V is on the perpendicular to the interface from R. Within the crystal, the section gPSH of the ellipsoidal wavefront (center A), whose semi-major and semi-minor axes are respectively AP, AS (with AS the axis of rotation), determines the refracted front μC, whose corresponding refracted ray is AC. The line Lμ, perpendicular to the incoming front and touching the interface,

Fig. 6.2 Huygens' construction

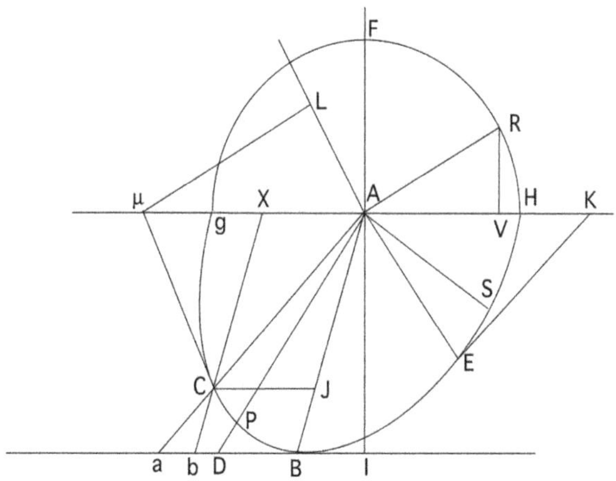

[15] Dijksterhuis, pp. 205–7.

[16] One that cuts perpendicularly down through any face along a line that bisects an obtuse facet angle.

represents the distance light travels in air in the same time that the surface gPSH is formed from the end of the front that is incident at A; point C on the refracted ray is determined by a tangent drawn through μ to gPSH. From C draw CX to the interface along a parallel to the refraction AB of a ray that is incident along the perpendicular FA (whose angle of refraction we shall call the normal deviation). Then, Huygens wrote above the diagram, the distance AX will satisfy the following relation:

$$\frac{AV}{AV}=\frac{L\mu}{AH} \textit{ which is the same as } AX=\frac{AH^2\sin\angle FAR}{L\mu} \tag{6.1}$$

Given (6.1), the procedure for computing the angle of refraction ($\angle IAC$) runs as follows, though Huygens specified the method only in his published account[17]—it does not appear explicitly in the earlier manuscripts:

$$\begin{aligned} AJ &= \frac{AB\sqrt{AH^2-AX^2}}{AH} \\ aB &= (AB)(AX)/AJ \\ aI &= aB + AB\sin(\angle IAB) \\ AI &= AB\cos(\angle IAB) \\ \angle IAB &= \arctan\left(\frac{aI}{AI}\right) \end{aligned} \tag{6.2}$$

Since Huygens had no specific need to compute a refraction from an incidence in 1677 it's possible that he had not by then developed (6.2). If he had, then he would have been able from it to reverse the calculation, and so to find AX from a measured refraction angle and the ellipse parameters as

$$AX=\frac{(AH)(AB)\left[\cos(\angle IAB)\tan(\angle IAC)-\sin\angle IAB\right]}{\sqrt{AB^2\left[\cos(\angle IAB)\tan(\angle IAC)-\sin\angle IAB\right]^2+AH^2}} \tag{6.3}$$

Given AX from (6.3), then (6.1) will yield either the corresponding angle of incidence, given Lμ (the radius of the sphere in air), or else $L\mu$ itself given the angle of incidence. Still, there is no evidence in the notebook that Huygens had (6.2), much less the ungainly expression (6.3) for AX.

The value of AH (the section of the ellipse by the interface) as well as that of $L\mu$ must be known to compute AX from (6.1). To find AH requires the orientation of the ellipsoid within the crystal as well as the angle of AB—the refraction of a

[17] Huygens, *Traité*, sec. 34.

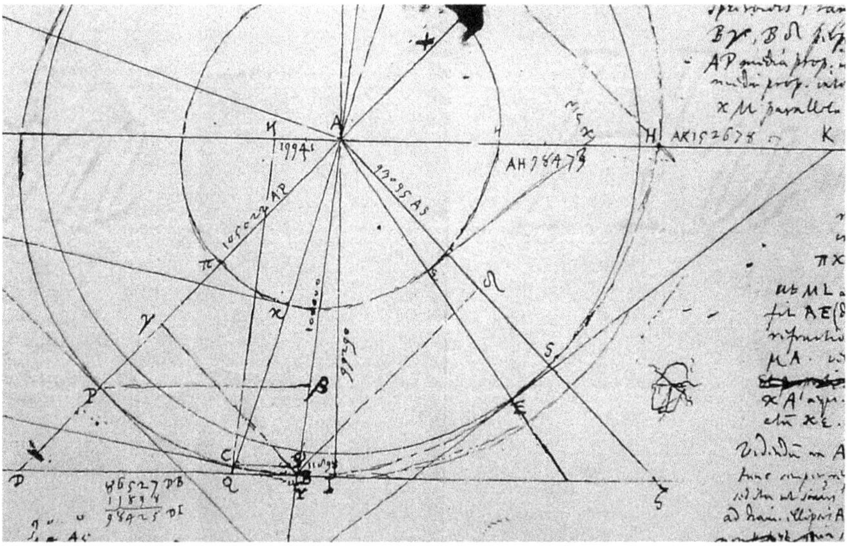

Fig. 6.3 Huygens' parameters in 1677

normally-incident ray—whose magnitude is set to 100,000 as a reference, thereby retaining six significant digits. Symmetry considerations dictate the ellipsoid's orientation in terms of the angle between crystal facets.[18] Only one other observation is then needed in order to determine Lμ, which must be found from some measurement in extraordinary refraction since it has to be specified in units of AB. This could be done by measuring both a refraction and its corresponding incidence; given both, (6.3) would yield AX and then Lμ from (6.1), but, again, there is no clear evidence that Huygens possessed (6.3). Note however (Fig. 6.3) that Huygens did depict a ray AC that goes straight through the crystal without deviation—the "radius recta penetrans", in his accompanying words. This undeviated ray intersects the tangent through B at point a (Fig. 6.2), and the corresponding value of AX is marked on the diagram. We will consider its significance below.

We can read many others of Huygens' values directly from the diagram and from the additional numbers below it and to the left (Table 6.1).[19] His figure and words together also specify the meaning of all of the marked points, excepting only one, namely I. However, the diagram places I visually along DB—the line through B and

[18] Since the refractions are the same in all three principal sections of the facets that meet at a solid obtuse angle the axis of revolution of the ellipsoid (AS) must lie in all three sections.

[19] Huygens' figure additionally depicts lines AE and EK. He describes AE as the refraction of a ray (from the left) that just scrapes the interface, while EK is the tangent to the spheroid at its intersection with AE. Considering the same (scraping) incidence of a ray AR from the right, we have tan($\angle$FAR) of incidence equal to Lμ/LA. As the incidence approaches 90° the interval LA must go to zero since Lμ is fixed, in which case Aμ becomes equal to Lμ (Fig. 6.2), and so (for scraping incidence from the left) Huygens' AK must also represent Lμ, the radius of the air sphere.

Table 6.1 Huygens' parameters

		On the 1677 diagram	In the *Traité*
AP	Semi-major axis	105,022	105,032
AS	Semi-minor axis	93,095	93,410
AH	Section by interface	98,473 (diagram) 98,470 (words)	98,779
AB	Normal deviation	100,000	100,000
AI	Vertical to tangent at B parallel to interface	99,290	99,324
BI	Interval from B to I	11,898	11,609
DB	Interval along line parallel to interface	86,527	Not used
DI	Interval from D to I	98,425	Not used
AK (Lμ—see note 16)	Radius of ordinary sphere in air	152,678	156,962
AX	Interval along interface for the undeviated ray	19,941	17,828

parallel to the interface. Moreover, in the summation to the left and below the figure Huygens has added a number to DB in order to generate DI, which makes sense only if all three of the intervals represented by that number, DB, and DI lie on the same line. Finally, and of greatest significance, the notebook contains two pages of computations immediately before the 'Eureka' (the first of the two includes an initial diagram; neither page is reproduced in the *Oeuvres*; see Fig. 6.4). And here we find BI given explicitly as 11,898, and the normal deviation (∠IAB) as 6°50′.[20]

Since DI/AI must be the tangent of ∠PAI, and so of ∠HAS, we can deduce the interfacial angle that corresponds to Huygens' 1677 parameters using the procedure that he later specified in the *Traité*. And from that we can find the angle that is made with the crystal base by a vertical edge that forms a side of two obtuse facet angles. Next, the numbers for BI, AI entail the normal deviation.[21] And finally, the angle of incidence for the undeviated ray (AC) can be deduced from Huygens' value for AX combined with ∠IAP, the normal deviation (∠IAB), and the values for AP, AS. The number for AX must have been computed from some value for the angle of incidence, which we can accordingly find from AX itself using (6.1). Note that Bartholin had the undeviated ray parallel to the vertical edge just mentioned (Table 6.2).

---

[20] Codex 9, pg. 46a for the top figure, pg. 46b for the other. Note that Huygens' diagrammatic value for AH (98,473 in Fig. 6.3) should be 98,470 given the other three radii, and he did write 98,470 in the remarks above the main figure.

[21] Huygens' procedure leads to the following expressions in terms of the interfacial angle (∠i.f.a.):

obtuse facet angle = ∠o.f.a = arccos[cot{∠i.f.a/2}cot{∠i.f.a}]

inclination of optic axis = 180° − arctan[sin{∠o.f.a/2}tan{∠i.f.a}]

The values for the obtuse facet angle and the inclination of the vertical edge, as well as for the interfacial angle, appear explicitly on pg. 430 in the *Oeuvres*, pg. 48 in Codex 9, immediately following the 'Eureka' page. In addition, the value ∠IAP appears on the page before the 'Eureka' diagram which was omitted from the *Oeuvres*.

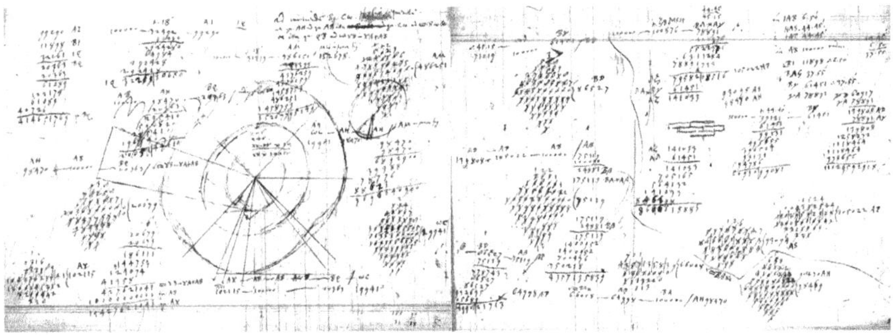

Fig. 6.4 Calculations from Huygens' notebook (these two pages are consecutive, but both are apparently numbered pg. 46 in the Codex)

Table 6.2 Bartholin's and Huygens' angles

	Bartholin in 1669	Huygens in 1677	Huygens in 1679
Inclination of optic axis to face	44°29′	44°45′	45°20′
Interfacial angle	103°40′	104°6′	105°
Obtuse facet angle	101°	101°18′	101°52′
Inclination of vertical edge	72°34′	72°	70°57′
Inclination of undeviated ray to crystal base	72°34′	72°	73°20′
Normal deviation	Unspecified	6°50′	6°40′

We see from the table that by the summer of 1677 Huygens had produced his own values for the crystal angles and for the deviation of a normally-incident ray, but that he continued to use Bartholin's claim that the undeviated ray parallels the acutely-inclined vertical edge of the crystal. The diagram has another number—the radius of the sphere in air (Lμ—that must have come from some measurement. It cannot result from the normal deviation, because that angle is used in conjunction with the reference AB and the crystal angle to set the ellipse radii. Neither can it come from a value for the ordinary index of refraction unless Huygens was prepared to assume at this point that the radius of the ordinary sphere in refraction is precisely the same as the semi-minor axis of the ellipsoid. But even if he were prepared to do so, the ordinary index would not likely have given him the number he wrote down. Though Huygens did not specify a value in 1677, Bartholin had given the ordinary index as 5 to 3. Using Bartholin's ratio combined with Huygens' 93,095 for AS would set Lμ (hereafter denoted AK for reasons given above, note 16) to 155,518, whereas Huygens' 1677 value is 152,678. Where then might AK have come from?

One possibility is that Huygens obtained AK from the presumed incidence of 18° for the undeviated ray (in which case he would not have measured it by means of a refraction since he took the assumption that it paralleled a crystal edge from Bartholin). To do that he needed first to find AX from the angle of refraction, and then AK from the (equal) angle of incidence using (6.1), but, as noted above, there is no indication that he at this point possessed (6.2) and so its consequence (6.3).

There is however another, simpler route to AX from the refraction using the algebraic expression for the ellipse to generate the radius AC along which the refraction lies. AX then easily follows (Fig. 6.2):

$$AC = \frac{(AS)(AP)}{\sqrt{AS^2\left(\cos\left(\angle IAP - r\right)\right)^2 + AP^2\left(\sin\left(\angle IAP ? r\right)\right)^2}} \tag{6.4}$$

$$AX = AC(\sin(r) - \cos(r)\tan(\angle IAB)$$

And then AK follows as:

$$AK = \frac{AH^2 \sin(i)}{AX} \tag{6.5}$$

Huygens' value of 19,941 for AX in 1677 indeed follows from (6.4) (or for that matter from (6.3)) for an 18° angle of incidence/refraction, i.e. for (as he then thought) a ray parallel to the acute vertical edge of the crystal. Huygens must therefore have produced one of (6.3) or (6.4) or their equivalents. Of the two, (6.4) is more easily found from the geometry of the ellipse and is accordingly the one that he most likely used. This is more than conjecture, because the sketch to the left of the main diagram (Fig. 6.5) shows an algebraic representation of the ellipse, with Huygens seeking an expression from it for what he labeled *x*. In the sketch he replaced AS, AP respectively with *a, b* and then introduced a pair *t, r* of variables, so that its final line reads:

$$\frac{AS^2 AP^2}{AS^2 + \frac{AP^2 t^2}{r^2}} \propto x^2 \tag{6.6}$$

The variable *x* represents some line that terminates in the end *A* of the semi-major axis, but Huygens has overwritten and obscured its other terminus. If we move in closer (rightmost image, Fig. 6.5) we see that he has drawn heavily over what might have been a *C*. In that case Huygens' *x* would be AC, and we can compare his (6.6) with our (6.4) by rewriting the latter as:

$$AC^2 = \frac{AS^2 AP^2 \left(\frac{1}{\left(\cos\left(\angle IAP - r\right)\right)^2} \right)}{AS^2 + AP^2\left(\tan\left(\angle IAP - r\right)\right)^2} \tag{6.7}$$

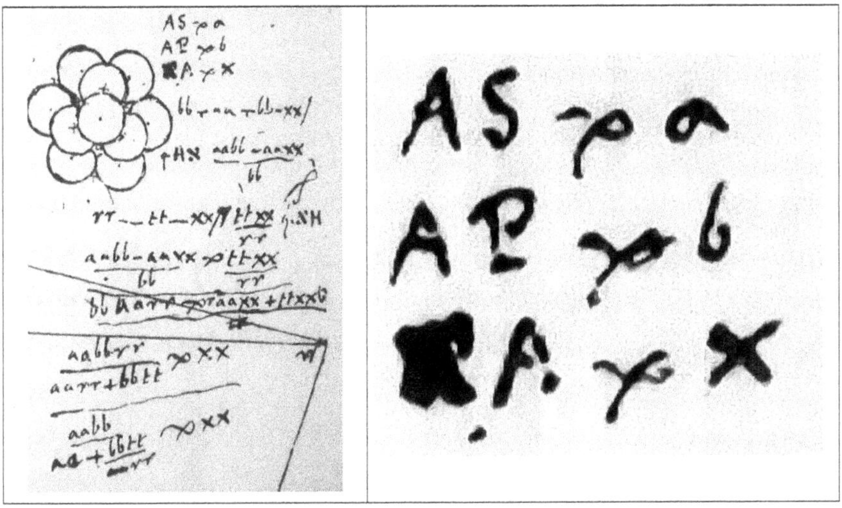

Fig. 6.5 Huygens' algebraic sketch

Note that $\angle IAP - r$ is the angle between the semi-major axis (AP) and the refraction; call it ε. The expressions $\sin^2(\varepsilon)$, $\cos^2(\varepsilon)$ would represent Huygens' *t, r* respectively, and it makes sense to express the relationships in terms of the angle between the ellipse's fixed axis AP and the refraction. Although the factor of AS^2AP^2 in (6.7) is obviously not constant, Huygens may have simply sketched a quick derivation before working out the details.[22] He needed the equivalent of (6.4) to calculate AX from an angle of refraction alone, and this algebraic fragment catches him in the act of deducing it.

Having obtained his value for AX from the undeviated ray, Huygens could then use (6.5) to calculate AK. However, this gives 150,259, not the 152,678 that he wrote on the diagram. Unless we assume that he erred to three significant digits out of six in a simple division and multiplication,[23] the discrepancy can mean only one thing: Huygens must have performed some sort of measurement in the summer of 1677. Is there any evidence as to what this might have been?

---

[22] It's also possible, but not probable, that Huygens' x represents AX, in which case our equivalent would be rewritten for comparison as

$$AX^2 = \frac{AS^2AP^2\left(\frac{\left(\sin(r)-\cos(r)\tan\angle IAB\right)}{\left(\cos(\angle IAP-r\right)}\right)}{AS^2+AP^2\left(\tan\left(\angle IAP-r\right)\right)^2}$$

Again the factor of AS^2AP^2 would not be constant.

[23] Which is unlikely given the careful multiplications that we find in Fig. 6.4, where we also see that Huygens did not use logarithms, no doubt to keep his accuracy to 6 digits.

It's certainly possible that Huygens directly measured some refraction angle, generated a corresponding AX from it *via* (6.4) (or (6.3)), and then used (6.5) to compute AK. However, not only is there no evidence from 1677 concerning such a measurement, but there are also no refraction angles (using the natural crystal) anywhere else in Huygens' notebooks or in the *Traité* except for the undeviated ray,[24] which in 1677 Huygens had not even determined for himself. If there's no evidence that Huygens measured other refraction angles, then is there any evidence that he had a different method for generating AK from observation?

Indeed there is, and we can find it by comparing manuscript fragments with his published *Traité*. The fragments were likely produced as Huygens began his new investigations during the summer of 1679. They read in translation[25]:

[Codex 9, p. 90v]

CR, CG $\propto$ a $\propto$98,473 fin. C. 6.50′ $\propto$ c $\propto$99,290, sin 6.50′ $\propto$ LM $\propto$ d $\propto$11,898. CM $\propto$ rad. 100,000

N to CG a bit greater than 8 to 5.

[Codex 9, p. 92v]

The regular refraction a little bit greater than [from] the last very exact observation [than = quam] 5 to 3, and it fits very well with [that (of) = ista] 500 to 293

The smallest elevation of the [bottom = fundi] [is] greater [−] observed many times [−] than was the case taking N at 153,456, and at the very least put[ting] N proportional to 158,928.

The [size of the] angle of the ray [that] from exact observation go[es] through without refraction was at the least about 73 gr.

N having been taken at 158,928 and the proportion of the regular refraction [having been taken] 500 to 293 [,] the ratio [of] N to the lesser axis of the spheroid is the same as that [of] N to the radius of the sphere of the extension of light in regular refraction.

According to the second sentence of the 92*v* fragment, Huygens had "many times" observed the elevation of a point in extraordinary refraction, from which he concluded that N (i.e. AK) must be 158,928 "at the very least". He had also performed a "very exact observation" of the ordinary index, finding it closest to 500/293. Immediately after this he tells us that the undeviated ray makes an angle with the horizontal of "at the least" 73°. These two measurements differ from the 152,768 and 72° of 1677, whereas all of the other parameters on the preceding fragment of 90*v* are unaltered. Note further that it is at this point that Huygens concluded the radius of the ordinary sphere must be the same as the semi-minor axis of the spheroid.

Sections 39 through 42 of the *Traité* reveal what Huygens was doing. There he describes, and then calculates, "a very singular effect", which is that the elevation of

---

[24] And the normal deviation.

[25] Huygens' Latin is loose at this point, so the translation, for which I thank Mac Pigman and Tony Grafton, can only be approximate.

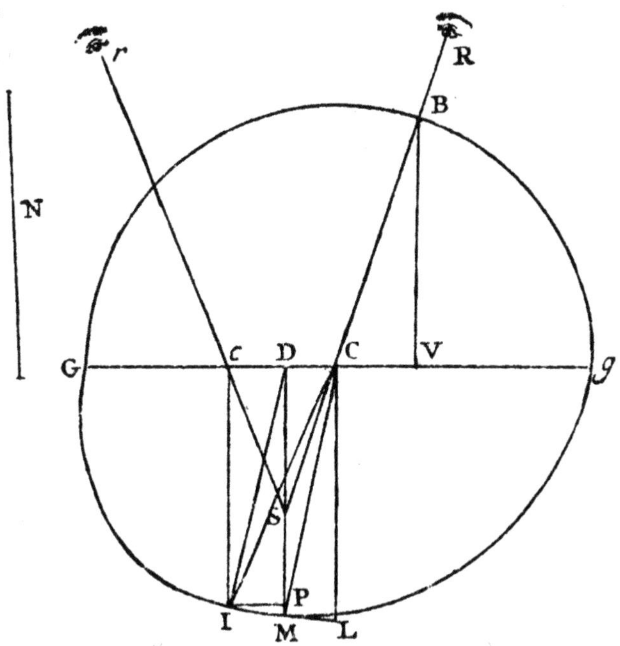

Fig. 6.6 Huygens' figure in the *Traité* for image depth observation

a point is different in extraordinary refraction depending upon which plane of incidence the eyes lie in, and that in all cases the depression of the image below the surface is greater than it is in ordinary refraction. In section 41 he computes the effect for a principal section.

The method requires placing both eyes in the principal section (Fig. 6.6) in such a fashion that the extraordinary rays from a point I that reach the two eyes (*cr* and *CR*) emerge at equal angles from the crystal.[26] The extraordinary image of I will lie at S, where the backward-projections of the rays intersect; S lies on the perpendicular line through D, which bisects the distance between the points *c*, *C* of emergence. If the eyes are sufficiently far from the surface that BV can be taken for CB, then the

[26] Huygens did not explain how to do so in practice. However, the conditions will be satisfied, he noted, if I is so marked that ID is parallel to CM, where M lies on the ellipse directly below D, and CM is the normally-deviated ray from C to M. This might be done in the following way. First set a point M below the crystal, and then mark D directly above it on the crystal surface by observing its ordinary refraction. From D, and in the principal section, mark C where the extraordinary image of M emerges along the normal by computing, in a very good approximation, the distance CD as, where *h* is the crystal height and δ is the normal deviation. Then, in a somewhat less accurate, but still good, approximation (which amounts to ignoring the distance ab in Fig. 6.2), mark I on the crystal bottom a distance CD from M. Set the eyes such that eye R sees I by extraordinary refraction in coincidence with the marked point C. The procedure requires considerable practice.

apparent depression DS of Y will be very nearly a third proportional between AK and AH (in terms of the parameters in Fig. 6.2):

$$\frac{AK}{DS} = \frac{DS}{AH} \tag{6.8}$$

Since AI represents the crystal height (Fig. 6.2), (6.8) gives the ratio of the image's measured depression *d* to the height *h* of the crystal as

$$\frac{d}{h} = \frac{DS}{AI} \text{ and so } AK = \left(\frac{d}{h}\right)^2 \left(\frac{AI^2}{AH}\right)$$

The combination of this result from the *Traité* with the fragment tells us how Huygens could obtain a value for AK without actually measuring any refraction angles at all. He took a straight rule, set it vertically on edge next to the crystal, and then marked on it the crystal's height and the apparent position of the raised point. He did this "many times", according to the fragment, and decided in the end that the best value for AK was 158,928. It's not an easy observation, which is why Huygens had to do it "many times", and why as well that the fragment sets AK to "at the very least" 158,928.

Although the 92*v* fragment almost certainly dates from 1679, the fact that it appears without any specification of the associated formula (6.8) may indicate that Huygens was already familiar with the method. If so, then we have a reasonable (indeed the only likely) source for his computation of AK in 1677. He would at that earlier date have measured a different value for the image height from the ones that he obtained 2 years later, which is hardly surprising considering his later remark that he had done the experiment "many times". Since it's quite clear that Huygens could not have used his 1677 value for the undeviated ray to compute AK, since there is no indication that he (ever) measured any other angle of refraction, and since we see that he did 2 years later refer to the method of image height for computing AK, it's plausible that this is what he did in 1677 as well.

Why would Huygens have worked with image height, which is at best an extremely difficult observation to make with any degree of accuracy? The answer is simple: it provided him with an independent, and uniquely specified, way to determine the critical parameter AK. Since Huygens always sought to produce the single best measurement that he could and never formed a resultant by combining multiple ones, he either had to choose a particular angle among all the possible ones to find AK, or else he had to find a different method that avoided the problem altogether, despite any other disadvantages. Because the method of image height requires a specific observational configuration, the problem of choosing an otherwise arbitrary angle of refraction evaporates. The notebook fragment 92*v* shows that Huygens was doing just this, certainly in 1679 and very likely in 1677 as well.

We return at long last to the comparison with Laplace. Both Huygens and Laplace certainly began with a microphysical model, albeit one in Laplace's case that he treated with the tools of the calculus. Those tools enabled him to wipe out the basic parameters of his model given his assumption that the forces act only over "insensible" distances. Huygens of course did not work with forces, and he did not go further than to assume certain symmetries for the presumptive disturbances that traversed his medium. Unlike Laplace, who threw the basic parameters of his model in the wastebin of a presumptively-vanishing integral and left the only relevant measurement to others, Huygens assiduously chased his parameters through measurements. In the late seventeenth century discrepant measurements were rarely used to generate a final result, and Huygens followed the usual procedure, which was to work at it until confident that one had done as well as possible.[27] But in so doing it was critical that the numbers generated should at least be reasonably close to one another, and indeed Huygens' manuscript fragments indicate that this was how he worked.

Where Laplace relied on a trio of reasonably-near measurements by Haüy and Thenard to sustain his claim regarding capillary rise, Huygens generated a plethora of results for his parameters. And while the capillary measurements bore at best a distant relationship to Laplace's presumed forces, Huygens' took their meaning directly from his geometric construction. That posed a problem for Laplace when Etienne Louis Malus unexpectedly produced detailed results confirming Huygens' construction after rendering it algebraically. Laplace invented a rationalization. No doubt the construction works, but not because of ellipsoidal waves. Not at all. Reverse-engineering Malus' algebra to generate an expression for a velocity that would *increase* on refraction in spar, as would be the case for a particle acted on by a force, Laplace triumphantly concluded that Malus had in effect confirmed a particles-and-forces model. But what the form of the forces might be or their range of activity and disposition, remained completely untouched, as had been the case with capillarity.[28]

Numbers, measurements, broad reaches of quantification in areas not touched in the seventeenth century certainly expanded rapidly in the early nineteenth. Yet the elaborate analytical calculations and approximations often deployed early in the century by Laplace and his cohort buried the foundations of their calculations. Huygens' geometric procedures, though notably limited in comparison with Malus' algebraic translation, nevertheless drew the basic elements of his theory in bold relief. Figleafing, it seems, was a uniquely French invention.

Jed Z. Buchwald Doris and Henry Dreyfuss Professor of History, California Institute of Technology. Before joining Caltech in 2001, he taught at the University of Toronto, directed MIT's Dibner Institute, and was Bern Dibner Professor at MIT. A MacArthur Fellow (1995) and fellow of both the American Philosophical Society and American Association for the Advancement of Science. His work spans the history of physics (Newton, Maxwell, Hertz), the interplay of theory and experiment and, more recently, the story of the Zodiac of Dendera and the decipherement of the Rosetta Stone.

---

[27] Buchwald, J.Z., "Discrepant Measurements and Experimental Knowledge in the Early Modern Era", *Archive for History of Exact Sciences*, 2007, 61:1–85.

[28] For a brief discussion see Buchwald, J.Z., *The Rise of the Wave Theory of Light*, 1989, The University of Chicago Press, pgs. 31–40.

Chapter 7
A Black Sheep in Isaac Newton's Family? The Haunts of Benjamin Smith, Jr.

Mordechai Feingold

Abstract Indubitably, the most colorful of Isaac Newton's younger relations was Benjamin Smith, grandson of Barnabas Smith, Hannah Newton's second husband. Most of what we know about him derives from anecdotes recorded by individuals who sought him out in his old age for information regarding his celebrated uncle—accounts that were, for the most part, unflattering. Consequently, Smith has been remembered, when he is remembered at all, as a 'wild character' and 'a reprobate', even 'a hellion who brooked no restrictions'. This essay reviews the surviving evidence and proposes a more nuanced portrait of this eccentric but cultivated *littérateur*.

Keywords Benjamin Smith · Isaac Newton

Shortly after Isaac Newton's death on 20 March 1726/7, John Conduitt asked Thomas Mason, rector of Colsterworth, Lincolnshire, to dispatch to London the person who stood to inherit Newton's estate there. That person was John Newton, grandson of Newton's uncle Robert Newton, and the son of John Newton who had been employed as Newton's gamekeeper. John Newton left immediately for London, carrying with him Mason's cover letter, which included an unflattering characterization of John Newton: 'I have according to yor. desire made Sr Isaac's Heir & Representative, who is ye Bearer of y^{s}, acquainted wth it, but God knows a poor Representative of so great a man, but y^{s} is a case y^{t} often happens.' Mason did not exaggerate. Within a decade John managed to waste his patrimony: he died in 1737 by breaking his neck, after 'falling down with a tobacco pipe in his mouth when he was drunk'.[1]

[1] King's College, Cambridge, Keynes MS 134, Thomas Mason to John Conduitt 23 March 1766/7; David Brewster, *Memoirs of the Life, Writings, and Discoveries of Sir Isaac Newton*, 2nd ed., 2 vols (Edinburgh, 1860), 2:425.

M. Feingold (✉)
Humanities and Social Sciences, California Institute of Technology, Pasadena, CA, USA
e-mail: feingold@hss.caltech.edu

S. Gattei (ed.), *The Incomparable Historian*, Archimedes 75,
https://doi.org/10.1007/978-3-032-25678-2_7

A more charming representative of 'so great a man' was Catherine Barton, noted as much for her intelligence as for her beauty. But certainly, the most colorful of Newton's younger relations was Benjamin Smith, grandson of Barnabas Smith, Hannah Newton's second husband. Smith's father, also Benjamin, was born in August 1651. He had two sisters: Mary, 4 years his elder—who married in 1666 Thomas Pilkington of Belton, Rutland—and Hannah, born in September 1652, who became the second wife of Robert Barton of Brigstock. Benjamin Smith married on 24 August 1693 Sarah Bishop, the daughter of John Bishop of Stickford. The couple had three children to reach adulthood: Hannah, born on 25 July 1694; Newton, born 24 May 1696; and Benjamin, born 25 January 1698. It appears that the family resided mostly in the neighborhood of Sarah's family: Hannah was born in Partney and the two other children in Halton Holegate, 2 miles south. Sarah was buried at the Church of Halton Holegate on 17 March 1699, and Benjamin Smith was interred there 8 years later.[2]

Benjamin Smith may have died intestate, for in 1708 the 12-year old Newton Smith, represented by William Wimberley of South Witham, sued his two siblings for ownership of properties in South Witham, Sewstern, and Buckminster.[3] From a 1712 letter written by Newton to William Wright of Spilsby, it appears that Thomas Brackenbury of Spilsby was Newton Smith's designated guardian, and Isaac Newton was entrusted with the guardianship of the two other children. In this letter, Newton intimated that he had asked William Wimberley to travel to Spilsby and inquire about the condition of Hannah and Benjamin and the state of their affairs.[4] Newton Smith was admitted pensioner at St John's College, Cambridge, on 6 July 1713, age 17, having previously attended the Oakham Grammar School. A year later he was admitted to Lincoln's Inn, describing himself as the son and heir of Benjamin Smith.[5] On 18 August 1715, he married Catherine Wimberley, daughter of William, and their first child was born the following year. Newton Smith inherited most of the sizeable estate that Barnabas Smith had bequeathed to Benjamin Smith, but it is unclear what legacy fell into the hands of Hannah and Benjamin Jr. Newton Smith drew up his will on 9 May 1728, 2 weeks before he was buried, bequeathing his entire estate to his wife and children.[6] No mention was made of his siblings.

Most of what we know about Benjamin Smith comes from anecdotes preserved by individuals who approached him in his old age for information regarding his famous uncle. Smith's reluctance to oblige seems to have rankled. Thomas Maude, for example—a poet who was employed by Harry Powlett, sixth Duke of Bolton, as

---

[2] A. R. Maddison, 'Notes Taken in the Church of Halton Holegate, Aug. 15, 1836), *Lincolnshire Notes and Queries* 11 (1911), 108–10, at 109.

[3] TNA C 5/268/23.

[4] William C. Wimberley, 'Original Letter of Sir Isaac Newton', *The Gentleman's Magazine* 192 (July 1852), 166. The letter is located now in the CUL, GBR/0012/MS Add.10235.

[5] *Admissions to the College of St John the Evangelist in the University of Cambridge*, eds John E. B. Mayor and Robert F. Scott, 3 vols in 2 (Cambridge, 1882–1903), 2:210; *The Records of the Honorable Society of Lincoln's Inn*, ed. W. Paley Baildon, 2 vols (London, 1896) 1:377.

[6] TNA PROB 11/622/261.

steward of his Yorkshire estates—was an admirer of Newton, intent on collecting as many details as possible about the great man. To that end, he diligently inquired 'both in and about his native parish, and among the very few of his surviving distant relations of half-blood', which included writing a solicitous letter to Benjamin Smith. The latter invited Maude for a meal, during which he allegedly railed against his parishioners, yet his guest was unable to 'obtain from him the smallest fragment' of what really interested him. Such disappointment accounts for Maude's caustic rendering of his encounter with Smith.[7]

Outright censorious was William Sheepshanks. Born in Linton on 18 March 1740/1; admitted to St John's College, Cambridge, in 1761; elected fellow in 1767; granted MA in 1769. By 1774 he had relinquished his fellowship and taken up residence in Grassington, a short walking distance from Smith's church at Linton. Having performed various ministerial duties during Smith's final years, including officiating over Smith's funeral service on 7 January 1777,[8] Sheepshanks was positioned to know Smith. And quite simply, he disapproved of the rector's character—indeed, regarded Smith unfit for the ministry—and didn't hold back when Thomas Whitaker approached him *c.* 1800 by for information about Smith:

> His temper was very unamiable: he always considered his situation at Linton as a species of banishment. He despised his parishioners, and took no pains to conceal his contempt for them. Their habits, their general poverty, and, above all, their dialect, were the perpetual objects of his derision. He called them 'baptized brutes;' and they, in return, regarded him with dislike, and treated him with disrespect.

Sheepshanks' critique was reinforced by his niece, Ann Sheepshanks, several decades later. Asked by Augustus De Morgan what she had heard from her uncle, she replied: 'Smith was one of the worst specimens of his order, even in those wretched times.'[9] Sheepshanks' disapprobation, I suggest, contributed to the garbled hodgepodge of information he furnished Whitaker, which the latter incorporated into his *History of Craven*—about Newton as well as about Smith. Certainly, Sheepshanks' animus caught the attention of a reviewer of Whitaker's book: many particulars relating to Smith are included in it, he wrote, 'but little to the credit of Mr. S[heepshanks]'.[10]

As for De Morgan, he took umbrage with Smith for a different reason. For several years he had obsessed over the precise nature of the liaison between Catherine Barton, Newton's favorite niece, and Charles Montagu, Earl of Halifax. Was she

---

[7] Thomas Maude, *Wensley-Dale; or, Rural Contemplations: A Poems*, 3rd ed. (London, 1780), 29; Thomas Maude, *Verbeia; or, Wharfdale, a Poem, Descriptive and Didactic, with Historical Remarks* (York, 1782), ix.

[8] *The Registers of the Parish Church of Linton-in-Craven, Co. York*, ed. F. A. C. Sharpe, 2 vols (Leeds, 1900–1903), 1:301–9, 329.

[9] Thomas D. Whitaker, *The History and Antiquities of the Deanery of Craven, in the County of York*, 2nd ed. (London, 1812), 463; Augustus De Morgan, 'Newton's Nephew, The Rev. B. Smith', *Notes and Queries*, 2nd ser. 3 (January–June 1857), 41–42, at 42.

[10] *Gentleman's Magazine* 85 (1805), 1144.

Halifax's mistress? Did Newton approve of their relationship? Worse still, might he have colluded in the affair for personal gain as Voltaire mischievously insinuated?

> I thought in my youth that Newton made his fortune by his merit I supposed that the Court and the city of London named him Master of the Mint by acclamation. No such thing. Isaac Newton had a very charming niece, Madame Conduitt, who made a conquest of the minister Halifax. Fluxions and gravitation would have been of no use without a pretty niece.[11]

Sheepshanks insisted that he sought to clarify the matter by asking Smith directly: when was Catherine's born? Smith purportedly responded: 'in the same year with himself'. Upon being informed of Voltaire's innuendo, Smith stated: 'when his uncle was made Warden of the Mint by King William, Mrs. Conduit was not born; and when he succeeded to the office of Master, she was only a child.' Convinced of Smith's intention to deceive, De Morgan sought to discredit the testimony by tarnishing the narrator's persona. Smith, he wrote, appears 'to have borne a character through his whole life which is entirely incompatible with his chief residence for nine years having been Newton's house'.[12]

Subsequent scholars uncritically perpetuated this reading of Smith. 'By all accounts', Frank Manuel wrote, Smith was a 'wild character' and 'a reprobate'. Westfall considered Smith to have been 'a racy profligate young man', and 'a hellion who brooked no restrictions'. Christianson dismissed Smith as a 'wild youth who consorted with the very type of reprobate element Newton had so despised'.[13] All this made me wonder: could a review of the surviving evidence yield a more nuanced portrait? What I had caught a glimmer of in my reading was an eccentric but cultivated man of letters, whose youthful penchant for high living greatly reduced his means and compelled him to accept a remote church living, where he attempted, the best he could, to harmonize his personal tastes with the demands of his new vocation.

Virtually nothing is known of Smith's upbringing and schooling. He undoubtedly attended one or more grammar schools but, unlike his elder brother, he was not sent to university. Smith intimated to Sheepshanks that when 'about eighteen years old, his uncle sent for him, and at his house he chiefly resided till the death of Sir Isaac in 1727.' Yet, no evidence exists to substantiate this. Based on Smith's failure to furnish any information about Newton not revealed by others, De Morgan concluded that Smith could not have enjoyed more than a few meals with his uncle. De Morgan was perceptive but probably too harsh. Catherine Barton, who helped convince Newton to settle a Berkshire estate worth upward of £4000 on the three children of her brother Robert Barton—who died before Newton and hence his

[11] Voltaire, *Mélanges de littérature, d'historie et de philosophie* (Geneva, 1756), 178, trans. Richard S. Westfall, *Never at Rest. A Biography of Isaac Newton* (Cambridge, 1980), 596. The anecdote first appeared in English in *The Works of M. de Voltaire*, trans Tobias Smollett et al., 33 vols (London, 1762–1781), 13:127.

[12] Whitaker, *The History of Craven*, 462n; De Morgan, 'Newton's Nephew', 41–42.

[13] Frank E. Manuel, *A Portrait of Isaac Newton* (Cambridge, MA, 1968), 35; Westfall, *Never at Rest*, 849, 857; Gale E. Christianson, *In the Presence of the Creator. Isaac Newton and His Times* (New York and London, 1984), 573.

children would not have had a share in his estate[14]—undoubtedly looked after her nephew Smith as well. Indeed, when in London, it is more likely that Smith boarded with Catherine and John Conduitt—who married in 1717, just as Smith turned 18—than with Newton. Still, he did interact with the great man and, consequently, it seems reasonable to accept the gist of several anecdotes attributed to him: that in his advanced years Newton became 'rather corpulent'; that Caroline, Princess of Wales, visited Newton when 'his age and infirmities' prevented him from waiting upon her—albeit not 'frequently' as Smith claimed; and that of all of his 'intimate friends' Newton 'had the greatest regard' for Samuel Clarke—'whom he called his Chaplain'. More questionable is Smith's recollection that Newton held Edmund Halley 'with disapprobation, on account of his infidelity and licentious conduct.'[15]

According to Sheepshanks, Smith entered Pembroke Hall, Cambridge, shortly before Newton's death. In fact, as we shall see below, this event occurred several years later. During the 1720s Smith instead pursued a lifestyle of a 'dilettante': an 'aesthete who hobnobbed' with men of letters and polite society. One of Smith's biographers cited Newton as having described his nephew as 'a man of wit, possessing great powers of conversation'; Smith was certainly endowed with both charm and intelligence, which, along with the well-known fact of his famous relations—both Newton and the Conduitts—undoubtedly facilitated his entry into London's literary circles. Some evidence for Smith's dabbling in poetry is recorded. Dixon claimed to have seen a couple of his compositions which, he believed, were quite good. At least one of his poems, a risqué construal on the Orpheus and Eurydice myth—perhaps written in the second half of the 1720s, and known to John Gay who refered to it in the *Beggars Opera* (1728)—was published in 1735, complete with a musical score:

Young Orpheus tickled his harp so well,
With a twinkum twankum twang,
That he gained his Eurydice out of hell,
With a twinkum twankum twang.
But had she been honest as she was fair,
'Tis a thousand to one she had never come there.
'Tis much to be feared she proved a scold,
And therefore the Devil had got her in hold.
But lest she should poison all hell with her tongue,
He straightway released her for an old song.
Which was twinkum, twankum, twinkum, twankum,
Twinkum, twankum, twang.[16]

[14] John Nichols, *Illustrations of the Literary History of the Eighteenth Century*, 8 vols (London, 1817–1858), 4:13.

[15] Whitaker, *The History of Craven*, 462.

[16] *A complete collection of old and new English and Scotch songs, with their respective tunes prefixed*, 2 vols (London, 1735), 2:139–40. The score was occasionally reprinted. See, *The Convivial Songster: Being a Select Collection of the Best Songs in the English Language: Humourous Satirical, Bachanalian, &c*... London 1784), 242–43. I cite the slightly revised (and improved) version of the ditty, copied by Sheepshanks, and published in Augustus De Morgan, 'Newton's Nephew II', *Gentleman's Magazine*, 2nd ser 3 (1857), 250.

During this period, Smith surely exhibited the profligacy and eccentricity that became his sole legacy. Sheepshanks, who inspected Smith's papers following the latter's death, states that he found several letters from Newton to his nephew, in which he 'pressed him to chuse a profession'. Testimony of Newton's displeasure with his nephew's lifestyle may be inferred from his couching his admonitions with 'some vulgar phraseology', which prompted Sheepshanks to burn the letters.[17]

Following Newton's death, Thomas Pilkington, Catherine Conduitt, and Benjamin Smith became the chief administrators of his estate. Totaling some £32,000, the fortune was divided equally among eight relatives. The sticking point proved to be the ownership of Newton's papers. Ultimately, since only Conduitt was wealthy enough to indemnify Newton's debts, it was agreed that he would inherit Newton's *nachlass* in perpetuity, save for a few manuscripts deemed fit to be published, which would be 'sold to the best advantage' of the beneficiaries. Conduitt agreed to sign a £2000 bond, 'to be answerable to the relations for all profit I should make by selling or otherwise disposing' of them.[18]

Each of the relatives appears to have also received a painting or a statue of Newton, in addition to other souvenirs and plate. Newton Smith received one of Kneller's portraits of Newton, which William Stukeley considered to be 'the finest original picture' of the great man. Smith's daughter, Sarah, inherited the painting and, following her death in 1786, it was purchased by the Duke of Rutland.[19] Thomas Pilkington acquired another painting by Kneller, as well as 'Medals, Pistols, Gun, and some plate'. Pilkington's impoverished granddaughter sold it *c.* 1805.[20] As for Smith, Maud stated that he 'left a small ivory bust of admirable workmanship, executed by that celebrated artist, Marchand, which from its elegance, similitude and placid expression, is truly valuable. It is said to have cost Sir Isaac one hundred guineas'. Maud's phraseology makes clear that the bust was still in Smith's possession when he died. Consequently, it cannot be that Le Marchand ivory bust which the astronomer Matthew Raper bequeathed to the British Museum in 1765.[21]

According to Sheepshanks, Smith told him that Newton 'left him about £500 *per annum*'. Since Newton died intestate, Sheepshanks was either misinformed or he misconstrued Smith's words. Smith's share of Newton's estate amounted to *c.* £4000; even if wholly invested in stock and land, such a legacy could not have yielded more than £250 *per annum*. Intriguingly, shortly after receiving his share, Smith received funds from an additional source. A 1735 legal settlement reveals that the Wragg Marsh estate in Lincolnshire—property of Richard Thompson—'was charged with an annuity of £200 paid to Benjamin Smith, of Stamford, Esq., for 2

[17] Whitaker, *The History of Craven*, 463.

[18] King's College, Cambridge, Keynes MS 127A.

[19] William Stukeley, *The Family Memoirs of the Rev. William Stukeley, M. D.*, ed. William C. Lukis, 3 vols (Durham, 1882–1887), 2:263; William C. Wimberley, 'The Genealogical Descent of Sir Isaac Newton', *The Gentleman's Magazine* 38 (July–December 1852), 271–73, at 273.

[20] W. I. R. V., 'Portrait of Sir Isaac Newton', NQ, 7th Ser. 7(January–June 1889), 347.

[21] Maude, *Wensley-Dale*, 80n; Milo Keynes, *The Iconography of Sir Isaac Newton to 1800* (Woodbridge, 2005), 72–73.

years'.[22] Thompson must have been either the grandfather or the uncle of Carrier Thompson, who married in 1716 Hannah Smith, Benjamin's sister. The payment might have been intended to compensate Benjamin Smith for relinquishing whatever share he was due in his father's estate.

Sheepshanks also confused details regarding Smith's sojourn on the Continent. Soon after Newton's death, he narrated, Smith had 'left England, and resided at Paris about two years; there he became acquainted with Mr. Philip Yorke, afterwards earl of Hardwicke, and Lord Chancellor. He then went to Rome, where he staid about three years more.' Such an extended journey abroad took place only several years later. Smith may have traveled briefly to Paris in the late 1720s. He became famous, or notorious, for his obsession with dancing, an obsession that purportedly took him to the French capital on several occasions, to acquaint himself with new fashions. He also lived wastefully. In early May 1730 a newspaper advertisement announced an extravagant wager he had laid:

> The Beginning of next Week Benjamin Smith, Esq; a Nephew to the late Sir Isaac Newton, sets forward from Stamford in Lincolnshire to Clerkenwel Church, which he is to walk in thirty hours: There are some thousand Pounds depending on the same. It seems one Homer is to attend him to see it done.[23]

I'm unable to find evidence of Smith's actually venturing to traverse the 90 miles in the time allotted. What is of interest, however, is Smith's reflection on the wager, which casts light on his reputation as a passionate classicist. Reflecting on the name of the person who was appointed to accompany him, Homer, Smith invoked Ovid: 'De mihi maeoniden, & tot circumspice Casus' ('Give me Homer and look around upon my many misfortunes; [his talent will be entirely destroyed by such great troubles]').[24] Whatever personal misfortunes Smith had in mind in 1730, graver ones lay ahead. A later anecdote recounts another wager, this time to the tune of 100 guineas, 'in walking between Stamford and Grantham against time'. Smith won the wager, being 'a great pedestrian … fitly made for the purpose, thin and tall, with a suitable frame of body. Such a feat earned him the name of "walking Smith"'.[25]

By 1732 Smith finally submitted to his uncle's wish that he choose a profession. On 23 September he was ordained and appointed deacon of All Saints Church in Stamford, whose rector was Newton's friend and admirer—and Smith's friend—William Stukeley. Eight months later Smith was ordained priest. By then steps were taken to procure for him a church living. Conduit and Martin Folkes had prepared for publication Newton's *Observations upon the Prophecies of Daniel and the*

---

[22] *Lincolnshire Notes & Queries* 16 (1921), 11.

[23] *The London Evening-Post*, 30 April–2 May 1730.

[24] Ovid, *Tristia*, I.i.47–48; S. G. Owen, 'On Some Passages of Ovid's Tristia', *The Classical Quarterly* 8 (1914), 21–32, at 21.

[25] 'Anecdotes of a Singular Character', *The Sporting Magazine* 8 (1796), 74–75, at 74. For the popularity of competitive pedestrianism in eighteenth-century Britain, see Earl R. Anderson, 'Footnotes More Pedestrian Than Sublime: A Historical Background for the Foot-Races in Evelina and Humphry Clinker', *Eighteenth-Century Studies* 14 (1980), 56–68; Bob Harris, *Gambling in Britain in the Long Eighteenth Century* (Cambridge, 2022), 116–22.

Apocalypse of St. John. Smith, who apparently contributed little, if anything, to the enterprise, was permitted to dedicate the book to the Lord High Chancellor, Peter King—one of the pole-bearers in Newton's funeral—in anticipation of a proper reward. The dedication was brief and somewhat constrained. Having justified the choice of the dedicatee on account of his long friendship with Newton, Smith concluded: 'I shall always reckon it one of the Advantages of my Relation to *Sir Isaac Newton* that it affords me an opportunity of making this publick acknowledgment of the unfeigned Respect of … Your Lordship's most obedient, and most humble Servant'.[26] The book was published in January 1732/3, and was reviewed in the February and March issues of *The Present State of the Republick of Letters*.[27]

Smith was promptly rewarded. On 15 June 1733 he was appointed rector of the second Mediety of Linton, Yorkshire. Smith was clearly disappointed—not only on account of the remote location, but the meagerness of the revenue. Sheepshanks recorded Smith's abject rendering of the manner in which the living was bestowed on him. Shortly after the publication of the *Observations upon the Prophecies of Daniel and the Apocalypse of St. John*, Peter King summoned Smith

> and addressed him as follows: A mediety of the Rectory of Linton, in Yorkshire, stated to me as worth £100 per annum, is now vacant: will you accept it?' Mr. Smith remained silent some minutes. The Chancellor repeated exactly the same words in a stronger tone of voice. [Smith] now saw this was the only thing he should ever have offered, and that, from the look and tone of his patron, he had nothing farther to expect. He, therefore, accepted the living; but always mentioned the interview and the offer as a cruel mockery, having fixed his own expectations upon £500 at least, and that in a situation more fitted to what he thought of his own taste and manners.[28]

Nor did some of Smith's Lincolnshire acquaintances approve of his new vocation. William Warburton, future bishop of Gloucester, reacted furiously when Stukeley informed him of Smith's ordination—an ordination Stukeley had sponsored. 'It is with the utmost concern I see you write that you gave [Smith] a title; purely on your own account', Warburton reprimanded on 10 February 1732/3, 'because I am afraid it should raise a terrible clamour against you. The news of his going into orders creates a furious scandal here'. After venting about the general malaise of the times, Warburton concluded:

> I have no notion of a man's running to the tremendous Altars of Jesus, reeking from the hot pollutions of a brothel, and covered over with all the stains of lewdness and impiety. We may write, and preach, and idly waste a midnight lamp, in defence of our sacred dispensation but, while such become our coadjutors, and sit with us as watchmen on the rampart of Faith, we shall be an eternal derision to our adversaries, and the torrent of Infidelity will still roll on.

Nine days later Warburton wrote again. Stukeley had assured him that it was not he but someone else who had signed the ordination 'testimonial'; he merely furnished

[26] *Observations upon the Prophecies of Daniel and the Apocalypse of St. John* (London, 1733), A2-A2v.

[27] *The Present State of the Republick of Letters* 11 (1733), 121–41, 165–77.

[28] Whitaker, *The History of Craven*, 463.

Smith with a 'title'—perhaps referring to Smith's characterization as 'eruditus' ('learned') in the ordination record. This was 'a matter that relates only to his support, not at all to his morals', Stukeley justified, adding as an additional excuse: recently Smith had nearly lost an eye. This tidbit prompted the somewhat mollified Warburton to quip: 'one would imagine the Bishop was in that condition when he laid his hand on [Smith], and so … saw but half his faults. But Providence will turn every thing to the best and, if the Gentleman reforms, I shall be exceeding glad.'[29]

Lacking an academic credential must have weighed on Smith. He was admitted fellow commoner to Pembroke College, Cambridge, on 18 July 1734—described as 'Presbyter'. Brazenly, in the Admission Register he recorded his age as 24, thereby shaving off some 11 years. The tutor assigned to him was William Trollope, a native of Casewick, located some 3 miles east of Stamford. Perhaps the two had known each other and conspired to provide Smith with an academic pedigree. By his own admission, Smith remained at Cambridge for only a short time. But even such a brief residence entitled him to acquire in 1746 a B.D. degree, as a '10 year man'.[30]

As mentioned above, Sheepshanks stated that Smith spent 5 years in France and Italy; another account suggested a longer stay abroad: upward of 12 years, which is probably a more accurate estimate. Smith appears to have left England in early Fall 1734, not taking up residence in Linton until 1743. The only solid evidence regarding his whereabouts is to be found in the travel diary of the physician Alexander Cunningham, later Sir Alexander Dick, third Baronet of Prestonfield (1703–1785), who accompanied the young aspiring artist Allan Ramsay on an Italian tour. The two left London on 24 July 1736 and after spending 5 weeks in and around Paris set out on 29 August for Italy. They arrived at Genoa on 23 September where it was their 'fortune' to meet Smith, who 'was desirous to be of [their] company to proceed through Italy by the way of Leghorn'. They set sail on 28 September only to encounter a terrible storm, which led them to conclude they were doomed. 'Smith and his man Tom, who were both stript in order to swim, had many grievous moanings betwixt them fearing a sudden separation for ever.' The wind ultimately died down and the felucca reached shore not far from Pisa, where 'Mr Smith who felt uncommon joy in recovering his man Tom, he gave us all his benediction in the warmest manner.'[31]

After visiting Livorno, Cunningham and Ramsay arrived at Florence on 5 October where they stayed till the 21st. Smith and his man Tom appear to have been there, too, for they traveled together to Rome, arriving there on 26 October. From a letter Cunningham sent his father in late March 1737, it appears that Smith was a constant and informed companion:

[29] Nichols, *Illustrations of the Literary History of the Eighteenth Century*, 2:20–21, 23–24, William Warburton to William Stukeley, 10, 19 February 1732/3.

[30] Pembroke College, Cambridge, 'Register of Matriculations', 327. I wish to thanks Elizabeth Ennion-Smith, the College Archivist, for her assistance in investigating Smith's Cambridge career.

[31] *Curiosities of a Scots Charta Chest 1600–1800*, ed. Atholl Forbes (Edinburgh, 1897), 195–6.

> [We] have been over all observing the grandeur and variety of Human glory, the particulars of which would fill volumes, among other things he made very curious and learn'd observations on the purity of the first centuries of the Church and the places the Apostles were martyr'd and the Catacombs which we saw where the Saints were buried. He made me observe the plainness and decency of the Church of England and this—and gave good reasons why a reform must inevitably happen soon here for the good of Christendom.

On one occasion, after touring the Temple of Bacchus, on the outskirts of Rome, they passed the coach of the Chevalier de St George, as James II's son was known. The sight of the 'Old Pretender' prompted Smith to reflect 'upon the unhappy fate of the old Chevalier, who had lost three kingdoms for his religion, and whose aspect had a very melancholy cast.'[32]

Smith did not follow Cunningham to Naples in late March 1737, but remained in Rome. By August he was in Siena from whence he sent a gloomy letter to Cunningham, in which he vented about the people of his native Lincolnshire town of Stamford:

> I doubt not in your passage through Stamford you was better informed than I am in what scurvy manner I have been treated, the worst of such usage is that it serves for an example to others when once a man is on the ground every Ass will be kicking. The money that I told you I expected to receive when you left Rome was not paid till this very day morning. That had not Belloni and another person supplied me I know not in what manner I could possibly have existed.[33]

A contemporaneous letter from Ramsay to Cunningham adds context to Smith's letter: 'You must know after the most humane treatment on my side Parson Smith has behaved to me', like the 'rascal' that he is. Four months earlier, 'when he was in great necessity and none of the English would assist him', Smith begged Ramsay for a loan of 20 crowns, promising to repay within a month. Ramsay obliged, 'believing him to be an honest unfortunate man'. However, Smith not only failed to repay his debt but

> had gone off to Sienna *en tres mauvaise odeure*, after having borowed £150 of an English man there who had got it left him in a legacy and a 100 crowns f^r Belloni for which Parker stands accountable. I have sent Parson Smith's letter to Belloni who has promised to see me pay'd as soon as Smith's money comes.[34]

The debts were ultimately paid, but Smith's reputation was tarnished.

Details concerning the last four and a half decades of Smith's life are few and far between, primarily anecdotes. According to a contemporary biographer, 'having spent most of his fortune' abroad, Smith returned home, accompanied with 'a genteel Italian servant—probably not the Englishman Tom, mentioned above—who was an excellent classic, and a dog named 'Sereno'. The dog was Smith's faithful companion, and in its last sickness the animal was attended by a physician 'and a

---

[32] Ibid, 128, 124.

[33] Ibid, 133.

[34] Ibid, 134–35.

nurse, who sat up with him several nights in the last stages of his illness'.[35] Archbishop Herring's 1743 ecclesiastical visitation furnishes a precise date for Smith's arrival at Linton. Responding in early summer, Smith intimated that he had only taken up residence 7 weeks earlier, and as the parsonage House was not 'in a Condition for [him] to inhabit', he ordered repairs. The responses to the Visitation also reveal that the parish was inhabited by some 160 families, served by two rectors, Thomas Gale and Smith. Gale did not reside in Linton and his curate, John Alcock, carried out his ministerial functions, for which he received £25 per annum. As for public services, they were performed 'twice every Ld's Day, & once Every Holyday, Wednesday and Friday in Lent.' Catechizing children and servants also took place during Lent.[36]

Smith expended considerable expense in renovating the parsonage. Once completed, it included a study, with an adjacent bedroom; a room for his 'female domestic'; and a closet for his learned attendant. The latter was undoubtedly the Italian servant whom Smith had brought back from Italy, and who read to Smith Greek and Latin works. For years, we are told, when Smith was unable to fall asleep, he would ring the bell and 'his reader then arose, procured a light, and read to him two or three hours, till he found himself disposed to sleep.' As for the 'female domestic'—who, it was widely rumored, was Smith's mistress—she was apparently a formidable figure: 'a grand madam for housekeeper, who dressed in an unheard-of fashion, and spoke a language which the simple villagers did not understand.'[37]

Smith's passion for dancing was the pursuit most commented upon. Thomas Maude furnished the earliest and fullest account: A 'scholar and a recluse', Smith was

> reputed one of the best dancers in England. He commenced the art at an early age, continued the practice with its refinements throughout life, and occasionally travelled to the continent to retouch, and report the last fashionable airs. He daily exercised in private at home, to a rustic fiddle, but with this peculiarity, as I have been informed, in the front of a looking-glass, with his musician's back turned towards him. However, one day the fiddler's curiosity being excited to the daring presumption of looking over his own shoulder at his master's steps, the reflecting mirror gave notice, and the poor man was accordingly mulcted, and occasionally suspended, for his imprudence.

Robert Southey wittily likened Smith to King David: 'a mighty dancer before the Lord', who had 'paid twelve guineas for learning one dance in France'. According to another anecdote, when riding to visit a friend in fine weather, 'he would sometimes alight, tie his horse to a gate, and dance a hornpipe or two on the road to the astonishment of any who happened to pass.' One commentator went so far as to make Smith 'a prominent figure among the dancing members of the Old Assembly

[35] 'Anecdotes of a Singular Character', 74. Sheepshanks concurred: 'compelled by necessity, he came to reside at Linton'. Whitaker, *The History of Craven*, 463.

[36] *Archbishop Herring's Visitation Returns, 1743*, eds Sidney L. Ollard, and Philip C. Walker, 5 vols (Cambridge, 1928–1931), 2:146.

[37] De Morgan, 'Newton's Nephew', 42.

Rooms, Leeds'—unaware that the establishment only opened its doors the very year in which Smith died![38]

The state of Smith's finances also appears to have been muddled by Sheepshanks. 'Circumstances now fell out', he wrote, 'which drove him to the necessity of a long course of law to recover his annuity'. During his 'seven years of poverty and distress', Smith repeatedly applied to Lord Chancellor Hardwicke 'for additional preferment; but always met with a refusal. These repulses he never mentioned but with great asperity and indignation.' The reason for Hardwicke's refusal, Sheepshanks conjectured, was his disapproval of Smith's 'conduct and character in early life'—and perhaps rightly so, Sheepshanks glossed: 'In no part of his life, so far as I know, had [Smith's] conduct been so regular as that a Patron, who was acquainted with it, could find any satisfaction in promoting him.'[39]

It is unclear which lawsuits Sheepshanks referred to. Smith was involved in two sets of litigations during the late 1720s and early 1730s. In the first, he was pitted as a defendant, alongside John Conduitt and Thomas Pilkington, against claims by John Warner, the husband of Catherine Conduitt's sister, that Newton owed him a considerable sum of money.[40] The second litigation, taking place between 1729–1731, involved a property dispute between Smith and his brother-in-law Carrier Smith which, as noted above, appears to have been resolved.[41] These suits do not correspond with Sheepshanks' description. Perhaps they relate to another tidbit of information Sheepshanks related. Upon returning to England, with his fortune diminished, Smith sold 'whatever interest remained in his estate to Sir Robert Clifton, a Nottinghamshire Baronet, for an annuity of £200 per annum, for their joint lives—a bargain which might have left him without bread to eat.' Robert Clifton, fifth baronet, died in 1767, and it is possible that his successor, Sir Gervase Clifton, sixth baronet, refused to continue to pay the annuity, so that Smith resorted to legal action. Lastly, there exists a 1761 'record of a fine' in the King's Silver Office, in which William Manners was the plaintiff, and Benjamin Smith, with his sister Sarah and her Husband Benjamin Brown were the 'deforciants', in the matter of 'messuages, cottages, and lands in South Witham.'[42]

Be that as it may, the anecdote highlights another addiction of Smith: gambling. As noted above, Smith was a pedestrian who often bet on his talent. He was also an

[38] Maud, *Verbia*, viii; Harker, *Rambles in Upper Wharfedale*, 141; Southey's Common-Place Book, Fourth Series, ed. John W. Warter (London, 1850), 415–16; John Doran, *Saints and Sinners: Or, In Church and about it*, 2 vols (London, 1868), 2:246–47; 'Anecdotes of a Singular Character', 74; Edmund Bogg, *Two Thousand Miles in Wharfedale* (London, 1904), 525.

[39] Nichols, *Illustrations of the Literary History of the Eighteenth Century*, 4:34. Peter King died in 1734. Hardwick served as Lord High Chancellor from 1737–1756. Sheepshanks claimed that Smith met Hardwick in Paris, but I found no evidence that the Lord Chancellor visited the French capital during Smith sojourn to the Continent. Perhaps Smith traveled to Paris in 1749, and met Hardwick's son, Philip, who was there in Fall 1749.

[40] TNA C 11/773/24, C 11/859/9, C 11/775/55, C 11/776/20, C 11/495/35.

[41] TNA C 11/1804/11, C 11/355/31, C 11/1929/28, C 11/1796/12, C 11/1690/6, C 11/1792/3.

[42] Wimberley, 'The Genealogical Descent of Sir Isaac Newton', 273.

avid card player. Upon encountering 'any poor person who was a good cribbage player', we are informed, Smith 'would maintain them three or four months only for the sake of playing with him.' Smith was also credited for being a 'fortunate as a lead-mine adventurer, and a share he bought for £80 he sold for an annuity of 120 guineas.' Above all, Smith delighted in betting on annuities. Sheepshanks recalled Smith's steadfast refusal to reveal when he was born; he had also falsified his age upon entering Pembroke College. The reason for this, one commentator explained, was 'for the purpose of making advantageous contracts in life-annuities'. According to his biographer, Smith, like most contemporaries, paid little attention to the growing body of literature which applied the mathematics of probabilities, based on mortality rates, to annuities—such as Abraham de Moivre's *Annuities Upon Lives* or Thomas Simpson's *The Doctrine of Annuities and reversions*—and they engaged in the practice largely as a game of chance. Smith seemingly won as often as he lost. 'When he had accumulated any considerable sum, he always purchased with it a life annuity'. One such transaction was with a wine merchant—perhaps Cuthbert Readshaw, mayor of Richmond—'with whom he had long dealt for wine'. However, after the contract was drawn, Smith 'could never be induced to take a drop of liquors, from apprehension that his friend might shorten his life.'[43]

The same source also cast doubt on Smith's abject poverty. Though his lengthy stay on the Continent had 'greatly diminished his fortune', in later years his yearly revenue from his rectorship and his various speculations was estimated to be between £500–£700 *per annum*, though much it had been 'yearly consumed among low parasites and fantastical projects'. Maude concurred: 'He was easy in his circumstances, and chiefly laid out what he acquired, in annuities for his own life.'[44] Smith's will further confirms his being a person of some means: he possessed various properties in and around Linton, in addition to unspecified monies.

What remains is to consider Smith's relations with his parishioners. Sheepshanks, we noted, claimed that Smith despised them, and they responded in kind. Dixon, in contrast, recalled conversations he had with 'several very old inhabitants' of Linton, whose memories of their former rector differed from Sheepshanks': 'they one and all spoke of him with respect and cast no reflection whatever on his moral character. They would say he was a queer gentleman, but not a bad one with those who knew how to humour him.'[45] Clearly, Smith had misgivings about the nature and location of his calling. To cite Dixon again, mid-eighteenth-century Linton, was not a place for 'a gentleman of extensive learning, who in his leisure hours pondered over the divine conceptions of the Greek Dramatists, and the orators and poets of glorious old heathen Rome'. Those who interviewed him in his old age might have heard him belittle his parishioners, but he also appears to have treated them with gentleness. Mindful of their poverty, he never took legal action against those who were late to pay their fees. His charity, as well as wit, may be inferred from yet another anec-

[43] Ralph Nevill, *The Merry Past* (London, 1909), 252–53.

[44] Maude, *Verbia*, viii.

[45] Dixon, 298.

dote: Having waved wedding fees from poor couples, Smith discharged them by saying: 'Go your way, poor devils, go your way; I have done you mischief enough already'.[46]

Smith drew up his will on 18 February 1776. Noteworthy is that it lacks the standard preamble pertaining to the testator's religious convictions. Smith left his entire estate to his housekeeper, Mary Spink, who may or may not have been the housekeeper whom Smith had brought with him to Linton in 1743. Also possible, she was Mary Leyland, baptized in Linton on 9 March 1728, who married on 25 September 1750 William Spink, with Smith officiating. If correct, it is unclear whether William died or Mary divorced him. Certainly, the will refers to her as 'Miss Mary Spink'; the administration of the will refers to her as a 'spinster'. One Mary Spink was buried in Linton on 12 March 1794. However, as no will appears to have survived, it cannot be determined for certain whether she was Smith's beneficiary.[47]

Thus passed away the last beneficiary of Newton's estate. Further research may uncover additional information on the life of this eccentric, cosmopolitan man of letters, who came to consider his rectorship as a place of exile, perhaps even his purgatory. Recalling his younger years, he commented: 'Instead of cultivated society, I have been driven to herd with baptized brutes, and when I was gaping for a pair of colours, I was thrust into the tub [pulpit].' Upon reflection, however, he would add that 'he believed Providence had placed him there to expiate for the sins of his youth'.[48]

Mordechai Feingold Kate Van Nuys Page Professor of the History of Science and the Humanities, California Institute of Technology. His publications include *The Newtonian Moment: Isaac Newton and the Making of Modern Culture* (Oxford, 2004) and two books co-authored with Jed Buchwald: *Isaac Newton and the Origin of Civilization* (Princeton, 2012), and *The Winding Trail to Newton's Principia Mathematica* (forthcoming).

[46] GM 2nd ser 3 (1857), 250.

[47] *The Registers of the Parish Church of Linton-in-Craven*, 1:248; 2:398.

[48] De Morgan, 'Newton's Nephew II', 250; Maude, *Verbia*, ix.

Chapter 8
The Astronomy of Thomas Hornsby, 1763–1772: History, Geography and Computation

Jim Bennett

Abstract This paper examines the early astronomical career of Thomas Hornsby during his tenure as Savilian Professor of Astronomy at Oxford (1763–1772), a period marked by an unusual fusion of historical scholarship, geographical inquiry, and extensive computational practice. Drawing on his six contributions to the *Philosophical Transactions*, the study highlights Hornsby's distinctive approach to determining the solar parallax through meticulous engagement with global observational records, from historical measurements by Flamsteed and Horrocks to contemporary data from the 1761 and 1769 transits of Venus. His analyses reveal both a remarkable command of archival sources and an enthusiasm for re-computing, correcting, and synthesizing disparate observations. Hornsby's ambitious proposals for the 1769 transit—supported by his use of terrestrial globes and voyagers' reports—demonstrate an unusually expansive geographical imagination within eighteenth-century astronomy. The paper further traces Hornsby's observational work on the 1769 transit, his careful documentation of the 'black drop' effect, his revised determination of the solar parallax, and his final major investigation into the proper motion of Arcturus and the obliquity of the ecliptic. Together, these studies illuminate an astronomer whose archival diligence, computational skill, and geographical curiosity generated a distinctive and self-sustaining scientific practice prior to his later transition to meridian-based observational astronomy at the newly founded Radcliffe Observatory.

Keywords Thomas Hornsby · Solar parallax · Transits of Venus · Eighteenth-century astronomy · Astronomical computation · Radcliffe Observatory

Author "Prof. Jim Bennett" died before the publication of this work was completed.

J. Bennett (✉)
Science Museum, London, UK

S. Gattei (ed.), *The Incomparable Historian*, Archimedes 75,
https://doi.org/10.1007/978-3-032-25678-2_8

Thomas Hornsby was appointed Savilian Professor of Astronomy at the University of Oxford in 1763, at a time when his qualifications for the post were not obvious. He had been elected to a fellowship in Corpus Christi College in 1760 and it was there that he began making observations with a 33-inch mural quadrant and a 44-inch transit instrument, both by the leading London maker, John Bird. He also observed occasionally from the traditional location for the Savilian Professor, a room at the top of the Mathematical Tower in the schools' quad. His major achievements as Savilian Professor were the foundation and equipping of the Radcliffe Observatory, and the observing programme he carried on there from 1772. Between December 1763 and December 1772, he presented six papers to the Royal Society, which were published in the *Philosophical Transactions*. They form rather an isolated and coherent grouping within his *oeuvre* and will be the subject of this paper.

Among the books written by John Heilbron, one of the most unusual and original in concept is surely *Geometry Civilized* of 1998. As both a geometry text and an eclectic history of geometry in culture, it adopts geometrical theory and practice as a guiding spirit through a narrative that relies on three media—text, diagram and image. The interplay of geometry, much of it practical, and cultural history celebrates a confluence of interests that is a construction of the work itself, unrestrained by the norms of historiographic convention. A further characteristic is the evident pleasure of the author, which the reader is always invited to share.

This book came to mind as I worked through Hornsby's early papers. In his case the confluence was threefold: history of an archival character, geography and astronomical computation. He too created an unusual and self-sustaining practice. He too clearly took pleasure in following wherever it led his enthusiasm. Whether readers shared his enjoyment is less certain.

The first paper was on the value of the parallax of the sun.[1] Hornsby immediately reveals his characteristic interest in the historical record of earlier astronomers by discussing the works of John Flamsteed, Edmond Halley, Jeremiah Horrocks, John Pound and James Bradley. He has particular praise for Horrocks, who as a modestly-equipped 24-year-old had come closer to the truth than many celebrated European astronomers. Neither is he limited to English observers, but considers the work of Johannes Kepler, Johannes Hevelius, Giovanni Domenico Maraldi, Giovanni Domenico Cassini and Jean Richer, drawing on observations of Mercury and of Venus.

Hornsby discusses Nicolas-Louis de Lacaille's attempt from the Cape of Good Hope to measure solar parallax by corresponding altitudes of Mars and Venus on either side of the equator, in collaboration with various observers in Europe. This draws in further astronomers in addition to Bradley: Eustachio Zanotti, César-François Cassini de Thury, Guillaume Le Gentil and Pehr Wilhelm Wargentin (with collaborators in Sweden Stronmer and Schemmark), whose results populate a table of 27 observations, yielding an average value of 9.842 arc seconds for the Sun's mean horizontal parallax. This compares with 9.575 from a table of 41 observations

[1] Thomas Hornsby, 'A discourse on the parallax of the sun', *Philosophical Transactions*, 53 (1763), pp. 467–495.

constructed by Lacaille himself. Taking into consideration observations of Venus at her inferior conjunction with the sun in 1751, made in Greenwich, Paris and Bologna, and gathering a conclusion from the most coherent among all the results he has, Hornsby offers a value of 9.92 for the sun's mean horizontal parallax, assessed prior to the 1761 transit of Venus, which 'engaged the attention of the curious of all nations.'[2] He then addresses the results available from the 1761 transit.

A method proposed by Halley in 1716 lay at the basis of the practice widely adopted. Observers at stations distant in latitude would time the visible transit and a sequence of trigonometrical steps would yield the parallax. While there was a broad consensus on the method, the choice of observing stations was both a more complex question, requiring much further work, and more controversial. Halley's proposal for 1761 was that the times should be noted at Hudson's Bay (Port Nelson) and the mouth of the Ganges. By repeating the computations, Hornsby finds significant mistakes made by Halley—mistakes already detected by Le Gentil and known to the French astronomers at the *Académie des sciences*.[3] After analysing these errors, Hornsby moves on to consider the stations that were chosen in preference: Toboliki (Tobolsk) in Siberia (observer, Jean-Baptiste Chappe d'Auteroches, or d'Auteroche), Bencoolen in Sumatra (Charles Mason and Jeremiah Dixon were the intended observers) and Rodrigues (Rodrigue; observer Alexandre Guy Pingré). Hornsby extends his enquiry to comparisons of the duration of the transit at Tobolski with other stations, in Scandanavia: Cajaneburg, Tornea (Torneá, Tornio), Upsala (Uppsala), Hernosand (Härnösand), Abo (Turku) and Stockholm; and in India: Calcutta and Madras. Hornsby is drawn into a major computational project, by deriving the solar parallax from each of the timed transit differences:

> I have ... made the necessary calculations, and compared the duration of the transit observed at several places with the duration as observed at Tobolski.[4]

He constructs a table of 12 results and the mean of the most coherent of them is 9.579 arc seconds.

Hornsby runs through a number of variant methods of finding the parallax from his data—too detailed to treat here but requiring much further computation. He then broaches the method due to Joseph-Nicolas Delisle,[5] where instead of comparing times of the duration of the complete transit, the difference is noted between the distant stations for the times of internal contact, at either ingress or egress respectively. For this method the longitudes of both stations must be very accurately known (within a few seconds), but it had the potential advantage of using stations where one observation had been compromised by poor weather or where only one contact had occurred during daylight. Hornsby accordingly launches into an extensive discussion of the grounds for accepting the longitudes he will use for his next

---

[2] Hornsby, 'A discourse', p. 470.

[3] Harry Woolf, *The Transits of Venus: A Study of Eighteenth-Century Science* (Princeton: Princeton University Press, 1959), 16–21, 50–60.

[4] Hornsby, 'A discourse', p. 474.

[5] Woolf, pp. 33–35.

set of calculations, including a particularly detailed consideration of the difference between Greenwich and Paris. This reveals another of his passions—a detailed command of the astronomical literature. In addition to the stations already cited, he now draws in St Helena, the Cape, Paris, Greenwich, Stockholm, Göttingen, Bologna, Florence, Rome, Madrid and Lisbon, together with a further set of observers.

In discussing the observations from Rodrigues, he alleges that Pingré has made a mistake, which is attributed to his subtracting his clock error instead of adding it. This was a matter that would feature in later discussions.

Hornsby then calculates the corrections for parallax at each of 14 stations, where he considers the longitudes sufficiently secure, and presents them in a table, omitting Rodrigues, of which there is further discussion. He relates that Pingré had noted a discrepancy of 1 minute, in comparison with the observations from the Cape, but not finding a source for an error, had preferred his own record over that of Mason:

> *... not because he could find no mistake in his own, but because he has proved that no mistake could possibly be committed.*[6] (Emphasis in original.)

Hornsby then presents a large table of the results of his calculations for 32 pairs of stations, yielding an overall mean value for the sun's parallax of 9.695 arc seconds.

In his conclusion, Hornsby raises the possibility that Mason and Dixon had indeed 'set down their observations 1 minute too soon'[7] and leaves the question somewhat open for the present. Further resolution and refinement can be looked for with the 1769 transit—'the subject of a future paper':

> Let it suffice at present to observe that it will in part be visible to the inhabitants of this island, as Venus will be seen wholly entered upon the Sun's disk more than half an hour before the time of sun-set at Greenwich.[8]

The abiding impression from this unusual paper is of Hornsby's impressive command of the observations scattered across the globe and across the historical record, together with his relish for astronomical computation. It is difficult to believe that he is not enjoying his geographical panorama, while at each turn he seems to embrace unflinchingly the explorations opened up by his calculations. The topic seems to spark an individual enthusiasm in its archival, geographical and computational possibilities.

A paper of 30 April 1764 (read on 10 May), reported his observation of a solar eclipse on 1 April,[9] but on 1 December 1765 (read on 13 February 1766) he fulfilled his promise to consider the prospect of the 1769 transit of Venus, in a paper he called a 'discourse', somewhat grandly 'inscribed' to the President, Council and Fellows of the Royal Society. It was printed over 19 pages of the *Philosophical*

---

[6] Hornsby, 'A discourse', p. 492.

[7] Hornsby, 'A discourse', p. 494.

[8] Hornsby, 'A discourse', p. 495.

[9] Thomas Hornsby, 'Observations on the eclipse of the sun, April 1, 1764', *Philosophical Transactions*, 54 (1764), pp. 145–149.

Transactions.[10] We may judge that this introductory flourish indicates the weight Hornsby grants to his work, as he well might, given the computation involved.

Hornsby begins with a consideration that had not featured in his work thus far, approaching the problem using gravitational theory, citing the work of Matthew Stewart and of Tobias Mayer. This indicated a larger orbit for the earth and a smaller value for solar parallax than Hornsby' previous conclusion. The 1769 transit will be an opportunity to resolve the anomaly.

Hornsby computes the elements of the orbit of Venus and after adjustments for aberration, the equation of time and parallax, finds the time of total ingress at Greenwich on 3 June will be $7^{h}. 24'. 40''$. in the evening, more than an hour before sunset. He then turns to a terrestrial globe ('Having rectified the globe to the declination of the sun in the middle of the transit = 22°. 26′. 40″, and also the times of the two internal contacts ...'[11]) and traces the visibility of the event around the world, as he rotates the globe, while his prose sweeps through the exotic geography of its passage.

He then moves on to suggestions for observing stations. The experience of the 1761 transit would discourage sites where the sun will be close to the horizon at either contact (the solar altitude should be at least 5°). He then has to consider the effect of parallax in latitude and longitude, differentiating between those places where the effect will lengthen the time of transit and those where it will be shortened. The object is to find two practicable stations where the difference is maximised. Where the duration will be lengthened, he mentions several possible stations in latitudes 73 or 74°N and is particularly attracted by Tornio ('Tornea') in Swedish Lapland, where he calculates that the transit will be lengthened by 12′ 58″. This sets Hornsby in search of an answering location where the transit duration will be shortened and he calculates that in close to the opposite meridian, at latitude 55°S, the difference from Tornea will be 24′ 33″. This, however, points to a location in the 'great South sea', and poses the question, 'in what parts of the South sea we may reasonably expect to find land.'[12] Hornsby's quest then turns to historical geography, as he searches for records of suitable stations.

Records of circumnavigations suggest that there are 'islands scattered here and there about the tropic of Capricorn, particularly the island or islands of St. Peter, in about 150° of W. longitude from Greenwich, and in about 21° of S. latitude.'[13] This moves Hornsby on to sets of calculations relating to a range of locations of different degrees of reliability. He cites the search for a southern continent by Pedro Fernandes de Queirós at the end of the sixteenth century. His report to the Spanish crown spoke of many islands

[10] Thomas Hornsby, 'On the transit of Venus in 1769', *Philosophical Transactions*, 55 (1765), pp. 326–344.

[11] Hornsby, 'On the transit', p. 331.

[12] Hornsby, 'On the transit', p. 334.

[13] Hornsby, 'On the transit', p. 334.

> … and particularly a large tract of land lying in or near the 15° of S. latitude, well peopled and well cultivated, the inhabitants generally of a peaceable disposition. The produce of this country is represented to be such as to render it a fit object to any commercial nation; consisting of gold, silver, pearl, spices of many sorts, and sugar-canes.[14]

Hornsby's source was the second edition of Harris's *Voyages*.[15]

He runs through further historic accounts, including of the Solomon Islands, the islands Abel Tasman named 'Amsterdam' and 'Rotterdam' in 1643 (islands of Tonga; 'the inhabitants represented to be of a civil and peaceable disposition, and to all appearance unacquainted with the use of arms; the lands well cultivated, and planted with all kinds of fruit-trees.'[16]) and a good few more. Hornsby draws also on accounts given by Joseph-Bernard de Chabert de Cogoli to the *Académie des Sciences* of four separate islands in the South Sea and he picks up many more from perusing Guillaume de l'Isle's 'map of the southern hemisphere'.[17] These then populate a table of 17 candidates, with their latitudes and longitudes, with greater or less security regarding their positions and even their existence. He then calculates the parallactic angles for those whose positions he considers most securely known among them, and the differences in transit durations from Tornio, which could be between 20 and 24 minutes: 'It appears, I think, not only possible, but highly probable, that observers may be stationed in the South seas.'[18]

If, after all, a location in the South Sea is not practicable, Mexico could yield a difference in duration from Tornio of almost 18 minutes, while the method of timing matched pairs of events at stations with secure longitudes draws in many further possibilities, including in France, Britain and Ireland, the Persian Gulf, the mouth of the Indus and of the Ganges, Madras and Calcutta; 'It may naturally be expected that the end of the transit will, if the weather be favourable, be observed at many of our own settlements in these parts.'[19] He computes, for example that the difference in egress between Madras or Calcutta and Mexico will be 11′ 44″.

It is pleasing to think of Hornsby settled comfortably as ever in Oxford with his maps, globe and large volume of voyages, together with other texts, deploying his computational facility in his very particular geographical, historical and mathematical project. He pursues it with such an expansive commitment that it is difficult not to think that he is enjoying himself in this virtual exploration.

His conclusion is to favour stations in Lapland and the South Sea: 'Upon the whole, the necessity of sending an observer into the South seas sufficiently appears',[20] while it should be possible for a southern observer to remain in residence

[14] Hornsby, 'On the transit', p. 335.

[15] John Harris, *Navigantium atque itinerantium bibliotheca. Or, a Complete Collection of Voyages and Travels*, London: T. Woodward, 1744–48.

[16] Hornsby, 'On the transit', pp. 336–337.

[17] Perhaps the *L'Hemisphere Meridional pour voir plus distinctement les Terres Australes*, Amsterdam: Covens and Mortier, 1742.

[18] Hornsby, 'On the transit', p. 339.

[19] Hornsby, 'On the transit', p. 342.

[20] Hornsby, 'On the transit', p. 343.

sufficiently long to obtain a good longitude determination by Jupiter's satellites. He urges the obligation for action, as there will not be another transit till 1874.[21] Finally he offers complementary inducements of a more worldly kind:

> How far it may be an object of attention to a commercial nation to make a settlement in the great Pacific Ocean, or to send out some ships of force with the glorious and honourable view of discovering lands towards the South pole, is not my business to enquire. Such enterprizes, if speedily undertaken, might fortunately give an advantageous position to the astronomer, and add a lustre to this nation, already so eminently distinguished both in arts and arms.[22]

Hornsby's continuing interest in fixing geographical positions is shown in his urging in January 1765 that the Board of Longitude should be charged with 'settling the Longitudes and Latitudes of the principal Capes & Forelands &c of the British Isles', as well as 'to procure & effect a determination of the exact situation of the most remarkable Ports to which the Ships of Great Britain usually resort.'[23]

Hornsby's next paper, of 14 June 1769 (read on 15 June), reports his own observations of the transit, made at Oxford on 3 June, and of those made at the Earl of Macclesfield's observatory at Shirburn Castle, with which Hornsby maintained a close cooperation.[24] Hornsby himself observed from the Mathematical Tower in the schools' quad. He also reports observations of the associated solar eclipse the following day from both observatories, together with a good many observations of the transit by others at sites across Oxford. His clear record of the visual effect afterwards known as the 'black drop',[25] when observing internal contact with a 7½-ft Dollond refractor, is a valuable description of 'live' observing by Hornsby, who

> … perceived that the Planet appeared to be wholly entered upon the Sun, though the limbs of the Sun and Venus were not actually separated; that part of the Sun's edge, where the ingress happened, being very sensibly obscured by a penumbra, and the limbs appearing to be united, by a kind of ligament of a considerable breadth. This ligament became narrower and narrower, and was at length reduced to a point, and actually broken at $7^h\,21'\,57''\frac{1}{2}$ mean time, or $7^h\,24'\,13''\frac{1}{4}$ apparent time. At $7^h\,24'\,23''$ apparent time, the thread of light between the edges of the Sun and Venus, which was, before compleated, now appeared to me of a very sensible breadth, and to equal ⅒ of the Planet's diameter. If I have estimated this breadth properly, the true internal contact must have happened considerably more than a minute sooner.[26]

---

[21] Hornsby, 'On the transit', pp. 342–344.

[22] Hornsby, 'On the transit', p. 344.

[23] Cambridge University Library, Royal Greenwich Observatory Archives, National Maritime Museum, Barrington Papers, BGN: 5:1–5:4. Online at https://cudl.lib.cam.ac.uk/view/MS-BGN-00000/17, accessed 24.9.2022.

[24] Thomas Hornsby, 'An Account of the Observations of the Transit of Venus and of the Eclipse of the Sun, made at Shirburn Castle and at Oxford', *Philosophical Transactions*, 59 (1769), pp. 172–182.

[25] Bradley E. Schaefer, 'The Transit of Venus and the Notorious Black Drop Effect', *Journal for the History of Astronomy*, 32 (2001), 325–336.

[26] Hornsby, 'An account', p. 176.

Although some observers of the 1761 transit had noted the effect,[27] the impression from Hornsby's record is that it took him by surprise in 1769, yet he remained remarkably controlled. The implications were significant: 'it will, I fear, occasion a much greater uncertainty in the quantity of the Sun's parallax deducible from these observations than was reasonably expected.'[28]

Hornsby followed this with a paper of 19 December 1771 (read the same day), on 'The Quantity of the Sun's Parallax', derived from reported observation of the 1769 transit.[29] Again he studies the records of travellers, but they are no longer distant historical figures but his own contemporaries, including James Cook, Charles Green, Daniel Solander and William Wales. He compares their observed values of transit duration with a computed value using an assumed solar parallax, so as to find a corrected parallax from the discrepancy. This yields nine values around a mean of 8.650 seconds of arc. He then describes at some length how he selected two timings in particular—Cook's value for ingress and the mean of his and Green's value for egress, both observed on 'King George's Island' (Tahiti)—and takes the difference of this duration from that measured by Wales and Joseph Dymond at Prince of Wales's Fort in Hudson's Bay. Using other available data he calculates values for the longitudes of both stations. He concludes a mean parallax of the sun of 8.78 seconds of arc and a distance of 93,726,900 English miles.[30] From this he calculates the distances of all the known planets.

We might think of Hornsby's 'virtual' explorations of the South Sea as an anticipation of Cook's voyage only a few years later, pulling together his historical, geographical and computational resources. In dealing with the observations from the 1769 expeditions he has become engaged with the results of an enterprise for which he furnished part of the context.

In passing, Hornsby announced that the observation made at the Cape of Good Hope in 1761 by Mason and Dixon was 'abundantly confirmed' and that of Pingré on Rodrigues was mistaken by 1 minute.[31] The dispute over whether Pingré or Mason and Dixon had been mistaken in this respect had survived from the 1761 transit and had received some notoriety. Nevil Maskelyne, for example, wrote to Hornsby on 13 June 1769, in anticipation of his revised determination of the parallax, that if Pingré had been correct

> … then Mason & Dixon must have mistook 1 whole minute, which I now somewhat suspect. If they did, they were block-heads or perhaps not quite honest in their accounts, for if they found they differed 1 minute from each other they ought to have sent their observations unaltered. The honor of our nation is concerned.[32]

---

[27] Harry Woolf, *The Transits of Venus: A Study of Eighteenth-Century Science*, Princeton: Princeton University Press, 1959, pp. 143, 148–149.

[28] Hornsby, 'An account', p. 177.

[29] Thomas Hornsby, 'The quantity of the sun's parallax as deduced from the observations of the transit of Venus, on June 3, 1769', *Philosophical Transactions*, 61 (1771), pp. 574–579.

[30] This is very close to the modern value for the solar parallax of 8.794 seconds.

[31] Hornsby, 'The quantity', pp. 578–579.

[32] Royal Astronomical Society, RAS MSS Radcliffe A.1.59.

Fortunately for the nation's honour, Hornsby felt able to vindicate the observers at the Cape.

Incidentally, Hornsby would later use an event reported in Wales's *Journal* of his voyage, published in the *Philosophical Transactions* in 1770,[33] in lectures delivered as Sedleian Professor of Natural Philosophy in 1785, and probably on other occasions.[34] The manuscript text of a number of his lectures has survived and on this occasion his subject was 'The different Kinds of Air'. In support of the assertion that uptake of water vapour from the sea surface alters the refractive properties of the lower layers of the air, Hornsby cited Wales's direct confirmation of a seaman being unable to see from the masthead, the low-lying land of Cape Churchill, clearly visible from the quarter deck of the ship. While the report itself was memorable, so was Wales's intrepid response:

> This appeared so extraordinary to me, that I went to the main-top-mast-head myself to be satisfied of the truth thereof; and though I could see it very plain both before I went up, and after I came down, yet could I see nothing like the appearance of land when I was there.[35]

The Royal Society must surely have approved of the adherence of their commissioned observer to their motto, 'Nullius in Verba'![36] Hornsby was able to adopt this engaging and instructive anecdote in his later lectures.

Last in the set of six is Hornsby's paper on the proper motion of Arcturus (23 December 1772; read 24 December). Here he uses observations of his own in 1767–8 (made with his 33-in quadrant and 44-in transit instrument) and observations from 1690 by Flamsteed. Hornsby attempts to correct the latter by means of an analysis of Flamsteed's mural arc.[37] He had become thoroughly acquainted with Flamsteed's *Historia Cœlestis* and was not content to accept the inaccuracies of Flamsteed's instrument but infers a complex pattern of errors from Flamsteed's own data, together with his remarks on the instrument itself. Comparisons are made with observations of nearby stars, which are checked where necessary for the possibility of their having their own proper motions. The whole exercise is impressively thorough, and the results are further checked by comparisons with a set of observations made at Shirburn Castle. Hornsby concludes that the proper motion of Arcturus is 1.205 seconds of arc (per annum) in right ascension westward and 2.005 seconds in declination southward, that no other prominent star has so large a value, that Arcturus is the nearest star to our system and that the search for an annual parallax among the stars might be concentrated thereabouts.

---

[33] William Wales, 'Journal of a voyage, made by order of the Royal Society, to Churchill River, on the north-west coast of Hudson's Bay; of 13 months residence in that country; and of the voyage back to England; In the years 1768 and 1769', *Philosophical Transactions*, 60 (1770), pp. 100–136.

[34] Bodleian Library, MS Radcliffe Trust d.12, f. 132*v*.

[35] Wales, 'Journal of a voyage', p. 115.

[36] 'Take no one's word.'

[37] Hornsby, 'An inquiry'. For comments by Hornsby on his paper, see a letter to an unnamed correspondent (internal evidence suggests the fifth Duke of Marlborough), Royal Astronomical Society, RAS MSS Radcliffe A.1.47.

Hornsby moves on to the implications of this result for the question of the alleged diminution of the obliquity of the ecliptic. While a progressive decrease had been accepted generally, this had been challenged by Cassini de Thury and by Pierre Charles Le Monnier. Hornsby deploys first a considerable set of observations from Shirburn Castle, then a set of his own in comparison with those of Flamsteed (again with instrumental corrections and taking account of the proper motion of Arcturus) to refute this revision, concluding that the obliquity has been declining at a rate of 58 seconds of arc per century. He then concludes his ambitious paper, filled with observation and computation, by reverting to the parallel considerations of natural philosophy: 'a quantity which will be found nearly at a mean of the computations framed by Mr. Euler and Mr. de la Lande, upon the principles of attraction.'[38] This last paper in the sequence of six shows emphatically that Hornsby had not lost his relish for computation.

This was Hornsby's final paper in the *Philosophical Transactions*, and the end of the series we have examined here. Remarkably, he then turned himself into a different kind of astronomical practitioner, namely an observer, while retaining all the qualities we have noted of assiduous application and rigour. The Radcliffe Observatory had been built and was being fitted out with a spectacular set of foundation instruments, as Hornsby described them, 'in all human Probability … not be equalled in the whole World'.[39] It would be his clear duty to observe and the meridian would now be his master. His disciplined application has been evident in the papers we have followed, but the astronomical agenda had been very different, revealing predilections for astronomical record, geography and computation and coming together to create an unusual engagement with fundamental practical astronomy.

A Postscript on the Use of the Globes

We conclude with a coda on the use of the globes, occasioned by Hornsby giving us a rare example of the application of a globe to an enquiry in professional, academic astronomy. These popular instruments, commonly encountered in matching pairs, celestial and terrestrial, were generally treated as educational items, or even as objects of prestige or display. While there were many treatises on their use, these fell within a broader enterprise, embracing lecture courses with demonstrations, tutoring, textbook writing and the commercial world of instrument making and selling. By contrast, we have seen that in Hornsby's 'discourse', as he called it, of 1765, where he anticipates the 1769 transit of Venus, he makes use of the versatility of the globe to predict the complete progress of the visibility of the coming transit and the globe makes an appearance in a major paper in the *Philosophical Transactions*. How might Hornsby have come to take this unusual and unexpected step? In the midst of a paper full of technical, computational astronomy he turns to an instrument identified more with popular lectures or polite domesticity.

---

[38] Hornsby, 'An inquiry', p. 125.

[39] Royal Astronomical Society, RAS MSS Ratcliffe A.7.

Hornsby had a foot in both these worlds. Among a number of positions he held contemporaneously in Oxford, in 1763 he was appointed 'Reader of Experimental'. Philosophy'. In this capacity he gave lecture demonstrations in the School of Natural History within the Old Ashmolean Building, following predecessors John Whiteside and James Bradley.[40] We can say that he joined the ranks of the popular lecturers, though he was very firmly on the academic side of the genre. We know that Hornsby purchased a pair of globes from the fashionable and entrepreneurial instrument maker George Adams, who first published globes of his own making in 1766, issuing versions of 12 and 18 inches diameter. This was in the period when Hornsby was enthusiastically engaged in his historical/geographical/computational project on the solar parallax, being just 2 years after his paper anticipating the 1769 transit of Venus, where he used a terrestrial globe to trace its visibility around the world.

Adams had been promising to issue new globes in several sizes as early as 1755 but had encountered difficulty and delay, not least a serious fire in 1757 at his premises—business and home.[41] Hornsby had a notebook, probably from relatively early in his lecturing career, where he kept track of items necessary for illustrating his lectures and recorded some of their movements.[42] There is a pattern in the notebook, where for each of his lectures, delivered over 26 days at the Old Ashmolean Building, there is a first page of entries marked 'Mine', then a page with the number of the day repeated but without 'Mine', and then a few blank pages. Items that are 'mine' were presumably acquired by Hornsby to be added to the instruments he had inherited from Whiteside and Bradley. The Adams globes were 'mine', as was a pair of 'old' globes. The former would have been a significant purchase: Adams's price for the pair ranged from 9 guineas to £26, depending on the stands, while the much-anticipated entry of the leading maker into the production of new globes would have justified the unusual step of naming him in Hornsby's list. The earlier globes might have naturally been identified as 'old' by contrast with the splendid new instruments and not necessarily been of a great age. The globe available for the procedure of 1764 would have been the 'old' terrestrial. Hornsby the astronomer would have been particularly pleased that his new celestial globe included the southern constellations identified by Lacaille as recently as 1763, appearing on an English globe for the first time.[43]

The use of the globes was a staple topic in this genre of lecture-demonstrations and it was included by Hornsby. We know this because a good number of the texts he prepared for his lectures have survived. Most of his teaching on the globes would

[40] Jim Bennett, 'The Lost Cabinet of Experimental Philosophy of the University of Oxford', in J. Bennett and S. Talas, eds., *Cabinets of Experimental Philosophy in Eighteenth-Century Europe*, (Leiden: Brill, 2013), pp. 69–78; Jim Bennett, 'Thomas Hornsby', in *Oxford's Sedleian Professors of Natural Philosophy*, edited by C. Hollings and M. McCartney (Oxford: Oxford University Press, 2023), pp. 73–91.

[41] John R. Millburn, *Adams of Fleet Street, Instrument Makers to King George III* (Aldershot: Ashgate, 2000), pp. 107–132.

[42] Bodleian Library, MS Radcliffe Trust e.11.

[43] Sylvia Sumira, *The Art and History of Globes* (London: The British Library, 2014), pp. 132, 141, 150.

have been covered by working through various operations and manipulations performed when solving problems using globes, so through performance rather than reading a text. Hornsby's notes for a lecture on the revolution of the earth, for example, contains the reminder to himself:

> Shew the Seasons upon the Orrery;
> And the different Lengths of the Day on the Globes
> And the reason why the Days lengthen & shorten so rapidly near the Equinoxes & so slow near the Solstices.[44]

Occasionally Hornsby's lecture text runs through the features and components of the globes, explaining their functions; more often they are simply listed as an *aide-memoire* when lecturing.[45]

In one notable instance of his use of a globe when lecturing (it is not certain whether from a Savilian or an experimental philosophy lecture),[46] he unpacks a passage from the Greek poet Hesiod's *Works and Days*, which he offers to his audience in Greek; the English translation is:

> When Zeus has finished sixty wintry days after the solstice, then the star Arcturus leaves the holy stream of Ocean and first rises brilliant at dusk. After him the shrilly wailing daughter of Pandion, the swallow, appears to men when spring is just beginning. Before she comes, prune the vines, for it is best so.[47]

Hornsby rectifies his celestial globe for the latitude of Hesiod's home, Ascra in ancient Boeotia (which he gives as 37°25′N), finds the position of the sun in the ecliptic for the given date (60 days after the winter solstice) and brings it to the western horizon, whereupon Arcturus is still well below the eastern horizon. But his instance is really aimed at demonstrating precession: once he makes this adjustment for an interval of 2681 years, he finds 'a very near Agreement'. He has room for some fudging, however: refraction will render the star visible in advance of the indication on the globe, whereas thick vapours near the horizon will delay visibility.

Jim Bennett (1947–2023) was Curator of Cambridge's Whipple Museum (1979–1994), Director of Oxford's History of Science Museum (1994–2012), and Keeper Emeritus at the Science Museum, London, from 2015 until his death. In 2009, he became Professor of History of Science at the University of Oxford. He is renowned for pioneering the study of scientific instruments as gateways to understanding practical science. His research focused on sixteenth–eighteenth century practical mathematics, astronomy, and scientific instruments, with key works like *The Mathematical Science of Christopher Wren* (Cambridge, 1982), *The Divided Circle* (Oxford, 1987) and *Church, State and Astronomy in Ireland: 200 Years of Armagh Observatory* (Belfast, 1990).

---

[44] Bodleian Library, MS Radcliffe Trust d.18, f. 2v; see also f. 22.

[45] Among many examples, see Bodleian Library, MS Radcliffe Trust d.16, ff. 62v–63; ibid, d.18, ff. 22, 101–101v, 105–105v, 137–138.

[46] Bodleian Library, MS Radcliffe Trust d.18, ff. 37–40.

[47] Translation by Hugh G. Evelyn-White, https://people.sc.fsu.edu/~dduke/lectures/hesiod1.pdf, ll. 564–570, accessed 31 January 2023.

Chapter 9
Telescopic Hindsight: Forging Galileo in the 1920s

Nick Wilding

Abstract Analysis of paper watermarks and other evidence reveals that one of the most famous manuscripts of the Scientific Revolution, the University of Michigan's autograph draft of Galileo's discovery of the satellites of Jupiter, is in fact a forgery, written on eighteenth-century paper. Contextual evidence extends this de-authentication to other supposed Galileo materials now located in the Morgan Library and Museum, New York, and the Biblioteca Maffi in the Archivio Storico Diocesano di Pisa, Italy. All these documents can be shown to be the work of the prolific Italian forger Tobia Nicotra, who produced and sold hundreds of documents in the 1920s, mainly for an American audience. These discoveries force a re-evaluation of historians' use of these fallen sources and the criteria upon which they accepted them; they also show the importance of the material study of documents for the history of science.

Keywords Galileo · Forgery · Tobia Nicotra · Watermarks · De-authentication

Reviewing John Heilbron's Quixotic Galileo against David Wootton's Hamletian Galilei, I struggled to nip at Achilles' heels.[1] One potential chink in the carapace was the omission from both biographies of what I then thought was a lovely and iconic

I would like to thank Pablo Alvarez, Curator of the Special Collections Research Center, University of Michigan Library, for his exemplary generosity, integrity and stoicism, without which this research would have faltered. Thanks also go to Silvia Nannipieri and Elisa Carrara at the Maffi Archive in Pisa, Philip Palmer at the Morgan Library & Museum in New York, James Rose and Michael Blanding.

'*Galileo: Watcher of the Skies* by David Wootton and *Galileo* by John Heilbron' in *London Review of Books* (second June 2011), 33, 11, pp. 31–22.

N. Wilding (✉)
History Department, Georgia State University, Atlanta, GA, USA
e-mail: nwilding@gsu.edu

S. Gattei (ed.), *The Incomparable Historian*, Archimedes 75,
https://doi.org/10.1007/978-3-032-25678-2_9

document housed in Ann Arbor, Michigan. Smugly I chided the two titans for their common lapse, placing what I thought was a palpable hit. This is the story of why Heilbron and Wootton were right and I was wrong, but not for the reasons then available.

Galileo's *Sidereus Nuncius* (Venice, 1610) is widely celebrated as a powerful, paradigm-shifting book, possibly the most important text in the history of science. A retrospective dossier of some of the original preparatory manuscript observations and drafts was compiled in the 1820s and now sits as Mss. Gal. 48 in the BNCF. Its relationship to the printed book has been brilliantly analyzed by Paul Needham.[2] Galileo was peculiarly conscious of the nature and status of his own archive, and so it is rare to find working materials, rather than letters or tracts, runted from the BNCF litter. One exceptional example proving the general rule resides in Special Collections Library at the University of Michigan (Fig. 9.1).

This is, at first sight, a wonderful artifact: the top portion of the single half-sheet is a draft of Galileo's letter presenting the telescope to the Doge (the sent original is in the Venetian State Archives, dated 24th August 1609)[3]; the lower part contains a set of Jupiter observations from January 7th–15th. Hang on. What are the chances of this two-for-one, the Alpha and Omega of Galileo's *annus mirabilis*? How did this thing get to be here, which is to say, what, indeed, is it? Now, lest you think me some kind of paranoid conspirator or exuberant iconoclast, bear with me here, ursine and tenacious reader: part of the point of this article is to demystify the métier, pull back the Oz-green curtains of History and show the piles of junk we sift through in our often dreary, mesmerizing research, to ask what kinds of questions we ask, and why; to labor on the *work* of historians.

Let's start by examining the document itself, before we ask questions about where it came from. What do we mean by that? Well, we can start with words, but we'll quickly hear them fail us. First, the handwriting: it starts off quite formally and neatly, as though this was supposed to be the delivered final version rather than an archived draft. Indeed, a comparison with the sent autograph original in Venice shows a similarity which may be uncanny. It's not exactly traced, as the line-breaks are carefully redistributed, but it's clear that what we're reading is a duplication of the letter in Venice, not a draft for it. Handwriting changes, and in Early Modern Europe, most professional writers, in whatever capacity, had more than two 'hands', to suit different purposes. The idea of handwriting as a singular expression of character was just then emerging, facilitated by cheaper schooling, cheaper paper and cheaper postal services, as well as a host of other debatable developments, which used to be called 'the rise of the individual.' Here, in both the Venetian and

[2] Paul Needham, *Galileo Makes a Book. The first edition of Sidereus Nuncius, Venice 1610, Galileo's O*, Vol. II, edited by Horst Bredekamp (Akademie Verlag, Berlin, 2011). Other material ended up in BNCF Mss. Gal. 50 & 70.

[3] Galilei, Galileo, *Le opere* [ed. Antonio Favaro], Edizione Nazionale, 20 volumes (Florence, 1890–1909) [cited as *GO*] X, 250–251; reproduced in facsimile in Antonio Favaro, *Per il terzo centenario dalla inaugurazione dell'insegnamento di Galileo Galilei nello Studio di Padova, VII dicembre MDCCCXCII*, (Firenze: Tip. di G. Barbèra, 1892), plate XXI.

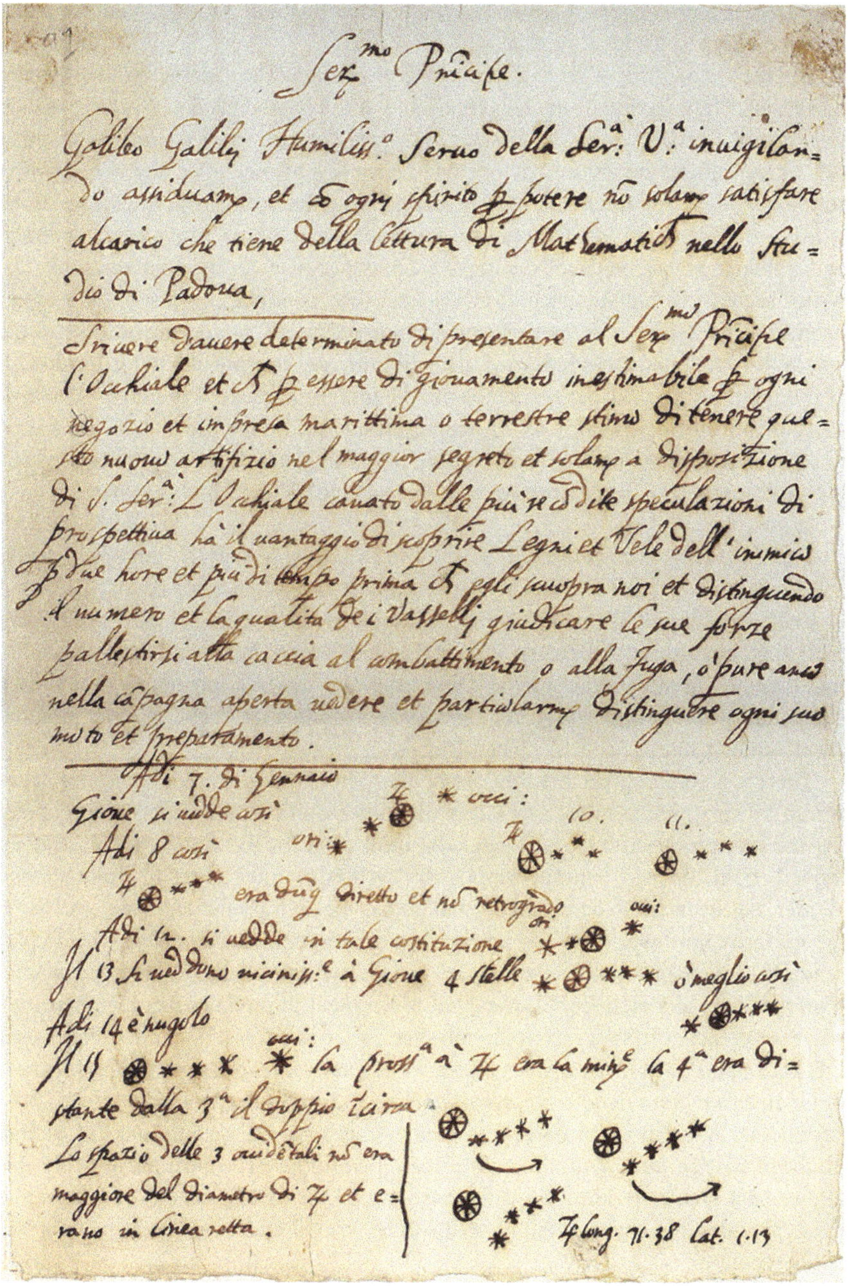

Ser.mo Principe.

Galileo Galilei Humiliss.o Servo della Ser.tà V.a invigilando assiduam.te, et con ogni spirito per potere non solam.te satisfare al carico che tiene della Lettura di Matematiche nello Studio di Padova,

Scrivere d'havere determinato di presentare al Ser.mo Principe l'Occhiale et per essere di giovamento inestimabile per ogni negozio et impresa marittima o terrestre stimo di tenere questo nuovo artifizio nel maggior segreto et solam.te a disposizione di S. Ser.tà L'Occhiale cavato dalle più recondite speculazioni di prospettiva ha il vantaggio di scoprire Legni et Vele dell'inimico per due hore et più di tempo prima che egli scuopra noi et distinguendo il numero et la qualità dei Vasselli giudicare le sue forze pallestirsi alla caccia al combattimento o alla fuga, o pure anco nella campagna aperta vedere et particolarm.te distinguere ogni suo moto et preparamento.

Adi 7. di Gennaio Giove si vedde così ✱ ♃ ✱ così: 10. 11.
Adi 8 così ♃ ✱✱✱ era dunque diretto et non retrogrado
Adi 12. si vedde in tale costituzione così:
Il 13 si veddono vicinissime à Giove 4 stelle ò meglio così
Adi 14 è nugolo
Il 15 così: la prossima à ♃ era la minore la 4.a era distante dalla 3.a il doppio in circa
Lo spazio delle 3 occidentali non era maggiore del diametro di ♃ et erano in linea retta.
♃ long. 71.38 Lat. 1.13

Fig. 9.1 Handwritten letter in Italian with notes and star-like symbols, discussing astronomical observations. The text mentions Galileo Galilei and refers to viewing and measuring stars, including their positions and movements. Several small star symbols and arrows illustrate star groupings and directions, likely representing constellations or celestial events. The letter appears to explain how to observe and interpret these stars, emphasizing their arrangement and changes over time. The purpose is to document and share detailed observations of the night sky for scientific study

the Michigander versions, Galileo has his Sunday Best on: a formal hand for a formal document to the head of state.

But then something strange happens. He messes up. The sheet is spoiled, but not discarded. The function of the text changes, but the hand does not: working now in ink-stained shirtsleeves his hand stays strangely formal. After a few lines Galileo (let's keep calling the writer that) seems to have lost his nerve, or flow. The writing starts wobbling, as though he's having second thoughts about the correct form of the words—does he teach 'Mathematiche' plural, with an 'h', as it says in his contract, or 'Matematica', as he usually writes? (and as he writes in the sent version). The writer breaks off mid-sentence, draws a line, then continues, in a slightly messier hand, perhaps at ease now the performance pressure is off, reordering the bulk of the text of the sent letter, though weirdly, using infinitives, 'to write of having decided…' as though drafting a self-pep-talk, turning the text from a letter to a mirror. The final section is even freer, with notes crowding around each other as though written on the fly. As soon as you pause from reading to looking, a little demon starts whispering its unfunny joke. What's wrong with this writing? There's nothing wrong. That's what's wrong.

The draft letter was composed before the end of August 1609; the observations are dated January 1610. So why is only one single consistent batch of rusty ink used to write the whole document? Why does it look so convincing, so uniformly, unremarkably, manuscripty? Because it's not really two texts, but one: it's a crafty splicing together of the Letter to the Doge and the observation notebook, written sequentially and at one sitting. As soon as you see this internal temporal paradox, the document shreds itself like a bad Banksy.

There's something remarkable about the penmanship: even though the letter forms are very similar in the Ann Arbor and other Galilean documents, the shape of the nib that produced them was not. The quill used for the Ann Arbor document, indeed, both top and bottom, supposedly written months apart in Summer and Winter, has been cut to an chisel tip, flat across the end, producing wide strokes on the north-by northwest axis and narrow strokes perpendicular to them. This is particularly noticeable in the drawing of circles, which bulge on both sides with plumper lines like incipient ellipses, but it's evident throughout the document. It's what we'd expect from a nice italic nib. Yet, as far as I can tell, Galileo never uses this kind of broad quill: he always cuts his to a fine point. Variations in the thickness of his lines are due solely to pressure, splaying the quill tines, not to the direction of the ductus. Whether in formal letters or shopping lists, his writing looks as though it's been written more with an marker than a calligrapher's quill; presumably this allowed for swifter writing. It's not a case of chronic anachronism, only anomaly: the Ann Arbor document certainly seems to be written with a real quill, just not one cut by Galileo.

Let's take a closer look. First, the handwriting isn't quite right. It's really close, but there are some odd doppelgänger twitches. Take 'Stella': when Galileo writes this word, the cross-stroke of the 't' dips as a ligature into the horizontal of the letter 'e'. I could only find one authentic example where this doesn't happen, and there it rolled into the top of the 'e'. But in the Ann Arbor manuscript, where, admittedly, the word 'Stella' appears only once, the cross-stroke of the 't' is on its own, and instead it's the base of the 't' that links to the 'e.' Take Jupiter, by Jove: Galileo depicts the planet on the first page of his logbook as either a six- or eight-spoked wheel, in equal measures, it then becomes, over the page, not a symbol but a circle, so that its diameter can be used to gauge the distances of the stars; in the Ann Arbor manuscript it's consistently, with only one exception, eight spoked, a spider, not a fly, and one caught in its own web. Next: Galileo almost always places a dot both before and after an Arabic numeral, as was normal practice, to denote a digit; Ann Arbor only does it after, and then only sometimes. Loops on 'd's don't know where they're heading in Ann Arbor; normally they're neat and efficient. Contraction marks, such as the flick at the end of 'no' to make it 'non' also meander uncertainly; 'ī' ('i[n]') is written as 'í': this may mean nothing to us, but it is genuinely meaningless for Galileo.

Perhaps most tellingly, in Ann Arbor, two incompatible forms are combined in the word 'recondite': an abbreviated 'n' and a looping 'd.' Contracting the 'n' is done by placing a bar or tilde over the 'o', like this: 'ō' or this: 'õ'. The frequency of contractions increases as space runs out, but they also speed up writing. Galileo had several choices for how to write a 'd', either with an upright ascender, or an open curve reaching back to the left, or a closed loop linking on to the next letter. The choice is determined by what comes before and after, and one thing you never see is the left-looping 'd' reaching back into the area already taken by the mark of a contraction. Kerning has rules. Galileo's writing may not be the neatest, but he aims for legibility, and the interpenetration of these two competing marks would just create confusion. In general, he avoids contractions before 'd's, but when he does use them, in a word like 'secūdā' (secundam), he always uses an upright 'd'. The example of 'recōdite' with a left-sprawling 'd' in the Ann Arbor manuscript is revealing: in the original letter in the Venetian State Archives, the word 'recondite' breaks at the end of the line, after a contracted 'recō=', followed by a left leaning (actually just flamboyantly looped) 'd' in 'dite.' The writer of the Ann Arbor manuscript has magically rejoined the word, but not realized that these forms are incompatible. This is not hand-writing, it's hand-copying.

A similar lack of familiarity with Galileo's graphology produces another mute howler: a word split after only one letter, 'e = rano', at the very bottom of the page. Much like hand-press compositors and the word processing programs that mimicked them, writers tended not to break a word like this; scanning right hand margins, I can find no other example of Galileo inserting a line break after a single letter. If he does not have space for at least two, he starts the word on the next line, however cramped for space he is. The knuckle-rapping of his monastic schooling stings at the end of each line, while forgers fail to flinch from their subject's subconscious.

Galileo actually has two quite distinct handwriting styles, one formal, one informal. In the formal hand his 'p's have a serif, like a reversed 'q'; in his informal hand, this tends to be more of a curve, like a reversed 'g.' In Ann Arbor, an imitation of the formal hand is consistently used even when the letter has proved too drafty to send and we've moved, 5 months later, to journalistic (or nocturnalistic) jottings. Galileo also uses space: if that's running out, he makes his writing smaller. But here we see the distance between the lines diminishing, so that descenders from one line fight with ascenders from another. Galileo is also pretty good, both in formal and informal writing, at keeping a straight left-hand margin, with a left-indent for the start of paragraphs. Here, we see several letters veer off into that marginal space.

There are more serious problems evident in the content: the text and the diagrams in Ann Arbor exhibit a profound misunderstanding of their source, whether that's astronomical or textual. The entries for January tenth and eleventh depict physically impossible satellite configurations. Galileo could never have seen these, and, if the Ann Arbor sheet is meant to be the first record of his observations, which were then transcribed into the logbook, as has been generally assumed, there's no way such a complete transformation of the observation could occur, especially as the logbook observations have been compared to historical reconstructions and shown to be surprisingly accurate. The copyist must simply have lost the plot, or started scatting, producing scat. And take what should be the triumphal entry for January 15th, which in the logbook is when the paradigm really starts shifting. There the point was made that 'they weren't in a totally straight line but like the example shows,' off-kilter, a crucial observation that Galileo is careful to depict as soon as he notices it, as he realizes it's the key piece of data for understanding what he's seeing, even if he doesn't yet understand it. In Ann Arbor this reads 'The 15th ✻ x x x x W[est] the closest to J[upiter] was the sm[alle]st the fourth was distant from the third roughly double The space of the 3 western was not greater than the diameter of J[upiter] *and they were in a straight line*' (translation and italics mine). Yet the diagram accurately places the moons not in a straight line, as it is copied from the logbook. This can't just be an observation made earlier in the evening of the 15th, as it goes against something Galileo had already remarked upon, two nights previously: 'nor were they in as exquisitely a straight line as before but the middle of the three western ones was a little raised up, or rather the most western one lower down.'

There follow, in Ann Arbor, three devious diagrams, which, if I'm reading correctly, are meant to show that Galileo was thinking of a model where one of the stars shifted off the line made by the other three, perhaps a rogue comet or planet. This explicitly contradicts the text next to it and also that night's observation in the logbook.

The Galileo scholar Stillman Drake, who first studied this document in 1962, via a facsimile in an 1934 auction catalogue, was the earliest to realize its potential import. He corrected the catalogue's willfully ignorant misdating from 1600 to 1609/10 and the recipient of the letter from Grand Duke Cosimo II (who only assumed that title in 1609) to Doge Donà. Excited by his find, he asserted, without seeing the original, 'The authenticity of the document appears to be unquestionable,' but also found it 'puzzling' and transferred this state onto Galileo, too, supinely suggesting that 'Galileo himself was puzzled by his successive observations.' Regarding the contradictions between the Florentine logbook and the Ann Arbor records, Drake was forced to conclude, 'I am at a loss to account for this odd discrepancy.'[4]

---

[4] Stillman Drake, 'Galileo Gleanings XIII: An Unpublished fragment relating to the Telescope and Medicean Stars,' *Physics*, 4 (1962), pp. 342–344, p. 344. The facsimile is on p. 96 of *The Library*

Returning in 1976 to the now-located original in Ann Arbor, Drake announced the solution to the problem. His caption for the reproduction of the document declares it to contain, somewhat ambiguously, 'The first known notations of observations by Galileo of Jupiter's satellites, for 13 and 15 January 1610 [...] Notes for positions on 7, 8, and 12 January were copied from another sheet, now lost. Positions numbered 10 and 11 do not relate to observations on those dates.'[5] Drake is liberated by the document's contradictions from the over-theorised epistemological strait-jacket into which French historians such as Alexandre Koyré had tightly strapped Galileo. Drake announced that Galileo was not yet a committed Copernican (despite Galileo's explicit statement that he was, in 1597, and again in 1610) and that, at the time of writing the Ann Arbor sheet, he thought that the stars were merely 'near', not orbiting, Jupiter. Anglo-American Empiricism was the core of science, not French theory. Drake illustrates this non-Copernicanism by offering a selection of vague options: the stars somehow 'oscillated back and forth' rather than orbited, or were somehow 'along Jupiter's epicycle between the planet and our eyes.'[6]

Drake's reconstruction requires an imaginative reordering of the composition of the observations, and also that the diagrams labels '10' and '11' be ignored as a 'mistake' by Galileo and in fact be understood as referring to the 13th (though in reverse order). The three diagrams in the lower right also have to be newly understood as 'looking down on the plane of the elliptic.' We have suddenly abandoned the brand-new telescopic observation mode and rocketed directly above Jupiter, without so much as a 'please fasten your seatbelt.' So much for empiricism. Historians generally do not take such liberties with their sources, and it's unclear where such editorial gymnastics get us, fun as they might be to perform. Drake asks us to imagine Galileo thinking that just one star was moving, around the other three, that they had nothing to do with Jupiter but proximity. Then, while drawing the third, lowest, arrowless, diagram, he realized, thought Drake, that this model was overly complex, and opted instead for the 'simple and unified hypothesis' of lunar orbits.[7] Occam's razor cuts the crap. Galileo then went, still on the night of the 15th, according to Drake, and drew the diagrams labelled '10' and '11', without the labels, as revised observations for the 13th (presumably either falsifying his data or relying on the now-lost, or invented, Ur-sheet). Then, at some later date, according to Drake, he went back and mistakenly gave them the wrong dates. Tsk! No primary source is objective, but it is not the historian's job to force sense with interpretative violence: this is not historiographical ingenuity, it's malpractice. Pardon my frank

---

of the Late Rev. Dr. Roderick Terry of Newport Rhode Island. Part One (American Art Association, Anderson Galleries, Inc. 1934).

[5] Stillman Drake, 'Galileo's First telescopic Observations' *Journal for the History of Astronomy*. 7 (1976), pp. 153–168, p. 162.

[6] Ibid. p. 163.

[7] Ibid. p. 165.

French, but we're not watching Galileo's thinking hand here, we're seeing Drake fake history.[8]

The desire to make sense, even where there is none, led two other Galileo experts, Owen Gingerich and Albert Van Helden, to reexamine the Ann Arbor manuscript in 2011. They follow Drake in ignoring Galileo's dates for the diagrams of the tenth and eleventh as inconvenient, but reject Drake's posited lost notes, instead claiming that Ann Arbor itself is the original record, at least for January seventh and eighth. On the eleventh, they imagine Galileo copying these notes into the Florentine notebook, which they had already observed, in 2003, seems to be written in the same ink for all the notes up to and including that night (they may well be wrong about this).[9] At that point, though, things get weird again: 'On 12, 13, and 15 January Galileo made entries in both A[nn] A[rbor] and F[lorentine logbook] each night. The AA entries are brief; those in F longer. This suggests that in general those in AA were made immediately after the observation, perhaps at the eyepiece, while the longer entries in F were the worked-out notes made later that same evening after a further check.'[10] The Ann Arbor document thus becomes the primary source, which makes its relationship to the Florentine logbook all the more troublesome, as this latter becomes displaced and decentered, a copy of itself. America wins the space race. Then Gingerich and Van Helden offer a really peculiar reading of Drake reading Galileo: 'at the bottom of the page are three doodles, which, as Drake pointed out, seem to show the Jupiter system from above but with the four satellites in a straight line revolving around the planet, clearly not independently but in lines almost as straight as a rail.' The one thing that is clear from the 'doodles' is that the four satellites are not in a straight line. One of them is moving: that's what the cartoon arrow is indicating. Drake's reconstruction reads 'The arrow shows the easternmost star going around the others, down and to the right.'[11] Again, the need to make sense has produced nonsense. The historians invent and describe an image that is not there, convincing themselves that bent lines are straight, and that this line of four moves in relation to an object when it does not.

Commitment to, rather than questioning of, the Ann Arbor document conjures up some baffling illogic: comparing Ann Arbor and the Florentine logbook for the former's final observation, 15th January, Gingerich and Van Helden state that in the former, the four satellites are drawn in a straight line, and that this matches Galileo's comment. But then, there is a peculiar glitch: 'Note that this agrees with the AA diagram, where all four satellites are in a very straight line. In editing this on the F sheet he specifically included the fourth star and, now sensitized to the idea that the satellites may not be in exactly straight lines, he changed his mind about the

[8] See also the similar discussion in Drake, Stillman, *Galileo at Work. His Scientific Biography* (University of Chicago Press, 1978) pp. 148–152.

[9] Owen Gingerich and Albert Van Helden, 'From *Occhiale* to printed page: the making of Galileo's *Sidereus Nuncius*,' *Journal of the History of Astronomy*, 34 (2003), 251–267.

[10] Owen Gingerich and Albert Van Helden, 'How Galileo Constructed the Moons of Jupiter,' *Journal of the History of Astronomy*, 43 (2001), 259–264. p. 262.

[11] Drake, 1976, p. 164.

linearity of the arrangement: "They did *not* form an entirely straight line, as the drawing shows" (our emphasis).'[12] The fourth star is already present in Ann Arbor, so the phrase 'specifically included' is misleading, at best. The Florentine notebook is supposed to be an expanded copy of the Ann Arbor 'eyepiece' observations, yet this Galileo is so committed to theory that in copying he counterfeits, directly contradicting both his diagram and his text. And then leaves documentary traces that directly contradict themselves. The documents are certainly problematic, but perhaps not as much as their historical readings.

Clearly, for the world's best historians to have to put so much effort into mastering and containing this document, to force self-contradictory and incoherent interpretations on it, to see things which are not there and then claim the confusion they have created is in fact Galileo's, some gaslight is at play. Let's get back to the document and see if we can approach it on different terms, avoiding the tyranny of coherence. Let's stop making sense.

It has been claimed, with reason, that one consistent giveaway of manuscript forgery is the lack of continuous movement in a stroke.[13] When signing a name, or drawing a freehand line, a hand will usually continue moving after it has left the page, like a tennis-player's follow through. When writing with a quill, this will result in a tapering line as the pressure on the page is reduced, and the nib de-splays for take-off. However, when the product, rather than the process, is being replicated, lines end heavily: the mark rather than the gesture is being copied. We see this effect quite clearly in the Ann Arbor document: the name 'Galilei' ends with a small club rather than a dagger: the writer has finished the word, then lifted the quill, rather than lifting while writing. Those cumbersome lines dividing up the manuscript horizontally exhibit the same heavy-handedness. Elsewhere, when Galileo makes such marks, there is always a trace of un-inked movement at one end of the line, at least while moving across the page from left to right. These ham-fisted lines are not kosher.

An even graver sin is anachronistic archery: the diagrammatic arrows do-si--doing the errant star around the fixed line appear a good century too early: in 1990 Ernst Gombrich fixed the first use of an arrow in a diagram to imply motion to 1737, where it indicated the direction of water flow; that chronology seems to have stood the test of time.[14] By outrageous fortune, the only arrows available in 1610 were material or metaphorical, not diagrammatic.

The last piece of evidence to consider before we start trying to piece together the provenance of this cultural artifact is the paper: it is old, with the telltale chain lines, laid lines and deckle edges of handmade paper. When light is shone through it, you can see a watermark, another potentially good sign. But the watermark is wonky: it doesn't match any of the paper Galileo used to write letters in the Summer of 1609,

[12] Gingerich and Van Helden 2011, p. 263.

[13] Kenneth Rendell, *Forging History: The Detection of Fake Letters and Documents*, University of Oklahoma Press, 1994.

[14] Ernst Gombrich, 'Pictorial Instructions' in *Images and Understanding: Thoughts about Images, ideas about Understanding.* Eds Horace Barlow, Colin Blakemore, Miranda Weston-Smith, Cambridge University Press, Cambrisge, 1990) pp. 26–45, p. 28.

when the document was started.[15] Usually, Galileo has a limited stock of different papers, and one can use watermarks fairly securely to assign a date to an undated document. Perhaps Galileo was on the road (he wasn't) and picked up a few sheets (and only this one survived), or had old stock lying around (but then threw the rest away). We could try to come up with reasons why this sheet, already so anomalous, also stands out of the reconstituted archive like a simian thumb. Or we could look hard at the evidence.

The paper's watermark has two monograms, the letters 'AS' inside a circle with a three-leafed clover above it, and the letters 'BMO' below. Nothing matching this mark shows up in any of the repositories of watermarks I've consulted, but that evidence is patchy. There is one that's quite similar, but with 'AC' instead of 'AS.' There are a series of others with 'AS' and 'BMO' but no circle and cloverleaf. But what's interesting is that all these close mismatches date from the 1770s, at the earliest. And even without a direct match we can make the general observation that the trend towards using monograms, especially two sets (excluding 'IHS', Jesus' monogram) only really kicks in during the second half of the seventeenth century, and becomes standard practice in the eighteenth. Old paper, but is it old enough?

In fact, paper is the silver bullet that slays this transmogrifying creature and reveals its monstrous essence. Indulge me with one further baroque detour, a trip to Texas and New York, and we'll get back to Michigan's Jupiter almost enlightened, or at least understanding more than anyone else ever has about these documents.

In June 2014, the PBS show *Antiques Roadshow* had a visit by a lucky individual who had inherited a framed Galileo letter. The show's expert, Francis Wahlgren, admitted initial skepticism, but then he heard the provenance: the piece came from a 1947 auction of the famous autograph collection of Philadelphian businessman John Gribbel. Wahlgren examined the document, authenticated it and valued it at $60,000 to $80,000. Not bad for a 1947 purchase price of $170, though actually quite a low estimate for a Galileo autograph. There was no mention of the document before Gribbel, and we don't know when or when he bought it. Favaro certainly didn't know about it during his lifetime, so it seems to have surfaced only after 1920; Gribbel died in 1936, so it must have been acquired before then. Soon after its glowing television debut the document was donated to the Morgan Library and Museum in New York. I'll refer to it as the Morgan letter.

The letter, though, is not what Wahlgren, Gribbel, PBS viewers or the Morgan believed it to be. One thing no-one had realized, because no-one checked, is that the text is almost identical to that of a letter published in the National Edition. There are minor differences, including the date, which has shifted 3 weeks later in the Morgan version.[16] That autograph letter lacked its opening salutation (it had also gone missing by the time Favaro printed it) though it had been reproduced in a detailed

[15] Needham documented and illustrated the four paper stocks used by Galileo between October 1609 and May 1610 on pp. 55–57 of *Galileo Makes a Book*.

[16] OG, X, 178–179. 24th August 1607. The Austin/Morgan document is dated 12th September 1607.

facsimile in 1828.[17] In the Morgan letter, the title, though not name, of the recipient, is included, and the missing text is patched over with an uncharacteristic opening, the word 'Sicome' (Because) which Galileo never used, not once, and certainly not to start a letter or sentence. Because that would be ugly. The text of the letter is slightly compressed, and instead of just signing his name, perhaps with some expression of joyful servitude, in the Morgan letter he is now 'lettore delle Matematiche' (Lecturer in Mathematics) a title he only once used in his sign-offs, in a letter to which we will return.[18] The forms of the letters are extremely close to those of the original, or at least to those reproduced in the facsimile. Stop, breathe, think: it clearly makes no sense for Galileo to send two nearly identical letters to the same person 3 weeks apart, but we'll let those kinds of arguments slide and papercut to the paperchase.

I want now to convince you simultaneously of two equally improbable things: that paper watermarks and library catalogues are *really* interesting, and I mean fascinating. Here goes. The Morgan Library and Museum in New York has not only an amazing collection of books, manuscripts and artworks, but, more important for our present inquiries, an astounding online public-facing catalogue. What does that mean? Well, rather than just recording the name of sender, recipient, date, place, those sorts of standard data, the catalogue records and makes public everything that's known about where the letter came from (including auction records). They provide information you, and perhaps they, didn't know might ever be useful, just on the off chance that it might be. And they even sometimes take an electron detecting version of an X-ray of the paper and include those in the item record. The might seem like overkill, but what a beta radiograph of a piece of paper can show you is precisely what's usually missing in our superficial, image obsessed approach to reproducing documents as mere two dimensional datasets: its three-dimensionality. This is not a fetish, and even if it were, fetishes, whether religious or psychological, are real, and therefore work.

So let's get dirty. Roll up your sleeves and plunge your arms deep into the papermaker's vat. This is full of wet pulp, called, wonderfully, 'stuff', the rotting underwear of long dead folk (if you're looking for a new *memento mori*, consider the recycling of textile into text). A paper making mould is swept and scooped into the vat and then lifted out like a gold prospector's or archaeologist's sieve. Watermarks are formed as the stuff forms little ridges around the wires of the paper mould while the water drains away, like sand around groynes on a beach or snow drifting around drystone walls. Or gunk in the shower plughole if that's what makes you sing. Unlike the rectilinear metal mesh of thick vertical wires that leave a broad parallel chain lines going up-down, and thinner laid lines going left-right in whole sheets of handmade western paper, these watermark wires serve no structural purpose. They are there so that you can read the paper, know where it came from, who made it,

---

[17] *Isographie des hommes célèbres* (Paris: A. Mesnier, 1828–1830) v. II, no pagination, arranged alphabetically.

[18] OG, X, 103.

perhaps what quality it is, know when you're changing stock, re-order from the same source, not get scammed. It might also be a marker of a particular papermaker's labour: proof that he, and no other, made these sheets: his mechanically signed work-stamp. They come in near-matching pairs and wear out within a few years, like your only pair of shoes. As with all graphic systems, watermarks vary over space and time, and enormous, sporadic, gloriously uneven efforts have been made to classify and catalogue them so that they can be used as a descriptive and analytical tool. This is a rabbit warren into which I do not want us to descend, as the tea-party never stops down there. But even when you can't find a direct match of a watermark, general trends are often discernible in the kinds and styles of marks used.

We've already noted that the Ann Arbor sheet has an uncharacteristic, or rather, anachronistic, double monogram dating it to at least the start of the eighteenth-century, or a hundred years too late to be the support for the text upon it. Guess what? the Morgan document has the same monograms: 'AS' and 'BMO.'

The forms of the watermark are different, though, meaning that the sheets were produced at different points in time (on different moulds) but most likely at the same mill. In the eighteenth-century, as paper production increased massively, it was no longer sufficient to employ a 'bull's head' or 'star' or 'foolscap' watermark to indicate size, quality, or origin of your paper. The increased use of letters, like an airport abbreviation, is part of a slow and uneven move towards bureaucratic standardization, with local authorities sometimes requiring mills to indicate their place of production. Indeed, in this case, 'BMO' denotes the site of production, Bergamo, a town near Milan in northern Italy. Other watermarks use 'BERGAMO' or 'BERMO'; we don't yet know what 'AS' refers to, most likely they are the initials of the papermaker or the paper mill owner. Briquet, the Godfather of paper studies, mentions a regulation drawn up in Venice in 1768 concerning the use of workers' initials, which actually gives the example of AS BMO' as the mark of one Antonio Sanzogno.[19] The form of the letters in the Morgan half-sheet is decidedly eighteenth-century. In fact, watermark 1112 of the titillatingly titled tome, *The ancient paper-mills of the former Austro-Hungarian Empire and their watermarks* (volume 8 of the series 'Monumenta chartæ papyraceæ historiam illustrantia' in case you missed it) gives us an exact match and, by seeing when it was used in dated documents, an archivally secure date: 1790.[20] This kind of revelation is one of the unintended consequences of the watermark-dating sub-industry. Their catalogues are to historians what ice core samples are for environmental scientists.

Now that we've fixed the date of the Morgan letter to 1790, and therefore proved that it is a forgery, let's see if we can learn more about the Ann Arbor paper. The

[19] Briquet, C.M., 'Les Filigranes: Ont-ils un sens caché?' *Le Bibliographie moderne*, (1909) 13, p. 322. The eighth August 1768 'Terminazione' is located in the Archivio di Stato di Venezia (ASV), Inquisitore alle Arti b. 23 and 24; the reference to Sanzogno is at ASV Cinque Savi alla Mercanzia, b. 465. I am grateful to Agostino Taboga for supplying this reference and a transcript of the document.

[20] Eineder, Georg, and E J. Labarre. *The Ancient Paper-Mills of the Former Austro-Hungarian Empire and Their Watermarks*. (Hilversum, Holland: Paper Publications Society, 1960).

monograms 'BMO' and 'AS' are not a common combination. The monogram 'BMO' appears in 33 recorded watermarks. All date from 1765 to 1792. There are six papers with the initials 'AS' that are also monogrammed with some combination of letters indicating Bergamo. None is a perfect match with the Ann Arbor watermark, but, again, they may be used to provide a general context. They all date from 1768 to 1792, that is, they are all within the 'BMO' window, which has now become an arrowslit. The paper of the Ann Arbor document was made in the late eighteenth-century.

There is simply no way any mould maker in 1607 could have shaped letters like that, or even used a double monogram; it's like finding a Coke logo in a Leonardo. To recap: the Morgan letter is undeniably forged on paper from 1790. The Ann Arbor half-sheet also comes from the very same paper mill and period, so that document too is forged. But not only is this an impossible coincidence, it's a clue.

Paper is heavy; moving it in bulk is costly. Imperial bureaucracies dictated and directed the tools they needed to run, so most Bergamese paper, famous for its excellence as a writing surface, went first to Milan (the capital of Lombardy, ruled by the Spanish Habsburgs), or then on to Spain. The area around Venice, especially around Treviso and Vicenza, but also up in Friuli, also made a lot of paper: it became the centre of the European book trade. The mighty Po was weaker than politics: Galileo got his paper from the Veneto, not rival Lombardy. These sheets are not only from the wrong time but also the wrong place. Abstract thought paradoxically requires a surprising amount of hardware, and this material for producing the supposedly immaterial has to be made, shipped, bought, used. Galileo thought globally, even cosmically, but wrote locally.

The shared watermark monograms tell us two things: first, that Galileo did not write either letter. But second, that a single forger did. Someone with access to paper from Bergamo. probably someone in Milan. Hold that thought—it's going to be important, I can feel that already.

The Morgan letter is ultimately of only incidental importance to us: it's a good forgery, and the Antiques Roadshow did what they could to treat it with caution. It was in a frame, mounted, so the giveaway watermarks weren't visible (though its three horizontal folds are not those of a real sent seventeenth-century letter, but more like the creases we are used to made by folding a letter to place it in a purpose-made envelope, which didn't exist until about 1840, and this should have raised even rooky eyebrows and questions). But it's crucially important in this story, because it provides a twin case to the Ann Arbor document, and shows, beyond doubt, that it, too is forged.

Galileo scholars been taken in for nearly a century by a very good forgery. It's been accepted without question as genuine, and profoundly affected how we have understood Galileo's discovery, putting a skeptical speed bump in the road to truth, or a dramatic moment of doubt in our hero's quest.

Why only a century, though? There's no mention of the document by Favaro or his ancestors, which is strange for a privately owned Italian autograph letter: Favaro was remarkably thorough and persistent at tracking down Galileo's materials; he knew all the private collectors of scientific autograph manuscripts in Europe, and very little escaped his notice. Might it, then, be the case that this document was produced *for* the market of post-robber baron American collectors? It was advertised as the first autograph letter by Galileo ever to be offered in the United States, and sold (in 1934, mid-Depression) for $975. Epistemological confusion was probably not the aim of the forger—it's more a cut'n'paste job than a strong content fabrication; presumably making new money rather than new philosophy was the aim.

It came to Ann Arbor in 1938 from the collection of a Detroit businessman and philanthropist, Tracy McGregor, whose agent Gabriel Wells had bought it for him at auction in 1934 at the sale of the library of Roderick Terry, who had died in 1933.[21] At that point, as described in Terry's auction catalogue, it was accompanied by a visiting card, now annoyingly lost, with a note attesting to its authenticity by an important figure in interwar science, Cardinal Pietro Maffi (1858–1931). Maffi was both an Archbishop and an astronomer, dual identities whose tension at this point is nicely illustrated by the refusal of his superiors of his request to replicate Foucault's pendulum in the cathedral of Pavia. In his mission to syncretize science and religion, and heal the open wound of Galileo's trial, he had proposed to put up a statue of Galileo in Pisa, eliciting backlashes from several directions.

Maffi's missing card, itself a peculiar medium for an authenticating attestation, apparently said that he had compared the letter with "other originals in the collection in Pisa." Now the only Galileo documents in Pisa at that time were a lone letter to Paolo Sarpi in the University archives, and two autograph Galileo letters unknown to near-omniscient Favaro, but acquired by Maffi in the 1920s, and subsequently included in the expanded reprint of Favaro's National Edition in 1939. As 'originals' is plural, 'the collection' must mean Maffi's personal collection. Maffi and Favaro had corresponded several times until the latter's death in 1922; Maffi went on to actively encourage the updating and reprinting of Favaro's edition in the late 1920s. Maffi's Galileo letters first surfaced in 1924, and were first reproduced in facsimile in 1930.[22]

Both are, to put it bluntly, extremely dodgy: the first, from Galileo to an unknown recipient addressed only as 'Marchese' (Marquis), assumed by the editors to be Galileo's patron Guidobaldo del Monte, from whom Galileo had not heard for 5 years, uses anomalous linguistic forms.[23] Galileo always writes 'all'incontro', but here we have an antiquated 'allo'ncontro'; instead of 'stimando' we find 'istimando.' It is in a hand that only superficially resembles Galileo's, and most strangely, offers up a copy of his new work, which he calls not the *Difesa*, but 'my book of the *Considerazione Astronomica*.' There's a real problem here: the letter is dated second February 1606; the *Difesa*, the only book to which it could possibly refer, was licensed at the end of July 1607 and published in August. The letter's

[21] See note 4.

[22] Lesca, Giuseppe, 'Reliquie inedite di Galileo Galilei,' *Nuova Antologia*, CCLXIX, serie VII, (1930) 54–70.

[23] *GO*, Appendices, II, Carteggio, eds. Michele Camerota, Patrizia Ruffo, with the collaboration of Massimo Bucciantini (Florence, Giunti, 2015), pp. 29–30.

editors in the appendix to the National Edition wonder whether Galileo sent a pre-print manuscript copy. But that also makes no real sense: the main book Galileo's *Difesa* attacked was not Capra's *Considerazione Astronomica* (published in 1605), but his *Usus,* which was only printed in March 1607. One historian has tried to make the chronology work by interpreting the phrase 'mio Libro della Consideratione Astronomica' as referring to Galileo's own annotated copy of Capra's book. This explanation, however, does not help the case, as Galileo's now lost copy of the *Considerazione* remained in Florence, its marginalia copied by Viviani. And that's also not the kind of book one might want to send to a respected patron, with a roasting of the author as a 'solo coglione' ('single bollock'). Moreover, the letter isn't folded as a sent letter would be, has no address page, nor wax marks from sealing. An impossible letter.

Maffi's other 'Galileo' is, conveniently, a fragment of a letter supposedly hidden beneath the rear pastedown of a composite volume of Boccaccio's *Laberinto* (1525) and *Ameto* (1534), with an incomprehensible note, also in something like Galileo's hand, at the end of the book about a bad story he'd never read.[24] Each piece of writing is supposed to vouch for the other. The volume also has an owner's note saying something like 'Collegii Trevisensis Societatis Jesu', which was presumably supposed to mean 'from the Jesuit College of Treviso,' a small town just North of Venice. The problem is that Treviso did not have a Jesuit College, though admittedly, the inscription is almost illegible, and may genuinely refer to another Jesuit college, such as Triers in Germany. The book is a small octavo, only 148 mm by 110 mm, and yet the letter, apparently the last of two or three pages, is on paper the same size and orientation as the book. Why Galileo would suddenly decide to cut up his paper to write a formal missive on a sheet about an eighth the size of his normal letter paper is a mystery. The letter is dated, astoundingly, wonderfully, chutzpahdically, 14th January 1610, that is, the cloudy night before Galileo's cosmological breakthrough with Jupiter's moons. I guess he had to be doing something that evening, but it seems strange he'd be writing a tiny letter saying nothing much to an unnamed recipient and mailing away one of his old annotated books.

There's more than boredom to work with, though: the apparently anodyne phrasing of the interminable sign-off very closely resembles that of a letter to the rectors of the University at Padua from 1603: both contain the phrases 'così resterò per sempre obligatissimo' and 'all quale prego da N[ostro] S[ignore] ogni felicità.'[25] Galileo uses neither phrase elsewhere in his extant, voluminous correspondence, and yet here they are together, twice. The autograph of that 1603 letter, like the autograph of the letter to the Doge used in the Ann Arbor document, is in the State Archives in Venice. For a while, this confounded me: would a forger really have gone to the trouble of visiting an archive, and having documents photographed? The answer is, he didn't need to: Favaro had reproduced both documents in facsimile in

[24] Ibid, p. 45–6.

[25] OG, X, 103.

1892 in a small book celebrating the tercentenary of Galileo's position at Padua.[26] The same book, with its generous flight of 25 facsimiles, also contains a pull-out reproduction of the first page of the Florentine logbook. A single pamphlet purchase could have produced the templates of this document and both portions of the Ann Arbor manuscript. When supplemented with the facsimile from 1828 which formed the basis of the Antiques Roadshow document, these four documents (Ann Arbor, Morgan, Maffi's two documents) can all be shown to be traced from easily accessible facsimiles. The very tools historians had embraced to guarantee, or sometimes test the authenticity of works from the late sixteenth-century on, the facsimile, also have the unintended consequence of facilitating the production of forgeries. Indeed, a forgery really only has to look like a facsimile, since that it what it will usually be compared against.

I've been casting aspersions like confetti. Now it's time for a good cry.

Maffi was given both these letters, along with the annotated Boccaccio, by a composer and musicologist, in 1924. The donor said he'd bought the Boccaccio book from a traveling salesman in Milan a decade earlier, who'd said it had come from Padua. The buyer perceptively noticed that there was a letter hidden under the glued down endpaper, and bought it for a song, so to speak. This, of course, happens all the time—a signed letter by one of the most famous people in the last five hundred years gets torn in half, with only the signed part surviving, and used as waste paper pastedown, in the very book it accompanied that also contains notes by Galileo. It's simply absurd.

The 'dal Monte' letter presumably has a similarly implausible origin story, but was also given to Maffi by the composer shortly after the gift of the botched Boccaccio. Why am I so suspicious about the source? The musicologist was called Tobia Nicotra. Not a household name, though in his day he was famous, then infamous, a character from a novel, though not a good one. It's a name I'd encountered previously, and seeing it in connection to these documents was like knowing the punchline to a bad joke.

Let's start with his downfall, which was also his rise to notoriety. In a curious coincidence, *Time* reported on Nicotra in the same dispatch as the Terry auction where McGregor acquired the Ann Arbor manuscript (though they mentioned only an even bigger ticket item):

> At an auction in Manhattan of the library of the late antiquarian Rev. Dr. Roderick Terry, Bookdealer Gabriel Wells paid $9,750 for a First Folio edition of Mr. William Shakespeare's Comedies, Histories & Tragedies.
>
> […] For an autograph which looked like Composer Wolfgang Amadeus Mozart's, Walter Toscanini, long-nosed son of Conductor Arturo Toscanini, once paid 2,700 lire ($229.50). Shortly Son Walter, a rare-book dealer, learned that the signature was forged, gave Milan police a tip as to who the forger was. They found the rogue, one Tobia Nicotra, in his workshop, busy making autographs of Christopher Columbus, Lorenzo de' Medici, Warren Gamaliel Harding, many another.

---

[26] Antonio Favaro, *Per il terzo centenario dalla inaugurazione dell'insegnamento di Galileo Galilei nello Studio di Padova, VII dicembre MDCCCXCII* (Firenze: Tip. di G. Barbèra, 1892), plates XI, XXI, XXIII.

> In court last week Accuser Toscanini testified that Nicotra visited the U. S. in 1932, posed as the late Musician Richard Drigo, was widely feted. Two years in jail was Nicotra's sentence. Said Dealer Toscanini: "I wanted to restore the faith of foreign collectors in Italian dealers."[27]

The scandal was widely reported, in *The New York Times*, *La Stampa*, *American Weekly* the *Diario di Noticias* in Brazil and a host of local newspapers. The impact on musicology has been especially serious, because Nicotra didn't just copy, he composed, in a wide variety of styles, supplying the demand deceased composers had failed to anticipate for autograph music. He stole period paper from libraries in Milan, sometimes chemically treating it to prevent ink feathering, and used quills and iron gall ink. Nicotra is now suspected of forging upwards of 500 manuscripts, mainly of collectible musicians, though he himself supplied, and possibly invented, this number. Only a dozen or so have been identified. His Haydns, Pergolesis, Mozarts and Wagners were purchased not only by amateurs, but also by institutions. As far as we know, Nicotra flourished, if that is the right word, from about 1923, when he sold a Pergolesi to the Library of Congress, to 1934, when the music stopped.

It is unclear how much prison time Nicotra actually served: he was back in New York a few months after sentencing and there are rumours that he was pardoned by the Fascists to work as an anti-opposition forger. He seems to have spent much of the rest of his life in Mexico as a professor of Italian, and died there in 1980.

Nicotra was arrested in December 1933: a search of his apartment, according to one early published report, revealed 'the entire armory necessary to forge autograph documents. The con-man made his reproductions of previously photographed documents on antique paper with a rare ability.'[28] Another notice from 1934 specifies that Nicotra's identified forgeries included Galileo.[29] Nicotra had also apparently received attention for removing old paper from 'the Milan library' (presumably either the Biblioteca Ambrosiana or Trivulziana). *The New York Times*' coverage claimed 'Many Americans Victims.'[30] The detective and criminologist Giorgio Florida wrote up a thinly veiled and riveting account in 1938, where the laconic imposter Nicotra has become Professor Cronao. The detective admits that 'the fake autographs were not only a technical miracle, but above all an artistic creation; classical music, fourteenth-century poems, political conceits bubbled from the creative brain of the forger, eccentric genius of the highest order.'[31] When combined with the badly damaged sentence document, Nicotra's M.O is revealed: india ink browned with rusty nails, paper stolen from various locations (the Milan State Archive, the

---

[27] *Time*, Monday, Nov. 19, 1934

[28] *La Bibliofilía* Vol. 36, No. 1/2 (GENNAIO-FEBBRAIO 1934), pp. 53–60. p. 58.

[29] *Philobiblon*, v. 7 (1934) p. 98.

[30] 'Autograph faker Gets Prison Term', *The New York Times*, tenth November 1934, p. 18.

[31] Florita, Giorgio, *Squadra Mobile: Tipi, Episodi, Racconti* (Corbaccio, Milano, 1938), 'I Falsi Autografi', pp. 227–248, 242.

Brera, the Royal Conservatory in Milan, the communal library of Como, the National Library in Rome); meticulous tracing of facsimiles; multiple identities.[32]

In 1935 the history of science collector Erik Waller wrote to Emil Starkenstein, mentioning that some time previously he had informed Sotheby's that a Galileo document they were offering was in fact a Nicotra forgery, and that they had withdrawn the lot. Sotheby's has no record of the pulled lot, and it may well be that this was a thwarted attempt to sell the Ann Arbor manuscript at auction.[33] Waller had already had a run-in with Nicotra a decade earlier: Nicotra had managed to incur a debt of 7000 lire concerning 'the autographs in question,' presumably refusing to reimburse Waller for the sale of an object found to be inauthentic, so Waller's antennae may have pricked when he heard the provenance of the object at auction.[34] How many Columbus, Luther or Medici letters are still out there undetected, let alone the scores of scores he produced, we may never know. The Galilean material, if it is limited to the quartet identified here, can now be understood for what it is, part of a wide-ranging decade-long criminal campaign, rather than a cardinal's sin.

Maffi received his doctored Boccaccio in 1924; the fake letter to del Monte was also definitely in his collection by 1930, and for him to write his note authenticating the Ann Arbor manuscript, he had to have handled it before 1931, the year he died, unless his calling card was actually also forged by Nicotra. Nicotra's letters to Maffi from 1924 survive in Maffi's archive, astonishingly signed 'Nicotra-Drigo,' acknowledging his dual imposter identity. They make it clear that the donation came with instrumental strings attached: a fawning plea from a struggling composer for a job also includes an explicit request for authenticating another document (it's not clear if this is the 'Marchese' letter, the Ann Arbor document, the Morgan letter, or something else). He asks about 'a writing that seems to me attributable to Galileo for the many graphological coincidences.'[35] 'Coincidences' might not be quite the right word.

Nicotra does not seem to have dealt in genuine documents along with his fakes: this is a shoddy approach that undermines the exquisite workmanship of the documents themselves. Given the similarity of these documents, their shared provenance and the relatively narrow timeframe of their appearance on the market, it seems

[32] Archivio di Stato di Milano, Sentenza Penale I grado, n. 2445/1934. Thanks to Silvia Nannipieri for allowing me to see a reproduction of the file.

[33] Georg Kinsky, 'Autographenfälschungen', *Börsenblatt für den deutschen Buchhandel*, 39 (February 16, 1932), pp. 119–121 mentions a previous Nicotra forgery of Handel's Jephtha, pulled from Sotheby's in 1930. The first print accusation of Nicotra seems to be Stefan Zweig, 'Sin gefährlicher Fälscher von Musik-Autographen,' *Philobiblon*, IV (1931) p. 296.

[34] Nicotra to Waller, Cantù, 1/30/1925, Erik Waller correspondence, Uppsala University Library. The Waller manuscripts include a small number of identified fakes. One of these seems to be a genuine Galileo autograph (Waller Ms. fake 00006). It is unclear on what grounds it has been deauthenticated: Favaro saw it (or, if it is a direct copy, its original) when it belonged to Count Paolo Galletti, deemed it genuine, and included it in the National Edition (OG, VI, 612). Perhaps Nicotra sold this genuine document to Waller, who then assumed it was fake?

[35] Nicotra to Maffi, 23/10/1924. "Carteggio" (1904–1931), n. 2 della serie 'Atti vari,' Biblioteca Maffi, Archivio Storico Diocesano di Pisa, Italy.

likely that Nicotra forged all four and targeted Maffi in his scam. Maffi spent nothing, and acquired two documents considered until recently by their archive to be authentic. The real target was the first U.S. buyer of the authenticated document, whether consigned directly by Nicotra or by a fence.

The Ann Arbor document has proved difficult to decipher. We can now understand why. Its epistemological anomalies are the forger's quirks and misunderstandings, rather than the creak of Galileo's constraining chrysalis as he tries to butterfly into a new paradigm. Historians' solutions have been to make the Ann Arbor an intermediary stage in the composition of the Florentine logbook, a rough draft whose cosmological kinks were straightened out in the fuller, righter, realer notes. This is precisely why they are such good material to forge: they place a tantalizingly human head-scratching 'er' at the start of 'eureka.' I'm profoundly sorry that the crown of that University of Michigan's Special Collections may have lost a jewel, or rather, found that one was paste. Heilbron did well to avoid it.

Nick Wilding Professor at the History Department, Georgia State University, Atlanta. He is a historian of Early Modern Europe, the history of science, and the history of the book. He is the author of two books: *Galileo's Idol: Gianfrancesco Sagredo and the Politics of Knowledge* (Chicago, 2014) and *Faussaire de Lune: autopsie d'une imposture, Galilée et ses contrefacteurs* (Paris, 2015), as well as a dozen book and journal articles. He has held fellowships at the Medici Archive Project, Stanford, Cambridge, Columbia, the American Academy in Rome and the New York Academy of Medicine, and teaches at the Rare Book School. His current research on book forgery has been featured in *The New York Times* and *The New Yorker*.

Part II
The Nineteenth and Twentieth Centuries

Chapter 10
Professor of Insects and Worms

Jessica Riskin

Abstract Jean-Baptiste Lamarck (1744–1829) was the Professor of Insects and Worms at the Museum of Natural History in Paris. Living through the storms of the French Revolution and Napoleonic Period, he founded biology, coining the term to name a new science devoted to all and only living things, and authored the first theory of evolution. Lamarck's science was foundational to modern biology, yet its radicalism—he usurped God's monopoly on Creation and re-assigned it to mortal, living beings—brought him and his ideas plenty of trouble. During his lifetime, Napoleon and his scientific inner circle hated Lamarck and did what they could to undermine him. Charles Darwin then adopted central elements of Lamarck's theory, but after Darwin's death, his most influential followers re-interpreted his theory to eradicate all traces of Lamarckism, rendering organisms once again the passive objects of outside forces, allowing room for an omnipotent God working behind the scenes. This conception of living organisms as passive in the evolutionary process has remained dominant since the turn of the twentieth century. In contrast, in Lamarck's theory, living beings were active, creative, self-making and world-making. Elements of this very different conception of living organisms have recently, gradually been returning to mainstream biology in fields such as niche construction and epigenetic inheritance.

Keywords Lamarck · Evolution · Biology · Invertebrates · Agency · Darwinism · French revolution

On June 10th 1793, Jean-Baptiste Lamarck—the Romantic naturalist who would later coin the term "biology" and develop the first theory of evolution—found himself at a curious pass. Going to bed the evening before, he had been a botanist, deeply engaged in the study and classification of plants; but the Revolutionary National Convention's Committee of Public Instruction now declared him a

J. Riskin (✉)
History Department, Stanford University, Stanford, CA, USA
e-mail: jriskin@stanford.edu

S. Gattei (ed.), *The Incomparable Historian*, Archimedes 75,
https://doi.org/10.1007/978-3-032-25678-2_10

zoologist. More specifically, the Committee appointed Lamarck to the lowliest of three positions in zoology in the reconfigured Paris botanical garden—what had been the Jardin du Roi became the Republican Jardin des plantes—in its new Muséum national d'histoire naturelle. At age 50, a widower with limited funds and six children, Lamarck had to retool overnight to take up the Professorship in Insects and Worms.[1]

In this humble position, Lamarck went on to teach and publish ideas that became foundational to modern biology. Yet his approach to living things was radically different from the dominant tradition in evolutionary biology that took hold by the end of the nineteenth century and that remains with us. In fact, Lamarck's theory has languished mostly in exile from mainstream science or on its fringes for almost two centuries, although some of his ideas have lately, slowly begun re-entering the mainstream in areas such as epigenetics, which studies factors outside the genome that influence gene expression.

Lamarckian biology differed from the main tradition of later evolutionary theory in a couple ways. First, he ascribed a fundamental agency to living organisms, describing these as continually creating and transforming themselves and the world around them. His was a proto-ecological vision of a world made by living beings, which were, Lamarck said, the only creative forces in the world. To live, as Lamarck defined it, was to create and compose.[2] The dominant tradition in evolutionary biology, in contrast, has rendered living organisms as the fundamentally passive objects of outside forces.[3]

Second, in keeping with his view of living things as actively making and remaking themselves and the world, Lamarck's mode of scientific explanation was narrative and holistic, dwelling in the desires and actions of living beings. He vehemently rejected the model of scientific explanation that his contemporaries were then establishing as the reductive account of a fundamentally passive, artifact world.[4] So, Lamarck's ideas offer a perspective on the development of biology and evolutionary theory, and of conceptions of living organisms, and also on science: what it is and isn't, how it came to be the way it is, and what it might otherwise be like.

When I studied biology in high school and college, my teachers and textbooks often took Lamarck's observations about the giraffe to epitomize his entire theory and its utter wrongness. Here's what Lamarck had to say about the giraffe in his magum opus, the *Philosophie zoologique* (1809):

---

[1] Décret de la Convention nationale no. 1020 du 10 juin 1793, l'an deuxième de la République, une et indivisible, *Relatif à l'organisation du Jardin national des Plantes et du Cabinet d'Histoire naturelle, sous le nom du* Muséum d'Histoire naturelle; Landrieu, *Lamarck*, pp. 53–54; Spary, *Utopia's Garden*, pp. 182–83. For the coining of "biologie," see BMHN, Ms. 742 and Lamarck, *Hydrogéologie*, p. 8.

[2] Lamarck, *Mémoires*, p. 255; *Recherches sur les causes*, Vol. 1: p. 1; Vol. 2: 1817, p. 289; 1864, p. 315; see also Vol. 2, 1841, p. 306; 1911–1912, pp. 349–50; 1951, p. 376.

[3] See, for instance, Dawkins, *The Selfish Gene*.

[4] Lamarck, *Réfutation*, pp. 357–69, 390–91, 409, 482; see also Lamarck, *Mémoires*, pp. 325–26 and 394.

> Concerning habits, it is curious to observe their results in the particular shape and size of the giraffe: we know that this animal, the largest of mammals, inhabits the interior of Africa, and that it lives in an area where the terrain, being almost always arid and devoid of grasses, obliges it to graze the leaves of the trees, and to struggle constantly to reach these. The result of this habit, maintained over a long time in all individuals of the race, is that the front legs have become longer than the hind legs, and the neck has elongated itself such that the giraffe, without standing up on its hind legs, lifts its head and attains a height of six meters (almost 20 feet).[5]

The key word in this passage is actually not "giraffe," but "habits." Lamarck wrote that animals capable of forming habits could, by doing so, transform their physical forms in tiny, incremental yet heritable ways. That claim provoked what must surely be one of the longest-lived controversies in the history of science. Over 200 years after Lamarck published the *Philosophie zoologique*, the idea that living organisms might transform themselves by their own agency in heritable ways remains a hot-button question in evolutionary biology.

In the first instance, Lamarck's claim that living beings could transform themselves heritably was controversial because it usurped God's monopoly on Creation. Instead of appealing to an external source of agency, as in the theological argument from design and the classical mechanist physiology that it informed, Lamarck said first of all that living organisms originated in spontaneous generation, which was triggered by forces such as electricity and heat acting upon inanimate matter. Then, once spontaneously generated, living matter tended to complexify itself by a force intrinsic to it that Lamarck called the "*pouvoir de la vie.*" This was just like other natural forces—electricity, magnetism, gravity—all of which followed certain laws: electricity followed conductors, magnets attracted and repelled one another, and living matter tended to complexify itself.[6]

As organisms became more complex, another cause of transformation began to act as well, namely responsiveness, an ability to react to the environment. In lower organisms, this responsiveness took various primitive and rudimentary forms, but from the level of birds and mammals, the responsiveness took the form of acts of will, through which animals formed "habits" and "ways of life" in response to their circumstances. So, not only did Lamarck place evolutionary agency within the evolving organism, but he described the process of transformation over time as a transfer of agency from outside to inside the organism, as nature "transport[ed] into the interior" of animals the forces of transformation, putting these increasingly at their disposal.[7]

This must be the case, Lamarck said, because if each creature owed its organization purely to an external source, then animals would be "totally passive machines,"

[5] Lamarck, *Philosophie zoologique*, 1: 256.

[6] Lamarck, *Recherches sur l'organisation*, p. 101; Lamarck, *Philosophie zoologique*, 2:95, 127; Lamarck, *Histoire naturelle*, 50, 134; Lamarck, *Hydrogéologie*, 188. On Lamarck's "ceaseless tendency" of living beings to compose and complexify themselves, a "continually active cause," see also his *Philosophie zoologique*, 1:132; 2:69, 100, 101, 104. For *monade*, see 1:285; 2:67, 212. "Animated point" is in Lamarck, *Histoire naturelle*, 16.

[7] Lamarck, *Philosophie zoologique*, 2:310–11.

lacking "sensibility" and "the intimate sentiment of existence," with no power to act, nor ideas, nor thought, nor intelligence: that is, they wouldn't be alive.[8]

In fact, Lamarck went even further: not only did living organisms create and transform themselves, but he claimed they composed the inanimate world around them. And here he gave the pivotal role to plants, which made sense for someone who began his career as a botanist. Actually, that's not quite right—he started off studying for the priesthood because his family, having drained their funds on the military careers of his older brothers, sent him off at age 11 to the Jesuit seminary in Amiens, which he hated. When his father died, Lamarck's first words (according to his son Auguste) were "so I won't be a priest!" He threw off the seminarian's collar and left the seminary without looking back.[9]

What he really wanted was to be a soldier like his brothers, and he managed to finagle a letter of introduction from a local countess whose brother was a high-placed general. With this letter, Lamarck at age 17 rode off on one of his mother's old horses to join a regiment fighting in Westphalia (this was in the midst of the Seven Years' War). He arrived on the eve of the Battle of Vellinghausen, in which the French were utterly routed, and managed to get himself taken into the regiment as a sub-lieutenant by the simple expedient marching off into battle with a company of grenadiers, then seizing command of the company when both the captain and the lieutenant were killed.[10]

And that's how he came to botany (how's that for a surprise twist?). While Lamarck was garrisoned with his regiment in Provence, he was enchanted by the flowers that blanket the Provençal landscape in a rainbow of colors. During a leave at home in Picardy, he traded some precious music notebooks to his brother (he loved music and played the viola da gamba), in exchange for a copy of a botanical handbook, Chomel's *Summary of the History of Common Plants*. Back in Provence, Lamarck took the handbook with him on rambles, collecting specimens and getting a local pharmacist to help him identify them.[11]

At age 24, Lamarck left the military. According to Georges Cuvier, who wrote the first official biographical essay about Lamarck, in the form of a remarkably hostile "eulogy" (more on that presently), he had to leave after one of his companions playfully attempted to lift him by the head, causing "a serious derangement in the glands of the neck." If Lamarck's life were a novel, this neck-stretching episode would of course foreshadow his giraffe-themed posterity. Perhaps, in fact, Cuvier used some poetic license, because according to Auguste, the neck problem was

[8] Lamarck, *Philosophie zoologique*, 2:310–11.

[9] IF, GC, Ms. 3156, no. 5; Cuvier, "Eloge," iii; Géoffroy Saint-Hilaire, "Discours," 209; Landrieu, *Lamarck*, 21; A. Lamarck—Eugène de Lamarck, 11 June 1865, 377.

[10] IF, GC, Ms. 3156, no. 5; Landrieu, *Lamarck*, pp. 15, 21; Cuvier, "Eloge," iii–v; Géoffroy Saint-Hilaire, "Discours," pp. 209–210.

[11] Note de Mlle de Lamarck sur son père, IF, GC, Ms. 3156, no. 9; Cuvier, "Eloge," V; Landrieu, *Lamarck*, 24.

actually scrofula. Anyway, Lamarck had to abandon the army and go to Paris in search of medical treatment.[12]

Once in Paris, Lamarck set about becoming a botanist in earnest, following courses in the Jardin du Roi. One day, on a botanizing jaunt with his brother, he encountered none other than Jean-Jacques Rousseau, who was also out botanizing. Rousseau was then about 60, living in Paris, and in the thick of his botanical phase; he often went botanizing on the outskirts of the city and condescended to co-botanize with Lamarck.[13]

Further inspired by his encounter with Jean-Jacques, Lamarck produced a three-volume work on *French Flora* in which he adopted elements of the Swedish botanist Carl Linnaeus's scheme of nomenclature and also endorsed Linnaeus's rather scandalous idea that plants had sex: that flowers were in fact sex organs, the stamens comparable to testes and the pistils to ovaries. Linnaeus wasn't the first to say so, though he was the most influential—it was almost a century since the English doctor Nehemiah Grew had first proposed the idea—but it remained controversial. Now Lamarck boldly said not only that plants had sex but that this was the very thing that distinguished them from minerals as organic beings. He waxed rhapsodic about the flowers' "sweet perfumes" and "laughing and lively" welcome to the festival of nature, saying that to the self-centered human observer, this all seemed "to exist only for him." But in fact, flowers were not meant to beguile humans; they were meant to beguile other plants.[14] Lamarck's *French Flora* caught the notice of Georges Buffon, who was then director of the Jardin du Roi, and who made Lamarck his protégé.

Around this time, Lamarck began developing his truly radical idea that living things actually created the inanimate world, composing it from elements. He wrote that all "minerals," by which he meant all inanimate, compound matter, arose from the actions of living beings. Plants, he thought, played the primary role in this process, because they could produce themselves out of just sunlight, air, and water.[15] A hyacinth or narcissus bulb suspended in a carafe of distilled water could grow an entire flowering plant. This, when weighed, would prove much heavier than the original bulb, clearly demonstrating the plant's ability to form itself as a compound material being by merging "elements" together.[16]

Lamarck described a similar experiment using a turnip: carve a small hollow in its side, suspended it in the air by its root, and keep the hollow filled with water. You

[12] Cuvier, "Eloge," V; Landrieu, *Lamarck*, 23; IF, GC, Ms. 3156, no. 5; see also Packard, *Lamarck*, 13. Lamarck's departure from the military may also have had to do with conflicts with other officers in the regiment; see Landrieu, *Lamarck*, 23 fn. 2.

[13] Bourguin, "Les grands," 195; Landrieu, *Lamarck*, 26; Packard, *Lamarck*, 17.

[14] Lamarck, *Flore française*, Vol. 1, pp. 2, 103–105.

[15] Lamarck, *Recherches sur les causes*, Vol. 2: 1822, p. 292; see also p. 305; and *Mémoires*, p. 298: "[P]lants, constantly composing their own substance by the action of their organs, and … nourishing themselves … with pure water, free particles of the earthy principle, air, and light, truly form primary combinations."

[16] Lamarck, *Recherches sur les causes*, Vol. 2: 1830, pp. 296.

will see it grow and create flowers, once again, just from water, sunlight, and air. Lamarck reported that he knew people who "in order to obtain greenery in their apartment in winter, amused themselves by raising peas in an earthenware dish, in which there was only pure water and cotton wool to support the roots."[17] At this point Lamarck himself lived in a small apartment near the Jardin with his wife Rosalie and two young children; I like imagine them in the dark, gray Parisian winter, watching decorative pea plants unfurl themselves from earthenware dishes, conjuring their delicate, pale green tendrils from thin air.

This ability to merge "elements" into compounds was unique to plants, Lamarck announced: no animal could subsist on "pure water, earth, fire and air."[18] Plants therefore originated the cosmic relay chain, bringing together the primitive elements and handing the combinations off to animals. Then animals, by eating and digesting, could turn plants or other animals into more elaborate compounds, and therefore shared with plants the defining ability of all living beings: to build, compose, and create.[19] Botany was the most fundamental science since it studied plants, the original sources of all things on Earth, whether living or inanimate.[20]

Alas, as we know, Lamarck was not allowed to remain a botanist, but had to remake himself at age 50 into a zoologist. As Professor of Insects and Worms, he was entitled to lodgings in the Jardin, so he was able to house his family in the apartment on the second floor of the fifteenth-century mansion Buffon had acquired to serve as the Intendant's residence.[21]

Later, Lamarck told his students he had feared there would be a lot less interest in studying insects and worms than great, magnificent creatures such as lions. "What interest, I said to myself, can be inspired by the sight of a mite that lives in cheese or one that feeds on skin ulcers," or a sea cucumber amid the maritime garbage along the shoreline? Lamarck's assigned creatures seemed to be mostly disgusting or at least tiresome household pests: the larvae that ate holes in clothing and furniture, or that devoured provisions of grain and vegetables; the slugs that were a gardener's nemesis. But then a couple new thoughts occurred to him. First, he realized that his purview included nine tenths of the known animal kingdom. Second, he reasoned that the simplest animals must have come first in the order of nature's production. Therefore, in simpler animals, one might witness the essence and foundation of animal life.[22]

Contemplating his insects and his worms with a faint gleam of hope, he divined an organizing principle and renamed his creatures "invertebrates," thereby inventing an entirely new taxonomic category. The invertebrates, Lamarck said, "conduct us

---

[17] Lamarck, *Recherches sur les causes*, Vol. 2: 1832, pp. 298–99.

[18] Lamarck, *Recherches sur les causes*, Vol. 2: 1827, p. 294–95.

[19] Lamarck, *Recherches sur les causes*, Vol. 2: 1911–1912, pp. 349–50.

[20] Lamarck, *Recherches sur les causes*, Vol. 2: 1842, p. 306.

[21] Hamy, *Débuts de Lamarck*, p. 181.

[22] Lamarck, "Discours d'ouverture pour le cours de 1816," BMHN Ms. 742, elements 78–88 and in Lamarck, *Inédits*, p. 27.

insensibly toward the inconceivable limit of animalization," toward those beings "barely endowed with animality" with whom "nature perhaps began." This "limit" or "minimum" or essence of animality was none other than the unbridled power of generation. Simpler creatures have fewer animal faculties, Lamarck explained, but those faculties they do enjoy are much broader in extent than in more complex animals, starting with the power of generation. The very simplest animals, such as polyps, have the most widespread generative faculty.[23] The life of each of their parts is independent of the other parts, so that they can fully regenerate from any section, as the Genevan naturalist Abraham Trembley had famously shown with his experiments slicing up polyps, or fresh-water hydras, then watching each part grow into a full organism.[24]

Other invertebrates, too, possessed extraordinary powers of generation and regeneration: the metamorphoses of insects, the regeneration of the head of snails, the legs of crustaceans, the rays of starfish; the ability of coral polyps to multiply by a perpetual budding, forming plant-like stems. Lamarck wrote that because these were the animals with the most extended faculties of generation and regeneration, the invertebrates far outnumbered the vertebrates, and he said we would likely never know their full extent because of the "infinite smallness" of the smallest among them. These tiniest and simplest animals displayed a "frightening fecundity" such that "it seems, so to speak, as if matter animalizes itself from every direction."[25]

"But what must be our astonishment," Lamarck continues, upon learning that the tiniest and simplest of creatures, the madrepore and millepora coral polyps, with just a little help from shellfish, have been the chief producers of the great calcareous mountain chains and enormous banks of chalk found all over the earth? These minuscule animals create islands, fill up bays and gulfs, clog ports and "completely transform the coastlines."[26]

The invertebrates became the basis of Lamarck's theory, not just that living organisms continually refashioned themselves and the world around them, but, even more radically, that they held a monopoly on creative agency. "For my part, without rejecting anything connected with religious belief," he wrote, "nor that it can be a consolation for the good man who persuades himself of it, I will say that this kind of consideration is absolutely foreign to my subject." Souls were not physically knowable, whereas the dynamic, creative action of living beings was everywhere apparent.[27] Life, to Lamarck, now became the manifest capacity to create in the face

---

[23] Lamarck, *Système*, pp. 11, 12 (the lower limit of animality, and see also p. 16: those enjoying a "minimum" of the "principal faculties of animality"); pp. 27–28 (simpler creatures have fewer but more widespread animal faculties).

[24] Lamarck, *Histoire naturelle des animaux sans vertèbres*, pp.156–57; Lamarck, *Système*, pp. 27–28.

[25] Lamarck, *Système*, pp. 26–27 (examples of invertebrate generation and regeneration); 20 (invertebrates outnumber vertebrates); 33 (tiniest invertebrates); p. 22 (frightening fecundity).

[26] Lamarck, *Système*, pp. 24–25.

[27] Lamarck, *Mémoires*, p. 255.

of all nature's forces of destruction. The science of life was the study of this capacity for heroic creation.

"There reigns throughout the universe an astonishing activity," Lamarck wrote. He meant living things doing the work of living—growing, reproducing, eating, digesting, moving, propagating life among themselves—but not only that. To live, Lamarck thought, was to engage in a constant work of construction and combination, merging elements into compound matter, in valiant defiance of the rest of nature's tendency to "destroy" all combinations.[28] There's a distinct note of tragedy in Lamarck's story, since of course each living thing is doomed to ultimate failure. It will someday lose the "astonishing principle" that has allowed it to resist the onslaught of nature continually dragging it "towards its ruin." The vanquished being is now "completely at the mercy" of nature's destructive power. When living things died, Lamarck explained, they surrendered all they had made in a lifetime—their bodies, secretions, and other productions—to the forces of nature, particularly fire, which then acted upon these, putting them through various transformations, producing a succession of minerals. "Thus, on all sides one perpetually sees nothing but an alternating succession of life and death, formation and destruction, … movement and rest."[29]

Lamarck's idea of living beings as tragic heroes, striving to create as much of the world as possible before surrendering to their fate, may sound like a Romantic notion, as indeed it was. Yet, two centuries later, the role of organisms in the formation of minerals has inspired renewed interest among geologists. According to the current scientific consensus, the first living beings, over 3.5 billion years ago, changed the chemistry of the atmosphere and the oceans, precipitating large-scale mineral deposits; and ever since, organisms have shaped the Earth's surface mineralogy. Geologists studying "mineral evolution" speculate, for instance, that biological activity may even have helped to stabilize the continents by increasing the rate of granite production.[30]

These geologists have suggested that without life, there would be much less free oxygen in the atmosphere, or perhaps none at all. Oxygenation associated with biological activity led to the "Great Oxidation Event" a sharp increase in atmospheric oxygen that took place a little over two billion years ago, when emerging microorganisms seeded the air with oxygen. This was probably the crucial turning-point in the proliferation of diverse minerals on Earth.[31] The free oxygen interacted with existing minerals to form thousands of new ones.[32]

---

[28] Lamarck, *Recherches sur les causes*, Vol. 1: p. 1 ("astonishing activity"); Vol. 2: 1817, p. 289 ("two powerful forces); Vol. 2: 1864, p. 315 ("without organic beings); see also Vol. 2, 1841, p. 306, 1951, p. 376.

[29] Lamarck, *Recherches sur les causes*, Vol 2: 1818, pp. 289–90 ("drags it towards its ruin"); Vol. 1, p. 2 ("alternating succession of life and death"). See generally Vol. 2: 1770–1792, pp. 273–275; and Vol. 2: 1808–1822, pp. 284–292.

[30] Hazen, et al. "Mineral Evolution," pp. 1693, 1704.

[31] Hazen, et al. "Mineral Evolution," pp. 1704, 1707, 1712.

[32] Kwok, "How Life and Luck Changed Earth's Minerals."

The Great Oxidation Event also brought multicellular life and the production of skeletal biominerals that, in their turn, transformed the surface mineralogy of the planet. During the early Cambrian, about 538 million years ago, a rapid rise in skeletal minerals—calcite, aragonite, magnesian calcite, apatite, opal—greatly increased the volume of rock on the Earth's surface.[33] Overall, people studying mineral evolution suggest that biochemical processes are responsible for most of the Earth's mineral species.[34]

Like Lamarck, these current scientists don't claim to know how the whole process began, but they believe that for as long as life has existed, organisms have played a crucial role in producing inanimate, mineral matter. "The origin of life depends on minerals," as one mineral physicist has put it, "but the origin of minerals depends on life."[35] Current work in geology prominently features the same idea that Lamarck proposed during the 1790's: that living beings are the creative, compositional forces in nature.

Lamarck's theory of living beings was also, as must be apparent, a theory of the composition of matter, and as such, it angered the authors of a competing theory of the composition of matter. In fact, when Lamarck tried to read a work presenting his theory at the National Institute, he was heckled so badly that he had to abandon the reading. His ideas were especially offensive to three influential people who would soon become favorites of Napoleon. One was the mathematician, astronomer, physicist, and statesman Pierre-Simon Laplace, whom Napoleon appointed Minister of the Interior and then ennobled as a count of the Empire.[36] And another was the doctor and chemist Antoine-François Fourcroy, whom Napoleon would appoint to his State Council, make him Director of Public Instruction, and again ennoble as a count of the Empire.[37]

Laplace and Fourcroy had each worked closely with Antoine Lavoisier, a chemist, tax collector and philosophical revisionist who lost his head to the guillotine in the spring of 1794, along with the other tax collectors. However, Lavoisier's new approach to chemistry—the "new chemistry" as people came to call it—remained very much alive. Lavoisier and his comrades had set out to do no less than remake the general understanding of the nature of matter. Every kind of matter, they had announced, was in fact, counterintuitively, reducible to different varieties of air. Water, for instance, was made of two different kinds of air, one highly flammable, the other excellent for breathing. These chemical revolutionaries demonstrated the composition of water and other substances in a series of decomposition and

---

[33] Microbes also foster clay mineral production; some geologists describe a "clay mineral factory" that took place during the Neoproterozoic Era between half a billion and a billion years ago. In addition to producing new minerals, the clay mineral factory would have further increased the availability of free oxygen in the atmosphere since clay absorbs and sequesters carbon. Hazen, et al. "Mineral Evolution," pp. 1693, 1709; Kennedy, et al., "Late Precambrian Oxygenation."

[34] Hazen, et al. "Mineral Evolution," pp. 1693, 1706.

[35] Kwok, "How Life and Luck Changed Earth's Minerals."

[36] Hahn, Pierre-*Simon Laplace*, Chap. 8.

[37] Querel, *Antoine de Fourcroy*, [pp].

re-composition experiments using an extremely precise balance to weigh the ingredients and products of each reaction. They assumed that matter was conserved, so the combined weights of the ingredients must equal those of the products. They then identified the lightest products they could make as the simplest units of composed substances. These airs or gases were the true elements of matter, they said, and not the ancient, Aristotelian tetrad of earth, air, fire, and water.[38]

Reduction was the new chemists' chief experimental procedure: in experiments, they reduced various kinds of matter to airy parts. Reduction was the defining feature of their worldview: they saw the material world as reducible to invisibly tiny, inert elements. Reduction, finally, was the central principle of their model of science: to them, explaining something scientifically meant breaking it down into its smallest (meaning lightest) parts.

Lamarck vigorously attacked what he called the "pneumatic chemists," including Laplace and Fourcroy, who had both co-authored articles with Lavoisier, and their reductive approach to science in a series of writings in the mid-1790's culminating in the *Memoirs* whose presentation at the National Institute provoked such heckling.[39] Generations of historians have shaken their heads in rueful mystification at Lamarck's apparently gratuitous antipathy for the new chemistry.[40] But it actually wasn't gratuitous: he had a powerful reason connected with the new science of life that he would soon name "biology." He was out to defend the creative agency and irreducibility of living things, which the new chemists denied. For instance, Fourcroy wrote that Nature merely lent certain materials to plants and animals for a brief time, ultimately taking these back to go on executing her perpetual cycle of composition and decomposition. Lamarck retorted that nature didn't compose anything: only living beings could compose.[41]

Against the new chemists' mechanical theory of composition by heat and pressure, Lamarck presented an entirely different theory of the composition of matter based on the agency of living beings and the principle that they were the unique force of composition. "The entire surface of the globe," he wrote with dramatic pathos, "the bosom of the waters and the whole atmosphere, are the vast field where nature ceaselessly destroys every composed substance which the principle of life does not defend or ceases to maintain."[42]

---

[38] See Riskin, *Science in the Age of Sensibility*, Chap. 7 for a discussion of the new chemical nomenclature.

[39] Lamarck, *Recherches sur les causes* (1794); *Réfutation* (1796); and *Mémoires de physiques* (1797). Lamarck had written the first of these, *Recherches*, almost 20 years earlier, in 1776, and presented it to the Academy of Sciences in 1780, but the commissioners charged with examining it had postponed its publication. A revised version was finally published in 1794. See Bange and Corsi, "Chronology"; Bourdier and Orliac, "Esquisse," pp. [ADD sources and explanations].

[40] See, for instance, Burkhardt, *The Spirit of System*, pp. 38–39.

[41] Fourcroy, *Philosophie chimique*, p. 272, ¶XIX; Lamarck, *Réfutation*, pp. 475, 482–83. See also Lamarck, *Mémoires*, pp. 340–41.

[42] Lamarck, *Réfutation*, p. 455.

At the same time, against the new chemists' reductive model of scientific explanation, analyzing all things including plants and animals into their lightest parts, Lamarck developed a very different kind of explanation. He argued that living organisms and the things they composed were not "susceptible to analysis"; they were irreducible. No analysis could break them down into their component parts because, Lamarck said, they had no pre-existing component parts: organisms synthesized themselves and the world around them just from sunlight, air, and water. What Fourcroy took for the *constituents* of organic matter were in fact the *products* of "organic action." Therefore, "instead of reducing everything to immediate principles," Lamarck concluded, let's simply say that vegetable and animal matters are the products of organic action characterized by distinctive "remarkable properties."[43]

Posterity has judged Lamarck to have been just plain wrong in rejecting the new chemistry, and Laplace and Fourcroy to have been right. Yet, in a certain sense, it really wasn't a matter of right versus wrong, but instead of competing ways of seeing things. If we define elements as the lightest parts of matter producible in laboratory experiments, it makes sense to say that hydrogen and oxygen are the elements that make up water. If, on the other hand, we define the "elements" of matter as the primitive substances out of which living beings compose themselves and the world around them, perhaps it makes sense to say that these elements are sunlight, air, and water. Indeed, according to this way of looking at things, it could even make sense to say that oxygen is not an element but a product of organic action, since its presence in Earth's atmosphere results from the activities of living beings.

Historically speaking, of course, Laplace's and Fourcroy's way of defining elements triumphed, along with their reductive definition of science. In the wake of their success, Lamarck's vision of nature, of science, and of life became, not just wrong, but un-thinkable, beyond the pale.

The third member of Napoleon's inner circle whom Lamarck antagonized was Georges Cuvier, who occupied the chair in comparative anatomy at the *Jardin des plantes* and whom Napoleon also appointed to a succession of ever higher administrative posts.[44] Cuvier was a devout Lutheran who categorically rejected the idea that living forms changed over time and instead upheld the fixity of species. His central anatomical principle, the principle of correlation of parts, was an assertion of rational design: he claimed that one could infer all the parts of a creature from any single part because they all followed from one another by rational necessity.[45]

Cuvier did believe in extinctions, but he took the fossils of extinct creatures to reflect a series of catastrophes, such as floods and fires, each eliminating the creatures of a given time and place, to be replaced later by brand new, equally fixed and

[43] Fourcroy, *Philosophie chimique*, pp. 125–28; Lamarck, *Réfutation*, pp. 357–69, 390–91, 409, 482; see also Lamarck, *Mémoires*, pp. 325–26 and 394.

[44] Pasquier, *Éloge de M. le baron Georges Cuvier* (1832), 15–19.

[45] Cuvier's successes at predicting fossil discoveries appeared to vindicate this principle, though he also applied an extensive knowledge of comparative anatomy to the task. Rudwick, *Meaning of Fossils,* 104–115.

unchanging creatures.[46] This theory implicitly supported the doctrine of special creation: that God had created each species for its niche in nature and had replenished the supply of beings following each catastrophe (such as the Great Flood) by new acts of creation. This could hardly have been better suited to Napoleon's purpose. Here was a scientific doctrine of revolutionary change that was at the same time profoundly conservative, each revolution imposed from outside rather than rising up from within, each a return to the same ultimate authority, and the whole thing guaranteed, not by the Church, but by Napoleon's own *savants*.

Cuvier's science, in other words, was cutting-edge, but at the same time, it bolstered rather than threatened an authoritarian order, first Napoleon's, and then that of the Bourbon Restoration (Charles X would make him a baron). It was emphatically not the science of the materialist, atheist revolutionaries from whom Napoleon took such pains to distance himself in his ongoing struggle to seize and retain the royalists' ground. But nor was it the older science of the Catholic Church. Cuvier translated the authority of religion into science, a powerful modern concoction.[47]

It's no surprise, then, that the Little Corporal loathed Lamarck. The astronomer François Arago told a story about this imperial antipathy that became legendary; it took place when Arago was just 23. On a frigid day in December 1809, the members of the Institut de France attended the Emperor in a chilly salon of the Palais des Tuileries to present to him their new publications. The savants and men of letters were dressed in green, the aides de camp in gold brocade. Standing next to the young Arago was Lamarck, now 65 and in ill health, though he would live another two decades. Lamarck held a copy of his *Philosophie zoologique*, his magnum opus in which he developed the powerful idea that living forms transform themselves continually in incremental, heritable ways, moving toward greater complexity and adaptation to their environments. Arago and Lamarck together wryly observed the apparent panic of each of the men around them that he might fail to receive the Emperor's notice.[48]

Napoleon at last entered the room and approached Arago as he would a new conscript, fixing him with a stare and barking "you're very young! What's your name?" Before Arago could answer, his neighbor on the left, eager to be noticed, replied "he cultivates astronomy!" Napoleon: "And what have you done?" Arago's neighbor on the right, not to be outdone, announced "he has just measured the meridian in Spain!" While Arago ruefully reflected that Napoleon must take him for a mute or an imbecile, the Emperor turned to Lamarck, who presented his *Philosophie zoologique*. Napoleon snatched it growling "I take [this] only out of consideration for your white hairs. Give it here!" Without looking at the book, Napoleon thrust it into the hands of an aid, while Lamarck, tears of humiliation rolling down his cheeks, tried to explain what was in it.[49]

---

[46] Cuvier, "Mémoire."

[47] See Rudwick, *The Meaning of Fossils*, pp. 104–115.

[48] Arago, *Mémoires*, XLVI.

[49] Arago, *Mémoires*, XLVI.

Had Napoleon known what was in the book, with its revolutionary understanding of living nature as continually creating and recreating itself from elements, he would surely have liked it even less. Lamarck's natural world of inherent active powers, its substance generated and shaped by its own living denizes, was uncongenial to authoritarians.[50] Cuvier was the victor who wrote the history, quite literally: as I've mentioned, he wrote Lamarck's official "eulogy," which was in fact an anti-eulogy, damningly describing Lamarck's theory as poetry rather than science. Cuvier's condemnation set the dominant tone for assessments of Lamarck's ideas for a long time to come.[51]

The fact that Lamarck's theory located evolutionary agency *within* the evolving organism, emerging as it did from Enlightenment French materialism, coming into print during and immediately following the Revolution, made Lamarck's approach smell to contemporaries of atheism, Jacobinism, social levelling, revolution, and regicide. These ill odors promised trouble not only for Lamarck and Lamarckism but for evolutionary theory more generally.

Charles Darwin struggled over his relations to both Lamarck and his own grandfather, Erasmus Darwin, who had been a doctor, naturalist, poet, and admirer of the revolutionaries in France, and had also described living organisms as transforming from generation to generation, sometimes in verse. Charles Darwin drew upon both Erasmus Darwin's and Lamarck's ideas, but he was torn between admiration for these Romantic precursors and embarrassment about them, and sometimes referred to their theories very scathingly. A page from Darwin's 1837 notebook shows him wrestling with his grandfather's and Lamarck's ideas, jotting things down and then crossing them out. A crossed-out line says "Man is derived from Monad each fresh" ("monad" was Lamarck's word for the simplest spontaneously generated living organisms)—but the sentence is both broken off and vehemently struck through.[52]

But Darwin did retain various forms of agency within the evolving organism, including the inherited effects of use and disuse of limbs and organs. It was really his followers in the period after his death, around the turn of the twentieth century, especially in Germany and American, who went about eradicating all traces of Lamarckism in their reinterpretation of his theory. The German evolutionary theorist August Weismann, in particular, offered interpretations of both Lamarckism and Darwinism that became standard in the twentieth century.

The "Darwinism" of Weismann's redefinition eliminated both of Lamarck's internal agencies, the complexifying life force and the role of habit or will. Weismann located evolutionary agency emphatically outside the organism, insisting that no kind of experience or behavior on the part of individual organisms could ever change the machinery of inheritance. Weismann posited an unbreachable barrier separating somatic (body) cells from germ (reproductive) cells. He insisted that no changes

---

[50] Lamarck, "Météorologie," 475.

[51] Cuvier, "Eloge," xx.

[52] Darwin, Notebook B, p. 18. For Lamarck's use of "monad," see *Philosophie zoologique*, Vol. 1, p. 285; Vol. 2, pp. 67, 212.

taking place in the somatic cells could ever be transmitted to the germ cells. This was the so-called Weismann barrier, eliminating the possibility of inheritance of "acquired" variations.[53]

The emblem of Weismann's theory was an experiment in which he chopped the tails off several generations of mice in order to show that their offspring had perfectly healthy tails. Weismann himself acknowledged that this experiment had no real bearing on Lamarck's theory, since it tested neither an intrinsic upward force of life nor the transforming effects of habit or will. But he said it would defeat the popular version of belief in inheritance of acquired characteristics.[54] It was more a spoof than an experimental refutation, but it was powerful and continues to be, still cited regular in biology textbooks.[55]

The "Weismann barrier" separating somatic cells from germ cells was in fact a barrier separating individual agency and evolutionary transformation. And it remained such—a denial of evolutionary agency to organisms—when it became a core principle of neo-Darwinism, when Francis Crick restated it as the "Central Dogma" of molecular biology: transcription can only work in one direction, from DNA to RNA to protein, and there can be no reverse transcription going the other way. Therefore, no bodily change in an individual organism can inscribe itself in the DNA and be inherited by the next generation.[56]

So, Weismann erected a lasting barrier between individual agency and evolutionary transformation. His doing so reflected the culture of the emerging German research universities—Weismann taught at Freiburg—in which the science and theology faculties had a tacit limited partnership arrangement.[57] The science faculties restricted themselves to demonstrations, in ever more specialized detail, of the mechanical complexity of natural forms, but this was a local and fundamentally passive mechanism, leaving a monopoly on questions of meaning, will and agency to the theology faculties. This was a kind of institutional application of the argument from design, which had banished agency to the province of a supernatural God, leaving behind a mechanical world that was infinitely intricate but fundamentally passive. Weismann himself emphasized that his mechanistic approach to biology, rendering living beings as passive machines, was predicated on the assumption of a divine presence."[58]

Weismannism came to dominate biological thinking in the United States in the early-to-middle decades of the twentieth century, crucially transmitted by someone who had come up within the same world of the German research university, Ernst Mayr. Mayr studied biology at the University of Berlin in the mid-1920's, receiving his PhD in 1926. His adviser was the ornithologist Erwin Stresemann, who greatly

---

[53] Weismann, *Germ-Plasm* (1893), xi.

[54] Weismann, "On the Supposed Transmission of Mutilations," pp. 431–434.

[55] See Futuyma, *Evolution*, p. 9.

[56] Crick, "Central Dogma."

[57] See Howard, *Protestant Theology*; and Zimmerman, *Anthropology and Antihumanism*.

[58] Weismann, "Mechanical Conception of Nature" (1882), 697, 710, 716, 708–709.

admired Weismann for having vanquished Lamarckism. Stresemann was a firm subscriber to the passive-mechanist view of living forms. He wrote, for example, "an animal does not act for itself, but under a higher commission: *animal non agit, sed agitur* [an animal does not act, but is acted upon]," and he endorsed the view that "[t]he animal does not think, does not reflect, does not establish aims for itself, and if it nevertheless behaves purposively, then someone must have thought for it." Notice in both of those expressions how the passive mechanism of the animal implies a higher power, an external Someone.[59]

When Mayr went to New York in 1931, he of course brought his training and influences with him. By then, at any rate, the United States had been importing the German model of the research university, its institutional culture and practices, particularly in biology, for more than half a century, since the founding of Johns Hopkins in 1876 on the template of the University of Berlin. American biology was a well-prepared soil into which to plant Weismann's ideas and the model of science they represented.[60]

The Weismannian interpretation of Darwin and repudiation of Lamarck thrived in the United States and elsewhere, remaining strong in throughout the twentieth century. As an example of how controversial Lamarck's theory remained two centuries after its first publication, consider the best-selling writings of Richard Dawkins and Daniel Dennette, both of whom describe themselves as "Weismannians," and have devoted a lot of ink to making biology safe from Lamarckism. They describe Lamarckism in extreme terms as "painful," "scary," "fatal," "devastating."[61] The surprising extravagance of their prose reflects the long history of evolutionary theory in which a tradition of writers from Weismann onward treated Lamarckism as a threat, not just to a particular scientific idea or theory, but to a whole worldview, to modern science itself. The modern science such authors seek to protect, though they apparently overlook its historical lineage, was founded upon a theological principle of rational design that dictated that science be the reductive account of a fundamentally passive-mechanical artifact world. Of course, none of these recent and current authors has a theological purpose himself, far from it; Dawkins is atheism's leading missionary.[62] Yet the theological origin of the model of science they've all defended with such force and power continues to inform their writing despite them.

Lamarck's model of science, in contrast, was rigorously naturalist and materialist to its core, ascribing all agency to mortal, living beings and outsourcing none to a supernatural god. Lamarck's approach differed also from what was already becoming the dominant model in being integral, narrative, and overtly interpretive whereas the emerging scientific ideals of the nineteenth century were analytical reduction and objectivity. Lamarck emphasized that knowledge resided in

[59] Stresemann, *Ornithology*, pp. 330–331, 339. Stresemann ascribes the view he endorses in the second quote, "the animal does not think…," to Bernard Altum.

[60] Howard, *Protestant Theology*, p. 5.

[61] Dawkins, Extended Phenotype, pp. 164–65; Dennett, *Darwin's Dangerous Idea*, p. 331.

[62] See Mayr, "How I Became a Darwinian," pp. 413–423; and Dawkins, *The God Delusion* (Boston: Mariner Press, 2006).

interpretive judgement and that "moral" things—habits, behaviors, willful actions—influenced physical things as well as *vice versa*.[63] This mode of thinking and writing was the one Cuvier scorned as poetic. As John Heilbron has shown with a lovely, Galilean flair, poetry and science had been intimately connected hitherto. Galileo's love of poetry, especially of the poet Ludovico Ariosto and his romance epic *Orlando Furioso*, shaped how he saw and responded to the world on every level and, in particular, informed what he saw when he looked through his spyglass at the Moon.[64]

The re-shapers of Darwinism in the later 19th and early 20th centuries insisted not only that theirs was the correct approach to evolutionary theory, but that causal reduction and passive objects were the distinguishing marks of modern science, setting it apart from all other modes of understanding, especially literary, humanist, and aesthetic ones, especially poetry. They promoted not only a theory of the development of life, in other words, but an idea of science as an absolutely distinct and exclusive endeavor, sharply constrained in its methods and approaches.

For instance, T.H. Huxley, Darwin's self-styled "bulldog," liked to insist on the difference between "scientific" and "unscientific" explanations; on the uselessness of literary or humanistic fields to scientific advancement; and that the "scientific method" was the unique path to truth. He also declared that living beings were purely passive automata and that consciousness was a mere epiphenomenon, like a locomotive's whistle. In this way, Huxley popularized not just a view of living things, but of science, as uniquely powerful in its reductive ability to cut through the illusions that other fields dwelt in, especially the whole humanistic world of volition and agency.[65]

Weismann too wrote for a popular audience, promoting with his revised Darwinism a standard for distinguishing science from non-science. He scoffed, for instance, that the inheritance of acquired variations "scarcely deserves to be called a scientific hypothesis." The internal formative agencies featured in earlier evolutionary theories, Weismann said, were utterly without value in science.[66] In this way, popularizations of evolutionary theory promoted a view of science as a thing apart, separate, and distinct from other modes of understanding in its objective, reductive account of a fundamentally passive natural world. This redefinition of science, which took place within 50 years around the turn of the twentieth century, setting science apart from all other forms of knowledge, meant that to offer a whole, integral, interpretive understanding of the world in the manner of Lamarck's science became—at least in many areas and in popular understanding—a thing of the past.

Recently, though, there has been a gradual return of Lamarckian ideas toward the mainstream as evolutionary biologists have been working to extend their theory far beyond the genome in areas such as epigenetics, niche construction and cultural

[63] Lamarck, *Philosophie zoologique*, 1:4.

[64] Heilbron, *Galileo*, pp. 150–154; see also pp. v, vii, 1, 2, 13, 12–22, 24, 84–86, 178, 229, 231, 250.

[65] Huxley, *Science and Culture*, pp. 17, 236, 316.

[66] Weismann, "On the Supposed Botanical Proofs," p. 415.

evolution.[67] This movement also includes many others, such as environmental philosophers contemplating the myriad forms of agency in the living world[68]; novelists imagining the inner lives of trees[69]; and researchers in the growing fields of "bioacoustics" and "biosemiotics" studying modes of communication among plants and animals.[70] Participants in these various projects are heirs to Lamarck's worldview, a world made by mortal, creative striving in its many forms, and to his science, which took that creative striving as elemental.

References

[Primary Sources]

[Archival]

Bibliothèque Centrale du Muséum national d'histoire naturelle [BMHN].
Institut de France [IF].
Papiers et correspondance du Baron Georges Cuvier [GC].
Auguste Lamarck—Georges Cuvier, 8 February 1830, Ms 3252/f. 358–359.
Auguste Lamarck—Georges Cuvier, 20 February 1830, Ms 3156/no. 5.
Mlle Lamarck au sujet de son père, 1830, Ms 3156/no. 9.

[Print]

Arago, François. 1854-1862. Mémoires de ma jeunesse. In *Oeuvres complètes de François Arago, 13 vols*, ed. J.-A. Barral, vol. 1, 1–102. Paris: Gide.

Cuvier, Georges. 1935. Eloge de M. de Lamarck, lu a l'Académie des sciences, le 26 novembre 1832. *Mémoires de l'Académie Royale des Sciences de l'Institut de France* 8: i–xxxi.

Cuvier, Georges. 1799. Mémoire sur les espèces d'éléphans vivantes et fossiles. In *Mémoires de l'Institut national des sciences et arts*, vol. 2 (An VII/1799). *Mémoires de la Classe des sciences mathématiques et physiques*, 1–22.

Darwin, Charles. 1837–38. Notebook B: Transmutation. CUL-DAR121. In *The complete work of Charles Darwin online*, ed. John van Wyhe. Accessed April 2023.

Fourcroy, Antoine-François. 1795. *Philosophie chimique: Vérités fondamentales de la chimie moderne, disposées dans un nouvel ordre*. 2nd ed. Paris: Du Pont.

Saint-Hilaire, Géoffroy, and Etienne. 1838. Discours prononcé aux funérailles de Lamarck. In *Fragments biographiques, Précédés d'Études Sur La Vie, Les Ouvrages Et Les Doctrines de Buffon*, 209–219. Paris: Librairie F.D. Pillot.

---

[67] Pigliucci and Müller, *Evolution*.

[68] Morizot, *Manières*.

[69] Powers, *The Overstory*.

[70] Hoffmeyer, *Biosemiotics*.

Huxley, Thomas Henry. 1888. Science and culture. In *Science and culture and other essays*, 1–23. London: Macmillan.
Lamarck, Guillaume Emmanuel Auguste Monet de–Eugène de Lamarck. 1907. 11 June 1865, reproduced in *Bulletin du Muséum national d'histoire naturelle*, 375–79.
Lamarck, Jean-Baptiste. 1778. *Flore française, ou description succincte de toutes les plantes qui croissant naturellement en France*. Vol. 3. Paris: Imprimerie Royale. [expanded edition published in 1805 in collaboration with Pyramus de Candolle.].
Lamarck, Jean-Baptiste. 1815. *Histoire naturelle des animaux sans vertèbres*. Vol. 1. Paris: Déterville.
Lamarck, Jean-Baptiste. 1802a. *Hydrogéologie, ou Recherches sur l'influence qu'ont les eaux sur la surface du globe terrestre*. Paris: Chez l'auteur.
Lamarck, Jean-Baptiste. 1972. In *Inédits de Lamarck*, ed. Max Vachon, Georges Rousseau, and Yves Laissus. Paris: Masson.
Lamarck, Jean-Baptiste. 1797. *Mémoires de physiques et d'histoire naturelle*. Paris: chez l'auteur.
Lamarck, Jean-Baptiste. Météorologie. In *Nouveau dictionnaire d'histoire naturelle, appliquée aux arts, à l'agriculture, à l'économie rurale et domestique, à la médecine, etc. Par une société de naturalistes et d'agriculteurs*, vol. 20, 451–477.
Lamarck, Jean-Baptiste. 1809. *Philosophie zoologique, ou Exposition des considérations relatives à l'histoire naturelle des animaux*. Vol. 2. Paris: Dentu.
Lamarck, Jean-Baptiste. 1794. *Recherches sur les causes des principaux faits physiques*. Vol. 2. Paris: Maradan.
Lamarck, Jean-Baptiste. 1802b. *Recherches sur l'organisation des corps vivants*. Paris: Maillard.
Lamarck, Jean-Baptiste. 1796. *Réfutation de la théorie pneumatique*. Paris: Agasse.
Lamarck, Jean-Baptiste. 1801. *Système des animaux sans vertèbres*. Paris: Deterville.
Pasquier, Étienne. 1832. *Éloge de M. le baron Georges Cuvier. Chambre des Pairs. Séance du 17 décembre 1832*. Paris: [s.n.].
Weismann, August. 1893. *The germ-plasm: A theory of heredity*. (trans: W. Newton Parker and Harriet Rönnfeldt). New York: Scribner's.
Weismann, August. 1882. On the mechanical conception of nature. In *Studies in the theory of descent*, ed. Raphael Meldola, vol. 2, 634–718. London: Sampson, Low, Marston, Searle, and Rivington.
Weismann, August. 1888a. On the supposed botanical proofs of the transmission of acquired characters. In *Essays upon heredity and kindred biological problems*, ed. Edward B. Poulton, Selmar Schönland, and Arthur E. Shipley, vol. 1, 385–417. Oxford, UK: Clarendon Press. 1889.
Weismann, August. 1888b. On the supposed transmission of mutations. In *Essays upon heredity and kindred biological problems*, ed. Edward B. Poulton, Selmar Schönland, and Arthur E. Shipley, vol. 1, 419–448. Oxford, UK: Clarendon Press. 1889–92.

[Secondary Sources]

Bange, Raphaël and Pietro Corsi. Chronology de la vie de Lamarck.
Bourdier, Franck and Michel Orliac. 2007. Esquisse d'une chronologie de la vie de Lamarck. 22 juin 1971. In *La vie et l'oeuvre de Jean-Baptiste Lamarck*, 23–43. https://archive.org/details/LamarckBiographie.
Bourguin, Louis Auguste. 1863. Lamarck. In *Les grands naturalistes français au commencement du XIXe siècle: Lamarck*, vol. 6, 194–221. Annales de la Société Linnéenne de Maine-et-Loire.
Burkhardt, Richard W. 1995. *The spirit of system: Lamarck and evolutionary biology*, 1977. Harvard: Harvard University Press.
Crick, F. H. C.. The central dogma. Nature 227 1970: 562–563.
Dawkins, Richard. 1982. *The extended phenotype: The long reach of the gene*. Oxford: Oxford University Press.
Dawkins, Richard. 1976. *The selfish gene*. Oxford: Oxford University Press.

Dennett, Daniel C. 1995. *Darwin's dangerous idea: Evolution and the meanings of life*. New York: Penguin.
Futuyma, Douglas J. 2005. *Evolution*. Sunderland, MA: Sinauer Associates.
Hahn, Roger. 2005. *Pierre-Simon laplace, 1749–1827: A determined scientist*. Cambridge: Harvard University Press.
Hamy, Ernest-Théodor. 1909. *Les débuts de Lamarck, suivis de recherches sur Adanson, Jussieu, Pallas, Géoffroy Saint-Hilaire, Georges Cuvier, etc.* Paris: E. Guilmoto.
Hazen, Robert M., Dominic Papineau, Wouter Bleeker, Robert T. Downs, John M. Ferry, Timothy J. McCoy, Dimitri A. Sverjensky, and Hexiong Yang. 2008. Mineral evolution. *American Mineralogist* 93:1693–1720.
Heilbron, J. L. 2010. *Galileo*. Oxford: Oxford University Press.
Hoffmeyer, Jesper. 2008. *Biosemiotics: An examination into the signs of life and the life of signs*. Chicago: University of Chicago Press.
Howard, Thomas Albert. 2006. *Protestant theology and the making of the modern research university*. Oxford: Oxford University Press.
Kennedy, M. J., M. Droser, L. M. Mayer, D. Pevear, and D. Mrofka. 2006. Late Precambrian oxygenation: Inception of the clay mineral factory. *Science* 311:1446–1449.
Kwok, Roberta. 2015. How life and luck changed Earth's minerals. In *Quanta Magazine*. https://www.quantamagazine.org/is-mineral-evolution-driven-by-chance-20150811/#. 11 August 2015
Landrieu, Marcel. 1909. *Lamarck, Le fondateur du transformisme*. Paris: Société zoologique de France.
Mayr, Ernst. How i became a Darwinian, In Mayr and William B. Provine, The evolutionary synthesis: Perspectives on the unification of biology. Cambridge: Harvard University Press, 1980.
Morizot, Baptiste. 2022. *Manières d'être vivant: Enquêtes sur la vie à travers nous*. Paris: Actes Sud. (English translation: *Ways of Being Alive*. Cambridge: Polity, 2022).
Packard, Alpheus. 1901. *Lamarck, the founder of evolution: His life and work*. London: Longmans.
Pigliucci, Massimo, and Gerd Müller. 2010. *Evolution: The extended synthesis*. Cambridge: The MIT Press.
Powers, Richard. 2019. *The overstory: A novel*. New York: Norton.
Riskin, Jessica. 2002. *Science in the age of sensibility: The sentimental empiricists of the French enlightenment*. Chicago: University of Chicago Press.
Stresemann, Erwin. Ornithology from Aristotle to the present, Hans J. and Cathleen Epstein, trans., G William Cottrell, ed., With a foreword and an epilogue by Ernst Mayr. Cambridge: Harvard University Press, 1975. [first published in German in 1951].
Querel, Alain. 2009. *Antoine de Fourcroy: savant, franc-maçon, homme politique*. Paris: Hermann.
Rudwick, Martin. 1972. *The meaning of fossils: Episodes in the history of Paleontology*. New York: Elsevier.
Spary, E. C. 2000. *Utopia's garden: French natural history from old regime to revolution*. Chicago: University of Chicago Press.
Zimmerman, Andrew. 2001. *Anthropology and antihumanism in imperial Germany*. Chicago: University of Chicago Press.

Jessica Riskin Frances and Charles Field Professor of History, Stanford University. She has also taught at MIT, Iowa State University and Sciences Po, Paris. Her research interests include early modern science, politics and culture and the history of scientific explanation. She is the author of Science in the Age of Sensibility: The Sentimental Empiricists of the French Enlightenment (Chicago-London, 2002) and The Restless Clock: A History of the Centuries-Long Debate over What Makes Living Things Tick (Chicago-London, 2016), as well as editor of Genesis Redux: Essays in the History and Philosophy of Artificial Life (Chicago-London, 2007) and, with Mario Biagioli, of Nature Engaged: Science in Practice from the Renaissance to the Present (New York, 2012). Her new book The Power of Life: The Invention of Biology and the Revolutionary Science of Jean-Baptiste Lamarck, was published by Riverhead Books in March 2026.

Chapter 11
Darwin, Merchants, and Famines

Giuliano Pancaldi

Abstract That the years around 1900 saw major changes in the life sciences is part of the historians' of biology common knowledge. This essay suggests that our understanding of those changes would benefit from enlarging the picture to include developments that, so far, have attracted little attention. These include what might be described as the rise of a new brand of civic life sciences. This development is illustrated here by focusing on Charles Darwin's late botanical works and on some merchant practices, encouraged and supported financially by Darwin himself, aimed at obtaining and distributing disease-resistant plants. Sources used in the essay include the seldom studied correspondence between Darwin and James Torbitt, a merchant and entrepreneur from Belfast who planned to obtain potato seeds resistant to the plant pathogen now known as Phytophthora infestans, responsible for the Great Irish Famine and subsequent outbreaks of the late blight disease.

Keywords Charles Darwin · Merchants · Famines · Disease-resistant plants · Agriculture · Ireland

11.1 Introduction

Historians of science agree that major changes occurred in the life sciences around 1900. Whether the focus is on the so-called "eclipse of Darwinism" (Bowler 1983; Largent 2009); Bellon 2011; the rise of Mendelism, including "social Mendelism" and genetics (Teicher 2020; Radick 2023); the "laboratory revolution" (Cunnigham and Williams 1992); developments connected with nationalism and colonialism (MacLeod 1971; Drayton 2000; Fa-ti Fan 2004); or a new "desire for morphology" (Tamborini 2020), the perception that the years around 1900 saw major changes in the life sciences is part of the historians' of biology common knowledge. This essay

G. Pancaldi (✉)
Department of Philosophy, University of Bologna, Bologna, Italy
e-mail: giuliano.pancaldi@unibo.it

S. Gattei (ed.), *The Incomparable Historian*, Archimedes 75,
https://doi.org/10.1007/978-3-032-25678-2_11

suggests that our understanding of those changes would benefit from enlarging the picture to include some developments that, so far, have attracted comparatively little attention. For brevity's sake, these developments could be described as the rise of the civic life sciences. These developments will be illustrated here by focusing on Charles Darwin's late botanical works and on practices, carried out in Britain during the last quarter of the nineteenth century, that bridged botany and agriculture in attempts to obtain disease-resistant plants.

It has been convincingly argued (Nyhart and Broman 2002; Turner 2005; Harwood 2006; Phillips and Kingsland 2015) that exploring the intellectual, social, and economic issues at the intersection between the life sciences and agriculture may shed new light on late nineteenth-century biology. The episode discussed here may contribute to an understanding of that intersection. It revolves around an initiative launched by James Torbitt, a merchant and entrepreneur from Belfast, who planned to obtain and distribute potato seeds that would be resistant to the plant pathogen now known as Phytophthora infestans, responsible for the Great Irish Famine and subsequent outbreaks of late blight disease. Following Darwin's science and his personal instructions, Torbitt claimed to be able to select disease resistant potatoes, that he wanted to distribute nationally with government's support. Torbitt secured Darwin's enthusiastic approval for the project, but despite Darwin's encouragement, intensive lobbying in London circles and private financial support, the plan ultimately failed.

The episode offers hints for interpreting the changes then under way in biology. Conclusions will focus on the transition from the gentleman tradition in the life sciences represented by Darwin (Desmond and Moore 1991; Browne 2002) to the emergence of new generations of practitioners, who shared an interest in Darwin's science while pursuing aims in line with the needs and ambitions of the rising middle classes. Such ambitions were nurtured by the benefits that several contemporaries attributed to what they presented as a much-needed turn towards what they called "practical Darwinism". A few concluding reflections will suggest the need to accommodate, in our reconstructions of fin-de-siècle life sciences, both the ideal of an individual scientist's "instinct for truth", as Darwin claimed to be pursuing with his work, and the goals and civic agendas pursued by merchants and entrepreneurs such as Torbitt.[1]

[1]This essay is one more attempt on my part to pursue several goals that John Heilbron pointed at in 1980 on becoming editor of *Historical Studies in the Physical Sciences,* after the editorship of Russell McCormmach. Heilbron wrote: "No historian now insists that ideas unroll independently of their institutional settings. But few try to discover the links that all agree exist; and fewer still have succeeded in making plausible connections between the content of science and the character of society." (Heilbron 1980, ii) I found these implicit challenges worth engaging and have since nurtured a desire to meet them.

11.2 Darwin, Variation, and Agriculture

The important role Darwin assigned to variation in his theory of evolutionary change is well known to historians (Endersby 2009). Late in the nineteenth century, however, among admirers and critics of Darwin's theory many viewed selection rather than variation as the main driver of Darwinian evolution. One reason for this view was that Darwin could not point with certainty to any decisive cause of variation.

What is not so well-known is that, after having devoted two bulky volumes to "variation under domestication" in the 1860s and produced a much revised second edition of the same in 1875 (Darwin 1875), Darwin, together with his botanist son Francis, conceived a new set of experiments aimed at delving into the possible chemical causes of variation. In the planned new research, the two Darwins intended to involve what was then the most prosperous, private agricultural research facility in England: Rothamsted. In those days research at Rothamsted focused mainly on chemical manures, the industrial production of chemical fertilizers being—together with land—the source of the funds needed to operate the expensive research facility owned by wealthy landowner, fertilizers' entrepreneur, and (from 1882) baronet Sir John Bennett Lawes.

Writing in 1876 to the chemist in charge of research at Rothamsted about the new research plan, Darwin premised that "no problem [was] … so perplexing as what causes almost every cultivated plant to vary" (CCD 1876, 53). The experiments Darwin envisaged in cooperation with Rothamsted were based on a bold hypothesis: that "the cause of variation must lie in different [chemical] substances absorbed from the soil… when their powers of absorption are not interfered with by other plants with which they grow … in a state of nature". The goal of the new experiments was to assess "whether variability depends at all on matter absorbed from the soil" (CCD 1876, 53–54). This topic and its ambition did not fail to impress chemists at Rothamsted, as well as Klimint Timiriazev, who as a young agriculturist from Russia visited Down house in July 1877 (Timiriazev 2006).

There is apparently no evidence that Darwin's collaboration with Rothamsted on this topic ever made it off the ground. Certainly, Darwin's interest in agriculture had not always been mainly theoretical as suggested by his exchange with Rothamsted. In the forties—the years of the Great Irish Famine (two recent assessments in Gray 2018, and Read 2022)—Darwin had published several articles devoted to practical agriculture, occasionally addressing issues of crop disease and plant breeding (see Wyhe 2009). In that same span of years Darwin had become the owner of two substantial farms in Lincolnshire and went on to devote to them the desultory sort of attention typical of a refined, "improving" but distant landowner (Desmond and Moore 1991, ch. 21).

What we do know for sure is that Darwin's attention to issues of practical agriculture was stimulated again early in 1876 when a new, unusual player entered the circle of his most frequent correspondents: James Torbitt.

11.3 James Torbitt, Merchant, Entrepreneur, and Agriculturist

Torbitt was a wine and spirits merchant who owned a dozen pubs in and around Belfast. He was also an entrepreneur of fluctuating fortunes and an agriculturist. He combined his commercial activities with an interest in the broad philosophical issues debated within the life sciences at the time, as well as long-pursued plans to launch some new manufacturing initiative connected to agriculture (Evans et al. 1996; DeArce 2008). Until the mid-Seventies, Torbitt's entrepreneurial interests had focused on the potato's starch content and nutrient profile. Beginning in approximately 1874, his plans included the idea of obtaining and selling—in his capacity as a mostly self-taught agriculturist and plant breeder—potato varieties resistant to the fungus that was causing the recurrent late-blight disease.

As we shall see, from 1876 onward Torbitt managed to obtain substantial support for his plans from Darwin, as well as some money raised by Darwin and his circle of wealthy relatives and acquaintances. For several years Darwin's support allowed Torbitt to conduct the slow and expensive crossing and selecting experiments aimed at obtaining and commercializing new potato varieties. Torbitt repeatedly asserted that he could not have carried out his experiments by relying only on the income from his merchant activities.

Torbitt's first goal when corresponding with Darwin—the two never met—was to obtain from Darwin, then at the apex of his public fame, statements asserting that Torbitt's plans were scientifically sound. And indeed, after few letters had been exchanged between the two, Darwin wrote to Torbitt:

> I cannot but think that the principle on which you are acting is right, and if you succeed you will have conferred an enormous benefit on the public. (CCD 1876, 96)

Darwin's statement amounted to a prestigious recognition of Torbitt's scientific credentials, heretofore pursued only erratically. Given Darwin's mention of the benefits to the public, his approval could easily be used by Torbitt to try to secure support for what he, and soon Darwin with him, presented as a "patriotic" undertaking of potential interest to millions of people and the government as well.

Torbitt immediately made use of Darwin's endorsement in a shrewd marketing campaign involving frequent ads in local and national newspapers. In the ads Torbitt offered his seeds for free to agriculturists willing to try them and report on the results obtained. Torbitt's campaign included letters to members of the House of Parliament, the House of Lords, and top Irish officials as well as, a few years later, public letters addressed to Prime Minister Gladstone. The letters were aimed at obtaining government's support for the long breeding and selection experiments Torbitt was carrying out in view of what he expected would become, in a few years' time, a commercial, nation-wide distribution of seeds that promised to propagate disease-resistant potatoes.

If these were Torbitt's chief goals when pursuing Darwin's support, what about Darwin's reasons for generously supporting Torbitt?

There was something interesting and unusual in the collaboration between the world-famous gentleman naturalist and the little-known merchant from Belfast with their shared ambition to save potatoes from a devastating fungus, and the poorest portions of humanity from recurring famines. The motives sustaining the two certainly deserve to be explored in depth.

Beginning in the mid-1850s Torbitt had tried to expand his trading business to include some manufacturing ventures. Inspiration for Torbitt's ambitions linked with agriculture came from the widely circulating *Commercial Dictionary* by Ricardian economist John Ramsay McCulloch (1839). According to Torbitt's own account (Torbitt 1875, 112), he found in that book an interesting apercu of the virtues and shortcomings of the potato from an economic point of view. (McCulloch 1839, 939–941). McCulloch identified some peculiar disadvantages of potatoes as a crop. Torbitt must have taken note of these as pointing indirectly to an unclaimed area, potentially open to a new commercial and entrepreneurial undertaking. For many practical purposes, McCulloch noted, it was impossible to preserve potatoes from one year to the next. According to McCulloch this amounted to a major disadvantage of potatoes when compared with corn, and implied—as he perceived well before the Irish famine of 1845—a recurrent danger for populations who relied overwhelmingly on potatoes for their nutritional needs (see also Salaman 2000).

In 1855 Torbitt applied for a patent to protect a process he had devised promising to be a new method for "preserving potatoes". This was by then a popular topic among patent seekers in Britain and the US. Torbitt described his invention as consisting of "improvements in the treatment and preservation of a constituent part of the potato": the part called "fibrine", an "element of nutrition identical with animal fibrine". The invention claimed to allow "the surplus produce of a luxuriant crop to be reserved to meet a subsequent scarcity", possibly "for an indefinite number of years". Torbitt's 1855 petition for a patent was ultimately rejected (Torbitt 1855).

A dozen years later, Torbitt was ready to try his chances as an entrepreneur and a patent seeker once again, and this time he gathered appropriate authorities in support. One was Lord Dufferin, an Ulster landlord who, when still a student at Oxford in the mid-1840s, had co-authored a dramatic report on the conditions of the Irish population he had observed during a visit to the country plagued by the famine (Dufferin 1847; Torbitt 1867). In the mid-1860s, having published a pamphlet on the land question criticizing John Stuart Mill's radical proposals for the pacification of Ireland (Dufferin 1868), many expected Dufferin to be appointed viceroy of the country.

Torbitt's new application for a patent was successful and it was "sealed" in March 1869. Specifications detailed a process like the one described in his earlier application. Torbitt now claimed, however, the "exclusive right to separate by mechanical pressure the liquid from the solid constituents" of the potato. The process aimed to enabling the manufacturing of a new commercial product, described by Torbitt as having "the consistency of paste or of Liebig's extract of meat, to which it bears a remarkable resemblance" (Torbitt 1868). Liebig's extract was an often-imitated commercial item in this period (Finlay 1995), and Torbitt claimed that the nutritional qualities of his "vegetable juice" were similar to those of Liebig's extract. To advertise his new product and prove the similarities, Torbitt visited Liebig in Munich. He also commissioned a chemical comparison of the two extracts from Arthur Hill Hassall, then Britain's leading food analyst (Torbitt 1875, 114–115).

Relying on the new patent, Torbitt set up an "experimental factory" in Belfast. In view of expanding his business to the continent, he also secured access to a potato starch factory in the Vosges, France. These were troubled years in Europe, however. The Franco-German war jeopardized Torbitt's ambitions on the continent, and in 1872 a new outbreak of the potato disease convinced Torbitt that the natural ingredient used in his factories deserved further study (Torbitt 1875, 114–115).

11.4 New Outbreaks of the Late Blight Disease

The 1872 outbreak stimulated an impressive array of research efforts in Britain (Jenkins 1874). In December Lord Cathcart, President of the Royal Agricultural Society of England (RASE), offered a prize of £100 for "the best essay on the potato-disease and its prevention". The competition attracted 94 contestants, but a special committee judged that none of their papers were deserving of the prize. In the end, the Society's council adopted a different scheme for the money, combining several different initiatives.

First, the Society decided to fund a foreign researcher—Anton De Bary of Strasbourg—in extending his earlier studies into the family of fungi (since 1863: *Peronosporeae*) to which the potato pathogen, known from 1876 as *Phytophthora infestans,* was thought to belong. De Bary's new study was concluded in December 1875 and published in the Society's journal the following year (De Bary 1876). Focusing on the "life-history" and morphological features of the fungus, however, De Bary's work did not claim to offer direct practical advice on how to prevent the disease.

Another initiative of the Society consisted in circulating a 25-item questionnaire to be sent to major potato producers throughout the country, focusing on their agricultural practice. One hundred questionnaires were returned (out of the three hundred sent) and carefully discussed by the secretary of the Society in a lengthy report. The report concluded with a cautious hypothesis asserting that potatoes' variable resistance to the fungus could perhaps be linked to the different crops being cultivated in the same fields where potatoes were subsequently planted.

In what was perhaps the most ambitious line of investigation launched by RASE in February 1874, six cultivators and dealers selling potato varieties claimed to be resistant to the fungus were invited to provide a ton of each, divided into twenty bags. These samples were used in a three-year experiment meant to compare the varieties' performance vis à vis the fungus while being cultivated—following instructions from and under the supervision of the Society—in twenty different locations in England, Wales, Scotland, and Ireland.

By the summer of 1874 experimental plots were being inspected by experts of the Society, including William Carruthers who acted as a consulting botanist and had played a role in formulating the Society's research agenda. Carruthers was keeper of botany at the British Museum, a member of the council of the Royal Society, and a rapporteur for the Agricultural Society's committee on the potato disease. He was inspecting the Society's four experimental fields in Ireland a few days before he joined the British Association for the Advancement of Science (BAAS) meeting in Belfast (Carruthers 1875), where Torbitt had decided to present his ideas for preventing the potato disease.

After the 1872 outbreak and the research efforts organized by the RASE, Torbitt's attention had been redirected. He now focused on the plant itself, its history, how it was cultivated, and the possible causes of the disease. This new focus led Torbitt to review the relevant botanical literature, including Charles Darwin's works. These may have nurtured in him the idea of winning over Darwin as a supporter of his new plans, following the same strategy he had adopted with Liebig when trying to launch his vegetable extract.

11.5 Torbitt's Path to Darwin

The path leading Torbitt to Darwin was far from straightforward, however. It involved readjusting Torbitt's ideas about life, reproduction, and aging in plants. It also involved bridging the gap separating a provincial merchant, aspiring entrepreneur, and amateur agriculturist from a gentleman of science who was then a world-famous naturalist and public figure.

Torbitt's decision to address the BAAS meeting in Belfast in 1874 was an early step in this process. On the occasion Torbitt submitted a paper that purported to combine a practical goal—preventing the potato disease—with a tentative theory explaining its causes.

The BAAS proceedings contain only a succinct abstract of Torbitt's presentation (BAAS 1874, 134). With typical zeal, however, Torbitt managed to have a local newspaper circulate an extended summary of his presentation, with a brief report of the discussion that followed (Torbitt et al. 1874). *The Times* in London also reported briefly on Torbitt's presentation and the ensuing discussion.

In the presentation Torbitt assumed that potato plants, if reproduced every year by tuber cuttings as was common practice, had a "term of life" of up to 30 or 40 years. On approaching that term, plants became "feeble" and could only be reinvigorated by crossbreeding, according to a view that Thomas Andrew Knight had circulated since the end of the eighteenth century (Knight 1795). Torbitt's proposal to fight the disease invoked the notion that "the operation of the laws of sexuality only was capable of producing new individuals" endowed with a renewed lifespan. In arguing for the latter, Torbitt mentioned in support Darwin's view according to which "it is a well-established fact that plants propagated by cuttings, or buds, all partake of a common duration of life" with the original plant. (Torbitt et al. 1874).

Torbitt's passing reference to Darwin at the BAAS was likely to please some and displease others. As is well known, John Tyndall in his presidential address in Belfast had celebrated a tradition of "higher materialism", culminating with Darwin, with the result of once again polarizing attitudes towards Darwin's science (Hesketh 2020). Darwin's critics at the meeting included the above-mentioned William Carruthers. He attended Torbitt's presentation and, during the discussion, thoroughly attacked Torbitt's ideas and the practical measures he recommended for preventing the disease.

During this row, Carruthers could rely on the extensive evidence available to him thanks to his role in the RASE's experiments. He claimed that experience was showing that vigorous, young, and healthy potatoes—including those reproduced recently from seeds—were attacked by the fungus just as old ones. In the end, both Torbitt and Carruthers neglected to mention the role that different varieties of the plant might play. The issue of variation, relevant for Darwin's recent works, was mentioned instead by a few botanists well acquainted with Darwin's work who joined in the discussion. They appeared to agree more with Torbitt's position than Carruthers. One of these participants was Samuel Alexander Stewart, an Irish botanist expert in mosses, who pointed out that several different varieties of potatoes were known to have "died out", being substituted by new ones. Another, Marmaduke Alexander Lawson, professor of botany and rural economy at Oxford, declared it "out of the question" that plants propagated by cuttings "should not live any longer than the original stock" (Torbitt et al. 1874). Comments by two other botanists were not detailed in the available reports, but what we do know about them suggests that they might have been more receptive to Darwin's views evoked by Torbitt. One was William Turner Thistleton Dyer, then professor of botany at the Royal Horticultural Society in London, who was corresponding with Darwin at that time on the movements of plants and insectivorous plants. Another botanist who joined in the discussion was Alfred William Bennett, who was corresponding with Darwin on the cross-fertilization of flowers.

What we certainly do know is that, prior to 1874, Carruthers as an expert in fossil plants and a geologist had displayed skepticism about Darwin's views on evolution (Carruthers 1869). We also know that a couple of years after the Belfast meeting Carruthers, based on his knowledge of paleobotany and in his capacity as president of the Geologist Association, went on to deliver a severe attack against evolution as promulged by Darwin, Haeckel, and their followers. His main argument was that "as yet" fossil plants had "disclosed nothing whatever as to *how* these forms originated." By this he claimed that he was not denying "development", but he certainly denied Darwinism. "Development – he stressed – is not the property of the evolutionist", and "the Mosaic narrative – the oldest scheme of creation –… represents the operations of that Creator as having been carried out in a series of developments…" (Carruthers 1876, 35). Carruthers had studied for the ministry in the Presbyterian Church for some time and he preserved an interest in Church history and theological literature throughout his life (Anonymous 1912, 196).

It took some time for Torbitt to further elaborate his ideas after the harsh confrontation with Carruthers at the BAAS. In the process, he strengthened his knowledge of Darwin's works, began to correspond with him, and convinced the world-famous naturalist to become a supporter of his initiative for preventing the potato disease.

Torbitt's further steps in this process was to submit his views, together with a report of his undertakings that far, to the Belfast Natural History and Philosophical Society. The presentation and publication that followed (Torbitt 1875) were aimed at reaching a broad circle of agriculturists "from Southampton to Dundee". There was still little strictly Darwinian in Torbitt's 1875 paper apart from the notion that sexual reproduction was superior to other forms of reproduction in terms of securing more vigorous seedlings. After the discussion at the BAAS, however, Torbitt may have thought that the limited convergence offered enough room for him to try to involve Darwin in a discussion over potato disease. Early in 1876 Torbitt sent Darwin an offprint of the Belfast Society paper, asking for his opinion on a theoretical issue: what should be regarded as an individual plant when dealing with plants obtained by cuttings? As Torbitt put the question to Darwin: "May not after all the sexual union be the only rejuvenator of life? The true and only starting point of the life of the individual?" (CCD 1876, 29).

Darwin's answer was cautious. It showed unwillingness to become involved in a discussion with a little-known correspondent on a delicate issue, that Darwin had touched upon recently (Darwin 1875, 405). Darwin's answer to Torbitt included a candid admission of ignorance concerning reproduction by cuttings: "How it is that a single cell, or a very few cells, suffice to give rise to a new organism will never be known until we can say what life is, and we are at present a long way off this goal." (CCD 1876, 30).

Darwin's caution notwithstanding, Torbitt found the response encouraging. In his view it showed that the great naturalist was paying attention to arguments concerning a topic of broad scientific import put forward by the little-known correspondent. Together with the attitude displayed by the Darwinian botanists who had avoided supporting Carruthers at the BAAS in Belfast, Darwin's letter must have encouraged Torbitt to reinforce the Darwinian elements in his proposal for combating potatoes' disease.

A new version of Torbitt's plan, formulated in the early months of 1876, emphasized the need for the agriculturists to *destroy* all the potatoes that succumbed to "the attack of the parasite". A few weeks after receiving Darwin's letter, Torbitt placed a paid advert in *The Times* conveying his plans in recognizably Darwinian language:

> EXTINTION OF POTATO DISEASE, with doubled or trebled crops. – Modus operandi – Growth from the seed. Exposure of plants to full force of infection. Destruction of those which succumb. Propagation of the rest by the sett. (In all places some plants will repel the attack of the parasite; in some, all). Seed supplied gratis. Address Robertson, Brooman, and Co. 165, Fleet street, London; or James Torbitt, Belfast, Ireland. (Torbitt 1876a)

Robertson, Brooman, and Co. were Torbitt's patent agents in London.

A new publication Torbitt was working on early in 1876, marked a further step in his adoption of Darwinian concepts and language. This text was to be his most ambitious publication and it coincided with a new, intense advertising campaign in which he was determined to make of his plan for fighting potato disease a recognizably Darwinian endeavor. Under the title *A treatise on the cultivation of the potato from the seed...*, (Torbitt 1876b) Torbitt's book promised to achieve both the extinction of the disease and a substantial increase of crop per acre. The frontispiece announced that copies of the book were being sent—"accompanied by a packet of seed"—to each member of the House of Lords, the House of Commons, and the principal landlords of Ulster."

To support his cultivation strategy, Torbitt continued to rely in part on old Knight's ideas, but now he also quoted a substantial extract from Darwin's *Variation* on "the difference between sexual and asexual generation", introduced by the flattering statement that "it will be generally admitted that there is no higher authority in the world [than Darwin]". (Torbitt 1876b, 57–58). Above all, he described his method of cultivation in an explicitly *selectionist* language:

> Let the plant have light and air, and let the Peronospora do its worst. Then every plant which is not immaculate when ripe, which does not resist the shock absolutely, destroy it. ... And those which resist the first year - destroy, destroy, destroy. *It is the principle of "natural selection" artificially carried out*, and it is done every day in animal life, and with results beyond, far beyond, what could possibly have been anticipated beforehand. If it were applied to the human race, how the matter of the better cultivation of a plant would fade into nothingness, by comparison. (Torbitt 1876b, 56–57, emphasis added)

To assess Torbitt's closing remark on selection in humans, it should be recalled that Thomas Huxley, writing for *Nature* in 1869 under the title "Darwinism and national life", had promulged the view that selection in humans was "a practical side of infinite importance" for Darwinism. (Huxley 1869).

Darwin received a copy of Torbitt's 1876 book in the first few days of April. On 4 April he wrote back to Torbitt with the glowing two-lines assessment already quoted, approving Torbitt's method and emphasizing its "enormous benefit on the public". Exhilarated, Torbitt replied by telegram: "Letter received with profound thanks. I can make it suppress the potato disease five years sooner, if published. May I?" Unusually, Darwin also answered by telegram: "You may publish what I have said" (CCD 1876, 92, 95–96).

A week later Darwin had assimilated Torbitt's plan and found it could fit with his botanical work well enough to offer Torbitt detailed scientific advice on how best secure its success:

> During the last 10 years I have been experimenting on crossing plants and shall publish the results in the autumn [Darwin 1876]. I would strongly advise you to intercross any two varieties, (and the more they differ in all respects the better) that is if you can get two varieties which are moderately free from the pest. I know that there is the strongest probability that seedlings raised from a cross of this kind would not only grow more vigorously, but would possess greater constitutional vigour, so as to be less viable to disease of all kind and death. (Darwin to Torbitt, 14 April 1876, in CCD 1882, Supplement to the Correspondence 1831–1880, 364)

Torbitt and Darwin's joint venture was now sealed. The following years there were ups and downs, however, and Darwin's motives for supporting Torbitt—as well as on rare occasions resisting him—must be considered in some detail.

11.6 Practical Darwinism

When assessing Darwin's response to Torbitt's 1876 publication one must consider that Torbitt's emphasis on selection in agricultural practice came at a time when Darwin's notion of selection was facing considerable skepticism. After Fleeming Jenkin's powerful critique of Darwin's theory in 1867, Darwin's supporters and friends had been placing growing emphasis on its possible practical applications to defend the theory. The most striking pronouncement on the matter was probably the one Joseph Hooker made on addressing as President the BAAS meeting in Norwich in 1868, where he articulated a memorable comparison:

> What Faraday's discoveries are to telegraphy, Mr. Darwin's will assuredly prove to rural economy, in its widest sense and most extended application. (Hooker 1868, lxviii)

Those were indeed years of epic achievements for transatlantic telegraphy. When suggesting the comparison, Hooker was defending Darwin against a nasty, anonymous attack published months earlier in the widely read *Athenaeum*. According to Hooker's polemical reconstruction, the author of the attack (Anonymous 1868), who was reviewing Darwin's *Variation* (first edition), had dared to assert that "Mr. Darwin's theory is a thing of the past, that Natural Selection is rapidly declining in scientific favour…" (Hooker 1868, lxx). By comparing telegraphy and the applications of Darwin's theory to rural economy Hooker was indicating a comparatively new vantage point from which to assess Darwin's merits, considering that debates since the publication of the *Origin* had focused mostly on scientific, religious, and philosophic issues.

Of course, Darwin's theory had been linked to "rural economy" since Darwin had established a parallel between natural and artificial selection (Secord 1985), and the parallel itself had raised skepticism among some interpreters. In Hooker's Norwich address, however, the notion of an "applied" or "practical" Darwinism evoked a different set of implications: Hooker with his parallel clearly focused on the *future* rather than the past of living forms.

The expectations evoked by Hooker soon reached botanists, agriculturists, and colonial planters. A 1869 article, published by *The Produce Markets' Review,* a journal known for offering trade information for grocers, and reprinted in the *Journal of the Society of Arts,* invited sugar cane planters worldwide to adopt "a little practical Darwinism", the goal being "the improvement of the sugar cane" (Anonymous 1869). The author, likely a member of the network of colonial botanists sponsored by Hooker through Kew, lamented that sugar planters were not "striving to improve their canes by selection", with the result that "the saccharine strength of the cane thus remains stationary, or even decreases, while the strength of the beet has been doubled by careful cultivation and selection, since the beginning of the century". (Anonymous 1869).

When Torbitt contacted Darwin a few years later to involve him in a campaign claiming to use Darwinian selection to obtain disease-resistant potato crops, and promising important benefits for agriculture and the public, Darwin was likely stirred as both a scientist and a citizen aware of the potential economic implications of his work.

In the early 1870s, Darwin himself had devoted considerable attention to potatoes as offering examples of graft hybridization, a topic from which he expected important theoretical implications for his hypothesis of Pangenesis and the science of heredity. In a substantial addition to the second edition of *Variation*, Darwin reviewed an abundant store of literature showing that "the potato affords the best evidence of the possibility of the formation of graft-hybrids." (Darwin 1875, vol. 1, 420–424) That new varieties, displaying great diversity of characters, could be obtained by grafting different parent forms without sexual generation, was often regarded with incredulity at the time. For Darwin, however, it showed "a highly important physiological fact". In Darwin's view, potatoes and other plants' graft-hybrids suggested that the elements contributing to the production of a new being "are present in the cellular tissue in such a state that they can unite without the aid of the sexual organs, and thus give rise to a new bud partaking of the characters of the two parent forms". Throughout 1875, Darwin had been exchanging views on the topic with George John Romanes, and as a result Romanes had carried out a series of detailed experiments aimed at obtaining genuine hybrids from potatoes by grafting (CCD 1875, letters from 14 January through 14 July). The plant's traits considered in these experiments were typically visible ones. On one occasion, however, Darwin mentioned also some more recondite characters such as precocity. Potatoes' vulnerability to disease, however, was never mentioned as an interesting topic for research in the correspondence between the two gentleman naturalists.

Indeed, as early as in the mid-1840s Darwin had been corresponding on potato disease with his mentor, the botanist and parish priest John Stevens Henslow, and with William Darwin Fox (Darwin's cousin), clergyman, landowner, and agriculturist. At the time they were considering how it might be possible to defend potato crops, and thus help the poor. Initiatives included assessing whether the descendants of wild potatoes from south America that Darwin had collected during his Beagle voyage proved more resistant to disease than recent European varieties. Observations carried out by Fox suggested not (Fox 1846; Ristaino and Pfister 2016), and the topic apparently did not attract Darwin's attention again until 1876.

It is also worth remembering that, since the late 1840s, Darwin had developed considerable skepticism about the notion that science should be valued for its practical applications. An exchange he had with Henslow in 1848 is revealing of his views in this regard. Well-known for his advocacy of science education for all classes of society and his attention to the poor, Henslow liked to stress the need to pursue a balance between science per se and its useful applications. "However delightful any scientific pursuit may be for its own sake alone – Henslow remarked in a popular lecture that, once printed, he sent to Darwin – "if it shall be wholly unapplied, it is of no more value than the very amusing occupation of building castles in the air." (Henslow 1848) Darwin objected strongly to this statement. Immersed in his long, and likely to remain "wholly unapplied" (his words) study of Cirripedia in that period, Darwin expressed to Henslow his deeply felt views on pure vs. applied science in the following, autobiographical statement:

> I believe there exists, & I feel within me, an instinct for truth, or knowledge or discovery, of something same nature as the instinct of virtue, & that our having such an instinct is reason enough for scientific researches without any practical results ever ensuing from them. (CCD 1848, 128; see also CCD 1858–1859, 269)

Considering the impact Darwin's *Origin* and *Descent of man* had had on public debates about broad scientific, philosophic, and religious issues for years, we may assume that in the 1870s Darwin was even more inclined to regard his own scientific persona as closer to the ideal of an "instinct for truth", harboring a comparative indifference towards the pursuit of practical applications. If this is so, Darwin's rapid and positive response to Torbitt's brand of "practical Darwinism" in the mid-1870s requires some additional explanation.

Surely, Torbitt's 1876 pamphlet is unlikely to have impressed Darwin on strictly scientific grounds. Torbitt added nothing to Darwin's recent botanical work, or on the causes of variation or on graft-hybrids. Torbitt's pages most appealing to Darwin must have been those in which the author presented his plan as directly linked to Darwin's notion of selection. Reading those pages, Darwin must have perceived an opportunity for relaunching, on the popular grounds of agriculture and the public good, his cherished notion of selection at a time when the notion was being subjected to frequent criticism.

This would also explain why Darwin was cautious about scientific differences with Torbitt. Tensions between the two emerged instead around mundane matters, such as their different attitudes when announcing intended goals or achievements to the public. Darwin, the gentleman scientist, advised caution, while the merchant preferred audacity. Another apparently delicate issue for Darwin was the need to assess the basic motives underlying Torbitt's experiments. Once Darwin asked Torbitt: assuming one would know whether in your search for disease resistant potatoes you are guided by "selfish" or by "patriotic" motives, what your answer would be? Torbitt's answer was frank: "My motives in prosecuting the subject seem to me to have been mixed partly selfish partly patriotic" (CCD 1878, 85). We are left wondering what Darwin might have made of this. Certainly, the emphasis on "patriotic" motives was gaining ground in those years as a new crisis was developing in Ireland, connected to the partial failing of the potato harvest (Jordan 1994, 203–205).

11.7 The Hunt for Disease Resistant Varieties

When assessing Torbitt's and Darwin's joint initiative it is important to keep in mind how elusive the notion of a disease-resistant plant was in the 1870s. A comparison with the better-known case of the hunt for Phylloxera—the insect that had been devastating southern France vineyards since the 1860s—can help.

Torbitt, the wine and spirits merchant, was of course aware of the news from France about this subject (Torbitt 1876a, 1876b, 5). Some French botanists were using Darwinian notions to deal with Phylloxera and the vines. In 1872 Pierre Marie Alexis Millardet, a botanist who would later play an important role in the study and treatment of Phylloxera, pointed at three concepts capable of guiding the plant pathologist: heredity, variation, and selection. (Millardet 1872, 31–32) Another French botanist, Jules Emile Planchon who had spent a few years in London working under William Hooker in the Kew Gardens and later studied American vines, ranked the latter according to their resistance to the insect (Planchon 1875). In the end Phylloxera was conquered by introducing new vines obtained by grafting French vinifera vines onto resistant American rootstocks (Paul 1996). In Torbitt's view, aging during long periods of reproduction without intercrossing was the common thread connecting the declining disease resistance displayed by grape, potato, and sugar-cane crops.

In daily practice, Torbitt's new, "Darwinian" search for disease-resistant potato varieties required all the usual, slow and uncertain procedures of plant crossing, selection, and breeding. Choosing seeds from two vigorous varieties that gave signs of being less affected by the fungus was the first step. Crossing was then carried out by shaking "the pollen on glass, to see under the microscope that it was perfect, and apply it early to the stigma by means of a hair pencil, repeating the operations several times" (CCD 1878, 257). Sowing seeds in appropriate soil and under favorable conditions was the next step, followed by putting them in a greenhouse (CCD 1880, 149), and making sure that accurate records of the original plant varieties and the seedlings were kept throughout. Then the new plants grown via this method had to be exposed to the "full force of the blight", and a new round of experiments was launched on those that had proved resistant.

As to the amount of work and care involved, by March 1880 Torbitt and Darwin estimated that 5000 seedlings had been treated each year by Torbitt (CCD 1880, 148), while the land being used for the experiments amounted to about 10 acres (CCD 1880, 159). Considering that from the beginning Torbitt had involved not only his own fields in the experiments but also dozens of farmers to whom he gave his seeds for free, provided they report on how they performed, the experiments also required a large number of trusted collaborators and reliable rapporteurs. The situation could prove critical, of course, but it allowed Torbitt to expand his experiments while giving him the opportunity to advertise his selected seeds among farmers in view of the business he planned to launch once the promised, disease-resistance performance of his seeds had gained credibility.

After a few years' worth of trials, Torbitt considered possibly relaxing the criterion of seeds-only reproduction. In November 1879 he cautiously argued that "the plant ought to be grown from its seed each fourth year", considering that "the yield of the plant in seed and tubers was, in the second and third years of life, much larger than in the first, fourth and fifth years". (CCD 1879, 500) Clearly, it was not easy for the agriculturist to keep considerations about yield separate from assessments of disease-resistance.

In fact, a doubt was making its way into Torbitt's mind:

> … given a variety which has been during four or five years free from disease, will it continue to be so forever? I have no doubt that it will not, and I have no doubt that it will not live forever. New varieties must be continually coming into existence, in my opinion, if the maximum capacity of the plant is to be made available. (CCD 1880, 310)

Perceptions of the relative progress made with the experiments oscillated accordingly. In an optimistic moment Torbitt was able to write to Darwin, who had asked whether he continued to be "fairly hopeful of complete, or even moderate, success":

> ... I have now to say, that as my judgement goes, complete success is not only fairly hopeful, but complete success has been obtained. I have no doubt whatever, that by means of cross-fertilization and selection, any number of varieties of the potato can be produced fungus-proof not only in the tuber, but also in the foliage. (CCD 1879, 455)

Assessments, however, were often connected to the resources available or expected to be at hand. The fluctuating success of Torbitt's commercial activities, which remained the main source of financing for his experiments, was often mentioned in his exchanges with Darwin. The independent and joint efforts the two enacted to obtain public support—from the government in London, Irish authorities, or public bodies like the RASE—were also a frequent topic of discussion, and frustrations were a recurrent note. The latter were often connected to the fact that Torbitt's plan encountered stiff opposition in London scientific and governmental circles fueled by the anti-Darwinian Carruthers.

As to the money involved in the experiments, in 1880 Torbitt and Darwin estimated that, over the years, the experiments had cost the Irish merchant about 1000 pounds (worth 98–99,000 pounds today). In the absence of public support, the two had already floated the idea of raising private funds in 1878 (CCD 1878, 101–102). In March 1880, when the chances of obtaining public financing looked as slim as ever, and even a loan from the government appeared unlikely (CCD 1880, 116), Darwin decided to help soliciting private subscriptions for Torbitt (CCD 1880, 118, 123). He recruited for this purpose first his friend Thomas Henry Farrer and James Caird, then some of Darwin's own relatives and friends, obtaining a total of 90 pounds that Darwin sent to Torbitt on 18 May 1881 (CCD 1881, 224). Together with the money Darwin himself had sent to Torbitt in March 1878, the total private funds made available to Torbitt through Darwin was in the order of a few hundred pounds. In January 1882, finally, Torbitt let Darwin know that, because his merchant trade was flourishing, he wanted to return the money he had received from the subscribers. On 10 January Darwin amiably answered: "As far as I am concerned, I am quite content to remain unpaid, as I gave the money for what I considered to be an excellent object." (CCD 1882, 23).

Darwin died on 19 April. Until then, despite their many, joint efforts Torbitt had not been invited to testify in front of the several government committees appointed to recommend measures for containing the recurrent potato crop failures. During such food production crises, the government spent millions to buy a competitor's potato variety to alleviate the crisis, rather than granting Torbitt the loan of a few hundred pounds he had been requesting for his experiments.

After Darwin's death, Torbitt kept in touch with his son Francis. In Darwin's *Life and letters* Francis left a sympathetic account describing how much Darwin had cared about Torbitt's plan. Francis also gave an update on the experiments the Irish continued to conduct. "Mr. Torbitt – he wrote – tells me that he still (1887) succeeds in raising varieties possessing well-marked powers of resisting disease; but this immunity is not permanent, and, after some years, the varieties become liable to the attacks of the fungus" (Darwin, Francis ed. 1887, vol. 3, 348–351). Torbitt also continued his frequent publicity campaigns through newspaper adverts (see DeArce 2008, 217–218).

In closing this paragraph, it is worth mentioning that the hunt for the potato pathogen, now known to be an Oomycete, and disease resistant potatoes is still in progress. An "open-access T-BAS phylogeny for emerging Phytophthora species" was announced in April 2023. It mapped "192 formally described species and 33 informal taxa in the genus Phytophthora" (Coomber et al. 2023). On the other front of the hunt, the highly complex genome of a tetraploid potato cultivar was decoded and published in March 2022 (Sun et al. 2022).

11.8 Conclusion

In the end, one may ask whether Torbitt's and Darwin's search for disease-resistant plant varieties was a genuinely Darwinian undertaking. The answer I think is yes, with qualifications.

Torbitt's plan, we know, was rooted in Knight's old notion of the "aging" of plant varieties. A notion that continued to play a role in Torbitt's later work, allowing him to claim a measure of originality after he had adopted the Darwinian concepts of selection and crossing. If we consider things from Darwin's perspective, we must also say that, before being contacted in 1876 by Torbitt already intent on his plan, Darwin was not devoting any special attention to the topic of disease resistant varieties. His occasional interest in the subject went back to the mid-1840s, the years of the great Irish famine. In the 1870s, however, before crossing paths with Torbitt, Darwin's reflections on Solanum tuberosum focused strictly on issues of heredity and grafting, in connection with his bold hypothesis of Pangenesis.

The endeavor that became Torbitt and Darwin's joint venture from spring 1876 onwards was thus made possible by the convergence of different paths and different research agendas, as well as different personalities and social profiles. One might say that the convergence materialized because Torbitt, after the 1874 BAAS meeting in Belfast, realized that there were many botanists in Britain paying attention to Darwin's botanical works and, unlike the anti-Darwinian Carruthers, willing to listen to Torbitt's arguments and consider his "patriotic" plan to extinguish the blight disease. One might equally say that, once he had become aware of Torbitt's plan which by 1876 emphasized the connections with Darwin's notions of selection and crossing, Darwin found good reasons to adhere to a call for "practical Darwinism" that some of his supporters had been advancing for years when defending him against his critics; even if this was a form of Darwinism towards which Darwin himself, apparently, continued to maintain a cautious stance.

Darwin's and Torbitt's paths were able to converge because their joint venture allowed Darwin to continue to pursue the gentlemanly ideal centered on what he called his "instinct for truth". At the same time, the venture also allowed Torbitt to follow his bourgeois conviction that science could go hand in hand with both "selfish" and "patriotic" goals.

Darwin's belief that life produces an incessant stream of unpredictable varieties or "inventions", favored his involvement in Torbitt's plan and their search for disease resistant plants. So did Darwin's believe that, with the art of the breeder, humans can intercept varieties and use them for their benefit when guided by Darwin's theory of selection, that he regarded as itself part of life's stream of inventions (Pancaldi 2019). Darwin conveyed such concepts lucidly when, intent on fund raising in Torbitt's behalf, he explained their plan to James Cairds, the doyen of British agriculturists, and to Thomas Henry Farrer, a high officer of the London Board of Trade (CCD 1880, 147–149).

Whatever the limits of the joint venture discussed here, and its effective failure during its protagonists' lifetimes, Torbitt and Darwin's collaboration showed that it was possible to pursue the goal of averring the "Newton of the blade of grass", that Darwin had evoked in the memorable last page of *Origin*, side-by-side with the commercial and civic goals that Torbitt pursued through his agricultural, marketing, and lobbying endeavors.

Clearly, towards the end of the nineteenth century there was a vast, multifaceted market in Britain for several forms of the "practical Darwinism" that many were invoking, unaware of the tragic detours the twentieth century had in store.

Abbreviations

BAAS	British Association for the Advancement of Science
CCD	Burkhardt, Frederick, et al. eds., *The Correspondence of Charles Darwin*, 1985–2023
PRONI	Public Record Office of Northern Ireland
RASE	Royal Agricultural Society of England

References

Anonymous. 1868. Review of Darwin. *Variation*, in *Athenaeum*, February 15, n. 2103, pp. 243–244.

Anonymous. 1869. The Improvement of the Sugar Cane. *Journal of the Society of Arts*, January 15, pp. 133–134.

Anonymous. 1912. Eminent Living Geologists: William Carruthers. *Geological Magazine*, New Series Decade, 9, pp. 193–199.

BAAS. *Reports*.

Bellon, Richard. 2011. Inspiration in the Harness of Daily Labor: Darwin, Botany, and the Triumph of Evolution, 1859–1868′. *Isis* 102:393–420.

Bowler, Peter J. 1983. *The Eclipse of Darwinism. Anti-Darwinian Evolution Theories in the Decades Around 1900*. Johns Hopkins University Press.

Browne, Janet. 2002. *Charles Darwin. The Power of Place*.

Carruthers, William. 1869. The Cryptogamic Forests of the Coal Period. *The Geological Magazine* 41:289–300.
Carruthers, William. 1875. Report on the Results of the Competition on 1874 for the Society's Prizes for Potatoes that should be free from Disease for Three Years in succession. *Journal of the Royal Agricultural Society of England, 2nd series* 11:376–395.
Carruthers, William. 1876. Address at the Opening of the Session 1876–77. *Proceedings of the Geologists' Association* 5 (1): 17–35.
Coomber, Alison, Amanda Saville, Ignazio Carbone, and Jean Beagle Ristaino. 2023. An Open-access T-BAS Phylogeny for Emerging Phytophthora Species. *PLoS ONE* 18 (4): e0283540. https://doi.org/10.1371/journal.pone.0283540.
Cunnigham, Andrew, and Penny Williams, eds. 1992. *The Laboratory Revolution in Medicine*. Cambridge University Press.
Darwin, Charles. 1875. *The Variation of Animals and Plants Under Domestication*. 2nd ed.
Darwin, Charles. 1876. *The Effects of Cross and Self Fertilisation in the Vegetable Kingdom*.
Darwin, Francis, ed. 1887. *Life and Letters of Charles Darwin*, 3 vols.
De Bary, Anton. 1876. Researches into the Nature of the Potato Fungus Phytophthora infestans. *Journal of the Royal Society of Agriculture of England, 2nd series* 12:239–269.
DeArce, Miguel. 2008. Correspondence of Charles Darwin on James Torbitt's Project to Breed Blight-Resistant Potatoes. *Archives of Natural History* 35 (2): 208–222.
Desmond, Adrian, and James Moore. 1991. *Darwin*.
Drayton, Richard. 2000. *Nature's Government: Science, Imperial Britain, and the 'Improvement' of the World*.
Dufferin, Frederick Temple Blackwood. 1847. *Narrative of a Journey from Oxford to Skibbereen During the Year of the Irish Famine*.
Dufferin, Frederick Temple Blackwood. 1868. *Mr Mill's Plan for the Pacification of Ireland Examined*.
Endersby, Jim. 2009. Darwin on Generation, Pangenesis and Sexual Selection. In *The Cambridge Companion to Darwin*, ed. Jonathan Hodge and Gregory Radick, 2nd ed., 73–95.
Evans, Alun, Louise Cooke, and Joseph Clint. 1996. Too Serious for Joking': James Torbitt, the Belfast Potato Breeder. *Ulster Folklife* 42:1–11.
Fan, Fa-ti. 2004. *British Naturalists in Qing China*.
Finlay, Mark R. 1995. Early Marketing of the Theory of Nutrition. The Science and Culture of Liebig's Extract of Meat. In *The Science and Culture of Nutrition, 1840–1940*, ed. Harmke Kamminga and Andrew Cunningham, 48–74.
Fox, William Darwin. 1846. Potato Disease. *Gardeners' Chronicle* 40:661.
Gray, Peter. 2018. The Great Famine, 1845–1850. In *The Cambridge History of Ireland*, ed. Thomas Bartlett and James Kelly, vol. 3, 639–665.
Harwood, Jonathan. 2006. Introduction to the Special Issue on Biology and Agriculture. *Journal of the History of Biology* 39 (2): 237–239.
Heilbron, John. 1980. Foreword. *Historical Studies in the Physical Sciences* 11 (1): i–iii.
Henslow, J. S. 1848. Address Delivered to the Ipswich Museum on 8th March 1848.
Hesketh, Ian. 2020. The Making of John Tyndall's Darwinian Revolution. *Annals of Science* 77 (4): 524–548.
Hooker, Joseph D. 1868. Norwich Address. In *BAAS Report, 1869*, lviii–lxxv.
Huxley, Thomas H. 1869. Darwinism and National Life, *Nature*, December 16, pp. 183–184.
Jenkins, H. M. 1874. Report on the Cultivation of Potatoes, with Special Reference to the Potato-disease. *Journal of the Royal Agricultural Society of England, 2nd series* 10:475–514.
Jordan, Donald E. 1994. Land and Popular Politics in Ireland: County Mayo from the Plantation to the Land War.
Knight, Thomas Andrew. 1795, April 30. Upon the Inheritance of Decay Among Fruit Trees, and the Propagation of Debility by Grafting. *Philosophical Transactions of the Royal Society*, pp. 81–84.

Largent, Mark A. 2009. The so-Called Eclipse of Darwinism. *Transactions of the American Philosophical Society, New Series* 99 (1): 3–21. Descended from Darwin: Insights into the History of Evolutionary Studies, 1900–1970.
MacLeod, Roy. 1971. The Support of Victorian Science: The Endowment of Research Movement in Great Britain, 1868–1900. *Minerva* 9 (2): 197–230.
McCulloch, John Ramsay. 1839. *Commercial Dictionary.*
Millardet, Pierre Marie Alexis. 1872. *La botanique.*
Nyhart, Lynn K., and Thomas Broman, eds. 2002. *Science and Civil Society* (Osiris, Vol. 17).
Pancaldi, Giuliano. 2019. Darwin's Technology of Life. *Isis* 110 (4): 680–700.
Paul, Harry W. 1996. *Science, Vine, and Wine in Modern France, 1750–1990.*
Phillips, Denise, and Sharon Kingsland, eds. 2015. *New Perspectives on the History of Life Sciences and Agriculture. Archimedes*, 40.
Planchon, Jules Emile. 1875. *Les vignes Américaines.*
Radick, Gregory, 2023. *Disputed Inheritance: The Battle over Mendel and the Future of Biology.*
Read, Charles. 2022. *The Great Famine in Ireland and Britain's Financial Crisis*, (People, Markets, Goods: Economies and Societies in History Book 19).
Ristaino, Jean Beagle, and Donald H. Pfister. 2016. What a Painfully Interesting Subject. *BioScience* 66 (12): 1035–1045.
Salaman, Redcliffe N. 2000/1949. *The History and Social Influence of the Potato.*
Secord, James A. 1985. Darwin and the Breeders. A Social History. In *The Darwinian Heritage*, ed. David Kohn, 519–542.
Sun, Hequan, et al. 2022. Chromosome-scale and Haplotype-Resolved Genome Assembly of a Tetraploid Potato Cultivar. *Nature Genetics* 54 (3): 342–348.
Tamborini, Marco. 2020. The Twentieth-Century Desire for Morphology. *Journal of the History of Biology, Special Issue* 53 (2): 211–216.
Teicher, Amir. 2020. *Social Mendelism: Genetics and the Politics of Race in Germany, 1900–1948.*
Timiriazev, Klimint. 2006. With Darwin at Down. With a letter by L. N. Bell. *Archipelago* 9:44–58.
Torbitt, James. 1855. Improvements in the Treatment and Preservation of a Constituent Part of the Potato, English Patents of Inventions, N. 971.
Torbitt, James. 1867. Torbitt to Dufferin, 26 September 1867, Ms. PRONI D1071/H/B/T/238.
Torbitt, James. 1868. Improvements in the Treatment, Preservation and Utilization of the Potatoe. English Patents of Inventions, N. 2941.
Torbitt, James et al.1874. The Potato Disease and Means of its Prevention, *The Belfast News-Letter*, August 21, 7.
Torbitt, James. 1875. Potato Cultivation. In *Proceedings of the Belfast Natural History and Philosophical Society, Session 1874–1875*, 112–125.
Torbitt, James. 1876a. Extinction of Potato Disease, Paid Ad in *The Times*, February 22.
Torbitt, James. 1876b. *Cras Credemus. A Treatise on the Cultivation of the Potato from the Seed, Having for Proposed Results the Extinction of the Disease*, etc.
Turner, R. Steven. 2005. After the Famine: Plant Pathology, Phytophthora infestans, and the Late Blight of Potatoes, 1845–1960. *Historical Studies in the Physical Sciences* 35 (Part 2): 341–370.
Wyhe, John van, ed. 2009. *Charles Darwin's Shorter Publications 1829–1883.*

Giuliano Pancaldi, Professor of the History of science, University of Bologna, retired. His research explores the interplay between the history of science and technology, biography, and society. His books include Darwin in Italy: Science across Cultural Frontiers (Indiana, 1991) and Volta: Science and Culture in the Age of Enlightenment (Princeton, 2003). He is currently working on the history of early biotechnologies; his first publication in this area is "Darwin's Technology of Life", published in Isis in 2019.

Chapter 12
Measures by Americans: The United States and the Establishment of International Electrical Standards, 1881–1897

Daniel J. Kevles

Abstract From 1881 to 1893, leading American and European physicists and engineers met in a series of international congresses to define the values of the fundamental electrical units, measures of resistance, current, voltage, capacitance, and inductance, all crucial infrastructural elements in electrical science and in the electrical technologies that together with chemistry were driving the Second Industrial Revolution. The delegates, initially in disagreement, steadily forged consensus definitions for the units. The young American physicist Henry Rowland played a leading role in the effort, as did the German Herman von Helmholtz and his British counterpart William Thomson. In 1884, members of the congress in Philadelphia, inspired by German plans, urged the U.S. government to establish a physical observatory that would include a bureau of electrical standards. While the recommendation failed, it adumbrated the establishment of the National Bureau of Standards, in 1901.

At the Chicago congress, in 1893, the physicist Thomas Corwin Mendenhall, superintendent of the U.S. Bureau of Weights and Measures, persuaded the leading national delegates to agree on a set of definitions for all but two of the standards. In 1894, Congress, asked the Academy to deal with the remaining two. It soon did, but the Academy's president infuriated Mendenhall by excluding the engineer Henry Carhart, an aggrieved voltage expert. Rowland tactfully and decisively intervened, and in 1895 the entire set of electrical standards achieved the force of law. The outcome signaled that the United States was now a robust presence in the internationally critical field of electrical standards.

Paper Prepared for the Memorial Volume in Honor of John Heilbron

June 2023
Corrected, Nov. 25, 2025

D. J. Kevles (✉)
Department of History, Yale University, New Haven, CT, USA
e-mail: daniel.kevles@yale.edu

S. Gattei (ed.), *The Incomparable Historian*, Archimedes 75,
https://doi.org/10.1007/978-3-032-25678-2_12

Keywords Carhart · Electrical congresses · Engineers · Industry · Mendenhall · National Academy of Sciences · Ohm · Volt · Ampere · Helmholtz · Thomson

In the late 1870s, the leading technological nations of Europe were at loggerheads over the definitions and values of the fundamental electrical units—the ohm, the unit of resistance; the volt, of electromotive force; and the ampere, of current.[1] All were crucial infrastructural elements both in electrical science and in the technologies of the new electrical industries—the generators, dynamos, lighting systems, and telephones — that together with chemistry were driving the Second Industrial Revolution.

Central to the differences was the ohm. Physicists and electricians, especially telegraph engineers, in England and Europe had been wrestling with its definition since the 1860s, driven primarily by the need to specify the characteristics of telegraph cables and wires for both operational design and commercial contracts. By the late 1870s, two different specifications for the ohm were in play. One, developed by British physicists, was at base theoretical; called the "absolute ohm," it rested on translation of electrical work or force into mechanical work or force, with the latter expressed in standard metric measures of length, mass, and time—for example, *cgs,* for centimeters, grams, and seconds. As the spearheads of a committee of the British Association for the Advancement of Science, William Thomson (later Lord Kelvin), James Clerk Maxwell, and the physicist/engineer Fleeming Jenkin successfully urged that the theoretical *cgs* value be given material definition in the form of a foundational electrical circuit. Dubbed the British Association ohm, or commonly the "BA ohm," it served as a standard for the production of coils of one-ohm resistances.[2]

The other definition was devised in Germany by Werner Siemens, a hard-driving electrical entrepreneur who had extensive experience with the laying of telegraphic cable. Impatient with conceptualizations, he preferred an instantiation of a standard resistance suitable for general practical use. He defined one ohm as the resistance offered to one volt by a one-meter-long tube of mercury with a cross-section of one square millimeter at a temperature of 0^0 centigrade. Siemens' definition was arbitrary in relation to the theoretical unit, but it was convenient for assessments of lengthy telegraph wires and was within a few percent of the scaled-up value of the absolute ohm.[3]

However, in the late 1870s several physicists experimentally tested the value of the BA ohm or Siemens' mercury standard and reported disquieting results: the outcomes not only varied from each other but also indicated that the BA ohm

[1] *LRL,* 61.

[2] Bruce J. Hunt, "The Ohm Is Where the Art Is: British Telegraph Engineers and the Development of Electrical Standards," *Osiris,* 2nd Series, 9(1994), 48, 56–57, 60–61*; LLR,* 8–10, 60–61.

[3] *LRL,* 8–14.

differed from the absolute unit of resistance by as much as 2 percent.[4] In 1880, eager to address the discrepancies the French government began organizing an international congress of physicists and electricians (the contemporary term for electrical engineers) that would convene in tandem with an electrical exposition to be held in Paris the following year.

In the fall of 1880, George Walker, a former vice president of the Western Union Telegraph Company and the current American consul-general in Paris, got wind of the initiative. He alerted the State Department to the upcoming events, adding the hope that the birthplace of [Benjamin] Franklin, [Samuel F.B.] Morse, and [Joseph] Henry, and now the home of [Thomas] Edison and [Alexander Graham] Bell, "will not neglect to participate in the proposed congress of electricians, and to impress upon it those scientific ideas … which have received such a remarkable and rapid development in our own country."[5]

The Department, aware of the significance of electrical technology to the nation's foreign and domestic commerce, arranged for the United States's representation in Paris by a delegation of "eminent specialists," as President Chester A. Arthur later called them in a report to Congress.[6] The delegation included the physicists George F. Barker, of the University of Pennsylvania, who had devised several electrical meters, including a reliable and inexpensive standard cell for measuring voltage; and Henry Rowland, of The Johns Hopkins University, whose specialty was electrical physics, particularly the ohm.

Rowland was the standout technical authority in the American delegation. Trained in civil engineering, he was a self-made physicist, ambitious, he told his family, for scientific reputation rather than "filthy lucre." While an instructor at his alma mater, Rensselaer Polytechnic Institute, he taught himself his favored subject by reading extensively, devising instruments, and finding an encouraging patron and confidant in the Harvard astrophysicist Edward Pickering. His reputation came early, in the mid-1870s, when he was in his mid-twenties, from deft theoretical and experimental analysis of the magnetic permeability of iron, steel, and nickel. His research was marked by supple experimental skill guided by sure theoretical knowledge. Aided by Pickering, he was recruited, in 1875, to an as yet undefined position at then-aborning Johns Hopkins University. In July, he left for Europe in the company of the Hopkins president, who wanted Rowland to purchase equipment for the university's laboratory and enrich his knowledge and acquaintance in European physics and physicists.[7]

---

[4] *Ibid.*, 62–63.

[5] George Walker, "International Exposition of Electricity": *Report by Consul General Walker of Paris,* October 27, 1880, *Commercial Relations of the United States: Reports from the Consuls of the United States on Commerce, Manufactures, Etc. of Their Consular Districts,* No. 4, February 1881 (Dept. of State; Washington, D.C.: GPO, 1881), 253.

[6] Chester A. Arthur, "Annual Message of the President," Dec. 6, 1881, *Papers Relating to the Foreign Relations of the United States, 1881* (Washington, D.C.: GPO, 1882), iv.

[7] Daniel J. Kevles, *The Physicists: The History of a Scientific Community in Modern America* (Cambridge: Harvard U. Press, 1995), 25–26; Samuel Rezneck, "An American Physicist's Year in

Soon on his own, Rowland visited the major laboratories of the day on an arc from England through northern Europe and on to Austro-Hungary. Maxwell greeted him enthusiastically in Cambridge, and he was welcomed at the BA meeting, where Thomson publicly touted his magnetic research and he dined almost every evening at the gathering's white-tie tables. He spent 4 months in Berlin, where, his reputation having preceded him, Helmholtz allocated one of the much-coveted spaces in his laboratory for his use. Rowland succeeded in an experiment at which Helmholtz had failed — showing that a statically charged rotating disk produced a magnetic field. Helmholtz, who had paid Rowland the rare compliment of visiting him at his bench almost every day, immediately presented the result to the Berlin Academy.[8]

Henry A. Rowland. Thomas C. Mendenhall, "Henry Augustus Rowland, 1848-1901", *National Academy of Sciences Biographical Memoirs* (1903), 116

In the weeks before his departure for Europe, Rowland had analyzed one of the variant experiments on the ohm and, as he told Pickering, discovered an error in its methods and was inspired to conceive a new method for determining the value remarkable "for its theoretical simplicity." During his visits to the continent's laboratories Rowland assessed their scientific instruments with a critical eye, writing to Maxwell, "so far it seems to me that the accurate measurement of resistance either absolutely or relatively is an English science almost unknown in Germany." He

Europe: Henry A. Rowland, 1875–1876," *American J. of Physics,* 30(1962), 878–879.

[8] Rezneck, "An American Physicist's Year in Europe," 880, 882; Thomas C. Mendenhall, "Henry Augustus Rowland," *National Academy of Sciences Biographical Memoirs,* (1903), 120.

made it also an American science after his return to the United States, in May 1876, as, by then, a professor at Hopkins. He conducted an experiment that eliminated multiple sources of error in the apparatus and methods that figured in the discrepant assessments of the BA ohm. His results, published in 1878: the BA ohm differed from the absolute ohm by about 1 percent; and at least one of the commercially produced sets of standard coils differed from Rowland's exacting determination of the value of the ohm by 0.15 percent.[9] The next year Rowland was nominated for membership in the National Academy of Sciences (NAS) and was a newly elected member when he arrived in Paris with the American delegation to the international congress, which opened in mid-September 1881.

The Paris congress that year turned what had been disparate national initiatives for the ohm into a series of international congresses that expanded beyond the ohm to consider the definitions and values of current, voltage, capacitance, and inductance; in succeeding years half were held in Paris, the rest in Philadelphia, Frankfurt, Edinburgh, and—a defining milestone—Chicago, part of the World Columbian Exposition organized in 1893 to celebrate the four hundredth anniversary of Christopher Columbus' arrival in the New World. Several of the gatherings convened in conjunction with electrical expositions where thousands of soigné men and women wandered through the exhibits in great halls aglow with electric lights, encountering the marvels of the electrical present and promises of its revolutionary future.[10]

---

[9] Rezneck, "An American Physicist's Year in Europe," 879, 881, 885; Henry A. Rowland, "Note on Kohlrausch's Determination of the Absolute Value of the Siemens Mercury Unit of Electrical Resistance," *Philosophical Magazine*, 50(Sept. 1875), 161–163; Henry A. Rowland, "Research on the Absolute Unit of Electrical Resistance," *American Journal of Science* (3), XV(1878), 291, 438–439.

[10] See, for example, Gérard Borvon, *Histoire de l'électricité. L'exposition Internationale d'électricité de 1881, à Paris* (9/12/2009) @https://seaus.free.fr/spip.php?article500.

The Court of Honor. Chicago World's Fair, 1893. Conceived by the prominent architect Daniel Burnham and the landscape designer Frederick Law Olmstead, it was popularly dubbed "The White City," denoting both the whiteness of the buildings by day and the dazzling electrical illumination at night. Kenneth M. Swezey Papers, Archives Center, National Museum of American History, NMAH-AC0047-0000009.jpg

The stakes in the congresses were high — for science, commerce, and, since each of the units required a universal name, for national standing in the international electrical enterprise. The meetings drew the sustained participation of the era's most eminent physicists and electricians—notably, from Britain, Thomson and William E. Ayrton, an admired physicist/engineer and inventor of a number of electrical instruments; from France, Nicholas Mascart, Professor at the College de France, Director of the Bureau Central Météorologique and the country's leading electrical expert; from Germany, Siemens, a baron of the electrical industry as well as of the state, and Herman von Helmholtz, an intimate of the government and, along with his good friend Thomson, one of the century's towering physicists.[11] Thomson and Helmholtz, already aware of Rowland's experimental prowess, ranked him as a coequal in the course of the electrical-unit congresses, most of which Rowland attended.[12]

From the outset Rowland gave technical weight to the American delegation, and its influence was enlarged over the course of the meetings by the involvement of members from other branches of the expanding electrical enterprise in the United States. Particularly important were Henry S. Carhart, a University of Michigan

[11] See the lists of participants in the several congresses in the publications of their proceedings cited hereafter; David Cahan, *Helmholtz: A Life in Science* (Chicago: U. of Chicago Press, 2018), 581.
[12] *NASM*, 4/17/79; 4/20/81.

physicist with a sizable presence in the instruments of electrical measurement; the physicist Thomas Corwin Mendenhall, the director of the U.S. Coast and Geodetic Survey and superintendent of its subordinate Office of Weights and Measures, which was the lead U.S. agency in the international drive for electrical standards; and representatives of the American Institute of Electrical Engineers.

In the late nineteenth century and since, American physics was faulted for excessive absorption with pursuing measurements to the next decimal place. But that devotion was highly valued in the international effort to achieve reliable electrical standards, and their propensity for precision physics advantaged its American participants. Together they exercised decisive influence in the forging of an international package of standardized electrical units and in the establishment of the package as policy and law in the United States.

In Paris in 1881, sentiment initially ran in favor of substituting Siemens' unit for the BA ohm. National pride as well as science was at stake. So were friendships. The Helmholtz and the Siemens families were close—one of Helmnoltz's daughters would marry one of Siemens' sons—and Helmholtz strongly supported Siemens' mercury-column standard on the merits. Thomson and Helmholtz, taxing their own friendship, squared off against each other in a heated debate over whether to retain as the standard the absolute ohm instantiated in some material form other than the coil or Siemens' mercury-column. Cooler heads called an adjournment and proposed a compromise. The absolute ohm would be retained as the theoretical standard but it would be represented materially for practical and legal purposes in Siemens' one-millimeter square column of mercury with its length adjusted so that its resistance matched the absolute ohm as closely as possible.[13]

The British and Germans agreed to the compromise, and so did the rest of the congress. Thomson acknowledged that the British coil was materially imperfect and that a geometrically defined column of mercury had the distinct advantage of being reproducible anywhere on the globe, without requiring the presence of a certified standard to compare it with. The agreement became one of seven resolutions concerning electrical units that the gathering endorsed. They covered the volt, ampere, coulomb, and farad as well as the ohm, but the ohm held a crucial place because at the time it figured directly or indirectly via Ohm's law, which holds current equal to voltage divided by resistance, in the definition of all the rest but the volt. The essential operational resolution was to call on the French government to join with other powers in establishing an international commission to oversee whatever research might be necessary to establish definitively all the units, but particularly the necessary investigations to determine the length of Siemens' column of mercury. The congress, which met in multiple sessions until it adjourned on October 5, would resume in Paris a year hence, in 1882, to receive and deliberate upon the results.[14]

---

[13] *LRL,* 20, 72–73, 78–80; Cahan, *Helmholtz,* 581, 595.

[14] *Thomson,* "[Third Session], Sept. 21, 1881, *Congrès International des Electriciens, Paris 1881,* 41–43, 46.

But the measurements reported at the next year's meeting, where Rowland and Barker once again officially represented the United States, were also inconclusive. At the insistence of Helmholtz, who was much concerned with the industrial as well as the academic need for a reliable standard, the congress unanimously urged the French government to request the sponsoring governments of each of the delegations to take whatever steps might be necessary to encourage their scientists to determine as accurately as possible the mercury-column equivalent of the absolute ohm. Once such reliable measures were available the congress would meet in Paris again to fix the ohm's practical value.[15]

From the time the first Paris congress closed in October 1881 Rowland had aimed to weigh in definitively on the length of the mercury column. In November, on his own return from Paris, Barker, also a member of the National Academy, informed the membership that the State Department had appointed Rowland and himself to the international commission for the ohm. At his urging, the Academy expressed the earnest hope in the form of a resolution that Congress would appropriate the funds necessary for its two commission members "to carry out the needed experimental determinations with credit to the country." The resolution, buried in the Academy's annual report to Congress, produced no immediate action, but sometime in the months following the 1882 Paris congress, the French government at the meeting's behest sent the State Department a request calling for participating nations to sponsor research on the ohm.[16] At the request of the Department, on March 3, 1883 Congress appropriated $12,500 ($400,000 in 2026 dollars) for the purpose and to cover the expenses of sending up to three experts to the next Paris conclave, scheduled for mid-1883.[17] Drawing on the federal allocation, Rowland turned a Hopkins facility outside Baltimore into a laboratory suitably insulted from disturbance and in June 1883 set out to determine the length of the mercury-column ohm through three independent methods.[18]

As it turned out, the 1883 congress was postponed to the next year because the mercury-column experiments in the various countries were slow to completion.

---

[15] Chester A. Arthur, "Annual Message of the President," 12/4/82, *Papers Relating to the Foreign Relations of the United States...* (Washington, D.C.: GPO, 1883), iv; *NASM*, 11/19/1881; *Conférence Internationale pour la Détermination des Unités Électriques, 16 Octobre—26 Octobre 1882: Procès-Verbaux* (Paris: Imprimerie Nationale, 1882), 8–10, 54–55, 67, 70, 154,157; Cahan, *Helmholtz,* 578, 582.

[16] Arthur, "Annual Message of the President," 12/4/82, iv; *NASM*, 11/19/1881; *Conférence Internationale pour la Détermination des Unités Électriques, 16 Octobre—26 Octobre 1882:* (Paris: Imprimerie Nationale, 1882), *Procès-Verbaux,* 8–10, 54–55, 67, 70, 154,157; Cahan, *Helmholtz,* 578, 582.

[17] *NASM*, 11/19/81; *Report of the Proceedings of the National Academy of Sciences,* [1881–1882], 20; U.S. Congress, Senate, *CR,* 47th Cong. 2d Sess., 3/1/83, 3503; *Statutes at Large, Ch. 143. 1883,* "An Act Making Appropriations for Sundry Civil Expenses …," 603; John Trowbridge, remarks, *Report of the Electrical Conference at Philadelphia in September, 1884,* 38–39.

[18] ["The Determination of the Ohm,"] Extract from a letter from Henry A. Rowland, n.d., *Conférence Internationale pour la Détermination des Unités Électriques, Deuxieme Session* (Paris, 1884), *Proces-Verbaux,* 37.

Rowland's exacting procedures prolonged his work even past the time the congress finally convened, which was for a week from the end of April 1884. Barker, Rowland, and the Harvard physicist John Trowbridge, whose specialties included electrical measurement, had all been appointed delegates to it, but for unexplained reasons none made the trip. Rowland instead sent a progress report saying that he expected to have full results by November," adding, "As these experiments will be made with the greatest care and in very favorable conditions, thanks to the generosity of the [U.S.] Congress, it is hoped that no decision concerning the value of the ohm will be taken before this time; in this way, the United States and other countries will be able to arrive at the definitive standard."[19]

Despite high interest in Rowland's investigation, however, the congress in Paris was too impatient to await his results. The mercury-column lengths reported there were scattered within a few percent of each other. No matter: Mascart, reporting that lawsuits were increasing with the ongoing expansion of international telegraphic and commercial relationships, emphasized that the industrial need for a standard could not be postponed. Guided by the mean of the reported results, 106.02 centimeters, the congress settled on the rounded value of 106, counting it close enough to the truth for practical purposes. The congress dubbed the 106-centimeter column length the "legal ohm" and, so as not to imply international legalization of the centigrade system, it specified the resistance as the measurement taken at the "temperature of melting ice." It urged that the French government recommend the adoption of the legal ohm by the participating states. Drawing on the practice with the standard meter, which was housed in Paris, it also called for the various governments to construct a primary mercury standard for the ohm and arrange for the production of solid alloys that would be regularly compared with it and with each other.[20]

Rowland and Thomson further cogitated upon the recommendations for both the ohm and the comparisons 5 months later, in mid-September 1884, at an International Electrical Conference held at the Franklin Institute in Philadelphia. Established at the initiative of the Institute as part of an exposition of cutting-edge electrical technologies — it was the first of its kind in the United States — the Conference comprised some 100 other scientists and electricians, some British but most American. The U.S. Congress had appropriated $7500 for the event and authorized President Arthur to appoint a scientific commission to run it on behalf of the United States. Arthur named seven scientists to the commission, including Rowland, Barker, the Princeton astrophysicist Charles Young, the renowned federal astronomer Simon Newcomb, and Monroe B. Snyder, who was a research-minded young faculty member at the city's Central High School and the supervisor of its astronomical observatory.

The Scientific Commission designated Rowland its president and William Thomson its vice president. The two men were of one mind on the essential importance of the conference's purpose, its significance symbolized by the enthusiasm of

---

[19] *Ibid.*

[20] "The Electrical Congress of Paris," *Nature,* 5/8/84, 26–27; *LRL,* 90, 95–96, 103–104.

the crowds—eventually more than a quarter of a million people—attending the nearby exposition. Rowland held that the electrical technologies to come required agreement upon the fundamental electrical units, especially the ohm. Thomson proclaimed, "It is as necessary … to measure electricity as it is … to weigh out quantities of tea and sugar, fish, meat, etc."[21]

Prompted by the French, the U.S. State Department wanted to know whether the United States should adopt the unit values recommended in Paris, including that of 106 centimeters of mercury for the ohm, and whether the U.S. should establish and maintain a primary standard of resistance for comparisons.[22] In his opening address, Rowland had pointedly remarked that while the Paris congress had thought the rounded value of 106 centimeters "desirable" for the mercury-column standard, his own results, now in hand, put the value at 106.278 centimeters—very close to Lord Rayleigh's result and correct, he and his research assistant felt confident, within one part in 3000.[23] Faced with Rowland's measurements, the conference declared the rounded value of 106 centimeters as only "provisional."[24]

All the members of the Scientific Commission but Snyder were members of the National Academy, and two, Newcomb and Young, were on a committee of the Academy that was advising a Congressional Commission—it was chaired by Senator William B. Allison, of Iowa — that was then conducting a major investigation of the scope and operations of the government's multiple scientific agencies. At the time, federal science was mainly concerned with the astronomical and earth sciences, but the discussions in Philadelphia inspired Newcomb and Young to think more expansively.[25]

Throughout the Philadelphia conference, Thomson and Rowland had been resolutely united on the need for governments to establish local agencies for the maintenance and comparison of electrical standards. The idea appears to have come from Helmholtz, who had told Thomson the year before that the German government wanted to establish a "physical observatory" in Berlin on land donated by von Siemens that would, as part of its research program, provide a resource for the

---

[21] "Address of Professor Rowland"; "Address of Sir William Thomson," *Report of the Electrical Conference at Philadelphia in September, 1884* (Washington, D.C.: GPO, 1886), 20, 26, 30.

[22] John Davis, acting sec. of state, to Trowbridge, n.d.,; letter headed "French Legation to the United States," to the Secretary of State, n.d., no author, *ibid.*, 38–39.

[23] *Report of the Electrical Conference,* 20, 39–41.

[24] *Ibid.,* 39, 42, 67; Edwin H. Hall, "John Trowbridge," *National Academy of Sciences Biographical Memoirs,* XIV(1930), 199.

[25] Henry A. Rowland and Monroe B. Snyder to Secretary of State Thomas F. Bayard, 6/13/85; Simon Newcomb, Opening Remarks, 10/8/84; T. F. Bayard to the President, 1/25/86; List of Participants; *Report of the Electrical Conference,* 3–5, 7–9, 12; U.S. Congress, *CR, House,* 47th Cong., 2nd Sess., 2/10/83, 2442–2443; 48th Cong., 1st Sess., 1/8/84, 286; 5/7/84, 3933; 7/2/84, 5881; *U.S. Statutes at Large,* 48th Cong., Sess., 1, Ch. 332, "An Act making appropriations for sundry civil expenses…," 7/7/84, 194; Jane Mork Gibson, "The International Electrical Exhibition of 1884: A Landmark for the Electrical Engineer," *IEEE Transactions on Education,* E-23 (No. 3; August 1980), 169–176; Thomas C. Mendenhall, "Some Metrological Memories," *Journal of the Franklin Institute,* 6(1918), 96.

establishment and maintenance of electrical standards. In a talk devoted to the subject, Snyder held that such an agency –a "Bureau of Standards," he called it — should maintain all the standards "necessary to the great business interests of the country," including length, mass, time, temperature, and electrical resistance, and furnish the public, particularly physicists and practical experts, with authoritative multiple copies.[26]

Newcomb, a vigorous enthusiast of the proposition, saw an opportunity to press it upon the U.S. Congress via the National Academy's engagement with the Allison Commission. Together with Rowland and other allies, he spearheaded a resolution urging action to call the Congress' attention to the "national importance" of establishing a central authority to maintain legal electrical standards, not least because they formed "the basis of contracts for the supply of electricity," and to verify all instruments for electrical measures. On September 11, 1884, the conference endorsed the resolution with a round of applause.[27]

Soon thereafter, Charles Young, fresh from the meetings in Philadelphia, brought the recommendation to the Congress via the National Academy's report to the Allison Commission, thus extending the report to encompass the expansion of federal science into new areas, including those related to the Second Industrial Revolution.[28] Borrowing from Helmholtz, the report advised the government to establish a "physical observatory" for research in the "exact sciences." Implicitly bespeaking the special interest of Young as well as of two other astrophysicists on the Academy's committee, the report tasked the observatory with investigating the laws of solar and terrestrial radiation and their uses in meteorology. Explicitly invoking the recommendations of "the recent Electrical Conference in Philadelphia," the report called for enlarging the work of the Office of Weights and Measures to "include electrical measures" and transferring the Office to the new physical observatory. There it would undertake the responsibilities of the bureau of electrical standards that the conferees in Philadelphia had in mind.[29]

Although the Academy's report was in the Allison Commission's hands by December 1884, the Commission gave no attention to the proposed multi-purpose physical observatory in any of its multiple subsequent hearings. The initiative did, however, become a target of the Alexander Agassiz, the son of the famed Louis Agassiz and himself a distinguished Harvard zoologist he was rich from turning a

---

[26] "Address of Professor Rowland"; Thomson remarks; discussion, *Report of the Electrical Conference,* 21, 47–48, 73; Cahan, *Helmholtz,* 617–619.

[27] Newcomb, Rowland, Charles K. Wead, and William A. Rogers, "Recommendations of the Committee," n.d.; discussion of the recommendation, *Report of the Electrical Conference*, 107–108, 110, 183.

[28] The *Congressional Record* from Dec. 1884 through July 1886 contains no mention of anything concerning a bureau or agency of standards.

[29] "Report of National Academy of Sciences on Organization of Scientific Bureaus of the Government…," in U.S. Congress, Joint Commission to Consider the Present Organization of the Signal Service, Geological Survey, Coast and Geodetic Survey, and the Hydrographic Office of the Navy Department, with a View to Secure Greater Efficiency and Economy of Administration of the Public Service …, *Testimony,* 1886, 49th Cong., 1 Sess., Sen. Misc. Doc. 82, Series 2345, *8.

failing copper mine into a success and was an advocate of limited government, including limitations on federal support of science outside his own discipline. In an unsigned article in *The Nation* in late December 1885, he attacked the call for the establishment of the observatory: "The Government should not keep a school for original research, and it is not advisable to see established at Washington laboratories of chemistry, physics, biology, etc. etc. intended only for such work as can be done elsewhere."[30] The Allison Commission, perhaps following Agassiz when it issued its report in the spring of 1886, did not mention either the observatory or the need for federal maintenance of electrical standards.

In truth, the Academy's committee had been handicapped in making its case. Thomson had noted in Philadelphia that before bureaus of electrical standards could be created the standards themselves had to be established, and in 1886 unquestioned electrical standards did not yet exist.[31] During the rest of the 1880s, scientists and electricians tended for practical purposes to use the legal ohm of 106 centimeters recommended in Paris in 1884 However, lacking adoption by any government, the legal ohm had no legal standing. By the end of the 1880s, new measurements of the ohm had been made in several countries and the seven most reliable ones — accomplished in Britain, France, Germany, and the United States — clustered closely around a mean of 106.28 centimeters for the mercury column, essentially the Rayleigh/Rowland values. At the BA's meeting in Leeds in 1890, its longstanding Committee on Electrical Standards accepted a rounded value of 106.3 centimeters.[32]

At the BA's meeting in Edinburgh, in 1892, the Committee achieved consensus on the ohm, the ampere, and the volt by a group of international experts. Among them were Helmholtz, who had been worrying that it might be difficult to measure accurately the mercury column's cross-section, and Henry S. Carhart, the University of Michigan physicist who was a leading authority on the Clark Cell, a battery-like device widely used as a standard for voltage. In the end, the gathering agreed to define the ohm as the resistance offered by a column of mercury at the temperature of melting ice that was 106.3 centimeters long, of uniform cross-section, and—incorporating a proposal from Helmholtz to substitute mass for the cross-sectional area—weighed 14.4521 grams. It defined the volt and the ampere in specific material instantiations that accorded with their absolute cgs values rather than by their Ohm's-law relationship to each other—for the ampere, by the quantity of silver deposited per second when a current of one ampere ran through a solution of silver

[30] [Alexander Agassiz], "The National Government and Science," *The Nation,* 12/24/85, 525–526.

[31] Rowland to Thomas F. Bayard, Secretary of State, n.d., received 6/13/85, *Report of the Conference in Philadelphia,* 4; T. F. Bayard to [the Committee on Printing], 3/22/86, U.S. Congress, Senate, *CR,* 49th Cong., 1st Sess., Senate, 7/26/86, 7516; House, 8/5/86, 8028–8029.

[32] *LRL,* 122–124.

nitrate; and for the volt, the level of voltage generated by a Clark Cell of certain specifications.[33]

The Edinburgh agreement amounted to an informal concordat. It awaited formal ratification by the international community of scientists and electricians and then official adoption by governments. Carhart, thrilled by the meeting and its outcome, hoped that the ratification could be achieved at the International Electrical Congress that was scheduled for Chicago in August 1893 in connection with the World Columbian Exposition.[34]

The principal promoter and organizer of the Chicago Congress was the American Institute of Electrical Engineers (AIEE). Since its establishment in the spring of 1886—several months before the Philadelphia meeting and, some said, out of resentment that the Philadelphia promoters seemed inhospitable to practical electricians—it had had become increasingly engaged with the shaping of policy in electrical matters. In 1890, now some 300 members strong, it took up the cause of uniform and reliable standards, mindful that the explosive expansion of the electrical industry into power and traction as well as light and communication was making imperative the need for establishing international standards not only for the volt, ampere, and ohm but also for the variables that figure in dynamos and motors — work, power, and self-induction. The AIEE managed to raise the profile of those values on the international-standards agenda, calling in 1891, at a meeting in Frankfurt, for getting at least one American name into the pantheon of names by dubbing the unit of self-induction the "Henry," after Joseph Henry, in recognition of his major contribution to the concept.[35]

President Grover Cleveland opened the Exposition on May 1, 1893, addressing a crowd reported to number 300,000 and then pressing an oversized gold telegraph key atop a flag-draped table that sent a signal to Machinery Hall, activating a huge assembly of giant steam engines and dynamos with a great whirring of wheels and belts. They sent power coursing through the exposition's 690 acres, more than a square mile, energizing the fountains in the nearby lagoon, their water columns rising 150 feet, the technological displays in the vast exhibition halls, and bathing interiors and walkways with thousands of candle powers of dazzling light. More than 27 million people were later said to have visited the Exposition, drawn by the exhibits from 46 countries that ranged from music and art to agriculture, mining, and transportation and by the amusements along the Midway that included George

[33] *LRL,* 141–143; *Report of the Sixty-Second Meeting of the British Association for the Advancement of Science ... Edinburgh in August 1892* (London: John Murray, 1893), 134–135.

[34] Henry S. Carhart to Thomas C. Mendenhall, 10/15/92, TCM; *LRL,* 124.

[35] Gibson, "The International Electrical Exhibition of 1884," 175; *LRL,* 34–40, 111, 144; Carl Hering, "Report of the Delegation to the Frankfort [sic] Electrical Congress," *Transactions of the American Institute of Electrical Engineers,* 10/27/91, 544–547; T. C. Martin, "History of the Congress," *Proceedings of the International Electrical Congress held in The City of Chicago, August 21st to 25th 1893* (New York: American Institute of Electrical Engineers, 1894), xxi.

Gale Ferris, Jr.'s giant wheel, which carried people in cabins seating up to 40 to a height of 240 feet.[36]

Ferris' giant wheel A History of the World's Columbian Exposition Held in Chicago in1893, 447

The Electrical Congress began, on August 21 amid a sweltering heat and the whistle-and-belching din of trains outside with a public session in Columbus Hall at the city's Art Institute before an audience of 500. In short order, Elihu Thomson, a founder of the Thomson-Houston Company and a recent president of the AIEE, presented Helmholtz, the Congress' honorary president, who bowed repeatedly as the crowd greeted him with a prolonged standing ovation. Now in his early seventies and somewhat frail, Helmholtz was the director of what in 1887 had been formally established as the outgrowth of the idea for a physical observatory—the *Physikalisch-Technische Reichsanstalt,* an expanding cluster of buildings on the land Siemens had donated that was largely devoted to testing and standards.[37] The Congress' registered participants, 150 in all, included far more than its predecessors men from schools, colleges, universities, scientific laboratories and inventive enterprises, many of them making, like Elihu Thomson, a Who's Who of scientists, electrical engineers, and inventors — the agents of what the inventor Elisha Gray, the formal chair of the Congress, called "a new era in our civilization … the Electric Age."[38]

[36] Erik Larson, *The Devil in the White City: Murder, Magic, and Madness at the Fair that Changed America* (New York: Vintage, 2003), 235–242, 251–255; "World's Columbian Exposition," *Wikipedia;* "Ayrton's Remarks," *Proceedings of the International Electrical Congress*, 8.

[37] Martin, "History of the Congress," xxii-xxiii; Mendenhall, "Some Metrological Memories," 103–104; Cahan, *Helmholtz,* 688, 707.

[38] List of participants and Elisha Gray's remarks, *Proceeding of the Congress,* vi–viii, 1–2.

The Exposition's throbbing reminders of the new age doubtless strengthened the assemblage's eagerness for an agreement on electrical units. The Congress assigned the work on the matter to a Chamber of Official Delegates, all appointed by their respective governments, five each from Britain, France, Germany, and the United States, and seven more from six other countries. On Gray's recommendation, the Secretary of State appointed as the U.S.'s official delegates Rowland, Carhart, Elihu Thomson, Mendenhall, and Edward L. Nichols, a physicist at Cornell. (To their "great chagrin," Carhart later complained, the government neither paid their expenses nor otherwise took any official notice of them or of the Congress). The Chamber designated Rowland as its president. Mendenhall, had long been concerned in his research with the intricacies of electrical instruments and measures and was now officially so in his capacity as superintendent of the Office of Weights and Measures; highly respected by both Rowland and Carhart, deft, dapper, self-assured, and well connected in Washington, he loomed large in the Chamber's deliberations.[39]

Delegates to the Chicago Electrical Congress, 1893. Front row, starting second from left: Ayrton, von Helmholtz, Mascart, Rowland E. Thomson Mendenhall. Second row, between Rowland and Thomson: Carhart. Proceedings of the International Electrical Congress, Frontispiece.

---

[39] Mendenhall, "Some Metrological Memories," 100; Henry A. Rowland et al., "Report [to the Secretary of State] of the Action of the International Electrical Congress held in Chicago, August 1893, in the Matter of Units of Electrical Measure," 11/6/93, in *Reports of the Committee on Electrical Standards Appointed by the British Association for the Advancement of Science...* (Cambridge: Cambridge U. Press, 1913), "Twenty-First Report – Oxford 1894," "Appendix I," 485–489; Mendenhall to Othniel Marsh, 11/19/94, TCM, B:3, f:10; Carhart to Mendenhall, 10/15/92; 12/30/92, TCM; Henry S. Carhart, "[Notice of] *Proceedings of the International Electrical Congress, Chicago, 1893 ..., Science,* II(10/18/95), 525–526; Rowland to Mendenhall, 11/24/94, TCM, B:3, f:10.

The Chamber proceeded smoothly in the first session, but then it turned bumpy in the second, on the afternoon of August 23, when it took up a set of definitions for the volt, ampere, and ohm that Mendenhall had submitted on behalf of the American delegation and that were apparently based on the Edinburgh agreements. Mendenhall recalled that "a long and by no means harmonious debate" ensued, some of it marked by "national prejudices and animosities" and threatening at one point to defeat the international agreement that all parties had come to Chicago to achieve. The fault line ran between, on the one side, the American and English delegates, who agreed with Mendenhall's proposals, and, on the other, the French and Germans, who objected to certain particulars (which the record of the meeting did not spell out but which appeared in part to involve naming). William Thomson, who had not come to Chicago, was unavailable to work out the differences with his friend Helmholtz. At Ayrton's suggestion, Rowland appointed a small committee to deal with the dispute that comprised Helhmholtz, Ayrton, Mascart, and Mendenhall, plus Rowland himself *ex officio*.[40]

In his hotel room late that night, Mendenhall redrafted the definitions in light of the criticisms and objections, writing them down with a pencil on three sheets of a yellow pad that happened to be at hand. While his definitions still closely followed those from the BA at Edinburgh, he prefixed each with "international"—thus, the "international ohm"—which was likely to neutralize nationalist objections to any of the terms. When he presented the sheets to the rest of the committee at its meeting the following morning, the other four members all approved his redefinitions with only a few slight alterations. As the committee was about to adjourn, it occurred to Mendenhall, he later recalled, "that it might be well to insure against possible subsequent change of opinion by having the memorandum signed by all of the members." All, he later said, signed it "cheerfully."[41]

[40] *Proceedings of the International Electrical Congress,* 15–*17; LRL,* 160; Mendenhall to Dr. R.B. Owens, *Journal of the Franklin Institute,* (Aug. 1914), 216, 220.

[41] Mendenhall to Dr. R.B. Owens, 216, 220, which includes a reproduction of the penciled definitions with the signatures of all the members, including Rowland.

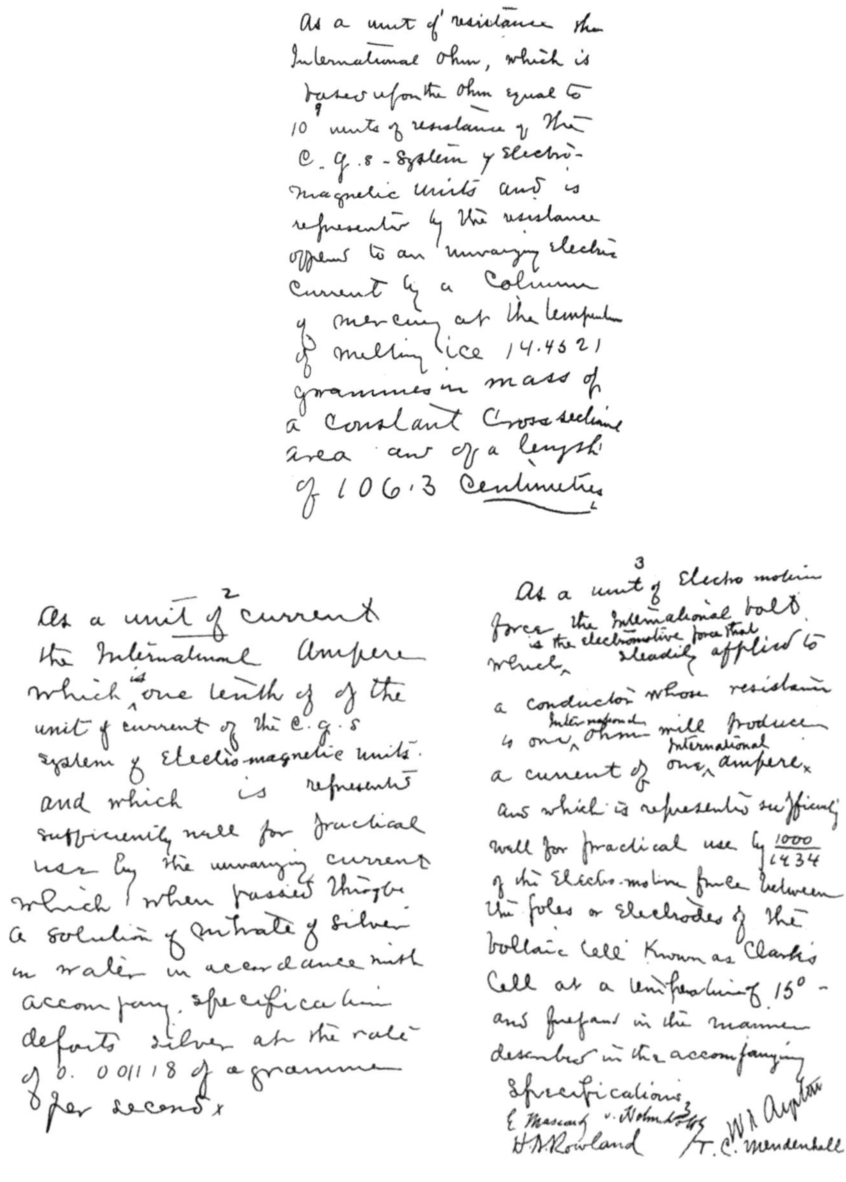
As a unit of resistance the International Ohm, which is based upon the Ohm equal to 10^9 units of resistance of the C.G.S. System of electro-magnetic units and is represented by the resistance offered to an unvarying electric current by a Column of mercury at the temperature of melting ice 14.4521 grammes in mass of a constant cross-sectional area and of a length of 106.3 Centimetres

2

As a unit of current the International Ampere which is one tenth of of the unit of current of the C.G.S system of Electro-magnetic units and which is represented sufficiently well for practical use by the unvarying current which when passed through a solution of nitrate of silver in water in accordance with accompanying specification deposits silver at the rate of 0.001118 of a gramme per second

3

As a unit of Electro motive force the International Volt is the electromotive force that steadily applied to a conductor whose resistance is one International Ohm will produce a current of one International ampere and which is represented sufficiently well for practical use by $\frac{1000}{1434}$ of the Electro-motive force between the poles or Electrodes of the Voltaic Cell known as Clark's Cell at a temperature of 15° and prepared in the manner described in the accompanying specifications

E. Mascart v. Helmholtz W. A. Ayrton
H. A. Rowland T. C. Mendenhall

Mendenhall's Memorandum on the Definitions of the "International" Ohm, Ampere, and Volt, Signed August 24, 1893 by Mascart, Helmholtz, Rowland, Ayrton, and Mendenhall

Mendenhall incorporated the three definitions into a memorandum that included in addition definitions for the "international coulomb," to signify the quantity of electricity, and the "international farad," to indicate the capacity of a condenser. Reporting to the Chamber that afternoon on behalf of the committee, he urged that the governments represented by the delegates adopt "as legal units of electrical measure" each of the memorandum's international units. The Chamber endorsed the recommendation unanimously. It also agreed to designate the "joule" as the unit of work, the "watt" as the unit of power, and, without dispute, despite earlier objections from the British and French, to name the unit of self-induction the Henry. Two days later, however, Helmholtz, evidently drawing on recent studies at the Reichsanstalt that found the German version of the Clark Cell superior to the British favorite, called for adding explicit specifications on its configuration and preparation. At Mendenhall's initiative, the Chamber empowered a committee of three — Helmholtz, Ayrton, and Carhart—to decide the issue.[42]

However, the committee was unable to complete its task because Helmholtz suffered an accident on his voyage home and had remained ill (he would die in September). Moreover, although the Chamber had specified the apparatus and methods for determining the ampere, Mendenhall evidently thought they needed to be reviewed and improved. Mendenhall, impatient, moved anyway to turn the other Chicago standards into electrical policy for the United States.

To accomplish his purpose, he broke with government's longstanding approach to standards of weights and measures. Although under the Constitution Congress was authorized "to fix" the standards, it had never mandated the value of any of them, except for the weight of coinage, on a uniform national basis. However, in 1832, to overcome large discrepancies in the collection of import duties, the secretary of the treasury ordered the fabrication of standard material instantiations of relevant weights and measures and their distribution to all of the country's custom houses. In 1836, Congress directed the secretary to deliver copies of the customhouse standards to the governors of all the states for use within their jurisdictions. Almost every state legislature soon adopted the standards as their own, thus achieving, albeit by voluntary state action, Congress's purpose of creating "a uniform standard of weights and measures" across the country. In 1866, Congress similarly instructed the secretary to furnish material metric standards to the states for their use and legalized the metric system. Mendenhall noted in April 1893 that "the metric system was thus the first, and thus far the only system made generally legal throughout the country."[43]

Mendenhall fractured these precedents when, in December 1893, in his capacity as superintendent of the Office of Weights and Measures and backed by the secretary of the treasury, he issued an official Bulletin promulgating the U.S. government's formal adoption of the Chicago electrical standards save for the pending volt and ampere. The definitions would, henceforth, bind all operations of the executive departments in which the fundamental electrical measures figured, including

[42] *Proceedings of the International Electrical Congress,* 17–19, 20–21; *LRL* 165–167; TCM, "Legal Units of Electrical Measure," *Science,* I(Jan. 1895), 12.

[43] Lewis V. Judson, *Weights and Measures Standards of the United States: A Brief History* (NBS Special Publication 447; Originally issued October 1963 1 updated 1976 by Louis E. Barrow; Washington, D.C.: Government Printing Office, 1976), 2, 5–6, 8, 10. "The Mendenhall Order of April 5, 1893," *ibid., "Appendix 3," 28.*

contracts.[44] However, his ultimate aim was to shatter precedent altogether by persuading Congress to mandate the use of the definitions everywhere in the country.

To that end, he turned to Representative Charles W. Stone, a business-minded lawyer, longtime Republican politico from western Pennsylvania, and member of the House Committee on Coinage, Weights, and Measures. Mendenhall drafted the bill with the assistance of Stone, who in the spring of 1894 shepherded it through the committee and ably explained the need for it to the House. Investment in the country's electrical enterprises now totaled between $100 million and $400 million and there was no "authoritative standard" by which disputes over contract fulfillment could be resolved, including for the government, which is "a great consumer of electricity." Stone added that the bill would make the United States part of a uniform international system of electrical standards and that it enjoyed the strong endorsement of the AIEE (which also lobbied Congress for its passage). The bill passed the House and then quickly cleared the Senate with only minor alterations, none of substantive consequence. Mendenhall, who had testified for the measure in both the House and the Senate, paid President Cleveland a visit to interest him in the matter, and on July 12, Cleveland signed the bill.[45]

However, the bill did not include the specifications for the Clark Cell or the ampere, both of which remained to be decided. Rather than hold up the bill, it occurred to Mendenhall that the government could call on the National Academy to finish the work on the specifications for both the ampere and the volt. He anticipated that the Academy would appoint a special committee for the purpose. At the Academy's meeting in April 1894, while the bill was still pending, he told the Academy's president, Othniel C. Marsh, that he would propose the idea to Stone if Marsh would appoint to the committee the five official U.S. delegates at Chicago. In his view, having brought the issue of electrical standards so far, they—and by implication the AIEE, which had done so much for the cause of the standards — should be given the opportunity to finish it. He emphasized to Marsh repeatedly that if he would not appoint the five, he would not recommend the Academy's participation to Stone. Marsh did not explicitly agree with the proposition, but he left no doubt in Mendenhall's mind that he supported it.[46]

---

[44] U.S. Coast and Geodetic Survey, *Bulletin No. 30;* "Office [of Weights and Measures] Report No. 5: Report of the Assistant in Charge ... for the Fiscal Year 1894," *Annual Report of the United States Coast and Geodetic Survey 1894*. No stranger to achieving his goals by executive action, Mendenhall had in 1892 issued an order, also with the secretary's approval, redefining the standards of weight and measure in terms of metric standards, which were now well established by international agreement, rather than in relation to the British standards for these units, which tended to discrepancies and variations. Judson, *Weights and Measures,* 12–17, 28.

[45] Mendenhall to R.B. Owens, *Journal of the Franklin Institute,* (Aug. 1914), 220; *CR,* 53rd Cong., 2d Sess., House, 6/8/94, 6/30/94, 5987–5989, 7038; Senate, 6/11/94, 6/22/94, 6/29/94, 6086, 6255, 6682, 6980–6988; Mendenhall to Marsh, 11/19/94, TCM, B:3, f:10.

[46] "Final and Official Recommendation of the Chamber of Delegates," 8/26/93, *Proceedings of the International Electrical Congress,* 20; Thomas C. Mendenhall, "Legal Units of Electrical Measure," *Science,* N.S. I(Jan. 4, 1895), 12; Mendenhall to Marsh, 11/19/94, TCM, B:3, f:10.

Mendenhall kept his part of the bargain, and the law tasked the Academy with the "duty" to specify and publish whatever details might be necessary for the determinations in practice of the ampere and volt. Marsh did nothing until just after the Academy's November meeting, when Mendenhall reminded him of the need for appointment of a committee on the matter, stressing that Carhart should definitely be a member not least because, as he recalled saying, "he had given more attention to the [Clark Cell] than any other American physicist."[47]

Marsh appointed the committee on November 9, designating Rowland as chair and as the rest of the members Barker, Mendenhall, and Trowbridge, all of whom were well versed in the deliberations on electrical units, and two other physicists and an engineer, all of whom were not.[48] Mendenhall, infuriated by the omission of even Carhart, sent Marsh a blistering letter, declaring that he had "not kept faith," reminding him of the expected quid pro quo under which he had brought the Academy into it. Mendenhall refused to join the committee, declaring that he was unwilling to be party to the transaction. Marsh, responding with only a few curt sentences of regret for the loss of Mendenhall's "valuable services," appointed the physicist Albert Michelson in his place.[49] Carhart confided to Mendenhall that he was "sore" about his exclusion from the committee, believing that appointment to it "justly belonged to me."[50]

Rowland stepped into the imbroglio as a solicitous statesman. After the Academy committee produced a full draft of the specifications, to Carhart's astonishment Rowland sent him the document with a request for his views. His first impulse was to have nothing to do with the matter, he told Mendenhall, but the committee's specifications for the Clark Cell followed the English form, which he considered decidedly inferior to the German version. In recognition that "the law of the land is at stake," he swallowed his pride and sent Rowland all his information and his judgment that the cell developed at the *Reichsanstalt* was superior and ought to be adopted. Rowland, finding Carhart's points meritorious, said he would urge their adoption, and they were incorporated in the committee's report. At a special meeting on February 9, 1895, the Academy approved the report unanimously, and, in accord with the Congressional act, authorized the official promulgation of the specifications. Marsh sent copies to both houses of Congress and initiated the publication of the report by the State Department, thus giving the specifications the force of law along with the statutory definitions of the rest of the electrical standards.[51] Together,

---

[47] *U.S. Statues at Large,* 53rd Cong. 2d Sess., Chap. 131, "An Act to Define and Establish the Units of Electrical Measure," 7/12/94, 102–103; Mendenhall to Marsh, 10/12/94, *Report of the National Academy of Sciences, 1894*, 40,and Note; Mendenhall to Marsh, 11/19/94, TCM, B:3, f:10.

[48] Marsh to Rowland et al., 11/9/94, *Report of the National Academy of Sciences, 1894*, 41–42.

[49] Mendenhall to Marsh, 11/14/94; Marsh to Mendenhall, 11/23/94, TCM, B:3, f:10; Barker to Rowland, 11/27/94, Henry A. Rowland MSS, Johns Hopkins University Archives, B:A—BAZ; *NASM*, 2/9/95.

[50] Rowland to Mendenhall, 11/24/94; Carhart to Mendenhall, 1/8/95, TCM, B:3, f:10; B:4, f:3.

[51] Carhart to Mendenhall, 1/26/95; 3/16/95, TCM, B:3, f:10; B:4, f:3; f:6; *NASM*, 2/9/95; Henry A. Rowland et al., "Specifications for the Practical Application of the Definitions of the Ampere

the two established the first federal policy that under the weights-and-measures clause of the Constitution mandated definitions of standards of any kind uniformly across the entire country.

Canada also adopted the Chicago electrical standards, but while in their own laws the governments of Britain, France, Germany, and Austria accepted one or more of them, they substituted standards of their own preferences for the rest. By the early twentieth century the result was a multinational system of disparate standards, with differences in values also introduced by the greater precision of measuring instruments; it was dissatisfying in principle and risked the unenforceability of commercial contracts. But by then each of the major electrical powers, including the United States, had created a national bureau of standards, and they took the lead in revising the Chicago definitions. They achieved the international adoption of the revision at a meeting held in October 1908 at the Royal Society in London that was attended by delegates from 24 countries and opened by Winston Churchill, then a member of Parliament.[52]

Yet the Chicago agreement had joined the electrical powers in a common framework of commitment and analysis. The Clark cell was gone from the standard for the volt; that was now defined as the product of the values of the international ampere and the international ohm. But the standards for those two remained, save for increased precision, essentially intact, including for the mercury column the Rowland-Thomson length. Arrived at on the shore of Lake Michigan, grounded in influential part in the experimental work of Rowland, and decisively shaped by Mendenhall, the Chicago agreement signaled that the United States was now a robust force in the electrical physics and engineering of the Second Industrial Revolution.[53]Abbreviations

B	Box
CR	*Congressional Record*
f	*Folder*
GPO	Government Printing Office
LRL	*L. R. Lagerstrom, Constructing Uniformity: The Standardization of International Electromagnetic Measures, 1860–1912* (Ph.D. Dissertation; University of California, Berkeley, 1992). Lagerstrom's dissertation,

and Volt," in "Special Meeting," 2/9/95, *Report of the National Academy of Sciences, 1895,* 7–13. Carhart learned later that Rowland had urged the Academy committee to credit him, but that Barker had blocked the recognition. Marsh threaded the needle by acknowledging Carhart's contribution in the report that went to Congress but omitting it from the version published by the Academy. Carhart to Mendenhall, 3/16/95, TCM, B:4, f:6; U.S. Senate, *Misc. Doc. No. 115,* 53rd Cong., 1895. Later that year Carhart, now a member of the AIEE, publicly applauded the report's treatment of the Clark Cell. Henry S. Carhart, "[Notice of] *Proceedings of the International Electrical Congress, Chicago, 1893…,*" *Science,* II(Oct. 18, 1895), 525–526.

[52] LRL, 224, 235–236, 243–244, 290–291; "International Conference on Electrical Units and Standards," *Nature*, 78(No. 2035; 10/29/08), 678.

[53] "International Conference on Electrical Units and Standards," *Nature*, 78(No. 2035; 10/29/08), 678–681; *LRL,* 169.

written under the supervision of John Heilbron, is a pathbreaking study of the development of electrical standards. I am much indebted to it.

NASM National Academy of Sciences, "Minutes of Meetings," Archives, National Academy of Sciences, Washington, D.D.

TCM Thomas Corwin Mendenhall Papers, Archives, Worcester Polytechnic Institute, Worcester, MA

Daniel J. Kevles, Professor of History Emeritus, Yale University, and Visiting Interdisciplinary Fellow, New York University Law School. His works include The Physicists (New York, 1978), In the Name of Eugenics (New York, 1985), The Baltimore Case (New York, 1998), and, as coauthor, Inventing America: A History of the United States (New York, 2002, 2006^2) as well as numerous articles, essays, and reviews in scholarly and popular journals. His current projects include a history of innovation, intellectual property, and the stuff of life from the Anglo-American colonial period to the present, and a collaborative history of the National Academy of Sciences.

Chapter 13
D'Arcy Thompson, Alan Turing and the Rise of Mathematical Biology

Brian D. Ross

Abstract D'Arcy Wentworth Thompson, a nineteenth Century biologist is the "Newton of the blade of grass." His magnum opus *On Growth and Form* (*OGF*) pioneered mathematical biology whereby he met Immanuel Kant's definition of a science. *OGF* influenced mathematician Alan Turing who created the equations of biological growth lacked by Thompson. By applying mathematics to the brain, Turing revealed the roots of consciousness to become the father of Artificial Intelligence (AI). We can adopt D'Arcy Thompson as its grandfather. D'Arcy Wentworth Thompson and Alan Turing were both giants in the World of biology, a World over which Darwin still rules.

Keywords Mathematical biology · D'Arcy Thompson · Alan Turing · Radiolaria · Phyllotaxis · Phrenology · Artificial Intelligence (AI)

The Book of Nature is written in the language of mathematics.
Galileo (1623), cited by D'Arcy Wentworth Thompson (1914) and John Heilbron (2010)

13.1 Biographical Details—Who Were D'Arcy Thompson and Alan Turing and How Are they Connected?

There is only one formal biography of Sir D'Arcy Wentworth Thompson (1860–1948) (D'Arcy Thompson, 1958). Published 10 years after her father's death, by his daughter, he emerges as a polymath with notable achievements in science and

B. D. Ross (✉)
Clinical Magnetic Resonance Department, Huntington Medical Research Institutes and Medical Engineering Division, California Institute of Technology, Pasadena, CA, USA

S. Gattei (ed.), *The Incomparable Historian*, Archimedes 75,
https://doi.org/10.1007/978-3-032-25678-2_13

the classics. That biography is supplemented by her second book (D'Arcy Thompson 1974). D'Arcy's mother was Anglo-Italian and it was her family that raised, educated and influenced him and ultimately determined his career choice as a zoologist. There is an archive of his work and correspondence, comprising 40,000 items. Held at University of St. Andrews, D'Arcy Thompson's intellectual home for 60 years the material has not yet been digitized. Accessible articles published around the centenary celebrations in 2017 for *On Growth and Form* (OGF) offer additional insights.

Lost from the view of his intended audience of biologists for decades, his name was sustained by the sculptor Henry Moore and architect Mies van de Rohe and dozens of other visual artists.

Much better known is Alan Turing (1912–1954) a mathematician. The descriptor 'enigma' used by his principal biographer (Hodges 1983) refers to his wartime work as a mathematician to crack a code. Turing's later efforts to define the origins of form and its changes in biology are inseparable from those of D'Arcy Thompson. A volume recording the centenary of Turing's birth (Cooper and Leeuwen 2013) provides more detail.

As a schoolboy Turning was interested in chemistry and in botany, topics in abeyance during a successful mathematical career. The need to construct and train a computing machine for the Enigma task provided the segue. The question now became how does the human brain 'learn'. His publication in 1952 provides complicated mathematics, concerned the changing shape of a fertilized egg into an embryo. It was firmly a matter for biologists and cites Thompson's OGF. Its successor, 'The design of a daisy', unpublished at the time was also biological. The die was cast. Turing was now a biologist.

Mind, brain and consciousness were territories which had been scrupulously avoided by Thompson: It was the juxtaposition of Turing's familiarity with math and computing which provided the advance towards a truly mathematical biology. Surely there is a case for inclusion of Alan Turing in any biography of D'Arcy Thompson.

Last and first words! Charles Darwin discovered D'Arcy Thompson while the latter was still a student at Cambridge, inviting him to translate Mueller's Fertilization of flowers (Muller 1873). This became Thompson's first published work (Thompson 1883). Darwin provided a laudatory Preface to D'Arcy Thompson's translation in what then turned out to be his own last published work. Thompson argued vigorously against Darwin's principal conclusions throughout his work, including *On Growth and Form* (OGF 1917/1942) that Darwin's Natural Selection was teleology, the theological argument from design in another form (Darwin 1859). Often seen as an attack on *Origin of Species*, OGF nevertheless remains a literary masterpiece. References, often in their original vernacular cover millennia of earlier science. Organized from the microscopic to the largest, the physical and chemical forces responsible for the shape of organisms are documented and quantified. It is also easy to follow Thompson's journey from comparative anatomist (his day-job) to questioning modernist. To the extent that the St. Andrews archive has been mined, both editions of OGF can be traced to original diagrams and equations culled from his correspondents. Thompson was a node in a Worldwide net-work of like-minded biologists (Esposito 2014).

Although no object in Nature was too lowly or too obscure to receive mention Thompson was reluctant to intrude on the human brain, specifically the 'mind'. This gap was filled by Alan Turing. Turing's work in biology also suffered a lengthy period of obscurity until mathematical biology re-emerged in the 1980s when OGF came back into print (Thompson 1998). By this time cybernetics and computers had become central concerns in the field now succinctly described as morphogenesis.

13.2 How Does a Biologist Become a Mathematician or a Mathematician a Biologist?

Many fields of the basic sciences contribute to the appeal of OGF but none as forcefully as mathematics. How could Thompson, a conventionally educated man of his era have accrued such knowledge? Reading the one existing biography and culling through the references in OGF reveals some of the answers whilst comparison with a Taxonomy (MSC 2020) which define the discipline of mathematics and adjacent sciences provides a quantitative framework. By the time D'Arcy Thompson was 50 years of age he was remarkably well prepared to compose OGF.

Even so, he often alludes to his own incompetence. The gap is spelled out on page 149 of OGF as follows "differential equations, easy to invoke, difficult to formulate". Whether Turing noted that comment, his biographer does not reveal. This was the major 'gap' which Alan Turing, the professional mathematician filled, to follow and then exceed him in the field.

A taxonomy of mathematics (MSC 2020) from the Mathematical Association of America (MAA, 2004) classifies the broad mathematical disciplines. Of 350 mathematical topics, Thompson identifies 145 (40%) as relevant to biology.

This formal analysis (Table 13.1) provides a quantifiable comparison with a taxonomy of imaging fields employed by biological scientists in the recent resurgence of the 'visual pursuit of science' (see below).

13.3 Mathematical Biology—History of a Science

More than 90 years after Thompson's contributions there were still no university departments of mathematical biology. When 'mathematical biology' eventually became a serious academic topic it acquired a new inventor; Nicholas Rachevsky was a physicist whose work called Mathematical biophysics (1938) had already been cited in OGF (Thompson, 1942). When it did finally emerge, mathematical biology

Table 13.1 A Partial Taxonomy of Mathematic Sciences. By MAA (2004)

1.0 Numbers and Computation

1. 1.1 Number Concepts
    1. 1.1.1 Natural
    2. 1.1.2 Integers
    3. 1.1.4 Irrational
    4. 1.1.6 Real
    5. 1.1.8 Famous Numbers
        1. 1.1.8.1 0
        2. 1.1.8.2 pi
        3. 1.1.8.3 e
        4. 1.1.8.4 i
        5. 1.1.8.5 Golden Mean
2. 1.2 Arithmetic
    1. 1.2.1 Operations
        1. 1.2.1.1 Addition
        2. 1.2.1.3 Multiplication

(Continued 3.0–10.0)

was attributed to Turing, rather than to Thompson by an applied mathematician, Professor J.D. Murray. Murray's successors in the Oxford Mathematics Department duly acknowledge and cite Thompson's pioneering work (Goriely 2018; Chirat et al. 2021).

13.4 Sociology of the History of Science

History of Science became a discipline early in the twentieth Century (Sarton 1928) attracting research into the origins of new ideas and their discoverers. It did not take long however for the very philosophy of science, its objectivity, methodology and means of dissemination to be challenged by sociology. The motives of inventors and discoverers became fuel for sociologists of science to uncover hitherto unrealized prejudices and schools of thought: which were to be explored and which abandoned? Notably scientific 'revolutions' in which huge shifts in human understanding might occur were now to be re-defined as 'paradigm shifts' (Kuhn 1962). Discoveries in science which were identifiable as revolutions became intensely rare. All else was routine science. While routine science still followed the rules by depending upon experimentation its results were interpreted as pursuit of the status quo.

In accordance with Kuhn's idea Daston and Galison (2007) draw the conclusion in their book titled appropriately *Objectivity*, that experts in science were susceptible to a kind of group-think. The authors suggest that this applies in particular to novel methods of imaging.

13.5 A Visual History of Biology

Beginning in the 1950s however there had already been a movement to invoke art and the visual as the driver of advances in biology (Ivins 1958). An article by Rudwick (1976) was followed by an influential book *Picturing the Book of Nature* by Kusakawa (2012). A conference at the Royal Society in London the same year (published in 2015, Henderson et al.) brought the subject to the fore and *Visual Science of Biology* now commands a presence within universities (see for example Hindle 2019). Instantly favored for its directness and ready interpretation without the intervention of the unfamiliar symbols and equations of mathematics, visual biology has found its niche in academe.

Mathematical biology thus finds itself in competition, not only with the non-sciences of biology, botany, zoology, embryology and comparative anatomy but with a renewed visual pursuit of biology. Thompson was not averse to the visual aspects of his science. OGF provides examples where mathematical and visual techniques overlap: a diagram of forces of the relationship between the movements of the horse and his jockey to optimize speed of travel, the flow of water over the hull of a ship, the surface of a fish and the helical twist in the horn of a narwhal.

It is legitimate therefore to re-examine his life and work within this new context. To formalize this, a taxonomy has been created to codify the ever-enlarging field of the techniques of image creation.

13.6 A Taxonomy of Imaging Techniques

While the history of 'imaging' is as old as man, beginning with cave paintings and scratchings on bones, it blossomed in the middle ages. To the manual arts of painting and lithography are to be added photography and the application of discoveries in physics and chemistry. A taxonomy of imaging techniques explores these old and new sources of the image to distinguish those which fulfill the ideal of 'objectivity' from those that may not.

Divided into eight classes, the Taxonomy of Imaging (Table 13.5) matches the MAA taxonomy of mathematics already presented in Table 13.1.

Taxonomy of Imaging—Primary technologies (Table 13.2)
Alphabet-soup of microscopy 1590–2020, began with light-microscopy and went through many modifications as the science of light and of lenses advanced. Major advances in magnification and spatial resolution, the use of specific stains applied in vitro or in vivo in enhancing tissue contrast and the discovery of fluorescence ensured a central place for microscopy in visual science. Improved length-scale precision opened the field to observation of biological features at the molecular level while ever faster imaging permits dynamical biological processes to be directly observed. The advances in light microscopy were paralleled by those in electron

Table 13.2 Taxonomy of imaging techniques

1590	Janssen	First compound microscope
1609	Galileo	Improved magnification system
1665	Hooke	First description of cells ("Micrographia")
1739	Leeuwenhoek	Observation of microorganisms
1858	Carmine dye	Staining of cell structures
1871	Fluorescence	Discovery of fluorescence for microscopy
1873	Golgi, silver stain	Nerve cell staining technique
1873	Köhler/Abbe limit	Optical resolution theory established
1883	Diffraction limit	Theoretical limit of light microscopy
1884	Gram stain	Differential bacterial staining
1900	Brightfield microscopy	Standard light microscopy method
1904	Fluorescent microscope	First microscope using fluorescence
1929	Epi-fluorescent microscopy	Fluorescence excitation from above
1930	Phase-contrast microscopy	Imaging transparent specimens
1941	Immunofluorescence	Antibody-based fluorescent labeling
1946	FRET	Förster resonance energy transfer technique
1955	Differential interference contrast (DIC)	Enhanced contrast using interference
1980	Calcium imaging	Monitoring intracellular calcium levels
1994	Confocal 4Pi microscopy	Enhanced axial resolution
1994	Green fluorescent protein (GFP)	Genetic marker for live cell imaging
1999	Red fluorescent protein (RFP)	Extended color palette for imaging
2000s	4D electron microscopy	Time-resolved electron microscopy

microscopy culminating in so-called 4-D (femto and atto) where the fourth dimension is time, overcoming yet another previously mooted limit.

The following sub-classification of images is proposed:

Primary—The list which includes all the known methods of image production (Table 13.5) can itself subdivided into eight classes:

1. F = images which display form only
2. f = images which display function only
3. c = images which display chemistry only
4. p = images which display physics only
5. Fc = images which display form + chemistry
6. Fp = images which display form + physics
7. Ff = images which display both form + function
8. U = images of undefined nature.

As early as 1875 Maxwell (Encyclopedia Britannica, Diagrams) offered another means of classifying the images Thus: i Diagrams of illustration; ii Metrical diagrams; iii Diagrams purely graphic and mixed graphic/symbolic; iv Diagrams in pairs; v Diagrams in kinematics; vi of configuration; vii of displacement; viii of velocity; ix of acceleration; x path and hodograph (per Hamilton); xi diagrams of stress xii reciprocal diagrams; xiii automatic description of diagrams; xiv indicator

Table 13.3 Taxonomy of illustration Menti, in Henderson (2015)

1. Watercolor
2. Instaglio techniques
3. Hatching
4. Cross-hatching
5. Stippling
6. Stipple engraving
7. Etching
8. Line engraving
9. Mezzotint
10. Aquatint
11. Coloring (of above) by hand
12. Color printing (3 plate method)
13. A la poupe (French method)
14. 'Secret method'

Table 13.4 Taxonomy of diagrams (per Maxwell 1877)

1. Diagrams of illustration
2. Metrical diagrams
3. Diagrams purely graphic and mixed graphic/symbolic
4. Diagrams in pairs
5. Diagrams in kinematics
6. Of configuration
7. Of displacement
8. Of velocity
9. Of acceleration
10. Path and hodograph (per Hamilton)
11. Diagrams of stress
12. Reciprocal diagrams
13. Automatic description of diagrams
14. Indicator diagrams
15. Boole-Venn diagrams (added 1880, after publication of Maxwell's article)

diagrams; xv Boole-Venn diagrams added 1880, after publication of Maxwell's article. The purpose which diagrams fulfill is the elucidation of results and the creation and rejection of hypotheses. Venn diagrams add logic.

Despite enormous advances in one of Galileo's principal interests, telescope design, their contribution to biology have been modest so the topic is included in Table 13.5 but not developed further.

Table 13.5 Taxonomy of Image Production

F: Images which display form only	Vision	Sonography	Light microscopy
	Camera obscura	Ultra-sound	Telescopy
	Camera lucida	Electron microscopy	Photography
	Cave painting	Scanning EM	X-Ray
	Sculpture	Tunneling EM	CT Computer tomography
	Silhouette	Atomic force microscopy	MRI
	Pottery		Multi-color MRI
	Ceramics	Flourescence microscopy	qMRI
	Engraving	Magnetic force microscopy	Endoscopy
	Expansion microscopy	FRET microscopy	Confocal microscopy
f: Images which display function only	Mesmerism	Transcranial magnetic stimulation (TMS)	Electrocardiography (EKG)
	Positron emission tomography (PET)	Electro-encephalography (EEG)	Electromyography (EMG)
	Single photon emission tomography (SPECT)	Magnetoencephalography (MEG)	
c: Images which display chemistry only	Molecular imaging	Immunocytochemistry	Nano-chemistry
	Magnetic resonance spectroscopy (MRS)	MRS Phosphrus	MRS carbon
	MRS Flourine	MRS nitrogen	Electron-spin resonance (ESR)
p: Images that display physics only	4D-EM	MR-relaxometry	Diffusion MRI
	Flat-lens design	DOSY magnetic resonance	
Ff: Images which display both form and function	Pottery	Phrenol-visual-math	X-ray fluoroscopy
	Art/painting	fMRI	Ph-MRI
	CVMR	PET-CT	PET-MRI
	Echo-cardiography	Ultra-fast photography	'Brainbow'
Fc: Images which display form and chemistry	CSI-MRS PASADENA -hyperpol	Optogenetics	DNP-hyperpolarization
		Optic-coherence	2-photon-laser
Fp: Images which display form and physics	DTI	MR-strain	
U: Images of undefined nature	Braille stethoscope	Haptics	Olefaction
		Phrenology	

13.7 Philosophy of Vision and Perception

Despite the enthusiasm for the new field, the concept of a visual science cannot go unchallenged. Since the time of Leonardo da Vinci and astronomer Kepler (1604) but particularly since the physiologist Helmholtz (1868) the actual mechanisms involved when external objects are seen and perceived may be entirely subjective. Indeed, a widely held view is that all images and all thoughts are 'theory laden' (Polanyi 1958).

The doubt goes to the root of the phenomenon of light. The earliest question was whether light arose from the observer or the object—extro-mission versus intromission. Leonardo da Vinci's experiment with a bull's eye supported the latter. When light enters the eye and passes through the lens, an image is inverted and cast in 2-dimensions on the retina. How we perceive the object the right way up, how the 3-dimensional outside World is correctly perceived has engaged physiologists and philosophers over the succeeding generations. Most favor the idea that outside objects are all constructs of the brain.

But none of the major proponents of art as science addresses Kepler or Helmholtz's concerns and the overwhelming opinion simply accepts visual images as real.

One French philosopher asked that the concept of 'theory-laden' images remain open, at least until the work of a psychologist, J.J. Gibson was fully proven (Latour 2014).

For J.J. Gibson (Ecological Perception 1979) images were no longer 'theory-laden'. Gibson criticized earlier psychologists for their unrealistic experimental conditions arguing that if the subject cannot move their head all sense of space (proprioception) is lost and the study fatally flawed. Gibson's publications adopted several hitherto unknown terms which have taken 50 years to reach the mainstream. The ecological and evolutionary approach on visual perception did eventually gain acceptance with a new brand of Gibsonians at the forefront (Jenkins 2008). Most recently an artificial intelligence (AI) technique developed to train computerized robots to maneuver through a crowded space perceiving and avoiding obstructions was named '*i*-Gibson' (2021) validating Gibson's idea of the reality of images. Gibson's work in turn, has been strenuously criticized by art historians (Gombrich 1971) and scientists (Fodor and Pylyshyn 1981) alike.

A direct comparison between mathematical and visual biology may be the only way to place the work of D'Arcy Thompson in context, since courtesy of Alan Turing the field of Mathematical Biology was already en-route to becoming artificial intelligence (AI).

We propose to accomplish this by means of three Case Studies. The object is to determine whether imaging or mathematics predominates in resolving an unanswered question in its biology.

13.8 The Mathematical Versus the Visual Pursuit of Science

Case Study #1 Radiolaria and 'Evolution' of a Micro-Organism
Radiolaria are a group of single celled animals, the size and shape of an amoeba found throughout the oceans of the world. Consisting of a spherical body each is covered by a skeleton of secreted calcium and strontium salts. The drawings and lithographs of vastly varying forms viewed through a light microscope were assigned a Darwinian evolutionary system by Ernst Haeckel (1862) (Fig. 13.1a, b, c, d, e).

Haeckel went beyond Darwin to coin the emotive terms 'survival of the fittest' and the 'descent of species'. He also included Man. He is credited with single-handedly gaining acceptance of Darwin's concepts throughout Europe.

Commencing with the illustration of his personal collection of more than one thousand of the unicellular organisms he later (Haeckel 1887) analyzed the much larger Radiolaria collection gathered by the Challenger International consortium. With the aid of a lithographer, he illustrated and identified 4138 unique species dividing them into hundreds of families. Half a dozen evolutionary trees were provided to sustain Haeckel's conception on the basis of their visual appearance. So popular was his work that other artists as well as designers and architects adopted and amplified Radiolaria as their models. A two-thousand fold magnified glass blown Radiolarian (Blaschke 1956, in Breidbach, 2006) and a fifty thousand fold magnified Radiolarian gateway to the Paris World Exhibition (Binet 1900), are the most extreme examples of his influence.

Haeckel's specimens were old and decayed. They consisted entirely of the skeletal remains after disappearance of the original body of the organisms. His pictorial reproductions of the angles of circles and hexagons cast in the calcium-strontium skeletons secreted during life were undoubtedly correct. In Haeckel's precise visual renderings D'Arcy Thompson found much to support an alternative, mathematical hypothesis. For Thompson the skeletons were too numerous and diverse and spread across all of oceans of the Earth, rather than geographically confined. This made them unlikely to define species that should have arisen by local competition. Instead the myriad forms were pre-ordained not by natural selection of any particular functions but by physico-chemical, mechanical and mathematical laws identical with those which govern the in-animate non-biological world.

In a lengthy chapter of OGF, he drew on the work of a physical chemist, Joseph Plateau (1873) to explain the recurring shapes. These, called bourrelets by Plateau, Thompson regarded as chemical precipitates created by surface tension across the thin films of protoplasm covering the spherical body of the unicellular Radiolaria. Thompson consulted Plateau's son Felix and, following the procedures initiated by the elder Plateau was able to reproduce many of the spectacular Radiolaria- like patterns. Thus it appears to be possible using only soap bubbles and hand-made circles of fine wire to reproduce in seconds patterns previously considered to have evolved in nature over millennia. Several examples appear in OGF. One (Fig. 13.1d) reproduces Haeckel's hand drawn Nesselaria the skeletal remains of which have the

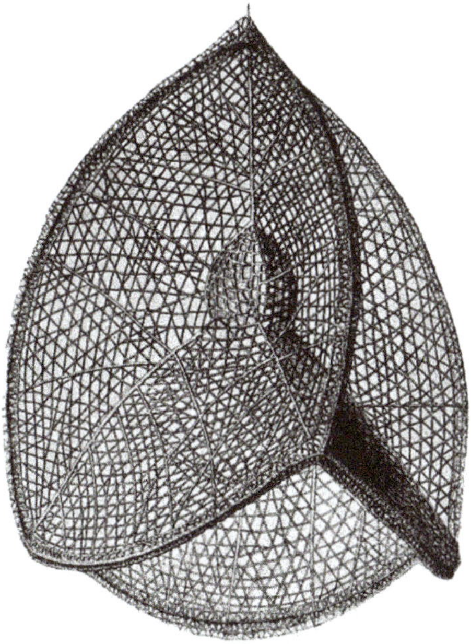

Haeckel Lithograph of Radiolarian (1862 and 1887)

Glassblown sculpture by Blaschke from Breidbach 2006

Fig. 13.1 (**a**) Radiolaria got their shape without the help of evolution. Reproduced by D'Arcy Thompson in his Dundee laboratory, using only a loop of wire and soap bubbles. (**b**) Blaschka's glass blown Radiolarian 1889, from Breidbach (2006). Inspired by Haeckel's drawing of Radiolarian Doratapis diodon approximately 1500 fold life size. (**c**) Radiolarian Gateway, Paris, 1900. Approximately 50,000 × life size the gateway to the Paris 1900 World Exhibition. Architect Rene Binet an admirer of Haeckel's art, likely used a lithograph of Rodiolarian Clathrocaniun reginae as his guide. Version of photograph in National Gallery, Washington. (**d**) Mathematical Rodalarian generated by a computer Using only the reaction diffusion equations (Richards, in Cooper and Leeuwen 2013). (**e**) Computer drawn Radiolarian with four radiating spines mimicking one of several naturally occurring organisms in the process of morphogenesis (Richard B, 1954; Cooper and Van Leeuwen, 2013)

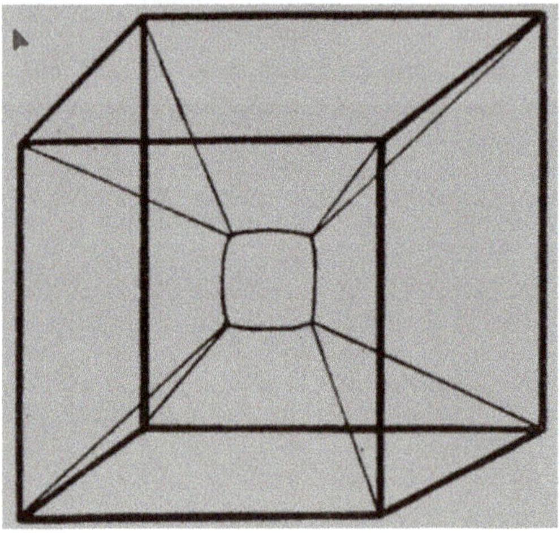

Fig. 13.1 (continued)

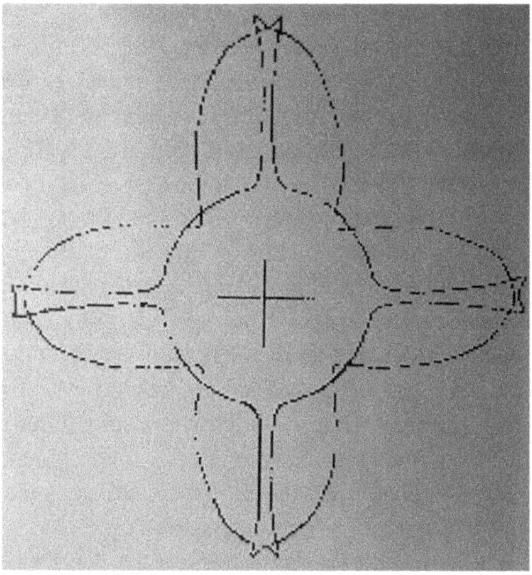

Fig. 13.1 (continued)

form of a spherical tetrahedron with a centrally placed bubble. Radiolaria have achieved their multifarious shapes through physical and mathematical events alone, without the intervention of evolution. Thompson declares victory for science, physics, chemistry and mathematics and claims to have 'lost faith in Haeckel's 4000 species of Radiolaria' (Thompson 1917/1942).

Proof of Thompson's alternative hypothesis would require a more direct demonstration of such a mechanism supported by actual mathematical equations. Alan Turing provided both the novel equations and the novel tool, a programmable computer.

Turing's concept of reaction-diffusion introduced laws of chemistry and the physics into biological research and remains current throughout several fields of biology and metabolic regulation (Epstein 2016; Zhang et al. 2018). Turing's last student Bernard Richards developed the simultaneous differential equations. He provided the additional mathematics, termed eigenfunctions and Legendre association functions, together with thirty pages of calculations and used the new Ferranti Mark 1 computer to create images of the 'unimaginable' shapes these reaction diffusion equations had generated. These were mathematical images of a handful of the Radiolaria observed down the microscope by Haeckel. As proposed by D'Arcy Thompson, organisms with minutely varying spherical bodies and projecting spines at the predicted locations around the Radiolaria surface needed no longer to be considered as distinct species.

Richards' unpublished thesis (1954) resolves the dispute that played out from 1862 to 1942 between Thompson and Haeckel over the multitudinous forms of Radiolaria and provides a crucial case study.

Case Study #2 Phyllotaxis and Symmetry in Botany

Nehemiah Grew (1680) offered the aphorism 'the study of plants draws a man into mathematics'. Phyllotaxis, literally the design of plants is the specialty which resulted from early attempt towards mathematization of botany. Thompson and Turing were both drawn to the mathematics of plant design. This subject had been known of since Babylonian times but was popularized in the nineteenth Century by brothers Bravais and Bravais, respectively a botanist and a crystallographer. Simply put, arrangement of leaves on stems, petals on flowers, scales on pine-cones and so on display visually striking patterns, logarithmic spirals with a numerical order they described as 'Fibonacci-like'. Thompson devoted a lengthy chapter to the issue in OGF but, after all he concludes that while the math is correct it is entirely incidental to plant design. Any other conclusion, he agreed would be 'nothing but unjustified Pythagorism'. A later editor of OGF eliminated the 'phyllotaxis' chapter entirely (Bonner 1961).

Turing on the other hand retained the Fibonacci series as the bedrock of plant symmetry (Turing 1952 unpublished; Saunders 1992).

Phyllotaxis has a unique and esoteric nomenclature. Angles between successive leaves became divergences; fundamental spiral arrangements along the stem when seen in cross section were named parastichies; the time intervals between their appearance during growth as plastochrones. and the number of leaves at each level is referred to as jugacies. The Fibonacci sequence arises when the observer applies numbers to successive leaves in adjacent spirals. See Fig.13.2b right hand numbers.

Among modern mathematicians, crystallographers, experimental botanists and geneticists the topic has remained controversial (Adler 1997). It is summarized in

Fig. 13.2 (**a**) Phyllotaxis is visible at Huntington Gardens 2011. Cycads illustrate Fibonacci ordered whorls. (**b**) If the surface is that of a cylinder (Fig 13.2a) we observe a series of spirals If flattened (Fig 13.2b) they present as a lattice The same spirals are then observed but numerically as Fibonacci series (to the right of Fig 13.2b) When the 'scales' are flattened and displayed as a lattice, the Fibonacci order is lost. (Thompson 1917/1942) 'Fibonacci is inexcusable Pythagorism'

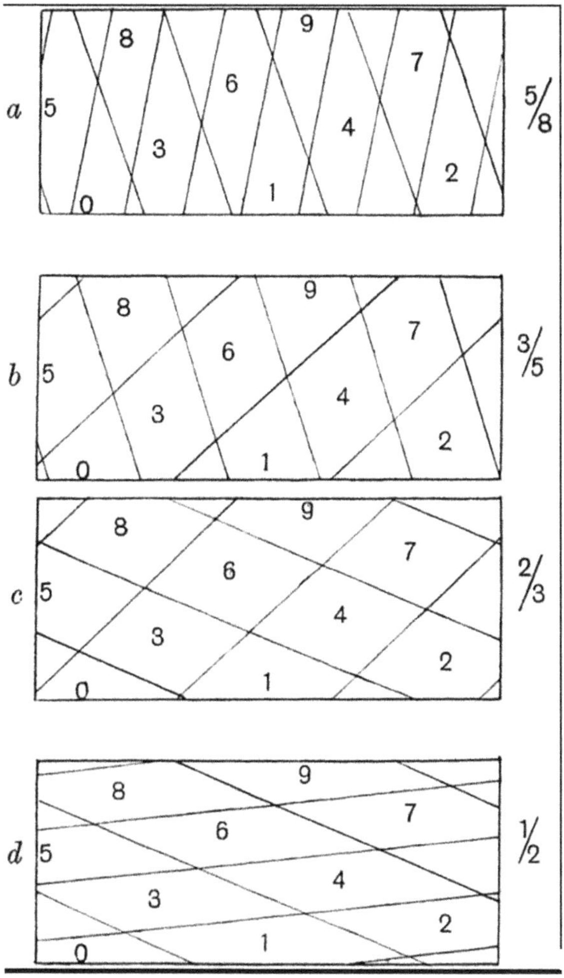

Fig. 13.2 (continued)

an 800 page tome (eds. Jean and Barabe 1998). *Huntington Gardens in California* displays 1500 Cycads, elegantly whorled plants, ancestors of today's conifers. The whorled appearance is the result of its numerous seeds arrayed in logarithmic spirals across the surface of the massive cones. The multiple logarithmic spirals are seen also in daisies, sunflowers, pine cones, pineapples and with some imagination can be discerned in the majority of the plants around the Gardens.

The logarithmic spiral was discovered by Rene Descartes (1596–1650) and confirmed as the root of the appearance of sea shells by Henry Moseley (1801–1872). D'Arcy Thompson picked up on this mathematical design theme in OGF and wholeheartedly supports its role throughout nature. But in his treatment of phyllotaxis, Thompson offers reasons for skepticism. Within the minute growing apex of plants the precise measurements needed to define a logarithmic spiral are not available. In addition he follows his Edinburgh University teacher P.G. Tait who found the angles between adjacent logarithmic spirals, divergences in the language of phyllotaxis, varied between 1/3 and 1/2, and concluded that such large differences would inevitably follow Fibonacci intervals. From this he decided that the recurring appearance of the Fibonacci sequence in botany was an accident of mathematics.

Alan Turing took the opposite view, according to a posthumous report by (Saunders 1992) who provides us with provide three different versions of Turing's account. The first drafted by a botanist, C.W. Wardlaw 'A diffusion reaction theory of morphogenesis in plants' reminds his readers that plants are among the most highly ordered of organisms. Across roots and stems, axial and radial structures recur which are difficult to explain on genetics alone. We are invited to agree that a chemical system somewhat akin to or identical with the postulated morphogen driven reaction-diffusion can generate recurrent patterns where none previously exists. By this time the existence of one plant chemical, auxin, which fits with Turing's description of a morphogen had been well-established.

Saunders' follows with a section headed Morphogen theory of phyllotaxis. Written in the first person, presumably by Turing himself, he defines the phyllotactic system as a group of congruences. This requires that he defines yet another three parameters: jugacy, plastochrone distance and divergence angle. When researchers display the helical coordinates on a flattened surface the spirals form a lattice. A lattice may be expressed in a theorem as three principal vectors with the third vector being the sum of the first two.

If, when you count the first two parastichy numbers (vectors) they fall into a Fibonacci sequence, then the third vector will automatically also do so. Turing thus agrees with that Tait had offered 80 years previously. But for Turing, Fibonacci was more acceptable than any of the alternative empirical laws which he found to be 'weird and magical'. It was another 60 years before direct microscopy of the growing apex in a plant, Arabadopsis allowed the accurate measurements of angles lacking to Tait, Thompson or Turing. The authors conclude that the reaction-diffusion model did apply. The reaction-diffusion equations deserve attention. The equations introduce an unanticipated element of instability. Form could change unexpectedly. His mathematics is cited as an early example of bifurcation, a cornerstone of modern mathematics, non-linear dynamics (Prigogine and Stengers 1984), and now too of biology (Saunders 1992; Cooper and Leeuwen 2013).

Patterns can be formed even in isolated chemical mixtures. One such, the B–Z reaction was devised in the 1950s to mimic oscillating biochemistry of the Krebs cycle, universal in biochemistry. The B-Z reaction is cited in manuals on Phyllotaxis (Jean and Barabe 1998) and re-appears in treatises on mathematical biology unrelated to phyllotaxis. But the B-Z reaction is just one of the many oscillating processes since identified in vivo. Turing's work is relevant to another important innovation, Using equations for the reaction—diffusion process devised by Turing all of the six patterns could be identified in an experimental study chaos theory (Lorenz 1963) published in 2014.

Morphogens necessary to Turing's theory have also developed a life independent of plants. The first to be formally identified was a gene "bicoid" in the fruit fly Drosophila; a Nobel Prize for the discovery followed.

In this Case Study mathematic biology predominates over the visual.

Case Study #3 Phrenology, Mind and Consciousness Define the Brain

Devised in 1791 by Joseph Gall, organology of the brain (later, phrenology) proposed that the brain substance rather than the ventricles contained all of the apparatus of the mind. Galen had postulated that it was the fluid content of the ventricles rather than the coverings of the brain which contributed all of its function. Gall's work therefore contributed significantly to the overturn of two millennia of Galenism. A less happy formulation of Gall's phrenology which followed was that the shape, size and functional capacity of 27 distinct neuropsychological activities of mind could be accessed from outside the skull by digital (literally) examination. The sizes of the organs of the brain were transmitted to the bony skull and direct access to personality required only the educated fingers of the phrenologist. Phrenology became a respectable haptic imaging tool. A century later, opinions changed again and modern dictionary definitions of phrenology usually include the disclaimer 'pseudoscience' (Fig. 13.3).

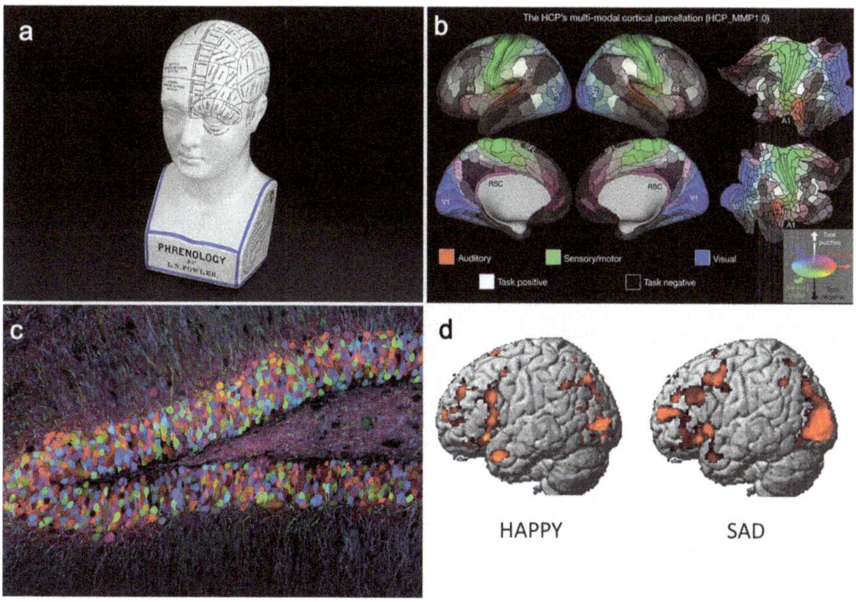

Fig. 13.3 (**a**) Phrenology to Novum phrenologium renovatum: starts with true digital imaging, Gall 1812. Plaster human heads, still ubiquitous in physician's offices emphasize their professional skill. (**b**) Modularity in human brain function continues to be supported by imaging, van Essen 2018. (**c**) (**d**) fMRI localizes specialized brain functions. From 1990 to present, recognizable fMRI images dominate the published Literature. (**e**) Neurotubules protect memories from quantum entanglement, Penrose 1994. Found throughout the brain, they suggest physics provides a mechanism for mind and offers a testable theory of global consciousness

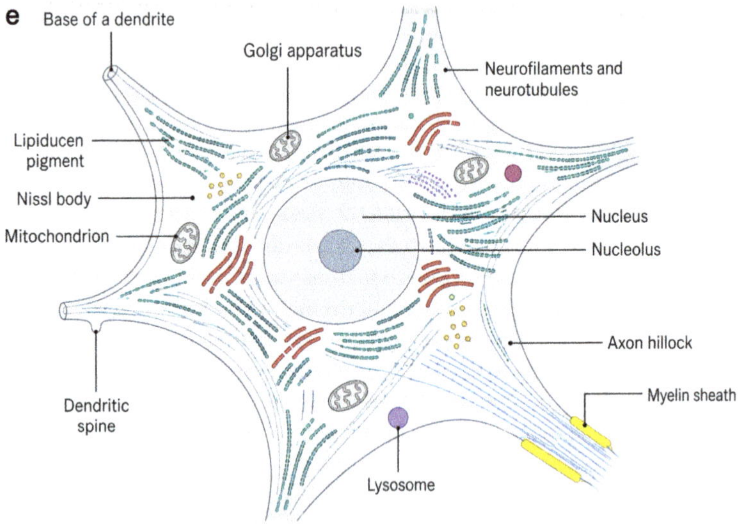

Fig. 13.3 (continued)

Why resurrect such an abused term? McFie (1961) indicated how useful was the name phrenology for the field hitherto defined as neuropsychology. Uttal chose the title The New Phrenology for his 2001 volume on visual analysis of human brain functions. And known today as modularity of mind (Fodor 1983), Gall's concept lives on.

The 1990–2000 was declared the Decade of the Brain tasked with exploring the matter further. Answering the question whether consciousness and other sophisticated functions of the brain were distributed or modular had become the dominant task of neuroscience. To paraphrase William Whewell's (1868) *novum organon renovatum*, The Decade of the Brain sought nothing less than a *novum phrenologium renovatum*.

Twentieth century medical historian Erwin Ackerknecht (1968) pointed out that Gall and Spurzheim's theory of human personality was as prevalent and influential in the nineteenth century as Freud's became in the 20th. It has been as persistent too: The Gall Society of London founded in 1827, was only disbanded in 1967.

Gall's academic world collapsed in ignominy in France after official approval of his modular brain concept was withdrawn. A theory of distributed and more global brain functions gained favor. That controversy remains a major component of the 'hard problem' of consciousness, identified as such by Dennett (1991).

The 2 questions for Case Study #3 are whether consciousness can be defined by visual or mathematical methods and whether focal or generalized across the entire organ?

Enthusiasm for Gall's modularity of the brain lives on. The number of recognized brain regions has increased from 27 to 360 (Essen, van and Glasser 2018). Nomenclature and above all technologies applied to the question have evolved from the phrenologists' haptic imaging through microscopy, biochemistry and the various modes of X-ray, positron emission tomography to magnetic resonance (MR).

Once completed the science of the form of the brain returned to re-defining its function. The same yardstick is employed as in the earliest methods of phrenology, comparing the new measures of form with some aspect of human personality. But if, as Uttal (2001) pointed out there is slipperiness in the definitions of psychology and furthermore if all images are theory laden, then any new phrenology based upon imaging and psychology might be doubly so.

13.9 Visual Phrenology

Progress in the neuroscience of phrenology during the twentieth Century has nevertheless been impressive with the great majority of suggestions coming from novel imaging techniques.

Microscopy has played a crucial role. Golgi and Cajal shared the Nobel Prize for medicine in 1906 based upon the histological appearance of brain cells using novel stains. The basis of their competing views was how, if at all, one cell can be connected with another to constitute a single organ. Light microscopy disclosed additional heterogeneity in shape, size and number of neurons to increase the number of brain regions from Gall's 27 to 52 (Brodman 1909/1994). The critical question which had divided Golgi and Cajal was resolved when electron microscopy (EM) revealed a space between neurons. Golgi's view was no longer compatible with this discovery (Palaye 1956).

The next innovation (see Table 13.2) confocal microscopy together with a combination of histochemistry, fluorescence and antibody labeling provided multicolored pictures appropriately named Brainbow (Livet et al. 2007) which demonstrated dozens of different cell types within each brain region.

With brain form now well established, visual methods of defining brain function became paramount. One such method is opto-genetics (Deisseroth 2006) whereby a light probe could deliver lasers to excite fluorescent indicators of genes within neurons. Optogenetics which has now been widely used in living experimental animals would not be ethical in a human.

But with the invention of non-invasive imaging technologies analogous studies did become possible during the latter half of the twentieth century. X-ray, the first such technology developed in 1895 was not alone useful to this end as it superimposed the images across the total thickness of brain. With the development of computerized X-ray tomography (CT) separate thin slices of brain anatomy could be imaged directly in awake humans.

The method proved immediately useful. Shrinkage of the hippocampus, the module of brain which was universally agreed to be associated with the neuropsychological tests and memory loss of Alzheimer Disease could be confirmed in individual patients by applying calipers to a CT- image.

Missing anatomical continuity, inter-neuronal connections across the synaptic space is made by special cell receptors and chemical neurotransmitters. Synthesis of specific radioactive tracers enabled such studies in human brain scanners. Positron emission tomography (PET) and nuclear imaging built upon these earlier advances.

Each of these technologies has pushed the frontiers of the new phrenology. But none has done so as dramatically and so publicly as magnetic resonance imaging (MRI). Discovered in 1946 in a physics laboratory and thereafter adopted by chemists, within 50 years nuclear magnetic resonance, NMR, had become magnetic resonance imaging, MRI.

But even this is not the visual technology that has made it to the front pages of the popular press. This level of enthusiasm has been reserved for functional MRI (fMRI). Modules most engaged in a particular process, be it speech, hearing, vision, movement or perception or in response to carefully tailored neuropsychological tasks light up on an fMRI image.

Philosophers, anthropologists, theologians and economists have all been tempted to evaluate the location and volume of the brain module they attribute to their given function. Anyone with the cost of access to a modern MRI device to spare ($500 per hour!), can tackle any question of human behavior. fMRI studies appear to confirm that the brain is a collection of individual organs.

Novel methods of magnetic resonance imaging brain water or assaying brain chemistry continue to be developed and are included in Table 13.2. MR chemical analysis termed MR spectroscopy (MRS) quantifies brain glutamate. One of the major neurotransmitters, metabolic turnover its consumption and regeneration, purports to measure the cost of thought. More powerful MR scanners and newer reagents promise further advances in visual phrenology.

13.10 Mathematical Phrenology

Mathematical theories of brain function date back to Pythagoras for whom number was the essence of the living world. Formal mathematical descriptions of brain function were unusual until the seventeenth Century. Beginning with algebra and Descartes a plethora of new mathematical devices was brought to bear on brain function, phrenology and consciousness into the twentieth century. Topology, non-linear dynamics, self-similarity and fractals, circadian theory, information theory, neural-network theory and most recently machine learning, artificial intelligence and computer modeling together developed a mathematics of phrenology as a vigorous counter-point to visual neuroscience. And finally, it has been productive to turn to quantum theory, as Thompson himself had suggested in OGF in 1942.

13.11 D'Arcy Thompson on Brain Function

Discussion of the brain was very limited in both editions of OGF. Thompson is recorded as stating that 'Consciousness is not explained by all the nerve-paths and neurons of the physiologist; nor do I ask of physics how goodness shines in one man's face and evil betrays itself in another'. His caution is echoed in an article entitled 'The unreasonable effectiveness of mathematics in the natural sciences' by a physicist, Eugene Wigner (1960). After noting the positive role of mathematics in the established sciences Wigner writes: "A much more difficult and confusing situation would arise if we could, someday, establish a mathematical theory of the phenomena of consciousness, or of biology, which would be as coherent and convincing as our present theories of the inanimate world. Mendel's laws of inheritance and the subsequent work on genes may well form the beginning of such a theory as far as biology is concerned".

13.12 Alan Turing Mathematizes the Mind

In his approach to phyllotaxis and the equations necessary to the design of a daisy and of Radiolaria, Turing was taking his turn into biology largely under the guidance of D'Arcy Thompson. No such connection is immediately apparent in his turn to neuroscience. For this Turing created his own ideas and tested them on the leading neuroanatomist of his day, J.Z. Young. Young responded with the results of his studies on brain function in octopuses, notably that the axons of nerves were able to seek out new connections and create neural networks.

Turing also tussled with his Manchester colleague, the philosopher Polanyi (1958). Turing devised an 'imitation game' whereby a machine might be constructed to deceive us as to its own nature precisely as another human can. Polanyi disagreed in accordance with his embrace of the concept of theory laden ideas. Turing's view and that of others in the new world of cybernetics was more positive and fostered the growing interest in artificial intelligence.

Rather than presenting each of the modern mathematical theories of consciousness listed above we should follow D'Arcy Thompson's meditation and focus on a quantum theory of novum phrenologium renovatum.

The most complete analysis is offered by a mathematician Roger Penrose in two books "The emperor's new mind (1989)" and "Shadows of the mind (1994)". A new physics is at hand provided by Godel's Entscheidungsproblem (1931), worked on also by Alan Turing (1934). Godel's concept was that mathematical theorems exist for which no proof can be provided. As a true science physics must have a mathematical explanation: ergo, according to Penrose, there must be a physics which lies beyond mathematics. This new physics of quantum theory Penrose suggests would be ideally poised to resolve a problem of such unimaginable complexity as the brain with its billions of individual constituent cells. It could apply to consciousness and other activities of the human brain.

Such a theory of mind remained somewhat empty until a territory could be found for it in the anatomy of the brain. Here Penrose added another observation from quantum theory. Summarized as entanglement this property prevents interactions between events at the quantum level and large object in their surroundings. For the continued existence of quantum events across the human brain quantum entanglement must be avoided. Only within minute, well isolated territories could such protection be envisaged. Here Penrose seeks his answers from beyond mathematics and physics.

The somewhat surprising idea that consciousness can be switched off and on without destroying any of its prior learned accomplishments or stored memories is familiar to anesthetists, sleep scientists. and a neurochemist (McIlwain 1958). A professional anesthesiologist Hammerhoff and Penrose (1996) identified a closed volume within the brain as one likely to limit entanglement. In Shadows of the Mind Penrose proposes that information is stored in the neuro-tubules, minute structures found throughout the brain. Here then is a credible mathematical theory of mind. We should then conclude that the burgeoning imaging techniques have been an artefact which has crowded out a mathematical law of consciousness.

In that most complex of multicellular organs, the human brain, with 86 billion inter-communicating cells the most recent visual studies suggest that as many as 360 modules (van Essen 2018) and up to 10,000 types of human brain cell might exist (Neuromorph 2023). One wonders whether a latter-day D'Arcy Thompson and mathematically innovative Alan Turing could rationalize these extraordinarily volatile numbers as they did for Radiolaria in Case study #1.

To satisfy the tenet of Karl Popper (1959) a means of falsification should be offered. Studies indicate that nanotubules contain water and that this water is subject to flow. A nuclear magnetic resonance imaging diffusion ordered radio-sequence (DOSY) allows water and its diffusion in nanotubules to be observed directly (Marega 2009). If such a brain function study in a conscious human could be achieved, the circle of phrenology would be complete.

13.13 Conclusions

The three case studies offered here appear to favor mathematical over the visual tools. New work repeatedly returns to D'Arcy Thompson as a pioneer in mathematical designs in Nature (Bouligand 2011; Hyde 2013).

Another novel driver in biological research which can be traced to Alan Turing and D'Arcy Thompson is that of Artificial Intelligence used in biology to describe 'Artificial Life' (Boden 2008).

Finally, authors have revisited the impression that Thompson in profoundly disagreeing with Darwin's *Origin of Species* bit the hand the fed him. But D'Arcy

Thompson likely foresaw the latest revision of Darwinism, the Extended Evolutionary Synthesis (EES) (Laland 2018).

Acknowledgements This work was conducted as a Reader at Huntington Library, San Marino California. Accordingly, I am grateful to the Trustees and Staff of the Library for their hospitality and encouragement. Among the staff were Daniel Lewis, Melissa Lo, David Zeidberg and Catherine Wehrey-Miller and several Readers, especially John Heilbron, Howard Weinbrodt, Stefano Gattei and Thomas Barron. This article is offered to thank John Heilbron for the generous guidance he provided during his sojourns at the Huntington Library. For figures and tables thanks are due to KAC Ross.

References

Ackerknecht, E. 1968. *A short history of medicine*. OUP.

Adler. 1997. The history of the study of phyllotaxis. *Annals of Botany* 80 (3): 231–244.

Binet, R. 1900. Monumental gateway Paris Esposition. L'Illustration, April 14.

Blaschke 1956: This can be found in, Breidbach, O. 2006. Visions of nature: The art and science of Ernst Haeckel. Munich: Prestel.

Boden, M. 2008. *D'Arcy Thompson, A grandfather of A-Life*. University of Sussex; MIT Press.

Bonner, J., ed. 1961. *On growth and form: An abridged edition*. Cambridge University Press.

Bouligand, Y. 2011. In *Morphogenesis: Origins of patterns and shapes*, ed. P. Bourgine. Berlin: Springer Verlag.

Breidbach, O. 2006. *Visions of nature: The art and science of Ernst Haeckel*. Munich: Prestel.

Brodman, K. 1909/1994. *Localization in the cerebral cortex*. London: Smith Gordon.

Chirat, R., A. Goriely, and D. Moulton. 2021. The physical basis of mollusk shell chiral coiling. *PNAS* 118 (48): 1–12.

Cooper, B., and J. van Leeuwen. 2013. *Alan Turing, his work and impact*. Elsevier.

D'Arcy Thompson, R. 1958. *D'Arcy Wentworth Thompson*. Oxford University Press.

D'Arcy Thompson, R. 1974. *The remarkable Gamgees*. Edinburgh: Ramsay Head Press.

Darwin, C. 1859. *Origin of species*. London: John Murray.

Daston, L., and P. Galison. 2007. *Objectivity*. New York: Zone Books.

Deisseroth, K. 2006. Next generation optical technologies. *Journal of Neuroscience* 26 (41): 10380–10386.

Dennett, D. 1991. *Consciousness explained*. Boston: Little Brown.

Epstein, I. 2016. Reaction-diffusion processes at the nano and micro scales. *Nature Nanotechnology* 11:312–319.

Esposito, M. 2014. Recovering the historical context of D'Arcy Thompson science of form. *Science in Context* 27 (1): 79–107.

Essen, van D and Glasser M. 2018. Parcellating cerebral cortex. *Neuron* 99:640–663.

Fodor, J. 1983. *The modularity of mind*. MIT Press.

Fodor, J., and Z. Pylyshyn. 1981. How direct is visual perception? *Cognition* 9:139–196.

Gall, F. J., & Spurzheim, J. G. (1810–1819). Anatomie et physiologie du système nerveux en général, et du cerveau en particulier: avec des observations sur la possibilité de reconnaître plusieurs dispositions intellectuelles et morales de l'homme et des animaux par la configuration de leurs têtes. Paris: F. Schoell.

Gibson, J. 1979. *An ecological approach to visual perception*. Houghton Mifflin.

i-Gibson, 2021, arXiv 02924. Author: Bokui Shen, Stanford Robotics. (This is an online Journal)

Gombrich, E. 1971. Exchange of letters (with J.J. Gibson & R. Arnheim). *Leonardo* 4:195–203.

Goriely, A. 2018. *Mathematics and mechanics of biological growth*. Springer Verlag.

Grew, N. (1680). Anatomy of plants. (Huntington Library USA) Proc Roy Soc UK.

Haeckel, E. 1862. *Die Radiolarien*. Berlin: Reimer.
Haeckel, E. 1887. *Challenger (14) Report on Radiolaria*. Edited by J. Murray. HMSO.
Hameroff, S., and R. Penrose. 1996. Orchestrated reduction of quantum coherence in brain microtubules. *Mathematical and Computer Simulation* 40:453–480.
Heilbron, J. 2010. *Galileo*. OUP.
Helmholtz, H. 1868. Handbook of Physiological Optics, Optical Society of America.
Henderson, F., S. Kusakawa, and A. Marr. 2015. Curiously drawn: Early modern science as a visual pursuit. *Huntington Library Quarterly* 78:141–415.
Hindle, S. 2019. *Symposium: Visual science: Huntington Library and division of humanities and social sciences*. Pasadena, CA: California Institute of Technology.
Hodges, A. 1983. *Alan Turing: The enigma*. OUP.
Hyde, S. 2013. D'Arcy Thompson's legacy in contemporary studies of patterns and morphology. *Interdisciplinary Science Reviews* 38 (1): 12–34.
Ivins, W., Jr. 1958. *Prints and visual communications*. London: MIT Press.
Jean, R., and D. Barabe. 1998. *Symmetry in plants*. Singapore: World Scientific.
Jenkins, H. 2008. Gibson's affordances; evolution of a pivotal concept. *Journal of Scientific Psychology* 12 (2008): 34–45.
Kepler J. (1604). Ad Vitellionem Paralipomena, quibus Astronomiae pars optica traditur.
Kuhn, T. 1962. *The structure of scientific revolutions*. Chicago University Press.
Kusakawa, S. 2012. *Picturing the book of nature*. Chicago University Press.
Laland, K. 2018. D'Arcy Thompson and the extended evolutionary synthesis (EES). In *St. Andrews Symposium 'The conceptual legacy of on Growth and Form*, ed. A. Love and S. Sarkar.
Latour, B. 2014. In *Representation in scientific practice revisited*, ed. C. Coopmans et al. MIT Press.
Livet, J., et al. 2007. Transgenic strategies for combinatorial expression (BRAINBOW). *Nature* 450:56–64.
Lorenz, E. 1963. Determinate non-periodic flow (chaos theory & attractors). *Journal of the Atmospheric Sciences* 20:130–141.
Marega, R. 2009. Diffusion ordered spectroscopy (DOSY) in functionalized carbon nanotubes. *Journal of the American Chemical Society* 131 (25): 9086–9093.
Mathematical Association of America (MAA). (2004). *Core subject taxonomy for mathematical sciences education*.
Maxwell, J. C 1877. "Diagram." In Encyclopaedia Britannica (9th ed., Vol. 7, pp. 164–177). Edinburgh: Adam & Charles Black.
Mcfie, J. 1961. Recent Advances in Phrenology. Lancet 278, 360–363.
McIlwain, H. (1958). "Neurochemistry and Therapeutic Endeavor." *AMA Archives of Neurology & Psychiatry* 80(3): 292–297.
Meli, Domenico. 2015. In *Curiously Drawn*, ed. Henderson et al.
Muller, H. 1873. *Die Befruchtung des Fruchte durch Insekten*. Berlin.
Neuromorph.org. 2023. George Mason University.
Palaye, S. 1956. Synapses in the Central Nervous System, Biophysics, Biochemistry. Cytology July 25, 193–202.
Penrose, R. 1989. *The emperor's new mind*. New York: Oxford University Press.
Penrose, R. 1994. *Shadow of the mind*. New York: Oxford University Press.
Plateau, F. 1873. *Experimental and theoretical statics of liquids subject to molecular forces only*. Brussels.
Polanyi, M. 1958. *Personal knowledge*. London: Routledge & Kegan Paul.
Popper, K. 1959. *The logic of scientific discovery*. London: Hutchinson.
Prigogine, I., and I. Stengers. 1984. *Order out of chaos*. Heinemann.
Rachevsky, N. 1938. Mathematical Biophysics, University of Chicago Press
Rudwick, M. 1976. Emergence of a visual language for geological science, 1760-1840. *History of Science* 14:149–195.
Sarton, G. 1928. *Introduction to the history of science*. Williams and Wilkins.
Saunders, P., ed. 1992. *Morphogenesis: Complete works of A.M. Turing*. North Holland.

Thompson, D' Arcy W. 1917/1942. *On growth and form*. Cambridge University Press.

Thompson, D 1883. Fertilization of plants, Macmillan, London.

Thompson, D' Arcy W. 1998. *On growth and form: The complete revised edition*. Dover.

Thompson Archive University of Saint Andrews 1886–1920 ms. 22329–22342: Letter exchanges with Felix Plateau.

Turing, A. 1952. Chemical basis of morphogenesis. *Proceedings. Royal Society of London* 237:37–42.

Uttal, W. 2001. *The new phrenology*. MIT Press.

Whewell, W. 1868. Novum Organon Renovatum. In *Volume II of the philosophy of the inductive sciences*. London: John Parker and Son.

Wigner, E. 1960. The unreasonable effectiveness of mathematics etc. *Communications in Pure and Applied Mathematics* 13 (1): 1–10.

Zhang, L., et al. 2018. Polyamide membranes with nano-scale Turing structure. *Science* 360 (6388): 518–521.

Brian D. Ross, After training in medicine and biochemistry at Oxford, he moved to California. Working at Caltech and the Huntington Medical Research Institutes, his team developed magnetic resonance techniques for diagnosis of human brain disorders, including Alzheimer's Disease, metabolic disorders of infancy and trauma. At the Huntington Library, he explored the history of mathematical biology and imaging, where John Heilbron provided valuable insights in work towards a biography of the Scottish biologist, D'Arcy Thompson.

Chapter 14
Hiroshi Sugimoto and the History of Science

Matthias Dörries

Abstract The 2023 London exhibition of the work of the Japanese artist and photographer Hiroshi Sugimoto, *Time Machine*, showcased his strong engagement with the natural sciences. Drawing inspiration from zoology, geometry, mathematics, electricity, and optics, Sugimoto refers to the history and philosophy of science and figures, well familiar to J.L. Heilbron, like Galileo, Newton, Franklin, Faraday, and Fox Talbot. Sugimoto values the craftmanship and artistry that goes into scientific investigations, pursues systematic experiments himself, and explores puzzling paradoxes from physics and geometry. Sugimoto's openness towards science is mirrored in Heilbron's delight in including culture and art in the history of the physical sciences.

Keywords Hiroshi Sugimoto · Physics · Geometry · Electricity · Galileo · Newton · Photography

On 26 October 2023, I was not—as originally planned—in Shilton (Oxfordshire), visiting John Heilbron, but in London, at the Hayward Gallery of the Southbank Centre. Built in 1968, it is a massive concrete building, Brutalist architecture of the 1960s. On approach it looks uninviting—if not depressing—with its dominant tones of grey, brown and beige. Nevertheless, its vast geometrical architecture made it a place perfectly suited to the exhibition I had come to see: *Hiroshi Sugimoto: Time Machine*. Sugimoto is a Japanese artist, born in 1948, whose mostly black and white photographs have found world-wide recognition. The exhibition was neatly divided into nine sections, each stacked into one of the modules of the building. Four concerned directly the natural sciences: *Dioramas*, *Conceptual Forms and Mathematics*, *Lightning Fields* and *Opticks*; two dealt indirectly with science: *Seascapes* (the elements of water and air) and *Architecture* (geometry). Science and technology play a double role in Sugimoto's work: they not only serve as inspiration, but also enter his

M. Dörries (✉)
AHP-PReST (UMR 7117), University of Strasbourg, Strasbourg, France
e-mail: dorries@unistra.fr

S. Gattei (ed.), *The Incomparable Historian*, Archimedes 75,
https://doi.org/10.1007/978-3-032-25678-2_14

photographic practice. Sugimoto strives to revive past scientific and photographic techniques going back as far as the seventeenth century, using them to refine and perfect his work. Names like Isaac Newton, Benjamin Franklin, Michael Faraday, and William Henry Fox Talbot, appear frequently in his texts and interviews. Sugimoto values especially the craftmanship and artistry that went into their scientific instruments and investigations. His darkroom serves as a scientific laboratory, where he experiments with new ideas and recovers past scientific and photographic knowledge. While Sugimoto does not fully subscribe to the "post-Renaissance rational scientific regard on the natural world," he relentlessly explores science's potential to enter his art: "Art resides even in things with no artistic intentions."[1]

14.1 Dioramas

Sugimoto started out in the 1970s with photographing the full-size dioramas of the American Museum of Natural History in New York. Dioramas represent natural events in an artificial setting. Sugimoto set out for the inverse: his photographs aimed to make the artificial settings look like natural ones (Fig. 14.1). He wrote, tongue in cheek: "However fake the subject, once photographed, it's as good as real."[2] Despite the considerable technical and artistic skills that went into their production, the photographs are still ultimately imperfect and disturbing. Sugimoto's photographs approach reality—they may deceive the viewer into thinking that the animals are alive—but in the end, they also willingly admit failure, just more representations in the ongoing search for reality. Fake and/or real, dead and/or alive, natural and/or unnatural: Sugimoto plays with opposites and looks for ways to represent them "simultaneously,"[3] making viewers uneasy about whether they can trust what they see. His photographs partake of a strange and paradoxical feature that can also be found in quantum mechanics, which—in violation of common logic—claims simultaneity of two logically exclusive states, famously exemplified by Schrödinger's cat which is dead *and* alive.[4] Setting up a tension and possible resolution between two opposites characterizes many of Sugimoto's works, which also explains his preference for black and white photography. His seascapes equally possess a binary structure: divided by a distant horizon, water and air each occupy half of the photographs, both often deceivingly alike, sharing certain features, or merging into each other. Water can look like air, and air like water—or not. They are photographs of the sea taken in specific places at specific times, but they also evoke

---

[1] Hiroshi Sugimoto, *Conceptual Forms*, https://www.sugimotohiroshi.com/new-page-24 (accessed 22 Feb 2024).

[2] Hiroshi Sugimoto, *Dioramas*, http://www.sugimotohiroshi.com/new-page-54 (accessed 22 Feb 2024).

[3] Hiroshi Sugimoto, "Unnatural nature" in *Hiroshi Sugimoto, Dioramas* (Bologna, New York: Damiani, Matsumoto Editions, 2014), 3–5, on 3.

[4] To my knowledge, Sugimoto has not referred to quantum mechanics in his writings.

Fig. 14.1 Installation view of Hiroshi Sugimoto, Polar Bear, 1976. Silver gelatin print. Photo: Mark Blower. Courtesy the artist and the Hayward Gallery

timelessness, eternity. For Sugimoto, the scenery of the sea stands for his very first memories as a child, but also as a "vision" of the earth that modern humans may share with their earliest ancestors.[5] Both the diorama and the seascape photographs confront the spectator with history and the deep past, a common theme in Sugimoto's work, hence the title of the London exhibition: *Time machine*.

14.2 Conceptual Forms and Mathematics

In 2002 Sugimoto turned towards science and mathematics after seeing the collection of nineteenth-century German plaster mathematical models at the University of Tokyo Graduate School of Mathematical Sciences. He chose a different approach from the American artist Man Ray, who in the 1930s had photographed mathematical models at the Parisian Institute Henri Poincaré. Struck by their beauty, Sugimoto set out "to photographically *remodel* the mathematical models."[6] The photographs

[5] Hiroshi Sugimoto, *Interview, Between Sea and Sky*, 11 September 2018, https://channel.louisiana.dk/video/hiroshi-sugimoto-between-sea-and-sky (accessed 22 February 2024).

[6] Hiroshi Sugimoto, "Surface of the Third Order," in *Hiroshi Sugimoto. Conceptual Forms and Mathematical Models*, ed. Klaus Ottmann (Ostfildern: Hatje Cantz, 2015), 5–6, on 5.

of the models, much larger than the originals, hint at the contrast between their embodied ideal of mathematical perfection and their actual physical imperfection, as the delicate plaster bears the marks of decades of use in teaching (Fig. 14.2). Sugimoto was particularly struck by one specific model: an acute hyperbolic shape, resembling a "vertically drawn-out Mount Fuji, its twin equation-curved slopes asymptotically approaching an infinite vanishing point."[7] While in modern terminology this specific geometrical body figures as a "Surface of Revolution with Constant Negative Curvature" or, more generally, a pseudosphere, it was first named "acute hyperbolic solid" in 1641 by the Italian mathematician Evangelista Torricelli, Galileo's successor as Mathematician to the Grand Duke of Florence.[8] Torricelli's study of a solid with *actually* infinite length puzzled seventeenth-century mathematicians and philosophers: How could it be that a solid of infinite length had a finite volume? The result was perceived as paradoxical and counterintuitive. It raised

Fig. 14.2 From left to right: Helicoid: Minimal surface; Surface of Revolution with Constant Negative Curvature; Kuen's surface: a surface with constant negative curvature. Installation view of Hiroshi Sugimoto, Conceptual Forms and Mathematical Model 006. Gelatin silver prints, aluminum and steel. Photo: Mark Blower. Courtesy the artist and the Hayward Gallery

[7] Ibid.

[8] Evangelista Torricelli, *De solido hyperbolico acuto,* in *Opere di Evangelista Torricelli*, ed. Gino Loria and Giuseppe Vassura, 4 vols. (Faenza: Stabilimento Tipo-Litographico Montanari, 1919–1944), vol. 1, part 1, 191–221. For an English translation of parts of this text, see: D. J. Struik, *A Source Book in Mathematics, 1200–1800* (Princeton, N.J.: Princeton University Press, 1986), 227–231. On Torricelli and the philosophical discussion around his solid, see the excellent article by Paolo Mancosu and Ezio Valiati, "Torricelli's Infinitely Long Solid and Its Philosophical Reception in the 17th Century." *Isis 82* (1991), 50–70. His solid figured later also under the name 'Gabriel's horn,' in reference to depictions of the archangel's trumpet.

discussions among philosophers about whether it is possible or impossible to have knowledge about infinity? It also called the ontological status of geometrical entities into question: Was it still possible to talk about space for a geometrical shape that is not closed? Was it still possible to use mechanical concepts, like center of gravity, in geometry? Should geometry be bound by physical, material constraints?

Akin to this seventeenth-century discussion, Sugimoto has mused over the ontology of a geometrical line, noting its paradoxical existence between being visible and imaginary:

> The notion of length with no width is very curious indeed, the pencil line I draw being only an approximation of an invisible mathematical line. Endeavors in art are also mere approximations, efforts to render visible unseen realms.[9]

Following up on this curious paradox, Sugimoto worked on an *Architecture* series of photographs of prominent buildings from the late nineteenth century and twentieth century, among them the Eiffel Tower, the World Trade Center, and the Guggenheim Museum (Fig. 14.3). These modernist buildings of a Machine Age had renounced all decoration. Sugimoto enhances their characteristic and easily recognizable geometric shapes by taking highly blurry pictures, "completely melting away many of the buildings in the process." However, the buildings survived his "onslaught of blurred photography."[10] Ironically, this very blurriness enhances the unique geometrical shape of the building; no finely drawn line is necessary. Sugimoto's *Architecture* photographs not only play with real and imaginary lines, but also catch a building's whole history from conception to realization, from imaginary or real outlines of first ideas, to the ultimate physical realization.

Beyond photographing mathematical models or buildings, Sugimoto set out to explore the limits of what was materially possible when manufacturing models derived from mathematical formula. His stainless-steel models of a "Surface of Revolution with Constant Negative Curvature" surmount the modest size of the historical models, and play with the tension between finite and infinite, visual and abstract, geometrical physical shape and mathematical formula.[11] These models are as perfect as currently technically possible thanks to the use of state-of-the-art computer and machinery. They possess a spotless, shiny surface and extend towards the heavens as far as the material allows, with a tip of just one millimeter in diameter. What physically is impossible, imagination must complete to achieve perfection. Just like the apparently natural number zero, infinity is an invention, "extracted

---

[9] Hiroshi Sugimoto, *Conceptual Forms*, https://www.sugimotohiroshi.com/new-page-24 (accessed 22 Feb 2024).

[10] Hiroshi Sugimoto, *Architecture*, https://www.sugimotohiroshi.com/new-page-5 (accessed 22 Feb 2024).

[11] On the link of Sugimoto's work with mathematics, see: Margaret Wertheim, "Mathematics and Material Form," in *Hiroshi Sugimoto: Time Machine* (London, Berlin: Hayward Gallery Publishing and Hatje Cantz, 2023), 124–125.

Fig. 14.3 Installation view of Hiroshi Sugimoto, Chrysler Building, 1997. Gelatin silver print. Photo: Mark Blower. Courtesy the artist and the Hayward Gallery

from the human imagination."[12] Sugimoto writes: "In my mind's eye I can see two fine lines extend from either side of the one-millimeter tip and meet at an infinite vanishing point."[13] The largest model *Point of Infinity* was installed in 2023 on a hilltop on Yerba Buena Island in San Francisco, extending some 69 feet into the sky (Fig. 14.4). Referring to the Tower of the Sun of the 1939 World Fair in San Francisco on Treasure Island, the model also functions as a sundial, indicating the position of the sun in the vernal and autumnal equinoxes.[14]

14.3 Lightning Fields

The series *Lightning Fields* was an outcome of historical and scientific investigations in photographic techniques. After decades of experiments to improve his developing process, Sugimoto had succeeded in obtaining perfect tonalities for his photographs. However, despite all efforts, he could not get rid of the "little sparks,

[12] Hiroshi Sugimoto, *Post Vitam* (Kyoto: Heibonscha, 2020), 175.

[13] Sugimoto, "Surface of the Third Order," (ref. 6), on 6.

[14] Julie Baumgardner, "Moving the needle: San Francisco to unveil Hiroshi Sugimoto's towering sundial monument," *The Art Newspaper*, 13 January 2023. https://www.theartnewspaper.com/2023/01/13/san-francisco-public-sculpture-hiroshi-sugimoto (accessed 22 February 2024).

Fig. 14.4 Hiroshi Sugimoto, Point of Infinity: Surface of Revolution with Constant Negative Curvature, 2023. Glass fiber, reinforced concrete and stainless steel, 69' / 21 meters, Installation, Yerba Buena Island, San Francisco, 2023. Courtesy of the artist, Photo credit: Sugimoto studio

glowing like fireflies"[15] on the developed film caused by static electricity. From 2006, Sugimoto chose no longer to consider these sparks as a nuisance, but to study them systematically for their aesthetic potential, thus turning the "demons" of static electricity into "angels."[16] A film, "Memories of Origin," shows him in a white coat in the darkroom, recording discharges on a film posed on a negatively charged metal plate.[17] Sugimoto first relied on a nineteenth-century Wimshurst generator, and later a twentieth-century Van de Graaff generator with voltages as high as 400,000 volts. He systematically varied parameters like humidity, length of charging time, and metals (iron, stainless steel, aluminum, brass, copper, titanium). To better understand how the type of metal affected the shape of discharge, Sugimoto also studied Michael Faraday's thirty-five *Deflagrations of Gold, and other metals, on vellum.*[18]

---

[15] Hiroshi Sugimoto, "Let there by lightning," in *Hiroshi Sugimoto: Nature of Light*, ed. Koko Okano (Nohara: Izu Photo Museum, 2009), 79–85, on 79.

[16] Ibid., 81.

[17] See the film: *Memories of origin* (2011), director: Yuko Nakamuro.

[18] Michael Faraday, "The Bakerian Lecture: Experimental Relations of Gold (and Other Metals) to Light," *Philosophical Transactions of the Royal Society of London, 147* (1857), 145–181. See Ryan D. Tweney, "On Replicating Faraday: Experiencing Historical Procedures in Science," *Archives of* Science *58* (2005), 137–148.

Ultimately, he shifted from discharges on metal plates to discharges in salt water, as this led to more satisfying results (Fig. 14.5).

Discharge experiments picked up in the late eighteenth century, following up on the American inventor and statesman Benjamin Franklin, who had demystified lightning by demonstrating that it was a phenomenon of electricity that could be domesticated with a rod. Still, electricity itself remained an elusive phenomenon. In 1777, the German physicist and writer Georg Christoph Lichtenberg in Göttingen realized that electric discharges on an insulator caused fine powder to arrange in characteristic patterns, now referred to as *Lichtenberg figures* (Fig. 14.6).[19] A few years later, in 1784, the Dutch director of Teyler's Museum in Leiden, Martinus van Marum, enjoyed the use of the most powerful generator so far, whose 500,000 volts could produce sparks of a length of 2 feet. Van Marum studied discharges on various metals, and had the colorful results copied by the artist Jan Christiaan Sepp for wider distribution.[20] These experiments were part of the "non-representational

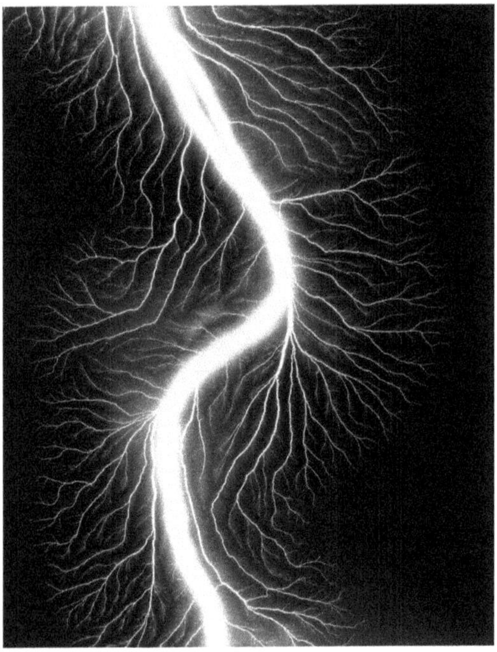

Fig. 14.5 Hiroshi Sugimoto, Lightning Fields 225, 2009. Gelatin silver print, 58–11/16 × 47 in. / 149 × 119.4 cm. Courtesy of the artist

[19] G. Chr. Lichtenberg, *De nova methodo naturam ac motum fluidi electrici investigandi, commentatio prior* (Gottingae: Joann Christian Dieterich, 1778). Yuzo Takahashi, "Two Hundred Years of Lichtenberg Figures," *Journal of Electrostatics 6 (1)* (1979), 1–13.

[20] See for colorful drawings of the deflagrations of metals and the two-feet spark: Martinus van Marum, "Première continuation des experiences faites par le moyen de la machine électrique teylérienne," *Teyler's Tweede Genootschap, Verhandelingen 4* (1787). The colorful deflagrations of lead and gold can also be found in J.L. Heilbron, "Fisica e astronomia nel settecento," in *Le scienze fisiche e astronomiche*, ed. W.R. Shea (Torino: Einaudi, 1992), 318–443, on 402–403 (*Storia delle scienze*, 2, ed. Paolo Galuzzi).

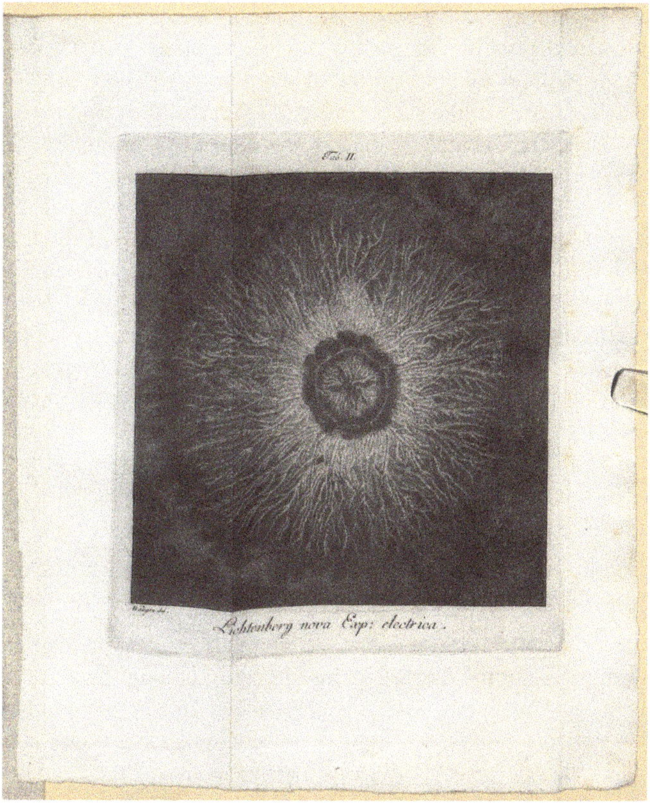

Fig. 14.6 Lichtenberg figure. Drawing by Friedrich L.H. Waagen (1750–1822). In G. Chr. Lichtenberg, *De nova methodo naturam ac motum fluidi electrici investigandi, commentatio prior* (Gottingae: Joann Christian Dieterich, 1778). Tab. II

imaging that emerged in the last decades of the eighteenth century … aimed at registering various forms of energy, vibrations, oscillations and excitations of matter through optical means."[21] For example, the German physicist and inventor of musical instruments Ernest Chladni studied how various modes of vibrations formed distinctive patterns of sand on a metal plate, now called Chladni figures. Lichtenberg and Van Marum were driven by the hope that the visual study of discharges would allow for a better understanding of the puzzling phenomenon of electricity. Though this research led nowhere, it clearly had an aesthetic aspect, and both marveled at the surprising patterns.[22]

---

[21] Tomáš Dvořák, "Reconsidering cameraless photography," *Philosophy of Photography 13* (2022), 3–15, on 7.

[22] For Lichtenberg's research making visible invisible phenomena, see: Florian Sprenger, "Sternenstaub. Zur Anschaulichkeit elektrischer Phänomene Ende des 18. Jahrhunderts," in *Zwischen Sichtbarkeit und Unsichtbarkeit. Visualität in Wissenschaft, Literatur und Kunst um*

Sugimoto's lightning pictures align with these early experiments, with the difference that now a photographic film serves as a recording medium. The British scientist and inventor William Henry Fox Talbot had done ground-breaking work in this line of research. Already a few years before the French Louis Daguerre, starting in 1833, Talbot had experimented by placing an object, for example a flower, on a surface covered with photo-sensitive material, like silver-nitrate, obtaining what he called a "photogenic drawing." A few years later, Talbot developed the *calotype* process, using a camera obscura and paper negatives from which positive prints could be obtained. The study of Talbot's work brought Sugimoto back to the historical origin of photographic techniques.[23] Eager to preserve and revive historical sources, he set out to produce photographs from Talbot's earliest paper negatives, from which even Talbot had not made positive copies. As these were highly fragile, Sugimoto saw himself working like "an archeological explorer excavating an ancient dynastic tomb."[24]

While the scientific and technical investigation of discharges had led Sugimoto to the origins of photography, the *patterns* of these discharges led him to origins of life:

> I wanted to get to the bottom of why an inorganic phenomenon such as electromagnetic discharge should produce shapes that looked so much like protoplasmic lifeforms.[25]

Having a meteorite with proto-organic molecules in his collection, Sugimoto looked at scientific theories on the origins of life on earth, among them the panspermia hypotheses.[26] Electromagnetic discharge phenomena, like lightning, may have triggered life by setting off chain reactions. In these moments of origin, the inorganic and the organic, the lifeless and alive, the instantaneous and the eternal, coincided. Flashes of light set inorganic material into chemical reactions that ultimately made life on earth possible. The lightning recordings can capture this instantaneous moment and restage the mystical moment of creation of life on earth. The photo of Sugimoto in the darkroom, provided by the artist himself (Fig. 14.7), makes him take on the persona of a scientist in a white coat, facing the mortal danger of high voltage. The lightning, the spark of the discharge, is visible, but not the eyes of the

---

1800, eds. Jürgen Kaufmann, Martin Kirves, Dirk Uhlmann (Paderborn: Wilhelm Fink, 2014), 39–64. For the history of deflagrations and cameraless photography during the nineteenth century, see Geoffrey Batchen, "Like Lightening," in *Sugimoto: Time Machine* (ref. 11), 140–141.

[23] Sugimoto also did some historic experiments with Talbot's original Electrostatic Discharge Wand from the Bodleian Library to produce discharges on film, see Batchen, "Like Lightening," (ref. 22).

[24] Hiroshi Sugimoto, *Photogenic Drawing*, https://www.sugimotohiroshi.com/photogenic-drawing (accessed 22 Feb 2024).

[25] Sugimoto, "Let there by lightning," (ref. 15), on 85. On Sugimoto's search for origins, see Batchen, "Like Lightening," (ref. 22).

[26] Recent research has emphasized the importance of lightning: Benjamin Hess, Benjamin L., Sandra Piazolo, and Jason Harvey, "Lightning Strikes as a Major Facilitator of Prebiotic Phosphorus Reduction on Early Earth." *Nature Communications 12, 1* (2021), 1535. https://doi.org/10.1038/s41467-021-21849-2."

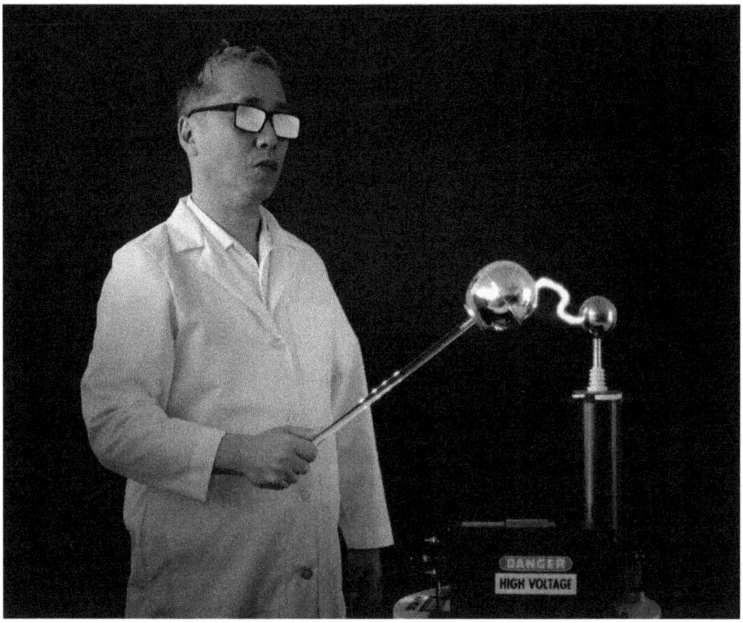

Fig. 14.7 Hiroshi Sugimoto, Courtesy of the artist. Photo credit: Sugimoto studio

experimenter. The irony here is that the scientist doesn't see, but photography and art can *see* lightning and make the strange phenomenon visible, while science remains blind to the objects and phenomena it has created. In his search for what science misses, Sugimoto turned once again to the human history of coping with the violent and fearsome phenomenon of lightning. In exhibitions of the *Lightning series*, Sugimoto has included a statue of Raijin, the god of thunder in Shinto religion, as a reminder that these photographs are an outcome of his own struggle with the demons to domesticate the disturbing electric sparks in the darkroom.

14.4 Opticks

From 2009, Sugimoto restaged the prism experiments of the 1660s by the English natural philosopher Isaac Newton, described in his 1704 book *Opticks*, of which Sugimoto had acquired a copy. Working in a prism chamber with white plaster walls, facing east to catch the morning sun, Sugimoto developed an instrument with an adjustable mirror to follow the moving sun and started to systematically observe and photograph the multitude of polarized colors, obtained by a ray of the sunlight shining through the prism into a dark room. 2009 was also the year when the American company Polaroid decided to stop the production of the instant film, introduced in 1948. Sugimoto, always keen to preserve traditional photographic practices, bought the remaining stock of film, and took hundreds of pictures of colors with a Polaroid camera over the next decade.

Working with prisms meant working with glass. For Sugimoto, glass is "sacred" and has a "commanding presence" due to its transparency and the effect that it seems to shine "outward from within." Over the last decades, he has collected and studied transparent glass-like stones, for example pallasites from meteorites and tektites from Libya. He also conceived a glass tea house, making use of the effect of glass that it can simultaneously "delineate and efface a plane within space."[27] Finally, following in Galileo's footsteps, he had Murano glass makers in Venice reproduce an ancient Sasanian Persian tea bowl that had made its way to Japan via the Silk Road.[28]

Glass is the medium through which Sugimoto as a photographer has contemplated the world: "My intellect and my emotions owe an enormous debt to the glass windows, known as lenses, that form part of this black box." Galileo's telescope, Sugimoto writes, has changed "our way of looking at the world."[29] Galileo and Newton laid the groundwork for all subsequent art and science. However, Newton had distinguished only seven colors, a fact, which left Sugimoto puzzled, as he observed a multitude, if not an infinity, of colors during his experiments.[30] Searching for an alternative interpretation of colors, Sugimoto turned to the German writer and naturalist Johann Wolfgang von Goethe and his *Farbenlehre* of 1810. Goethe was concerned with the effects of light and colors on the senses, and considered colors to be "acts of light, acts and sufferings." In the end, Sugimoto rejected both "Newton's impassionate arithmetic gaze on nature," and "Goethe's warm poetic reflexivity," opting for a middle way (Fig. 14.8). He set his sight on capturing "the colors themselves," and explored the nature of light in its ambivalence between materiality and immateriality, particle and wave:

> What I wanted to do was to capture not a form but the colors themselves…. It took me fifteen years of research before I could create prints that satisfied me. The works are closer to painting than photographs, except that what Mark Rothko tried to express with paint, I tried to express directly with light itself. Paint is a physical substance, and I envisioned fixing light itself without the mediation of paint.[31]

---

[27] Hiroshi Sugimoto, *Post Vitam* (Kyoto: Heibonscha, 2020), 173.

[28] On the history of glass and Galileo, see: Sven Dupré, "The Nature of Glass: Technologies of Transparency, Materials on the Move," *Isis 114* (2023), 393–399.

[29] Sugimoto, *Post Vitam* (ref. 27), 174.

[30] Ibid., 177-178. "Why must science always cut up the whole into little pieces when it identifies specific attributes? The world is filled with countless colors, so why did natural science insist on just seven? I seem to get a truer sense of the world from those disregarded intracolors. Does art not serve to retrieve what falls through the cracks, now that scientific knowledge no longer needs a God?" Ibid., 178.

[31] Hiroshi Sugimoto, "A Second Silent Spring," *The Yomiuri Shimbun*, 15 April, 2020. https://www.odawara-af.com/en/news/a-second-silent-spring/ (accessed 22 February 2024).

Fig. 14.8 Installation view of Hiroshi Sugimoto, Opticks 163, 2018–2022. Chromogenic print. Photo: Mark Blower. Courtesy the artist and the Hayward Gallery

Making use of up-to-date technology, Sugimoto digitized his Polaroid photographs, increased their original size and erased the "visual noise," that had been caused by the chemicals during development. Proceeding this way, Sugimoto realized that he was no longer acting as a photographer but rather as a painter "using photons."[32]

14.5 J.L. Heilbron—In Memoriam

Walking through the exhibition, I could not escape seeing parallels to John Heilbron's work in the history of science. There were the obvious ones: his studies of the history of electricity, of towering scientific figures like Galileo, Newton, and Franklin, his penchant for geometry, his predilection for studying historic instruments and experiments, combined with a wariness of metaphysical speculation.[33] Heilbron's focus on the early modern period suited his preference for exploring the

[32] Sugimoto, *Post Vitam* (ref. 27), 213.

[33] "Some reviewers have found it difficult to accept the finding that not metaphysical commitments but new instruments gave the main impulse to the development of electrical theory. … Our case history shows that the metaphysics of the paradigms and research programs supposed to guide scientists are seldom close enough to experimental work and theory construction to order them in useful ways." J.L. Heilbron, *Elements of Early Modern Physics* (Berkeley: University of California Press, 1979), ix.

physical and visible side of scientific work. This preference for the concrete and proven also showed in his vintage writing technique, working with index cards, all of which had minutely hand-written text on them. Hundreds of them stood next to the latest computer in his office.

But the exhibition also reminded me of something else: Heilbron's books are lavishly illustrated. Paintings, drawings, and photographs convey a message that the text alone cannot deliver. In his books, experimental and geometrical drawings may serve didactic ends, providing a rapid grasp of proportions and scales, or the cause of amusement, informing the reader of an approximate "mathematics of Hell."[34] But there is so much more: whole galleries of portraits of philosophers, academicians, cardinals, and writers; dangling figures in eighteenth-century public electric experiments; an abundance of putti;[35] striking geometrical patterns from various cultures; all kinds of frontispieces with allegorical drawings; illustrations of instruments with lonely hands pointing to key features; reproductions of Galileo's skilled drawings of the moon; photographs of meridian lines with their medallions and zodiacal signs in Italian churches. They depict historical objects of all kinds, providing puzzles, or in Heilbron's terms "hieroglyphs"[36] that require decipherment and explanation. But these pictures also serve to send messages to the reader. The book *Geometry Civilized*, for example, starts out with the frontispiece of an early eighteenth-century edition of Euclid's *Opera omnia*, showing the Greek philosopher Aristippus with his companions, shipwrecked on a beach in Rhodes. They discover geometrical drawings in the sand and Aristippus exclaims, "Bene speremus! Hominum enim vestigia video." ("Let's hope for the best! For I see traces of humans."). The frontispiece sets the tone: Heilbron unreservedly subscribes to "geometry as evidence of civilization." But characteristically, he also qualifies such a bold assertion in the preface: "this doctrine confuses culture with quantitative thought, the whole with the part."[37]

John Heilbron took an improbable artistic turn in *Galileo* and *The Ghost of Galileo*, wide-ranging cultural studies of Galileo's life and of the reception of his work in England during the Civil War. Abhorring speculation in historiography,[38] Heilbron indulged in literary dialogues *à la* Galileo, when his sources failed unambiguously to explain the emergence of Galileo's new science of motion or the presence of Galileo's *Dialogue* in an English painting. This device allowed Heilbron—like

[34] J.L. Heilbron, *Galileo* (Oxford: Oxford University Press, 2010), 31.

[35] See: J.L. Heilbron, "Domesticating Science in the Eighteenth Century," in *Science and the Visual Image in the Enlightenment*, ed. William Shea (Canton, MA: Science History Publications, 2000), 1–24.

[36] J.L. Heilbron, *The Ghost of Galileo* (New York: Oxford University Press, 2021), 1.

[37] J.L. Heilbron, *Geometry Civilized: History, Culture and Technique* (Oxford: Oxford University Press, 1998, sec. ed. 2000). Vitruvius, *De Architectura*, liber 6, preface. Heilbron routinely solved geometrical problems during long university administrative meetings.

[38] "To obtain an answer in this muddle of uncertain dates, perplexed attributions, and reciprocal influences without claiming greater certainty than our information allows, a Galileian dialogue, which permits digressions and approximations, might suit." J.L. Heilbron, *Galileo* (ref. 34), on 128.

Galileo—to stake out his own interpretation of a mysterious matter, and—unlike Galileo—to escape the Inquisition.

Sometimes, art can approach closer to reality than history can. John Heilbron travelled far during his long intellectual life: from physics to the history of physics, to the institutional and cultural history of the physical sciences, to cultural, political, and religious history, to history *tout court*, and ultimately to art. The last lines of John Heilbron's *Ghost of Galileo* aptly summarize Heilbron's own path and apply to Hiroshi Sugimoto as well: "May the image of Galileo the humanist-scientist inspire humanists who know no science and scientists disdainful of the humanities to widen their horizons!"[39]

14.6 Postscript

In 1958, the American painter Mark Rothko (1903–1970) gave an improvised talk at the Pratt Institute in Brooklyn, his last public intervention. He came up with the following "recipe of a work of art—its ingredients—how to make it—the formula," which echoes both Hiroshi Sugimoto's art, and John Heilbron's history of science:

1. There must be a clear preoccupation with death—intimations of mortality … […]
2. Sensuality. Our basis of being concrete about the world. It is a lustful relationship to things that exist.
3. Tension. Either conflict or curbed desire.
4. Irony. This is a modern ingredient—the self-effacement and examination by which a man for an instant can go on to something else.
5. Wit and play … for the human element.
6. The ephemeral and chance … for the human element.
7. Hope. 10% to make the tragic concept more endurable.

I measure these ingredients very carefully when I paint a picture. It is always the form that follows these elements and the picture results from the proportions of these elements.[40]

Acknowledgements Many thanks to Laura Gosney of the Southbank Center, Hayward Gallery, London, and Catherine Bailey of the Marian Goodman Gallery, New York, for their kind help in getting access to the photographs, and Hiroshi Sugimoto and the Sugimoto studio for providing photographs for this article.

Matthias Dörries, Professor of history and epistemology of science at the University of Strasbourg, France. He was a post-doc at the Office for the History of Sciences and Technology (OHST), University of California at Berkeley, in the early 1990s and has returned regularly since then. His work focuses on the history of the (geo)physical sciences and more recently the history of climate change and geoengineering.

[39] Heilbron, *The Ghost of Galileo* (ref. 36), 387.

[40] Mark Rothko, "Address to Pratt Institute, November 1958," in Mark Rothko, *Writings on Art* (New Haven: Yale University Press, 2006), 125–126.

Part III
Styles in the Historiography of Science

Chapter 15
Science and the Jewish Question

Tal Golan

Abstract This essay traces the intertwined histories of modern science and the "Jewish Question" from the Enlightenment to the rise of Zionism. It argues that the Jewish Question emerged as a distinctly modern problem, one that required scientific rather than theological explanation and solution. Beginning with eighteenth-century political economy and anthropology, European thinkers redefined Jewish difference in naturalistic, moral, and utilitarian terms, casting Jewish "improvement" as a task of rational administration. Nineteenth-century transformations—industrialization, emancipation, and the professionalization of science—further secularized the debate, replacing religious prejudice with socio-economic and biological theories of human difference. By the end of the nineteenth century, evolutionary and racial sciences had recast Jewish destiny in hereditary and geopolitical terms, preparing the conceptual ground for both antisemitic politics and Zionist responses. The essay concludes by showing how Theodor Herzl and his contemporaries internalized scientific and technocratic ideals, envisioning Jewish regeneration through the rational reconstruction of Palestine—the deliberate, technoscientific transformation of land, society, and governance through planning, engineering, and expert administration. In this sense, the Jewish Question was not only illuminated by modern science but also served science's own moral and political ambitions: to reform humanity, rationalize society, and demonstrate that reason and technology could remake both people and nature.

Keywords Jewish question · Enlightenment science · Political economy · Race theory · Emancipation · Zionism · Technocratic modernity

The Jewish Question got its name from the 1840s exchange between two young Hegelians—Bruno Bauer and Karl Marx—who embraced their master's dialectical method and pondered whether the political emancipation of the Jews could lead to

T. Golan (✉)
History Department, University of California San Diego, La Jolla, CA, USA
e-mail: stefano.gattei@unitn.it; tgolan@ucsd.edu

S. Gattei (ed.), *The Incomparable Historian*, Archimedes 75,
https://doi.org/10.1007/978-3-032-25678-2_15

their human emancipation and social assimilation.[1] Thus coined, the Jewish Question was a modern question that could not have been posed the same way in an earlier epoch that organized itself by faith and birth. In God's universe, everyone had their preordained place, including the Jews. The Jewish station in Christendom may not have been enviable, but it was steadied by Christian and Jewish dogmas and legalities.[2] It was only after the *Philosophes* secularized European society beyond redemption that the Jewish Question emerged regarding the place of the Jews in the newly secular, industrial, and eventually national order of things.[3] As a modern question, the Jewish Question had to be answered scientifically, and as science evolved so did the Jewish Question and its solutions. This essay explores these dynamics between science and the Jewish Question—from the eighteenth-century sciences of man and society that pondered the possibilities of Jewish emancipation to the fin-de-sicle sciences of evolution and race that undermined the assimilation project and informed the rise of Zionism.

15.1 Eighteenth-century Science and the Jewish Question

The eighteenth-century sciences of man and society—Locke's empirical psychology, Montesquieu's historical anthropology, Buffon's geographical biology, Quesnay's political economy—have all taught that human beings are born more or less equal and are shaped by their social and physical environments. Were the Jews not human, shaped by their experience? And if so, could they not be improved by emancipation and education? The answer seemed clear, a matter of deduction from first principles.[4] Yet, none among the *Philosophes* was publicly hopeful.[5] They could no longer accept theological explanations of Jewish differences that were rooted in the doctrines of damnation and redemption but remained uncertain about what could counter the tight grip Jewish religion had over its followers.[6]

Jewish society seemed impervious to the new scientific spirit that animated the Enlightenment project. The fabled accomplishments of the Scientific

[1] Bauer, Bruno, *Die Judenfrage* (Braunschweig, 1843). Marx, Karl. "Zur Judenfrage." In Deutsch-Französische Jahrbücher, edited by Arnold Ruge and Karl Marx, (Paris: 1844), 182–214.

[2] Baer, Yitzhak, *Galut*, trans. Warshow, Robert. (New York, Schocken, 1936).

[3] Funkenstein, Amos, "The Political Theory of Jewish Emancipation from Mendelssohn to Herzl," in Grab, Walter, ed., *Deutsche Aufklärung und Judenemanzipation* (Tel Aviv, Jahrbuch des Instituts für deutsche Geschichte, 1979) 13–28. See also Auerbach, Rena, ed. *The "Jewish Question" in German Speaking Countries, 1848–1914, A Bibliography* (New York: Garland, 1994).

[4] Warman, Caroline, ed. *Tolerance: The Beacon of the Enlightenment* (Cambridge: Open Book Publishers, 2016) 90–92.

[5] Hertzberg, Arthur. *The French Enlightenment and the Jews* (New York: Columbia University Press, 1968). Schechter, Ronald, *Obstinate Hebrews: Representations of Jews in France, 1715–1815*. (Berkeley and Los Angeles: University of California Press. 2003).

[6] Baron d'Holbach, *L'Esprit du judaïsme ou Examen raisonné de la loi de Moïse* (Paris, 1770).

Revolution—the mechanical philosophy, the experimental method, the empirical and quantitative spirits, Newton's clockwork universe, and the gentlemanly culture of science—did little to alter the premises of Jewish society, which promptly excommunicated its only philosopher who dared to wed the old deity with the new, and entrenched itself in its isolated districts.[7] No greater insult existed to those who fought to liberate human thought from the shackles of religious superstition. Consequently, the *Philosophes* came to consider Jewish religion as the epitome of arbitrary and irrational authority, and Jewish society as primitive, superstitious, and inferior in every way.[8]

The fledgling science of political economy, which taught how the enlightened state should optimize its resources, added a new and practical dimension to the bookish discussion. Sharing the enlightened appetite for natural causes, the *Économistes* investigated how such causes could be addressed by the state in its efforts to reform the Jews into useful citizens.[9] The *Physiocrats* taught that all wealth is extracted from nature, mainly through agriculture. Their tripartite economic model included the 'proprietary' class of landowners, the 'productive' class of agricultural laborers, and the 'sterile' class of artisans and merchants. This scheme easily lent itself to moralizing. The farmers were celebrated for their affinity with nature, urban culture was damned for its artificiality, and the Jews were disparaged for their focus on trade and finance, which had a draining influence on the national economy and morality.[10]

The Physiocrats failed to gain political power.[11] Nevertheless, their analysis, which tied the discussion of Jewish emancipation to the language of political economy and its indices of utility and efficiency, established a new standard by which to evaluate the integration of the Jews. One could see this standard at work in the late eighteenth-century policies of France's two enlightened neighbors, Prussia and the Holy Roman Empire, as they moved to improve the new and extensive territories they had torn off the dying Polish Kingdom, which included unprecedented numbers of Jews.[12] "It is a very good and advantageous acquisition, both from a financial

[7] Nadler, Steven, *Spinoza: A Life* (Cambridge University Press; 2001).

[8] Sutcliffe Adam, *Judaism and Enlightenment* (Cambridge University Press, 2004). Hertzberg, Arthur, *French Enlightenment and the Jews* (New York: Columbia University Press, 1968).

[9] *Cf. Journal des Économistes*, founded in Paris, in 1841.

[10] Arkel, Dik van. T*he Drawing of the Mark of Cain: A Socio-Historical Analysis of the Growth of Anti-Jewish Stereotypes* (Amsterdam University Press, 2009).

[11] Charbit Yves and Arundhati Virman, *The Political Failure of an Economic Theory: Physiocracy*, Population (2002) 57(6): 855–883.

[12] In 1772, during the 'first partition' of the Polish-Lithuanian Commonwealth, Prussia annexed 52,000 Square kilometers (20,000 sq. miles) and about 600,000 inhabitants of northwestern Poland. The Holy Roman Empire annexed 83,000 km^2 (over 32,000 sq. miles) and 2,650,000 inhabitants of southwestern Poland. Russia gained 92,000 km^2 (36,000 square miles) and 1,300,000 people of eastern Poland. Two additional partitions took place in 1793 and 1795, adding more land, people, and Jews to Prussia, the Habsburg Empire and Russia, and leading to the disappearance of the Commonwealth. See "Partitions of Poland," *Encyclopedia Britannica*, https://www.britannica.com/event/Partitions-of-Poland. Accessed 1 May 2024.

and a political point of view," wrote the Prussian monarch, Friedrich II, to his brother. "In order to excite less jealousy I tell everyone that on my travels I have seen just sand, pine trees, heathland, and Jews. Despite that, there is a lot of work to be done; there is no order, and no planning, and the towns are in a lamentable condition."[13]

It was in this context that the Prussian Cameralist and administrator Wilhelm Christian Dohm published in 1781 a detailed proposal to reform the Jews into productive citizens.[14] Dohm, who had taught finance and statistics at the Collegium Carolinum in Kassel before joining Friedrich's administration, was concerned less with the corruption of the Jews and more with the corruption of rational politics. A rational administration that seeks to order, discipline, and maximize its available resources, argued Dohm, must concern itself with the improvement of the Jews. He acknowledged the prevalent views of the Jews as morally and politically degenerate but contended that this may be largely the result of their treatment by Christian society, and posited that rehabilitation was possible by disassociating the Jews from their parasitic roles and returning them to the soil. Eager to make the Jews productive citizens, Dohm proposed a set of policies that would grant Jews civil rights, require them to perform military service, expose them to modern education, and move them East to the newly acquired territories, away from trade and usury and into the more productive fields of agriculture and the crafts.[15]

Dohm's proposal had little effect on the policies of his enlightened king who considered the Jews as unproductive as heathland but far less improvable.[16] Friedrich renamed his newly acquired territories West and South Prussia and planned to relocate hundreds of thousands of German colonists there. The Prussian administration quickly moved to improve the infrastructure of these territories. Massive reclamation projects provided new fertile land, the administrative and legal systems were reformed, and hundreds of schools were opened—all applied lessons of *Kameralwissenschaft*, the German science of administration; a mix of ethics, economics, finance, and legislation that was nourished at the Prussian universities of Frankfurt an der Oder and Halle.[17] Alas, no similar efforts were invested in the improvement of the Jews. Friedrich was ready to tolerate them on the Polish border

---

[13] MacDonogh, Giles, *Frederick the Great: A Life in Deed and Letters* (New York: St. Martin's Griffin, 2001) 363.

[14] Dohm, Christian Wilhelm von. *Ueber die bürgerliche Verbesserung der Juden* (Kaiserslautern, 1781).

[15] Hess M. Jonathan, "Rome, Jerusalem and the Imperial Imagination: Christian Wilhelm Dohm and the Regeneration of the Jews," in *Monstrous Dreams of Reason: Writing the Body, Self, and Other in the Enlightenment*, ed. Laura Rosenthal and Mita Choudhury (Lewisburg, PA: Bucknell University Press, 2002), 132–48.

[16] Blackbourn, David, "'Conquests from Barbarism': Taming Nature in Frederician Prussia", in Christof Mauch, ed., *Nature in German History* (Berghahn Books: New York, 2004) 10–30.

[17] Rüdiger, Axel, *Staatslehre und Staatsbildung: Die Staatswissenschaft an der Universität Halle im 18. Jahrhundert* (Tübingen: Niemeyer 2005). Recktenwald, Horst Claus, "Cameralism." In Durlauf Steven N. and Lawrence E. Blume, eds. *The New Palgrave Dictionary of Economics.* (New York: Palgrave Macmillan, 2008) 1236–1238.

where they performed useful trade but made sure that their numbers inland did not increase, especially in the growing urban centers, where "they formed cliques, deal in contraband and get up to all manner of rascally tricks which are detrimental to Christian burghers and merchants."[18]

Dohm's proposal gained more interest from the other great Enlightenment monarch, the Holy Roman Emperor Joseph II, whose newly acquired Galician territories included a large, poor, and deeply religious Jewish population. During the 1780s, the Josephinian administration launched a sustained campaign to make the "numerous members of the Jewish nation more useful to the state."[19] Improvement was to be achieved via educational and economic reforms. The educational reforms required all Jewish children to acquire the basics of secular education, namely reading, writing, arithmetic, and German language—either in their traditional schools or, free of charge, in new public schools that were now opened to Jews. The economic reforms removed constraints on various occupations to channel Jews away from their "characteristic usury and deceitful trade" toward productive occupations such as agriculture, crafts, and manufacturing.[20]

Drafted by the eminent Joseph von Sonnenfels, the first professor of *staadstwissenschaften* at the University of Vienna, a senior adviser to the Emperor, and the son of a baptized Jew from Moravia, the Josephinian policies were the first systematic effort to integrate the Jews into the modern public sphere.[21] Marketed as "Edicts of Tolerance," these policies included military conscription, restriction of corporate autonomy, and the demotion of the rabbinic courts to civil arbitration. Fearing Rabbinical opposition, Moses Mendelssohn, the Berliner figurehead of the Jewish Enlightenment, appealed in 1782 to the Jewish leaders to take their part in the experiment initiated by their Emperor: "If you desire protection, tolerance, and sufferance from others, then protect, tolerate, and suffer each other. Love and you will be loved."[22]

Mendelssohn was not the only one concerned with the resistance of the Jewish establishment to the enlightened efforts to 'improve' them. The Prussian biblical scholar Johann David Michaelis commended Dohm's ambition of reforming the Jews but cautioned that the Jewish religion, whose "special purpose is to maintain the Jews as a people almost completely separated from other people," will forever

[18] Giles (2001) 347.

[19] *Joseph II's Edict of Toleration for the Jews of Lower Austria (January 2, 1782)* https://germanhistorydocs.ghi-dc.org/sub_document.cfm?document_id=3648. Accessed 1 May 2024. See also, Blitz, C. Rudolph, "The Religious Reforms of Joseph II (1780-1790) and their Economic Significance". *Journal of European Economic History* (1989) 18: 583–586.

[20] Grodziski, Stanislaw, "The Jewish Question in Galicia," *Polin* (Oxford, 1999) 12: 61–71. Karniel, Josef, *Die Toleranzpolitik Kaiser Josephs II.* (Gerlingen, Ger., 1986).

[21] Habermas, Jürgen, Sara and Frank Lennox, "The Public Sphere," *New German Critique*, 1974(3) 49–55.

[22] Mendelssohn, Moses, Preface to Vindiciae Judaeorum, in *Moses Mendelssohn: Writings on Judaism, Christianity and the Bible,* Michah Gottlieb, ed. (Lebanon, NH: Brandeis University Press, 2011) 40–54. See also Feiner, Shmuel, *The Jewish Enlightenment* (Philadelphia; University of Pennsylvania Press, 2004) 105–162.

prevent the Jew from becoming "a full citizen with respect to love and pride in his country."[23] Michaelis, like many other intellectuals and administrators of his generation, supported the emancipation of the Jews but opposed giving them political rights unconditionally. To that end, a distinction between human and civil rights was borrowed from natural law doctrines—human rights were natural, inherent, cognizable, and inalienable; civil rights were a matter of positive law.[24] As long as the Jews live among us, wrote Fichte in 1793, their rights as human beings must be protected. "As for giving them civil rights, I see no remedy but that their heads should be all cut off in one night and replaced with others in which there would not be a single Jewish idea."[25] Fichte was not advocating mass decapitation, but he was certainly venting the Enlightenment's frustration with Jewish obduracy—and he shared the prevalent skepticism regarding their improvability.[26]

Translated into French in 1782, Dohm's program played a role in the French revolutionary debate over the status of the Jews.[27] There were not many Jews in late eighteenth-century France, so the debate was framed by the principles of liberal theory rather than the indices of political economy. The familiar concerns were aired in the National Assembly during the debate preceding the 1789 *Declaration of the Rights of Man and of the Citizen*. The spokesman for the minority of liberal nobles, Comte de Clermont-Tonnerre, summed up the concerns: "The Jews, they say, are not sociable. They are commanded to loan at usurious rates; they cannot be joined with us either in marriage or by the bonds of social interchange; our food is forbidden to them; our tables prohibited; our armies will never have Jews serving in the defense of the fatherland." In response, the moderate Comte put Dohm's teachings forward: "Men who have nothing but money can only work with money: that is the evil. Let them have land and a country and they will loan no longer: that is the remedy." Clermont-Tonnerre proposed that the new republic offer the Jews a deal: to trade their collective identity as a Jewish people, scattered among the nations, for citizenship in a modern state that treated all its citizens equally. "But, some will say to me, they do not want to be. Well then! If they do not want to be citizens, they should say so, and then, we should banish them."[28]

The resultant *Declaration of the Rights of Man and of the Citizen*, voted into law by the National Assembly in 1789, during the early phase of the Revolution,

---

[23] Michaelis, Johann David, "Herr Ritter Michaelis Beurtheilung," in *Dohm, Verbesserung der Juden* (Berlin, second extended edition, 1783) Part 2: 31–71.

[24] Hutcheson, Francis, *A System of Moral Philosophy* (London,1755) 261–262.

[25] Lauth, Richard., ed., J.G. *Fichte: Gesamtausgabe der Bayerischen Akademie der Wissenschaften* (Stuttgart: Frommann, 1964),1(1), 292–93.

[26] Sweet, Paul R. "Fichte and the Jews: A Case of Tension between Civil Rights and Human Rights," *German Studies Review* (1993) 16(1):37–48. Shortly after Joseph's death in 1790, most of the restrictions on the Jews were revived.

[27] Dohm, Christian Wilhelm von, *De la réforme politique des Juifs* (Paris: 1782).

[28] Clermont–Tonnerre, "Speech on Religious Minorities and Questionable Professions" (23 December 1789), in Hunt Lynn, ed., *The French Revolution and Human Rights: A Brief Documentary History* (New York: St. Martin's, 1996), 86–88.

specified the terms of the deal. The first and second articles of the Declaration stated that all men are free and equal by nature and must be treated as such by their government. With equal rights came equal obligations, and the third article of the *Declaration* demanded that no public authority be allowed which does not proceed directly from the nation. Many remained skeptics of the willingness of the rabbis to abide by this third article. A distinction was added in interpreting the *Declaration* between active citizens who shared tangible interests in the nation and deserved full rights, and passive citizens who did not and deserved not. The Jews were included among the second group, together with women, children, and foreigners. It was only during the closing days of the National Assembly, in late 1791, that a decree of complete emancipation was finally passed on the ground that it was impossible to give all men equal rights, except for the Jews.[29] But the questions remained open. Will the Jews accept the deal? Will their rabbis relinquish control over Jewish public life to the secular state? Can they give up their Messianic dreams for the secular promises of liberté, égalité, et fraternité?[30]

15.2 Nineteenth-century Science and the Jewish Question

The nineteenth century saw the emancipation of both Jews and science. Buoyant on the tides of industrialization, secularization, urbanization, education, and professionalization, Jews and scientists alike broke out of their isolated quarters and developed brilliant careers in the academy, the arts, and the economy. The Jewish emancipation spread early in the nineteenth century on the boots of Napoleon's soldiers. Napoleon, like Friedrich II, was not known for his love for the Jews but his policies, like those of Joseph II, reflected the viewpoint of an enlightened despot who treated religious and ethnic differences as an obstacle to his efforts to pacify and optimize the utility of his multi-ethnic populace.[31] "By putting them [the Jews] upon an equality with Catholics, Protestants, and others, I hoped to make them become good citizens, and conduct themselves like others of the community," Napoleon explained in his memoirs. "My intention was to render everything belonging to the state in the Constitution, purely civil and independent of any religion. I

[29] "Admission of Jews to Rights of Citizenship," 27 September 1791, In Hunt (1996) 99–101.

[30] Schechter, Ronald, "The Jewish Question in Eighteenth-Century France," *Eighteenth-Century Studies* (1998) 32:84–91.

[31] In private correspondence Napoleon referred to the Jews as "the most despicable of mankind." See the letter to his brother Jerome Napoleon dated 6 March 1808, in *New letters of Napoleon I* (Paris, 1898). See also Weider, Ben, "Napoléon et les Juifs," Paper given at *Congrès de la Société Internationale Napoléonienne*. Alessandria, Italy, June 21–26, 1997. https://www.napoleon-series.org/ins/weider/c_jews.html. Accessed 1 May 2024.

wished to deprive their priests of all influence and power in civil affairs, and to oblige them to confine themselves to their own spiritual matters and nothing else."[32]

Much like with the metric system, the Napoleonic Empire substituted a unified French civil code, newly revised in 1804 to reflect the universal principles of the revolution, for the previous patchwork of local traditions. The Napoleonic Code, as it came to be known, allowed freedom of religion, forbade privileges based on birth, and required government jobs to be allocated to the most qualified. Stressing uniform and accessible law, the code allowed a growing number of Jews to move out of their secluded quarters and enter the economic, social, and political life around them.[33] Once started, the process proved hard to reverse. The powerful European coalition assembled to stop Napoleon found it easier to defeat his mighty army than his popular civil code.[34] The growing apparatus of the nation-state—free and often compulsory education, economic and health regulation, legal restructuring, bureaucratic and professional certification—eroded the control of the Jewish establishment over its populace. The lifting of restrictions on choices of place and livelihood, the increase of physical and social mobility, and the expansion of commerce, industry, and finance allowed a growing number of Jews to move to the cities and prosper.[35]

The scientists, too, broke out early in the century from their exclusive laboratories, societies, and academies, and prospered. The Industrial Revolution demonstrated the long-promised benefits that could be had from the harnessing of nature. Universities added the task of creating new scientific knowledge to their core mission; engineering and medical schools multiplied to produce the cohorts of experts that could answer to the growing ambitions of the moderns; and national bureaus of standards were established to coordinate it all. New types of scientific practitioners emerged who relied on specialized training, hardware, and methods to promote higher standards of knowledge and performance. Disciplinary science was established, professional societies and journals for individual scientific disciplines appeared, and national associations agitated for the advancement of a society based on scientific principles.[36]

Early nineteenth-century British and French social reformers imagined society as a self-regulating system, reducible to its elements. They fancied themselves achieving for it what Newton had achieved for the physical universe. They analyzed

---

[32] O'Meara, Barry E., *Napoleon in exile; or, A Voice from St. Helena. The opinions and reflections of Napoleon on the most important events of his life and government, in his own words* (London: 1822) I: 183–84.

[33] Derenberg, Walter J. "The Influence of the French Code Civil on the Modern Law of Unfair Competition." *The American Journal of Comparative Law* (1955) 4(1):1–34.

[34] Napoleon referred to the Code as a greater achievement than all his victories: "Waterloo will erase the memory of so many victories; what nothing will erase, what will live on forever, is my Civil Code." Quoted in "Hellegold," GRUR (The German Association for Intellectual Property Law) 1931, S. 1301.

[35] Katz, Jacob. *Out of the Ghetto: The Social Background of Jewish Emancipation, 1770–1870* (Cambridge: Harvard University Press, 1973).

[36] Morrell, Jack and Arnold Thackray, *Gentlemen of science: early years of the British Association for the Advancement of Science* (Oxford: Clarendon Press, 1981).

human nature to its elements and synthesized stable social institutions by the careful balance of the universal forces of attraction and repulsion that controlled these elements.[37] Unfortunately, this Newtonian science of society provided little insight into the Jewish Question. Endorsing the Enlightened theories of human malleability and inevitable progress, Robert Owen campaigned for equality for the Jews, while Charles Fourier cautioned against a sudden and unconditional emancipation that would corrupt the French rather than reform the Jew. Henry St. Simón kept his views to himself, perhaps because many of his leading followers were assimilated Jews who were attracted by his vision of a new social order based on scientific administration and technocratic expertise.[38] The same could be said of Saint-Simon's former student and secretary, August Comte, whose mid-century social physics swallowed up these earlier systems. Like his former employer, Comte had little to say directly about the Jews, although his system associated Judaism with the primitive stage of human development when men blindly followed what they were taught by their ancestors.[39]

That was not the case east of the Rhine, in the German-speaking lands, where the question of Jewish emancipation continued to agitate the public sphere for the rest of the century. Important early nineteenth-century statesmen, such as Wilhelm von Humboldt and Karl August von Hardenberg, continued to promote Jewish emancipation as administratively rational, but their views ran against the growing conviction in the existence of German organic unity, incompatible with Jewish nature.[40] The notion of an organic society greater than the sum of its parts was undergirded by a philosophy that rejected mechanistic science as barren and sought a reflexive contemplation of natural phenomena via historical, aesthetic, and moral principles.[41] The pinnacle of this contemplative science was Hegel's comprehensive system that dominated German imagination during the middle third of the nineteenth century. Hegel's dynamic science of history taught that every people has a historical task to perform, and once a people performs its task it loses the reason for its existence and exits the world stage. The Jewish people fulfilled its mission by producing its antithesis—Christianity, the most perfect embodiment of divine reason. Next, the

[37] Manuel, Frank E. *Utopian Thought in the Western World* (Cambridge: Harvard University Press, 1979) Part 6: "The union of Labor and Love," 581–697.

[38] Silberner, Edmund, "Charles Fourier on the Jewish Question," *Jewish Social Studies* (1946), 8(4): 245–266. Szajkowski, Zosa, "The Jewish Saint-Simonians and Socialist Antisemites in France," *Jewish Social Studies* (1947) 9(1): 33–60.

[39] Comte, Auguste, *The Positive Philosophy of Auguste Comte* (London: Kegan and Trübner, 1893, 3rd ed) 2:210–12. Manuel, Frank E., and Fritzie P. Manuel. "The Battle of the Systems." In Manual (1979) 581–589.

[40] Arendt, Hannah, "Privileged Jews," *Jewish Social Studies* (1946) 8:3–30. Baron, Salo, Die Judenfrage auf dem Wiener Kongress: auf Grund von zum Teil ungedruckten Quellen dargestellt (Wien: Löwit, 1920). Lawrence, Paul Rose, *Revolutionary Antisemitism In Germany From Kant to Wagner* (Princeton University Press, 1993).

[41] Poggi, Stefano; Maurizio Bossi, eds. *Romanticism in Science: Science in Europe, 1790–1840* (Boston: Boston Studies in the Philosophy and History of Science, 1994).

Jewish people was expected to disappear by assimilation. Alas, not only did the Jews not disappear, but they seemed to thrive.

The anomaly was picked up in the early 1840s by a pair of young Hegelians who embraced their master's dialectical method but rejected his suggestion that the evolution of the human spirit achieved perfection in the modern Prussian state. The first was Bruno Bauer, a vocal champion of religious freedom in Prussian society who nevertheless supported the efforts to stall Jewish emancipation. In a pamphlet titled *The Jewish Question*, Bauer argued that Judaism and the modern state were inherently incompatible. To be free, argued Bauer, a society must separate religion from its public affairs and turn it into a private matter. Prussia, in its inevitable march towards perfection, will be able eventually to do so; however, Judaism, being based on a revealed law, could never be reduced to a private matter. Jewish emancipation was therefore two dialectic steps away—the Jews must assimilate, and the state secularized.[42]

Bauer's analysis of the Jewish Question prompted a sarcastic reply from the young Karl Marx who chastised Bauer for fighting yesterday's war.[43] Marx agreed with Bauer that no German was yet free, and that the Jews would do best to drop their self-centered quest and join their gentile brethren and emancipate themselves as human beings rather than as Jews. However, the Enlightenment war against religion was over. The industrial age had dawned, and the new fight for the soul of humanity was now with capitalism. The son of a converted Jewish family, Marx took sarcastic pleasure in arguing that in capitalistic Europe, Jewish emancipation was so successful that instead of the Jews turning Christian, the Christians turned Jewish—greedy, egotistic, and alienated. In Marx's view, it was society that needed to rehabilitate itself, not the Jews. By repenting its unholy capitalistic mentality, which allowed the Jews to prosper in the first place, the Jewish Question will disappear by itself.[44]

Marx's analysis reflected the changing mid-century terms of both science and the Jewish Question. On the one hand, the emancipated Jews were increasingly identified with the sins of capitalism instead of religion. Having entered the economic life around them, many Jews found it the easiest to secure an occupation in the thriving commercial and financial markets that enjoyed a longstanding Jewish presence and a rising demand from the rapidly expanding mid-century economy. Consequently, having overcome their religious segregation, the Jews found themselves newly

---

[42] Bauer (1843). See also Bauer, "Die Fähigkeit der heutigen Juden und Christen, frei zu werden," in Georg Herwegh, ed., *Einundzwanzig Bogen aus der Schweiz*, (Zürich, 1843), 56-71. For secondary sources, start with Rotenstreich, Nathan, "For and against Emancipation: The Bruno Bauer Controversy," *Leo Baeck Institute Year Book* (1959) 4:3–36.

[43] Marx, Karl. 1844. "Zur Judenfrage." In *Deutsch-Französische Jahrbücher*, Ruge Arnold and Karl Marx, eds. (Paris, 1844.) 182–214.

[44] Funkenstein (1979) 13–28. Peled, Yoav, "From Theology to sociology: Bruno Bauer and Karl Marx on the question of Jewish Emancipation," *History of Political Thought* (1992) 13(3): 463–485. See also Slezkine, Yuri, *The Jewish Century* (Princeton: Princeton University Press, 2004) 40–104.

recast as followers of another false god—Mammon—and judged, again, as morally degenerated and socially harmful.[45] On the other hand, social science too was found guilty by mid-century for its abstract tendencies and idealistic notions, which prevented any progress in the real world. The remedy was found in the empirical methods of disciplinary science that harnessed the human imagination and pointed it in the right direction. Marx, for example, borrowed Hegel's dialectics of state and society and furnished it with a concrete historical subject—the industrial proletariat—and with concrete empirical content taken from the economists' and socialists' literature. To do so, according to Marx, was to advance the quest for the developmental laws of society from a utopian discourse to a real science.[46] August Comte could not have agreed more.

Other social critics turned to other sciences for guidance. Linguistics, physical anthropology, comparative anatomy, embryology, and the theories of evolution and race have all offered evidence of historical continuities and hereditary links that constituted potent alternatives to the mechanistic interpretations of society. The scientific interest in race could be traced back to the failed efforts to catalog God's children. The great eighteenth-century taxonomist Karl Linnaeus failed to establish a clear morphological distinction between the human races and resorted instead to the anthropological classification of four 'varieties' of the *Homo sapiens*, defined primarily by geography: the American, the European, the Asian, and the African. As a natural historian, Linnaeus was careful to avoid moral analysis, but his prejudices often showed, for example when he described the Africanus as "crafty, sly, lazy, cunning, lustful, careless" and "governed by caprice."[47]

Linnaeus' French archrival, Comte de Buffon, no longer considered nature a divine creation nor its species fixed. Fascinated by the vision of a dynamic nature evolving in time, Buffon had no problem calling the human varieties 'races,' and putting them to work in explaining the development of human difference and diversity.[48] Buffon believed that the human races all degenerated from a single and superior Caucasian race and were differentiated by environmental factors such as weather, diet, and culture. The German naturalist Johann Friedrich Blumenbach also promoted the degeneration theory and claimed to establish with his superior

---

[45] Penslar, Derek Jonathan, *Shylock's Children: Economics and Jewish Identity in Modern Europe* (Berkeley: University of California Press, 2001); Katz, Jacob, *From Prejudice to Destruction: Anti-Semitism, 1700–1933* (Cambridge, Mass.: Harvard University Press, 1980).

[46] York, Richard, and Brett Clark. "Marxism, Positivism, and Scientific Sociology: Social Gravity and Historicity." *The Sociological Quarterly* (2006) 47(3): 425–50.

[47] Broberg Gunnar, "Linnaeus's Classifications of Man," in *Linnaeus. The Man and His Work*, Frängsmyr, Tore, ed. (Berkeley: University of California Press, 1983). Müller-Wille, Staffan, "Linnaeus and the Four Corners of the World", in *The Cultural Politics of Blood, 1500–1900*, Kimberly Anne Coles, Ralph Bauer, Zita Nunes and Carla L. Peterson, eds., (London: Palgrave MacMillan, 2015) 191–209.

[48] Georges Buffon, "Variétés dans l'espèce humaine," in *Histoire naturelle, générale et particulière, avec la description du Cabinet du Roy* (Paris: L'Imprimerie royale, 1749), t. 3, pp. 371–530. Farber, Paul L. "Buffon and the Concept of Species," *Journal of the History of Biology* (1972) 5(2): 259–284.

measurement techniques a clear anatomical distinction between the five human races.[49] Other naturalists supported the unity of the human species but argued that the races have evolved rather than degenerated. Others yet taught that the races had no common origin but were rather created separately, at different points of time, by spontaneous or divine creation.[50]

Fraught with inner debates, the nineteenth-century discourse on race developed in a somewhat incongruous fashion. While the origin of the races remained contested and their classification slippery, their attributes were increasingly called upon to undergird ethnic, cultural, and political theory.[51] By mid-century, influential accounts such as those of the historian Ernest Renan and Comte Arthur de Gobineau featured the human races as sui generis agents that drive human history, and a full-blown racial hierarchy with the white race on top of the other races, and the Aryans on top of the white race.[52] Racial interbreeding, with Semites in particular, was identified as the new and imminent danger facing the superior European civilization. The anonymous 1861 *Jews and the German State* was probably the first work to apply race theory directly to the Jewish Question. Protesting new statutes that allowed Jews to enter the civil administration, the author added to the familiar theological, philosophical, and economic arsenal, an evolutionary analysis of the innate immorality of the Jewish race, chronicled in the Bible and cultivated by centuries of religious and social isolation.[53]

Still, the 1860s were the heyday of liberal theory and Jewish emancipation, and the demand to consider the Jewish Question as a racial question remained limited to

[49] Blumenbach, Johann Friedrich, "On the natural variety of mankind," (1795), in Bendyshe, Thomas, trans and ed., *The anthropological treatises of Johann Friedrich Blumenbach* (London: Longman, 1865), 145–276.

[50] Paley, William, *Natural Theology or Evidences of the Existence and Attributes of the Deity* (London, 1802); Lamarck, Jean-Baptist, *Zoological philosophy* (Paris, 1809); Priestley, Joseph, "Observations and Experiments relating to equivocal, or spontaneous, Generation," *Transactions of the American Philosophical Society* (1809) 6: 119–129. Moore, Randy; Decker, Mark, and Cotner, Sehoya. Chronology of the Evolution–creationism Controversy (Santa Barbara, Calif.: Greenwood Press, 2010).

[51] Harris, Marvin, *The Rise of Anthropological Theory: A History of Theories of Culture* (Walnut Creek, California: Alta Mira Press). Brown, Ricardo, *Until Darwin, Science, Human Variety and the Origins of Race* (London: Pickering & Chatto, 2010).

[52] Biddiss, Michael D., *Father of Racist Ideology: The Social and Political Thought of Count Gobineau* (London, Weidenfeld and Nicolson, 1970). Renan, Ernest, *Le Judaïsme comme race et comme religion* (Paris, 1883) and *L'avenir de la science: pensées de 1848* (Paris, 1890). Gobineau, Arthur de: *The Moral and Intellectual Diversity of Races* (Philadelphia, Lippincott, 1856); Arvidsson, Stefan, *Aryan Idols: Indo-European Mythology as Ideology and Science* (Chicago: University of Chicago Press, 2006); Priest, Robert, "Ernest Renan's Race problem," *The Historical Journal* (2015) 58(1):309–330. Almog, Shmuel, "The Racial Motif in Renan's Attitude to Jews and Judaism", in Almog, Shmuel, ed., *Antisemitism Through the Ages* (Oxford: Pergamon Press, 1988) 255–278.

[53] The *Jews and the German State was* published in Berlin in 1861 under the pseudonym H. Naudh.

the radical wing of the conservative camp.[54] Physical anthropology was promoted as the study of human diversity, and cell theory was recruited to support the vision of a society organized by individual merit and voluntary union.[55] Lamarckianism celebrated the potency of the individual will; Spencer's cosmic laws explained how society improved via the processes of differentiation and specialization, and Darwinism, born in class-based England, was introduced into the flourishing German university system by progressive thinkers for whom biology was a weapon against superstition and inherited privilege.[56]

For Jews who grew up in Western and Central Europe at the height of Jewish emancipation, science became a vehicle for social mobility and assimilation. Jewish presence in the academy and the scientific professions quickly grew in the last third of the century. Prussia emancipated its Jews in 1866, and Germany's unification in 1870 brought full civil and political rights to all German Jews. By the 1880s, about one in five Garman Mathematics professors was jewish, and in Berlin and Breslaw—where Jews had the best showing—one of every four University teachers was Jewish. The numbers of Jewish faculty in medicine, chemistry, and physics were smaller but still significantly greater than their share percentage of the population. The same was probably true in the growing scientific industries as well. The embrace of science produced mixed results for the Jews, though. Their numbers grew in the universities and the professions, but remained concentrated in the lower ranks, most often in poorly paid positions. Ironically, the lingering academic reluctance forced Jewish scientists to the margins of nineteenth-century science, where many of them thrived and made important contributions.[57]

15.3 Fin-de-siècle Science and the Jewish Question

By the end of the nineteenth century, both science and Jews had accumulated enough cultural capital to be implicated with the spiritual and physical decay of the European society that hosted them.[58] As technical progress subverted the old order,

---

[54] Prussia emancipated its Jews in 1866, and Germany's unification in 1870 brought full civil and political rights to all German Jews.

[55] Spencer, Herbert, *The Principles of Biology* (London: William and Norgate, Vol. 1, 1864); McNeely, Ian F., *Medicine on a Grand Scale: Rudolf Virchow, Liberalism, and the Public Health* (London, Wellcome Trust, 2002).

[56] Sheehan, James, *German Liberalism in the Nineteenth Century* (Chicago: University of Chicago Press, 1978); Weindling, Paul, *Health, Race and German Politics between National Unification and Nazism, 1870–1945* (Cambridge: Cambridge University Press, 1989).

[57] Efron, Noah, *Judaism and Science: A Historical Introduction* (London: Greenwood, 2006), 157–174.

[58] Heilbron, John L., "Fin-de-sicle physics," in *Science, Technology and Society in the Time of Alfred Nobel*, Bernhard C.G, Crawford E., and P. Sorbom, eds. (Oxford, Oxford University Press, 1982) 51–73. See also Staley, Richard, "The fin de siècle thesis," *Berichte zur Wissenschaftsgeschichte* (2008) 31(4):311–330.

the champions of yesterday joined forces with the poets to declare modern science morally bankrupt and culturally dangerous. They argued that the mechanistic philosophy robbed humans of their soul and Nature of her mystery. That by attacking the precepts of religion and the evidence of the soul, and by destroying belief in purposeful creation, science menaced morality. The theory of evolution robbed humans of their moral compass and demoted them to the animal kingdom, red in tooth and claw.[59] Human history no longer marched toward perfection, and the individuals were no longer born *tabula rasa* and malleable but rather hardwired by the history of their race. The notion of society as an organism facilitated a lively synthesis of biological and social theories. Embryology became an extension of archeology, criminology of anatomy, and the Jewish Question of race theory. Some worried that if ethnic and moral characters are fixed in the blood by centuries of evolutionary pressures, then Jewish assimilation may take too long, if at all. Others worried to the contrary, that Jewish assimilation has spread too rapidly for their host societies to adapt and urged the authorities to clamp down on Jewish assimilation and its degenerative course.[60]

Confidence in the assimilation project was further eroded by the escalating Jewish crisis in the Russian empire, home to the overwhelming majority of the Jewish people. Facing the sudden addition of several million Jewish subjects following the late-eighteenth-century partitions of Poland, the Tsarist regime established the so-called *Pale of Settlement* that stretched from the Baltic to the Black Sea in the western part of the empire, and forbade its Jewish subjects to live and work outside the Pale without a special permit. Early in the nineteenth century, some Tzarist policies still aimed to integrate the Jews in the Pale into Russian society by the enlightened means of reform and education, but these half-hearted efforts were overwhelmed by popular discontent and official allegations of Jewish corruption and economic exploitation.[61] The second half of the century saw a surfeit of legal restrictions that perpetuated Jewish segregation and made the proposed integration difficult, if not impossible.[62] By the 1880s the Pale of Settlement faced a devastating combination of population explosion and extreme poverty. Close to a million Jews left Eastern Europe by the century's end, mostly to the New World, and the several millions who stayed behind faced increasingly virulent forms of persecution fueled

---

[59] Tennyson, Alfred. In Memoriam A.H.H. London: Edward Moxon, 1850. Canto 56.

[60] Marr, William, *Der Sieg des Judenthums ber das Germanenthum* (Bern, 1879); Dühring, Eugen, *Die Judenfrage als Racen, Sitten und Culturfrage: Mit einer weltgeschichtlichen Antwort* (Leipzig: Reuther, 1881); Katz, Jacob, "The Preparatory Stage of the Modern Antisemitic Movement (1873-1879)," in Jacob Katz and Almog, Shmuel, eds., *Antisemitism Through the Ages* (Oxford: Pergamon Press, 1988) 279–290.

[61] Klier, John Doyle, *Russia Gathers Her Jews: The Origins of the "Jewish Question" in Russia, 1772–1825* (De Kalb: Northern Illinois University Press, 1986).

[62] Demidov, Pavel Pavlovich, *The Jewish Question in Russia* (London: Darling, 1884). Klier, John Doyle, "The Jewish Question in the Reform Era Russian Press, 1855-1865." *The Russian Review* (1980) 39(3): 301–319.

from below by the longest economic depression in Europe's history, and from above by popular nationalist ideology and race theory.[63]

The escalating Jewish crisis in Eastern Europe furnished the Jewish Question with a new sense of urgency. "Shut your eyes and hide your head like an ostrich," cried out in 1882 a popular Jewish pamphlet, following the latest round of pogroms in Russia. "There is to be no lasting peace unless in the fleeting intervals of relaxation you apply a remedy more thorough than those palliatives to which our hapless people have been turning for 2000 years."[64] The author, the Jewish physician Leon Pinsker, gave up on his assimilation dream, i.e. the Russification of the Jews. Calling upon his modern education at the University of Moscow, Pinsker diagnosed antisemitism as an inheritable mental disease, the pathology of a common human phobia, the fear of ghosts. Having no land of their own, the Jewish people was like a spirit without a body, antisemitism naturally followed, and no remedy was to be found but the creation of a new body in the shape of a Jewish nationality, living on its own soil.[65]

The biggest Jewish philanthropist of the day, the Bavarian Jewish Baron Maurice Hirsch, also gave up on Europe but not on the assimilation project. One of the most innovative financiers of his time, Hirsch amassed a vast fortune in the booming mining and railway economies and dedicated much of this fortune to promote jewish assimilation.[66] In 1870, Hirsch settled in Paris and proceeded to develop from there an international empire of Jewish philanthropy on the fringes of Europe that supported modern schools and vocational training in the Ottoman cities, the French colonies of North Africa, the Russian Pale of Settlement, and the Habsburg provinces of Galicia and Bukovina. Hirsch considered the traditional forms of Jewish charity as only perpetuating Jewish degeneration. Instead, Hirsch aimed to reform his unfortunate coreligionists into productive citizens by teaching them skills and settling them on the land as small farmers. By the end of the 1880s, the rise in political persecution and legal restrictions in Eastern Europe led Hirsch to shift his focus to the New World, where good land was plentiful and "the laws and religious tolerance permitted the Jews to carry on the struggle for existence as noble and responsible subjects of a human government."[67]

---

[63] In 1881 the Jewish population in the Russian empire had reached four million, 4% of the total population. One million lived in the Kingdom of Poland, where they made up 13% of the inhabitants; 2.9 million, or 12.5% of the total population, lived in the fifteen Russian provinces of the Pale of Settlement, with the heaviest concentration in the nine former Polish provinces. See Polonsky Antony, "The Position of the Jews in the Tsarist Empire, 1881–1905," in *The Jews in Poland and Russia* (Liverpool: Liverpool University Press, 2010) 2: 2–39.

[64] Pinsker, Leon, *Auto-Emancipation: An Appeal to His People By a Russian Jew* (1882). Schoeps, Julius, *Pioneers of Zionism: Hess, Pinsker, Rülf: Messianism, Settlement Policy, and the Israeli-Palestinian Conflict* (Berlin: de Gruyter, 2013) 33–58.

[65] Ibid.

[66] Lehmann, Matthias, *The Baron: Maurice de Hirsch and the Jewish Nineteenth Century* (Stanford: Stanford University Press, 2022).

[67] Baron de Hirsch, "My Views on Philanthropy," *North American Review* (July 1891) 153: 1–4.

In 1891 Hirsch established the *Jewish Colonization Association* [JCA] to support Jewish emigration from Eastern Europe to the New World. Hirsch registered the JCA in London as a charitable organization but ran it tightly, in a businesslike fashion, from his Parisian residency. The JCA trained selected immigrants on large, self-sustaining farms. Then, once the candidates had become accustomed to agricultural life, the JCA settled them on lands it purchased in the New World, mostly in Argentina. The JCA provided the settlers with long-term leases, expert guidance, and sufficient credit to build a farm and stock it with the means necessary to become independent farmers and productive members of their new nations. In return, the JCA expected the farmers to fully pay back their loans with interest. All obligations and timetables were clearly stated and signed in personal legal contracts. Those who failed to keep up were quickly written off. Such strictness, Hirsch insisted, served a higher purpose—to build up a healthy Jewish farming class who till their own land, love their freedom, and are patriotic citizens of their new country.[68]

Notably, Hirsch kept the JCA out of Palestine, which he considered unsuitable for large-scale Jewish colonization. In 1891, he refused to support the plans of Pinsker's Russian organization, the *Lovers of Zion*, to establish Jewish colonies in Palestine, cautioning that "religious memories and ancient traditions" were not a solid basis for establishing a new homeland.[69] Four years later, in 1895, Hirsch rejected a similar request from the 35-year-old Viennese journalist Theodor Herzl. Herzl's plan owed much to the example set by the JCA of organized immigration to a new territory, away from Europe, where the Jews could reinvent themselves as modern subjects. The problem, according to Herzl, was that Hirsch did not think big enough. "How many specimens can you transport over there anyway? Fifteen to twenty thousand? More Jews than that live on one street in the *Leopoldstadt*."[70] No, a different scale of action was needed to solve the Jewish Question—an enormous immigration project, unprecedented in modern history, expertly planned and executed. "The Exodus under Moses," wrote Herzl, "bears the same relation to this project as a Shrovetide Play by Hans Sachs does to a Wagner opera."[71]

In Herzl's analysis, the enormity of scale required a new approach to the Jewish Question. To start with, no land, not even Argentina, was vast enough and no society was tolerant enough to assimilate the Jews *en masse* without triggering a new cycle of antisemitism. Jewish society would need then to find a new territory where it could refurbish itself and become a modern nation. Secondly, no amount of money in the world—not even Hirsch's legendary fortune—was enough to buy the Jewish masses one by one. To mobilize the Jewish masses, one needed to appeal to their collective imagination and passions by indirect means: "patriotic songs, the

---

[68] Norman, Theodore, *An Outstretched Arm: a history of the Jewish Colonization Association* (Boston: Routledge, 1985).

[69] Quoted in Matthias Lehmann, "Baron Hirsch, the Jewish Colonization Association and the Future of the Jews," *Jewish Studies Quarterly* (2020) 73–102, on p. 91.

[70] *The Complete Diaries of Theodor Herzl*, Patai, Raphael, ed. (London: Yoseloff, 1960) 21.

[71] Ibid, 38, June 7, 1895. The Shrovetide Play was a popular street-theatre form in early modern Germany.

Maccabees, religion, heroic stage plays, honor, etc."[72] Herzl tried to convince Hirsch that only Palestine, the mythic homeland, had the symbolic capital required to mobilize the Jewish masses and that only he, Herzl, could channel the human energy induced into an efficient political force. Hirsch thought otherwise. Herzl's proposal to establish a Jewish international organization that would press for the massive colonization of a national Jewish territory seemed to Hirsch a dangerous fantasy that would endanger, among other things, the discreet negotiations required to facilitate Jewish immigration and colonization in the New World, where assimilation was perhaps still possible.

As an elitist who did not trust what he termed "the irrational psychological dynamism of mass politics," Herzl preferred to initiate his mass movement from above.[73] It was only after the top Jewish philanthropists turned him down that he shifted his effort to mobilize his mass Jewish movement from below. "My plan does not depend on you alone," he wrote Hirsch. "There are others. There are, alternately and above all, the Jewish masses, and I shall know how to get across to them. This pen is power."[74] Half a year later, in February 1896, Herzl published *The Jewish State: An Attempt at a Modern Solution of the Jewish Question*. The assimilation project failed, trumpeted Herzl in the preface. Fin-de-siecle Europe had become a dangerous place for all Jews and the only remedy was to find a new place, far away from Europe, where they could regenerate and rejoin the trunk of humanity.

The bulk of the *Jewish State* outlined Herzl's plan for the Jewish colonization of Palestine. The plan closely followed the playbook of Victorian colonialism.[75] The first step was to secure a charter from the Ottoman Sultan for a land development company to operate in Palestine. With a charter in hand, a Zionist organization would proceed to secure investment capital. Once the capital was secured, an administration would be established to execute a survey of Palestine's natural resources and develop a rational plan for land improvement and distribution. Jewish immigration would then proceed on a large scale over several decades, in accordance with the best scientific principles. Starting afresh in a new land, unrestricted by previous investments in obsolete political and technological infrastructures, the Jews would be free to unleash the unprecedented powers of modern science and technology and shape a new and exemplary Jewish society.

Herzl's doctorate was in law, but his vocation was writing—from journalism to theatrical plays to science fiction which earned him the nickname 'the Jewish Jules Verne'.[76] *The Jewish State* showed his many talents. Herzl interlaced the disciplined

---

[72] Ibid.

[73] Schorske, Carl, "Politics in a New Key: An Austrian Triptych," *Journal of Modern History* (1967) 39: 345–86.

[74] *The Complete Diaries of Theodor Herzl*, 38, June 7, 1895.

[75] Galbraith, John S. *Crown and Charter: The early Years of the British South Africa Company* (Berkeley: UC Press, 1974). Kayyali, A., "Zionism and Imperialism: The Historical Origins," *Journal of Palestine Studies*, 1977, 6(3) 98–112.

[76] *The Complete Diaries of Theodor Herzl*, 287. February 1, 1896. Jules Verne was a famous French novelist who wrote what one might refer to as fin-de-siècle science fiction. To this day, Verne is still

prose of legal and financial planning with romantic appeals to mythical longings and ancient customs, including a detailed description of the new Jewish flag—white, with seven golden stars. The combination was extremely successful. *The Jewish State* was an overnight sensation, and its author became a household name in the Zionist circles. Quick to capitalize on the momentum of *The Jewish State*, Herzl organized the first World Zionist Congress in August 1897, in Basel, Switzerland. There, before more than 200 ecstatic Jewish delegates from all over the globe, Herzl announced the establishment of the World Zionist Organization [WZO] and invited the European powerhouses to support the Zionist plans for the large-scale Jewish colonization of Palestine.[77]

Herzl's critics compared him to the notorious Sabbatai Zevi, the false seventeenth-century Messiah who brought much pain on the Jews. In reply, Herzl claimed that the tremendous technoscientific development since the seventeenth century made a second exodus possible—mobilized once again by Jewish misery but led this time by the modern deities of capital and science. Herzl conceded that Jewish nature had been corrupted by centuries of abuse, but was confident that with proper design and management, the accumulated Jewish despair would drive the wheels of the Zionist movement like the steam drives those extraordinary "slaves of modernity," the machines. Herzl noted that Palestine's nature was also corrupted by centuries of abuse, but was equally confident that with adequate capital, scientific planning, advanced machinery, and expert supervision, the great modern experiment of replanting Jewish society in Palestinian soil could be successful. "The word 'impossible,'" he wrote at the end of *The Jewish State*, "ceased to exist in the vocabulary of technical science. Wherever the moderns appear with our inventions, we transform the desert into a garden."[78]

15.4 Zionism, Science, and the Jewish Question

Unlike other late-nineteenth-century ideologies such as Marxism, positivism, and eugenics, Zionism was never promoted as a science. The Zionist Movement was too practical to align itself with a specific ideology. Instead, it was conceived as an urgent project, to provide a workable solution to the Jewish Question.[79] Nevertheless, science and technology were central to the Zionist project from the very start. By applying the powers of science, the Jews would prevail over nature, refurbish their old homeland, and become self-determined again. By sharing the universal values of science, the Jews would prevail over their inner nature, free themselves of their

---

the most translated French author in the world.

[77] Straus, Nellie, tr., *The Congress Addresses of Theodore Herzl* (New York: Federation of American Zionists, 1917) 7–8.

[78] Herzl, Theodor, *The Jewish State* (New York: Dover, 1988) 90.

[79] Funkenstein, Amos, "Zionism, Science, and History," in idem, *Perceptions of Jewish history* (Berkeley: University of California Press, 1993) 338–350.

oppressive religion, and reinvent themselves as modern subjects. These two redemptions—of Palestine and Jewish society—were co-dependent. Both were necessary if the Jews were to reinvent themselves and rejoin the main trunk of humanity as a modern nation.

Science also furnished the Zionist project with a moral narrative, as the agent of progress to the Holy Land and the uncivilized Levant. This moral narrative stood on two legs. The first was the picture painted by archaeologists, geographers, historians, and biblical scholars of a prosperous biblical civilization later ruined by centuries of abusive Arab rule. The second was the Victorian ladder of civilization constructed by evolutionists and anthropologists who ranked the Arabs far below the Western-minded Jews. Put together, these two paradigms allowed the Zionists to argue that any hope for a better future for Palestine lay solely with the returning Jews and the capital, science, and energy they would bring with them.[80]

Thirdly, and importantly, science offered Herzl more than a specific solution to the Jewish Question alone. Herzl considered Jewish history to be shaped by the same evolutionary forces that controlled the rest of humanity, and the unfortunate state of the Jews as the inevitable outcome of a certain chain of historical circumstances—political, social, economic, legal, and religious—which may be reversed if understood well enough. The Jewish Question, he believed, was but a branch, early canalized, of the greater social question. The solution to the Jewish Question could therefore point to a more general solution for the troubles of fin-de-siècle Europe at large. La déité du jour, *l'e'lectricité*, wrote Herzl in the preface to *The Jewish State*, "was not invented for the purpose of illuminating the drawing-rooms of a few snobs, but rather for the purpose of throwing light on some of the dark problems of humanity. One of these problems, and not the least of them, is the Jewish Question."[81]

Any general solution for the troubles of fin-de-siècle Europe had to deal with the implications of universal suffrage and the rise of populist parties that sought political power through the ballot box. For Herzl, the problem was doubly pressing. First, as a liberal. Second, as an assimilated Jew who considered the great Jewish mass that lay to the East, at the Pale of Settlement, as a primitive, superstitious, and dogmatic society; a prime example of the physical and spiritual atavism caused by centuries of wretchedness.[82] Herzl was keenly aware that his solution to the Jewish Question would have to contend with these atavistic tendencies of the Jewish masses. "I will not agree that the Jewish commonwealth I wish to found will be small, orthodox, and illiberal," he wrote Baron Rothschild. "For three years I have labored on a coherent response to these and similar misgivings."[83]

---

[80] Said, Edward, *Orientalism* (New York: Pantheon, 1978). Hentsch, Thierry, *Imagining the Middle East* (New York: Black Rose Books, 1992). Norris, Jacob, *Land of Progress: Palestine in the age of Colonial Development, 1905–1948* (Oxford: Oxford University Press, 2013). Gilbert, Martin, *Churchill and the Jews* (London: McClelland, 2007).

[81] Herzl, *The Jewish State*, 41.

[82] Kornberg, Jacques. "Theodore Herzl: A Reevaluation." *The Journal of Modern History* (1980) 52(2): 226–252.

[83] *The Complete Diaries of Theodore Herzl*, 1347, August 22, 1902.

Herzl detailed his response in his 1902 utopian novel *Altneuland*. The novel portrayed a future Jewish society in Palestine as an implacably rationalistic society that solved the problem of the degenerated Jewish masses through the suspension of mass politics. Altneuland was ruled by a technocratic elite that undercut the political power of the masses. The Altneulanders elected a congress and a president, but the congress sat only a few weeks per year and the presidential post was purely honorary. The key decisions were taken by politically neutral experts who replaced professional politicians at the centers of power. Labor, production, and consumption were all rationally organized by making credit freely available, allowing workers' associations access to interest-free capital, and by state-controlled industry. Hunger, want, unemployment and disease were all eliminated, and security provided to the young, the old, and the sick. Each man had the right to the full fruits of his labor, and a state-regulated mixture of capitalism and socialism provided incentives and rewarded merit but avoided gross inequalities, and, with it, also the dangers of both socialism and Jewish orthodoxy.[84]

Not all Zionists warmed up to the Herzlian vision of a technocratic Jewish society. Many were ready to embrace the instrumental and moral powers of science, but not its ideology and cultural supremacy. Most acerbic among them was Achad-Ha'am, the intellectual leader of the *Lovers of Zion*, who ridiculed *Altneuland* and Herzl for blindly mimicking the Gentiles and leaving nothing that was authentically Jewish in their utopian society.[85] The German Jewish intellectual Martin Buber similarly lamented Herzl's failure "to realize how mistaken it was to consider technical progress, even the greatest, a Renaissance."[86] These critiques put on display the new dialectics that evolved within the Zionist movement between those who aspired for a national rebirth based upon authentic Jewish revival, and those who agreed with Bauer and Marx that the Jews would do better to emancipate themselves as human beings rather than Jews. Imagining Jewish sovereignty, the Zionists came to face the conflict that has long resided at the heart of the Jewish Question, between the ancient creed of Jewish uniqueness and the modern desire for normalcy. Could a Jewish state resolve this conflict? More than a century after the birth of Zionism, the Jewish Question remains undecided.

Tal Golan, Associate Professor of History of Science at the University of California, San Diego. His areas of specialization are science policy, and the relations between science, technology, and law. He is currently engaged in two research projects: Zionism and science, and the history of statistical evidence.

---

[84] Herzl, Theodor, *Altneuland* (Leipzig: Hermann Seemann Nachfolger, 1902).

[85] Achad-Ha'am, *Altneuland*, Ha'shelach (1903) 10:6. In Hebrew. Accessible online: https://benyehuda.org/read/5527. Goldstein, Y., "Eastern Jews vs. Western Jews: the Ahad Ha'am-Herzl dispute and its cultural and social implications," *Jewish History* (2010) 24:355–377.

[86] Buber, Martin, "If Herzl Were Still Alive," *Ha'aretz* (May 17, 1940) 9.

Chapter 16
Applied History of Science and the National Academy of Sciences

Peter Westwick

Abstract John Heilbron urged the application of history of science scholarship to science policy. The US National Academy of Sciences has had a major influence on science policy in the US, and studying its history illuminates the process of science policy formation and might thus help answer Heilbron's call. This paper shows how historians of science occasionally served on policy studies for the Academy, and how the Academy for a brief time in the 1970s elected historians of science to membership. Some of those historians challenged the Academy's ideal of objectivity and its faith in the fact/value dichotomy. In the process they helped the Academy come to grips with a general shift in its advisory philosophy, from a belief that its advisory committees were composed of disinterested experts to acknowledgment that scientific experts also had interests, and that the Academy's job was not to provide neutral experts, but rather to balance the interests within an advisory committee.

Keywords Science Academy · Science policy · Objectivity · John Heilbron · Thomas Kuhn

In his Distinguished Lecture to the History of Science Society in 1986, John Heilbron urged historians of science to apply their understanding to benefit science policy: "It is time to build the channels through which relevant and relevantly packaged research results of historians, philosophers, and sociologists of science and technology may flow to policymakers." The U.S. National Academy of Sciences, John perceived, offered a particular avenue to apply the history of science.[1]

Another type of applied history of science—a genre at which John tried his hand—is the history of a scientific institution, sometimes commissioned by the

---

[1] J.L. Heilbron, "Applied History of Science," *Isis, 78* (1987), 552–563, on 561. Archival citations in the notes are to the NAS archives, with specific file locations noted in parentheses.

P. Westwick (✉)
University of Southern California, Los Angeles, CA, USA
e-mail: westwick@usc.edu

S. Gattei (ed.), *The Incomparable Historian*, Archimedes 75,
https://doi.org/10.1007/978-3-032-25678-2_16

institution itself. Such applications can aim to celebrate of the institution in question, but another purpose should be critical, in the belief that the institution's scientists and leaders—and the general public—might benefit from learning its history. In the spirit of both celebration and critical reflection, the National Academy has commissioned histories of itself at 50-year intervals, and the third, honoring its 150th anniversary and undertaken by Ruth Schwartz Cowan, Daniel Kevles, and this essay's author, is now nearing completion. In addition to serving as its own example of applied history of science, the project sheds light on earlier instances of the Academy's applications of the history of science to science policy.

16.1 Historians of Science in the Academy

The U.S. National Academy of Sciences (NAS) was created by an act of Congress on March 3, 1863. Its model was the European science academy, which had emerged and proliferated two centuries earlier in the Scientific Revolution, as savants rejected the medieval universities that propagated scholastic philosophy. The rationalism and elitism of academic natural philosophers had meanwhile aligned with the bureaucratic values of the emerging modern state, and many academies accepted and benefited from official sanction. In the ensuing "Age of Academies," which lasted through the eighteenth century, the academy became the leading home for science, providing a bridge between the medieval university and the modern research university.[2] The creation of the U.S. Academy in 1863, after the Age of Academies had ended, might thus appear as an anachronism—like a young child trying to mimic its older siblings, not realizing that they had already moved on to new fashions.

The American academy, however, developed new ways to serve both science and the state. Its congressional charter included both honorary and advisory functions. The original act of Congress named fifty scientists as charter members of the Academy and allowed them to elect new members, an honor intended to elevate the profile of American science and promote its growth. The act also declared that the Academy "shall, whenever called upon by any department of the Government, investigate, examine, experiment, and report upon any subject of science or art."[3]

Over the next century the Academy grew substantially in the size of its membership and in the scope of its activities. Amid the mobilization of science for World War I the Academy created, at the direction of President Wilson, the National Research Council (NRC). The NRC's study committees were not confined to Academy members but rather could include any scientists and engineers from universities, industry, and the government, greatly expanding the range of expertise

[2] Roger Hahn, *The Anatomy of a Scientific Institution: The Paris Academy of Sciences, 1666–1803* (Berkeley, 1971), 2, 7–10, 43–45, 275; James E. McClellan III, *Science Reorganized: Scientific Societies in the Eighteenth Century* (New York, 1985), xix-xxiii, 254–259.

[3] Rexmond C. Cochrane, *The National Academy of Sciences: The First Hundred Years, 1863–1963* (Washington DC, 1978), act of incorporation on 595–596.

available to advise on federal policy. After World War II the Academy through the NRC became an engine for science policy, providing science advice to virtually every federal agency as well as to Congress.[4]

A prominent question ran through the Academy's history: how to define the "sciences" in its title and activities? The congressional charter did not specify; on the contrary, the directive to advise on "any subject of science or art" left a wide-open door. Whom should the Academy admit as members? Whom should it consult on matters of policy advice? Did engineers, physicians, or social scientists qualify? Science administrators? Or even, God forbid, historians of science?

The Academy first found historians of science useful in its advisory mission. The fledgling discipline of history of science grew in the postwar period in tandem with the study of science policy and was soon enlisted in its support. In 1963 the Academy created a Committee on Science and Public Policy (COSPUP) out of an earlier Committee on Government Relations. One of the first COSPUP reports, *Federal Support of Basic Research in Institutions of Higher Learning* (1964), employed A. Hunter Dupree as a consultant and included an extended historical account that drew on Dupree's recent book, *Science in the Federal Government*; the report thanked Dupree as well as the political scientist Don K. Price for "valuable contributions."[5] Two years later, a COSPUP proposal to Congress for a broader study of federal support of basic research listed Thomas Kuhn, I. Bernard Cohen, Cyril S. Smith, and Derek Price as potential consultants, but Congress declined to fund the project.[6] A decade later, an Academy review of US-Soviet science exchanges leaned heavily on the expertise of Loren Graham, a historian of Soviet science.[7]

The Academy proved quicker to consult historians of science than to admit them as members. At times it had elected members sympathetic to history of science: Joseph Fruton, for example, had been elected in 1952 for his scientific work before turning to history of science; similarly, Smith was elected in 1957 for his research in metallurgy. The Academy's president from 1963 to 1969, Frederick Seitz, had an interest in history of science, which may help explain COSPUP's use of Dupree; after retiring Seitz wrote biographies of scientists and a brief history of the Academy.[8]

---

[4] Daniel J. Kevles, *The Physicists: The History of a Scientific Community in Modern America* (Cambridge, Mass., 1995), 102–116; Cochrane, *History*, 209–216.

[5] NAS, Committee on Science and Public Policy, *Federal Support of Basic Research in Institutions of Higher Education* (Washington, DC, 1964), front matter. Dupree had testified to Congress in 1959 on science policy in response to Sputnik: Zuoyue Wang and Naomi Oreskes, "History of science and American science policy," *Isis, 99:2* (2008), 365–373, on 367.

[6] Harvey Brooks to Philip B. Yeager, 23 Nov 1965 (Committees & Boards: Com on Sc & Public Policy, General 1965).

[7] NAS Board on International Scientific Exchange review panel, draft minutes, 16 Sep 1975 (IR: Bd on Internatl Scientific Exchange: Panels: Review of US-USSR Exchanges & Relations 1975).

[8] Frederick Seitz and Nikolaus Riehl, *Stalin's Captive: Nikolaus Riehl and the Soviet Race for the Bomb* (Philadelphia, 1993); Seitz, *The Cosmic Inventor: Reginald Aubrey Fessenden* (Philadelphia, 1999); Seitz, *A Selection of Highlights from the History of the National Academy of Sciences, 1863–2005* (Lanham, MD, 2007).

Robert Merton was elected in 1968, part of a broad push in the late 1960s and early 1970s to include more social scientists—in part on the notion, backed especially by the next Academy president, Philip Handler, that the Academy's scientific advice to the government increasingly involved social issues such as health care, environmental pollution, and energy, and that the presence of social scientists would help the Academy render sound advice. This included the election of the first economists and sociologists and the creation in 1971 of a new membership group, Class V, for the behavioral and social sciences.[9]

The Academy, however, had not elected any members strictly for scholarly accomplishments in history of science. Its annual election process presented an obstacle to electing members in new fields. The Academy consisted of a few broad classes, such as physical and mathematical sciences, biological sciences, and behavioral and social sciences, with each class divided further into several sections. Each section nominated several candidates, which the class members then ranked and submitted to the overall membership. The Academy had an overall quota, set at 60 in the mid 1970s, with each class assigned a certain quota within that. With far more nominees that election slots, the annual voting was an extremely competitive process, with much politicking. Any field not represented by an existing section was thus effectively shut out of elections. The Academy had recognized this obstacle to new research frontiers and provided a loophole through so-called Temporary Nominating Groups, ad-hoc panels for certain fields—such as emerging interdisciplinary specialties—that could inject a few nominees into the pool of a particular class.

In 1975 Class V proposed creating a Temporary Nominating Group in history and philosophy of science.[10] The Academy's Committee on Membership Affairs voted it down. NAS vice president Saunders Mac Lane observed that "these are two separate fields with considerable controversy between them, so they could not easily be combined." Mac Lane added, "there probably are not as yet many people in this field who should now be elected to Academy membership."[11]

The NAS Council, however, also had the right to nominate a few candidates for membership, and it appointed an ad hoc committee under the mathematician Mark Kac to serve as a Temporary Nominating Group within the Council.[12] The Academy's archives do not provide access to election records, but the ad hoc committee seems to have nominated several historians and philosophers who subsequently won election, judging from the fact that several historians and philosophers of science became Academy members in those years: in 1977, Martin Klein, Otto Neugebauer, and Willard Quine; in 1978, Ernest Nagel and Patrick Suppes, with Joseph Needham elected as a foreign associate; in 1979, Thomas Kuhn. Even Mac Lane, who served

[9] Janice Goldblum, "Behavioral and social sciences within the NAS," c. July 1987 (NAS Council: Meetings: Agenda Aug 1987).

[10] James G. March to Allen Astin, 1 Jan 1975 (NAS Council: Agenda: Feb 1975); Ward Goodenough to Astin, 13 Feb 1975 (NAS Council: Meetings: Minutes Apr 1975).

[11] NAS Council minutes, 22 June 1975.

[12] NAS Council minutes, 11–12 Aug 1975 and 9–11 Aug 1976.

on the committee, seemed to warm to the idea of including historians and philosophers of science. He noted "considerable enthusiasm" for the field among some Academy members and "growing interest" outside the Academy, and he added that such scholars "would bring a neglected field of expertise" to the Academy.[13]

Temporary Nominating Groups, however, were just that: temporary. The Kac committee expired in 1979, at which point Fruton petitioned to create a permanent Academy section for History and Philosophy of Science. He originally included "Sociology" in the title of the section, but Merton thought that would dilute the "meager strength" of sociologists in their existing section, so Fruton confined it to history and philosophy. He informally surveyed 33 NAS members and one foreign associate (Needham) and got 28 replies. 22 of them agreed that the Academy should create a section; 3 said no (Abraham Pais, Tracy Sonneborn, and Eugene Wigner), and 3 were undecided. Of those who agreed, 12 said that they would transfer into a new section if created.[14]

The NAS Council acknowledged the "growing number of younger scientists in this area" but thought there were not enough to justify a new section, and it voted down Fruton's petition in April 1979. A straw vote of the entire membership at the annual meeting that month affirmed the Council's denial.[15] After Fruton stood up at the annual meeting and protested the negative vote, the Council, out of respect to him, revisited the decision in June—and reaffirmed it.

The problem, Handler explained to Fruton, was in the Academy's organization and election process. A section for history and philosophy of science would logically belong in Class V with the social sciences, but that class had a small allotment of slots in the annual quota and its nominating committee was "dominated by economists, anthropologists, sociologists, and psychologists," so candidates from the new section were unlikely to succeed. Handler argued instead that if a scholar made important contributions to the history or philosophy of a specific scientific discipline, he or she could just be elected by the relevant section—so a historian of astronomy, say, would go in the astronomy section. This ignored the fact that astronomers—or physicists, chemists, or biologists—were no more likely to give up one of their election slots to a historian than the social scientists were. That was Fruton's point: in order to get significant numbers of members of a new field into the

---

[13] NAS Council minutes, 25 Feb 1979.

[14] Fruton to "fellow-member of the NAS," 13 Nov 1978, and Fruton to David Goddard, 2 Jan 1979 and 9 Mar 1979 (NAS Council: Meetings: Agenda Apr 1979). Those who were willing to transfer into a new section: C.M. Brooks, S.S. Cohen, A. Cournand, J.T. Edsall, M.J. Klein, H.B. Glass, E. Nagel, J. Needham, O. Neugebauer, W.V. Quine, C.S. Smith, O. Temkin, plus Fruton. Eight of the thirteen were members of Class II, for biological sciences. Those who favored a new section but were not willing to transfer: S. Goudsmit, R. Guillemin, F. Haurowitz, G.E. Hutchinson, V.A. McKusick, E. Mayr, R.K. Merton, F. Seitz, O.H. Wangensteen, B. Witkop.

[15] NAS Council minutes, 21–22 Apr 1979.

Academy, the field needed its own separate section to provide nominations. Fruton, frustrated, gave up.[16]

History of science may have suffered over the battles fought earlier in the decade to elect more social scientists to the Academy; fatigue had set in, and the NAS Council was reluctant to wade once more into the fray. But history of science differed from social science, which many NAS members believed was a "soft" science unworthy of inclusion alongside natural science. History of science in fact had substantial sympathy from NAS members, perhaps because many of its practitioners—people like Fruton, Klein, Kuhn, and Neugebauer—had trained or practiced as "hard" scientists before turning to history or philosophy. History of science also of course had the potential to document science's achievements, perhaps to include those of Academy members. The problem was simply the unwillingness of NAS members to give up election slots that would otherwise go to scientists. The Academy earlier in the decade had gotten around this for the social sciences by increasing the annual quota for a few years, but it was not about to do so again for a small, marginal field.

16.2 Facts and Values

History of science proved itself useful to the Academy a few years later. The general issue was what counted as a scientific fact. The immediate issue was whether to legalize pot.

The increasing presence of science and technology in American life, together with the growing size and reach of the federal government, greatly expanded the Academy's advisory role in the decades following World War II. Academy studies on contentious policy issues, from national security to regulation of food, drugs, and environmental pollution, forced Academy members and leaders to confront the sausage-making reality of the policy process, as well as the realization that even scientists could have conflicting political interests and views.

The Academy had always assumed that its advisory committees were composed of disinterested experts. This Gradgrindian faith in science and reason came under challenge in the 1970s, not just from consumer advocates and policymakers who identified conflicts of interest among scientists, but also from academic scholars who pointed out the difficulty of separating scientific facts from social and political values. The election to the Academy of historians and philosophers of science both reflected these emerging concerns and reinforced them.

In 1982 an NRC advisory committee recommended changing federal law to prohibit only marijuana sales, thus decriminalizing marijuana possession. Frank Press, Handler's successor as Academy president, thought the report overstepped its

[16] NAS Council minutes, 24 June 1979; Fruton statement at NAS business session, 24 Apr 1979, Handler to Fruton, 4 May 1979, and Fruton to Handler, 14 May 1979 (NAS Council: Meetings: Agenda June 1979).

bounds by considering non-scientific factors, such as the cost of enforcement. Furthermore, Press wrote, existing scientific data was "insufficient to justify on scientific or analytical grounds changes in current policies .... I am concerned that the Committee may have gone beyond its charge in stating a judgment so value-laden, that it should have been left to the political process."[17]

Thomas Kuhn, who was not on the study committee, responded to Press with a long critique of the fact/value dichotomy. Kuhn noted that the "conceptual separation of scientific facts from values and the correlated insistence that NRC committees are fully competent only with respect to the first are standard parts of Academy discourse." However, Kuhn argued, the government usually asked the Academy's advice on topics "at the very edge or over the edge of current scientific research .... Committee members must rely on judgments (educated guesses, if you will), and it is likely that even the most expert will disagree even if they all share relevant values." Scientific research, Kuhn continued, involved many personal factors: professional history ("where and by whom trained, past research experience"); personality ("introvert vs. extrovert, optimist vs. pessimist, risk taker vs. conservative"); and other values, such as "preferring precision to broad scope, or cognitively fundamental results to socially useful ones."[18]

This line of reasoning, Kuhn observed, had two troublesome implications. One was that "disagreement is often inevitable, that it need not derive from values, and that, when it does, the role of values may not be eliminable." The Academy "promises clients a sort of neutrality which is often in principle unobtainable and thus exposes committee members to demands that strain integrity." The second was that scientists could claim they just supplied the means, and society chose the ends. "To invoke value neutrality or freedom from individual idiosyncrasy where it cannot exist is to deny personal responsibility and thus shield oneself from the consequences of a decision or recommendation."[19]

Press liked Kuhn's letter so much that he shared it with the Academy Council and invited Kuhn to give a seminar on the topic at the Academy. Press agreed with Kuhn that the fact/value dichotomy was often false, but he argued that the marijuana study had so little scientific content as to be almost entirely value-driven and thus inappropriate for the Academy. Kuhn replied with sympathy, but he added, "how on earth is one to draw the line?"[20]

Press seems to have absorbed Kuhn's admonition about value-laden facts. As for drawing the line, that was an issue of pragmatic judgment. Five years after his exchange with Kuhn, Press warned the NAS Council that certain recent NRC

---

[17] National Research Council, Committee on Substance Abuse and Habitual Behavior, *An Analysis of Marijuana Policy* (Washington DC, 1982), including transmittal letter of Press to William Pollin (director, National Institute on Drug Abuse), 21 June 1982.

[18] Thomas S. Kuhn to Press, 18 June 1982, emphasis in original (NAS Council: Meetings: Agenda Aug 1982).

[19] Kuhn to Press, 18 June 1982.

[20] Press to Kuhn, 24 June 1982, and Kuhn to Press, 11 July 1982 (NAS Council: Meetings: Agenda Aug 1982).

studies (such as reports on adolescent pregnancy and on AIDS) were "laden with social, ethical, and political value." The Council concluded that the Academy had pursued a "reasonable course" but that "continuing vigilance" was needed.[21]

History of science made subsequent, limited inroads at the Academy. Press had concluded to Kuhn, "I used to cite the example of measuring the charge of an electron as an activity free of values until one of our members proposed that he had observed a fractional charge in a controversial experiment that has not yet been repeated." Just a few years earlier, Gerald Holton had published an article on the presuppositions permeating Millikan's famous oil-drop experiment that had clinched the scientific debate in favor of unitary electron charge. Holton's article evidently had not reached Press.[22]

The Academy did, however, enlist Holton—who was not an Academy member—in the writing of a white paper on research ethics, "On Being a Scientist." The short pamphlet, intended for wide circulation to current and prospective scientists (especially graduate students), responded to the controversies at the time over scientific fraud, highlighted by accusations of scientific misconduct against the biologist David Baltimore, a Nobel laureate and Academy member. Holton provided historical examples of the scientific method, and the pamphlet—published in 1989, with 200,000 copies distributed to undergraduate and graduate science students—became one of the Academy's most popular publications. Its third edition, the most recent, was published in 2009.[23]

Holton's input, like Kuhn's intervention, was a good example of applied history of science. It also indicated that such applications did not depend on the Academy admitting historians of science. The National Research Council, the institutional innovation of World War I, enabled the Academy to enlist non-members to serve as experts on its advisory studies, and several historians of science besides Holton have contributed in recent decades to Academy studies on subjects such as the use of models in regulatory policy (Naomi Oreskes) and science and the law (Daniel Kevles).

16.3 History and Science

The Academy has not repeated the experiment of the 1970s and elected historians of science. It appears to have revisited the idea occasionally only to run into the same old problem: scientific sections were unwilling to give election slots to historians over scientists.[24] It may not help that in recent years it seems that fewer histo-

[21] NAS Council minutes, 13 Feb 1987.

[22] Gerald Holton, "Subelectrons, presuppositions, and the Millikan-Ehrenhaft dispute," *Historical Studies in the Physical Sciences, 9* (1978), 161–224.

[23] NAS, Committee on the Conduct of Science, *On Being a Scientist* (Washington DC, 1989); NAS Council minutes, 18 Oct 1989.

[24] John Brauman interview with Ruth Schwartz Cowan and Peter Westwick, 16 Mar 2017.

rians of science have PhDs in science than the Kuhnian cohort, which could make them less palatable to Academy scientists. The tradeoff—more historians of science trained in history than in science, at the price of exclusion from the Academy—might be acceptable. Other scientific academies are less exclusive. The modern Accademia dei Lincei embraces history and other liberal arts, Israel's academy includes humanities alongside the sciences, the Chinese and Russian academies of science have institutes for the history of science, and the Royal Swedish Academy of Sciences has elected as members Tore Frängsmyr and one J.L. Heilbron.

The US Academy's ongoing engagement with historians of science realized two Heilbronian desiderata. First, it helped historians of science to practice applied history, giving them an avenue to inform the Academy's attitudes towards science advising, à la Kuhn, and the scientific method, à la Holton, in addition to advising on the policy issues studied by particular Academy committees.

Welcoming historians into the work of the National Academy of Sciences also helps to restore an organic connection. John highlighted the common roots of the natural sciences and history in the early modern period, when new methods of doing science and of writing history challenged and upended the medieval worldview. These common roots included the collecting of information, distinguishing between primary and secondary qualities or sources, comparing data and establishing connections and correlations, testing hypotheses to arrive at theories about the workings of the world, and the role of good taste in deciding between competing theories.[25]

Science academies helped effect the subsequent separation of history from science. The Florentine Academy under the Medici in the fifteenth century took an inclusive approach that encompassed philosophy and the humanities, but its successors, starting with the Accademia dei Lincei, for the most part narrowed their focus to natural philosophy amid the Scientific Revolution. The Paris Academy of Sciences, for example, derived from competing plans: one utilitarian, to help humans control their environment, with members drawn from practical arts such as agriculture, hydraulics, and metallurgy; the other cultural, with savants bringing glory to the King through their erudition, to include writers, philosophers, and historians. Colbert initially fused the two concepts, but he then left out the humanists after some belle-lettrists resisted cooptation.[26]

It might thus be appropriate, if not redemptive, for science academies to contribute to the reconciliation of history and science. Historians weren't residents of Solomon's House, the scientific paradise described in Bacon's *New Atlantis* that provided a model for science academies.[27] Neither were they evident in the land of Balnibarbi, described by Swift in *Gulliver's Travels*, where academicians conducted experiments to extract sunbeams from cucumbers and in general prized theorizing

---

[25] Heilbron, "Physics and history: Forged in the Baroque," UC Berkeley Graduate Council, Hitchcock Lecture, 7 Nov 2007.

[26] Hahn, *Anatomy*, 11–15, 50–54.

[27] Heilbron, "Why revisit Solomon's House?" in Tore Frängsmyr, ed., *Solomon's House Revisited. The Organization and Institutionalization of Science* (Canton, Mass., 1990), 331–342.

over common sense and contemplation over action. Historians of science might have the good sense, and the historical perspective, to inform the National Academy of the distance it falls short of Bacon's utopia, and of the hazards of Balnibarbian indulgence.

Peter J. Westwick received his doctorate in history of science from UC Berkeley and is professor of the practice in history and thematic option at the University of Southern California. He is the author and editor of several books on the history of science and technology; he is currently writing a history of science and technology in California since the Gold Rush and is coauthor, with Daniel Kevles and Ruth Schwartz Cowan, of a forthcoming history of the US National Academy of Sciences.

Chapter 17
Citation styles in history of science: A critical discussion and a proposal

Olivier Darrigol

> *McCormmach counselled contributors [to HSPS] to write with clarity and distinctness. To these may be added economy of expression and an occasional smile. The ridiculous in the human condition touches scientists as well as their historians.*
>
> John Heilbron , "Foreword," *HSPS, 11* (1980), i–iii, on iii.

Abstract As a writer and as an editor of the journal HSPS, John Heilbron paid great attention to format and style as he regarded them a basic condition for the scientific communication of historical studies. In particular, he favored two well-defined styles for the citation of sources in journals and in books. This essay offers a critical discussion of the most frequent citation styles in current history of science, as well as recommendations for publishers and authors in harmony with the principles of completeness, concision, practicality, and elegance on which John implicitly based his own styles.

Keywords Citation · Styles · Conventions · Footnotes · References · Edition

When I first came to Berkeley in the fall of 1982, John Heilbron was my host at the egregious Office for History of Science and Technology. Although I had freshy defended a *Thèse de troisième cycle* in history of science, my training was mostly in physics, and I knew almost nothing about citation styles. I learned my trade as a historian mostly from John, during the first wonderful years I spent at UC-Berkeley. In addition to the broader principles of history writing, he taught me the austere art of proper citation and referencing, especially when I published my first article in English in *Historical studies in the physical sciences*. During his tenure as the chief editor of this journal, he personally edited the submitted manuscripts with utmost

---

O. Darrigol (✉)
UMR SPHere, Paris, France
e-mail: darrigol@univ-paris-diderot.fr

S. Gattei (ed.), *The Incomparable Historian*, Archimedes 75,
https://doi.org/10.1007/978-3-032-25678-2_17

care and precision. At an astonishingly rapid pace, he could read and memorize an entire article, detect internal inconsistencies and redundancies, and correct them by negative and positive surgery—to an extent that occasionally irked the author. Most of the patients, however, admired the surgeon's skills and were thankful for the improvements. John was often able to turn a not-to-well written article into a sound, polished piece of writing.[1]

Here is how John, in his "Swansong" of 1986, reflected on his and Russell McCormmach's ways of running *HSPS*:[2]

> Russ set a very high standard of editing, indeed, too high a standard for exact imitation. He insisted on intelligibility and directness in expression, and clarity and accuracy in the things expressed. Consequently, he was drawn into extensive correspondence with contributors who did not practice these virtues, and he ended rewriting their papers. ... No matter was too small or too large for his studied examination, for Russ was the copyeditor as well as the Editor in Chief of his magazine .... There is no doubt that every article in the first ten volumes of *HSPS* is the better for his care .... Although lacking the energy, patience, and insight to edit in Russ's manner, I tried to approximate it. ... One point of style on which I tried and failed to better Russ was lightheartedness. *HSPS* did not become a humor magazine. Nowhere in it will you find the puns that serve as arguments in some recent writing on the history and sociology of science.

I could not resist citing this exquisite sample of John's wit. On the sober side, part of John's role was to make sure that all articles respected the *HSPS* citation style, which he had brought to perfection by a system of abbreviations, by rigorous attention to punctuation and font, and by eliminating unnecessary symbols and capitals as carefully as a mathematical logician would (see Sample 1). He understood and made others understand that the format of a historical publication largely conditions its objective and transmissible content. Yet, there were and still are huge differences between *HSPS* style and the citation styles adopted by some other journals and publishers. These differences often have deplorable effects. Even the most submissive authors suffer when they are forced to respect rules with which they disagree. They feel like a chef who would be forced to add one spoon of vinegar in any of his compositions. The readers have concomitant digestion troubles.

Are citation styles just a matter of subjective convention, and should we calmly accept the formats that are imposed on us? I do not believe so. Some formats have pernicious effects, and it should be our duty as historians to criticize them and to convince the responsible editors and publishers to change their system. Admittedly, there cannot be a unique, evidently superior citation style that everyone should adopt without thinking twice. The elected style necessarily depends on tradition, on the type of text, on the type of journal, and on the form of publication (paper vs. electronic). Yet, some styles are defective under any circumstance, and it is

---

[1] Russell McCormmach launched *HSPS* in 1969. John Heilbron became the general editor in 1980 when McCormmach retired from the job. The title of the journal changed to *Historical studies in the physical and biological sciences* in 1986. In 2007 John passed the editorship to a team directed by Cathryn Carson and the journal's name changed again to *HSNS*, that is, *Historical studies in the natural sciences.*

[2] Heilbron, "Swansong," *HSPS*, *37* (2007), 185–187, on 185–186.

important to reflect on the criteria that may rationally orient our choice toward a better style. This is the purpose of the following inquiry. I will conclude with a few tentative recommendations. For every mentioned style, samples are given in an appendix. I will not address broader aspects of historical writing style such as general punctuation, quotation, appendices, headings, table of contents, and index. These are reasonably well addressed in easily accessible manuals of style.

17.1 Principles

Any serious historian must cite the sources on which he or she is relying. This information may include reported facts, background information, quotations, and relevant secondary literature. It is as necessary to a historical study as supporting experiments are to a physical theory. It is the means by which we can judge the honesty of the author, the solidity and originality of her arguments, and the extent of her knowledge. Also, it can be an important documentary basis for scholars who will later work on the same topic. A historical study without cited sources is simply worthless. In contrast, the citation and reference apparatus has a value in itself, and it may even have more value than the text it is serving (as John once told me).

In order to be efficient, the source-citation system should[3] respect a few principles. The referencing should be as complete as possible, including all details that matter to historians. It should be concise enough not to interfere with the smooth reading of the main text: overabundant notes and exceedingly long notes are to be avoided, for instance. It should be precise and devoid of ambiguities. It should be practical: for instance, it should not require the reader to constantly move from one part of the text to another. It should be pleasant to the reader's eye: even for a scholarly text, the success of the author depends on the reader's pleasure. To paraphrase John: The elegance of a text touches scientists as well as their historians.[4]

There are basic requirements that should be respected in any reasonable citation style. The references to primary and secondary literature should be given in full, including the first names of the authors, the full titles of article and books, volume number, year, and edition (for a book), the initial and final page (for an article), the place of publication for a book (the publisher's name it desirable, though often considered optional), the pages of the portion relevant to the citation, and the precise page in the case of a quotation. Manuscripts and letters should be precisely identified by author, correspondent (for letters), and location in an archive or in published collections. In the references, the punctuation and font case should respect the

[3] Here and in the following, I use "should" as an abbreviation for "should in the best possible world as I see it," not as an absolute dictate.

[4] Alas, some of my colleagues seem to be sharing Ludwig Boltzmann's opposite view that "elegance is for the shoemaker or the tailor" (see Clemens Schaefer, "Ludwig Boltzmann," *Die Naturwissenschaften*, *33* (1946), 33–67, on 35).

conventions of the *Chicago manual of style,* at least roughly.[5] In particular, the title of books and journals should be italicized, and the title of articles should be in quotation marks (when they are given in footnotes). These typographical rules are now nearly universal, and no major journal in our field questions them. However, there still are journals that do not comply with the basic requirement of completeness. Science journals, philosophy journals, and hybrid journals usually do not provide the authors' first names; they should do so when they happen to publish historical studies. Indeed, first names are necessary for the precise identification of historical actors (and present authors). Lastly, science journals should give the titles of cited articles when they publish historical studies, although they need not do so when they publish pure science.

17.2 Five common citation styles

We will now consider the five citation styles most commonly used by journals that publish science-historical studies.

The first and most common citation style in history of science journals is a variety of the style called "Notes and bibliography" by the *Chicago manual of style* (see Sample 1). The references are listed in full in footnotes or endnotes, separated by semicolons when there is more than one reference in the footnote. References to secondary literature come after a period, usually introduced by "see" or "cf." (in English publications, "see" should be favored, because "cf." suggests a comparison rather than a reference). There is no separate bibliography in this variety of the first Chicago style. When a reference has already been used in an earlier footnote, a backreference to this footnote replaces the full reference (Alas, some European journal still use the highly inconvenient *op. cit.* without footnote number). *HSPS* used a variant of this first style under John Heilbron's reign, and *HSNS* still does.

The second style is what the *Chicago manual of style* calls the "author-date" style (see Sample 2). In this style, a parenthesis is inserted at the proper place in the main text, including the name of the author of the cited source (of primary or secondary literature), the date of publication, and a page number (if needed). Footnotes are not used for primary citations; they are for additional commentary to the main text (sometimes including secondary citations). The references are given in full, by alphabetic order, at the end of the article, usually in a format close to the one

[5] *The Chicago manual of style*, 17th edition (Chicago: The University of Chicago Press, 2017); https://www.chicagomanualofstyle.org/home.html for the online version. In one respect, John Heilbron wisely departed from the Chicago rules for capitalization within titles: this capitalization (originally a British custom) should be avoided because if interferes with the use of capitals for grammatical reasons (in German) or for emphasis (honorary title, social category, etc.), and because it necessarily introduces a disparity between the rules for different languages (in most languages other than English and German, there is no such capitalization; in German, capitalization is reserved to nouns).

recommended by the *Chicago manual of style*. This style prevails in philosophy of science journals, although it is also encountered in a few historical journals.

The third style is a hybrid between the first and the second, and it is the one I have used whenever the publisher or the journal allowed it (see Sample 3). In this style, all citations are given in footnotes (as they are in the first style), but they are given only in the author-date form (as they are in the second style), with an alphabetic list of references at the end of the article or book. Note that in this style, parentheses or square brackets are unnecessary around the author-date, and therefore should be avoided. Also, there should be no coma between name and date.

The fourth style is another hybrid of the first and the second, and it corresponds to a variety of the "note-bibliography" system of the *Chicago manual of style* (see Sample 4). In this style as in the first style, the citations are entirely given in footnotes, but the references are now in abbreviated form, and the full references are found in an appended bibliography. This style is or was most common in books, and it is the one John Heilbron favored in his own books.

The fifth style or IEEE style (Institute of Electrical and Electronics Engineers) is commonly used in scientific publications (see Sample 5). Each citation is sequentially marked within the text through a number in square brackets. The references are given in numerical order at the end of the article, with minimal information (just initials and full name of authors, nor article title, and no final page number). In an improved version of this style (see Sample 5bis), the references are given in full (more or less), and they are numbered according to the alphabetic order of the name of the first author (consequently, the citation numbers are no longer increasing along the text).

17.3 Selection

To which extent do these five styles satisfy the above-listed principles of completeness, concision, practicality, and elegance?

The first style, with full references in notes, is highly convenient because the reader can see the reference on the very page it is called for, granted that the notes are run as footnotes or as electronic notes. Journals employing this style should avoid endnotes, or they should offer an electronic link between the note number and the note. Another advantage of this style is that it allows for the grouping of citations by paragraph: in many cases, several sources are referred to sequentially in the same paragraph, and the references can be given sequentially in the same footnote at the end of the paragraph, without fear of ambiguity. This saves a lot of footnote space at the bottom of the page, and this avoids an ugly proliferation of footnote numbers in the text. The main inconvenient of this system is that it becomes awkward for long articles and books in which the same reference is likely to be used repeatedly. This leads to a large number of back references in the footnotes, thus forcing the reader to move back and forth within the text and troubling the author when footnote numbers and backreferences are modified in the course of writing or editing.

The second style, with author-date citations within the text, has the advantage of offering a reference list, which is very convenient for future work on the same topic. But I find it ugly and inconvenient, as it disturbs the flow of reading in the main text. In historical studies (not so much in philosophical studies), a handful of citations may occur in the same paragraph, and the text becomes an indigestible pea soup. Moreover, this format is hardly compatible with an explicit distinction between primary and secondary sources (provided by the "see" or "cf." in the first style), and it does not allow the author to qualify the way in which a secondary source is used. For instance, in the first style we can write: "On the importance of fluctuations in Einstein's works, see the authoritative analysis by Martin Klein, …"; in the second style, one would just have (Klein, 1967) floating somewhere in the broth of the main text.

The third style, with author-date referencing in footnotes, does not have any of the defects of the first and second. It allows for much shorter footnotes, and, like the first style, it is compatible with the recommended grouping of footnotes by paragraph. It is very practical for the author, because it avoids cross-referencing and generally simplifies the gestion of references. Like the second style, it offers the various advantages of a separate bibliography. However, this style has one drawback that may have determined the preference of many authors for the first or the fourth style: In printed form, it forces the reader to move back and forth between the reading page and the reference list. This defect disappears in electronically published articles, which are now most common. It persists in printed books, but I believe the aforementioned advantages largely compensate for it.

The fourth style, with abbreviated references in footnotes, pleases the eye of the reader and it best satisfies the principles of directness and practicality. It is again compatible with the grouping of footnotes, and its abbreviated titles avoid the long footnotes implied by the first style. It remains highly recommended in printed books, although it requires additional work on the author's part and although it is not as concise as the third style. It becomes unnecessary in electronic publications.

The original IEEE form of the fifth style should be avoided, because it severely hurts the principle of completeness, and because it conflicts with concision and practicality: In a historical study, the names of sources should be seen where they are cited, and the references should appear in alphabetic order so that they can easily be exploited for future research. In the improved form of the fifth style, the references are more complete but still shun the first name of authors. The reference list is in alphabetic order, but numbers in square brackets are used instead of author-date for citation. In an electronic file, this system does not differ much from the second (author-date in text) system, since the reference can be seen by simply placing the cursor over the reference number in the main text. It therefore has nearly the same defects as the second system. At any rate, it conflicts with the common practice of historians and therefore hampers the respectability of a publication among historians.

17.4 Recommendations

- Strictly avoid the IEEE citation style with sequential numbers. Science journals using this style by default should accept another style when they publish historical articles. Historical journals should exclude it. The improved version of this style is more tolerable, but still too remote from the legitimate practice of historians.
- Historical journals should avoid the second, author-date in text, style. Journals that persist in favoring this style because some of their readers are accustomed to it (philosophers for instance), should nonetheless accept the third style when the author desires it. This is technically easy to achieve since author-date referencing is at the core of both styles.[6]
- The best overall style is the third style, author-date in footnotes, and notes grouped by paragraph when no confusion is possible. The superiority of this style is most evident in electronic publications.
- The first and fourth styles, with references in footnotes, will remain serviceable as long as footnotes truly appear as footnotes in the published text.
- The previous recommendations are chiefly addressed to editors and publishers. To authors, my recommendation would be to favor journals and publishers that require or at least accept the first, the third, or the fourth style.

Although my judgements and recommendations draw on long professional experience and benefitted from the lessons of a towering figure in the historian of science, I do not wish to engrave them in marble. It does not seem desirable to impose a single citation style forever, and it would be better to keep alive a spectrum of reasonable citations style, so that the style(s) may later evolve according to the changing needs of the discipline and the advent of new publication techniques. My personal recommendations are only tentative, and I would rather see them initiate a discussion with colleagues who might hold different views. Hopefully, a partial and revisable agreement will emerge in which the interests of authors, editors, publishers, and readers will happily converge.

Acknowledgments I thank Jean-Baptiste Grodwohl for useful comments on a first draft of this essay.

[6] I have usually succeeded in convincing the editors of *Archive for history of exact sciences* and those of *Studies in history and philosophy of science* to accept articles in the third style, although their default is the second style.

Appendix: Samples of the most common styles

J.L. HEILBRON

The contributions of Bologna to Galvanism

1. THE SOIL

Università dei Studi

LIKE ALL OF his colleagues at the University, Luigi Galvani was born, bred, and educated in Bologna. Inbreeding had become not only a practice but a right: any citizen with a doctorate could expect appointment as a lecturer with a salary should he take the trouble to apply and should the municipal taxes assigned the University suffice. Around 1690, a century before Galvani published his news about frogs, some 150 lecturers and professors existed under this regime. Since the University then had only a few hundred students, each learner had, on average and on the books, a half or a third of a teacher at his disposal. Statute required that each lecturer give 100 classes per year; the calendar allowed fewer than 100 days for instruction after subtraction of saints days and vacations; the university's rooms and customs permitted a maximum of some thirty classes a day; whence it appears that there was neither time nor place for most of the 15,000 lectures expected of the 150 lecturers.[1]

The logical resolution of this embarras de richesses was to cut the faculty; the political and social reality was to keep it, and to condone non-teaching teachers. A compromise ensued: the number of lecturers fell to about 70 in the 18th century, most of whom lectured in the University only when pressed to by a few members of the dwindling

*J.L. Heilbron, Office for History of Science and Technology, University of California, Berkeley, CA. It is a pleasure to thank Prof. Giuliano Pancaldi for copies of Galvaniana not easily available in the United States.

The following abbreviations are used: AS, Academia delle Scienze dell'Istituto di Bologna; *ASC*, AS, *Commentarii*; *ASM*, AS, *Memorie*; *ASO*, AS, *Opuscula*, bound with *ASM*; *Materiali*, Bologna, Università, *I materiali dell'Istituto delle Scienze* (Bologna, 1979); *SMSUB*, Istituto per la Storia dell'Università di Bologna, *Studi e memorie per la storia dell'Università di Bologna.*

1. E. Costa, "Contributo alla storia dello Studio bolognese durante il secolo xvii," *SMSUB, 3* (1912), 1–88, on 49–55, and "La fondazione dell'Istituto delle Scienze ed una riforma dello Studio bolognese proposita da Luigi Fernando Marsili," *SMSUB, 5* (1920), 45–66, on 48; Luigi Simeoni, *Storia della Università di Bologna: L'età moderna (1500–1888)* (2 vols., Bologna, 1940), *2*, 86–89, 92, 99.

Sample 1 From *Historical studies in the physical and biological sciences*, *22* (1991), p. 57. Abbreviations are given in the first, starred note. Full references are given in footnotes. The first footnote groups all secondary sources that were used to write the first paragraph. The first names of authors are given (when available).

ELSEVIER

Contents lists available at ScienceDirect

Studies in History and Philosophy of Science

journal homepage: www.elsevier.com/locate/shpsa

Charles Darwin as a statistical thinker*

André Ariew

Department of Philosophy, University of Missouri-Columbia, Columbia, MO 65211-4160, USA

A R T I C L E I N F O

Keywords:
Statistics
Probability
Population thinking
Variation
Explanation
Confirmation
Evolution
Modeling
Species

A B S T R A C T

Recently historians and philosophers of science have been interested in the role of statistics and probability in investigating population variation. The focus is typically on investigators applying statistics and probability to explain large scale phenomenon that arise out of the collective behavior of numerous and varied individuals. The case studies that inform this work come mostly from molecular physics and 20th century genetical versions of evolutionary theory. Charles Darwin's work on evolution is rarely mentioned in this context except to point out his shortcomings—he made evolutionary theory "ripe" for statistical investigations, but he was not a statistical thinker. But this is a mistake, Darwin was a statistical thinker. In this essay I describe two instances where Darwin utilized statistical methods to investigate evolution. In the light of these cases, we ought to revise our views about Darwin's scientific methodology, in particular, how he came to develop his ideas about evolution and about the nature of his "population thinking". Furthermore, Darwin's cases provide us with an expanded view about what constitutes "statistical thinking" in the biological sciences. In the examples we will find Darwin using statistical measures of type frequencies to *detect* large scale ensemble effects, *confirm hypotheses* by comparing between expected and observed averages, and applying the astronomer's law of error to *explain* evolutionary trends.

1. **Introduction**

According to Ernst Mayr (1959), one of Darwin's most significant contributions to science was to introduce "population thinking" to biology, a catchy phrase, but what does it mean? In the literature that Mayr instigated, "population thinking" refers to two distinct kinds of ideas, one is metaphysical and the other methodological. The metaphysical conception of population thinking is usually attributed to the novelty of Darwin's theory of speciation. Darwin overthrew a prominent view that types are real, and variation is an illusion. He replaced it with the view that individuals are real, and types are illusionary.[1] Darwin's contribution to the other, methodological, version of population thinking, however, is a bit more controversial. It has to do with methods for analyzing large-scale phenomenon in terms of the collective behavior of numerous individuals. Darwin is rightfully credited for replacing creationism with natural selection. The harmonies of ecosystems and prevalence of exquisite adaptations are not the products of direct intervention by the creator. Rather, they are aggregate result of the lives, deaths, and reproductive activities of individuals operating in conditions of scarcity. For some philosophers (Ariew, 2008; Walsh, 2015) that qualifies Darwin as a population thinker in this methodological sense. Yet, there is another kind of population thinking that was gaining prominence in Darwin's century in which it is said that Darwin did not engage. It involves, as Theodore Porter puts it, "employing a mode of reasoning based on stable numerical frequencies" (Porter, 1986, p. 135). Perhaps a better phrase for this kind of population thinking is "statistical thinking". Most philosophers and historians deny that Darwin was a statistical thinker.[2] While he employed the concept of chance several times in the *Origin of Species*, he made no references to stable numerical frequencies, nor did he invoke the theory of chances to explain long-run evolutionary trends (Walsh, 2015). As Porter puts it, "Darwin never developed anything like a quantitative model of evolutionary change". So, while Darwin made the investigation of evolution "ripe" for statistics he himself was no statistical thinker (Porter, 1986, p. 134). Gigerenzer et al. concur:

* Thanks to Antonine Nicoglou, Travis Holmes, Greg Radick, Jon Hodge, Denis Walsh, Sukhvinder Shahi, Philippe Huneman, Adam Krashniak, Charles Pence, an anonymous reviewer, and philosophy audiences at University Bordeaux, Paris IHPST, and Leeds University. Funding comes from a University of Missouri, College of Arts and Sciences, Research Leave Grant 2019–2020.

E-mail address: ariewa@missouri.edu.

[1] For a detailed look on the origins of Mayr's distinction see (Chung, 2003). For a defense of typological thinking, see (Lewens, 2009). For a critique of Mayr's assessment of Darwin, see (Wilkins, 2009).

[2] *Pace* Sober 1980. For criticisms of Sober's reasons for attributing statistical thinking to Darwin see Ariew (2008) and Walsh (2015).

References

Ariew, A. (2007). Under the influence of Malthus's law of population growth Darwin eschews the statistical techniques of Aldolphe Quetelet. *Studies in History and Philosophy of Science Part C: Studies in History and Philosophy of Biological and Biomedical Sciences, 38*(1), 1–19.

Ariew, A. (2008). Population thinking. In D. Hull, & M. Ruse (Eds.), *Oxford companion to philosophy of biology* (pp. 64–86). Oxford U. Press.

Ariew, A., Rice, C., & Rohwer, Y. (2015). Autonomous statistical explanations and natural selection. *The British Journal for the Philosophy of Science, 66*(3), 605–634.

Browne, J. (1980). Darwin's botanical arithmetic and the 'principle of divergence,' 1854-1858. *Journal of the History of Biology, 13*(1), 53–89.

Chung, C. (2003). On the origin of the typological/population distinction in Ernst Mayr's changing views of species, 1942–1959. *Studies in History and Philosophy of Science Part C: Studies in History and Philosophy of Biological and Biomedical Sciences, 34*(2), 277–296. https://doi.org/10.1016/S1369-8486(03)00026-8

Cowles, H. M. (2020). *The scientific method: An evolution of thinking from Darwin to Dewey.* Cambridge, Massachusetts: Harvard University Press.

Sample 2 From *Studies in history and philosophy of science*, *95* (2022), pp. 215 (front page), 222 (list of references).

The author-date format is used, with most citations in the text (and some in footnotes for secondary literature). The authors' first names are not given.

Natural Magic (*Magia naturalis*, 1558 and 1589), from the Italian version of the *Three Books on Pneumatics* (*Tre libri de' spiritali*, 1606), and from the meteorological treatise *On Transmutations of Air* (*De aeris transmutationibus*, 1610).[7]

2 | DELLA PORTA'S WRITINGS AND HIS CONCEPT OF NATURAL MAGIC

The name of Giovan Battista Della Porta was well-known to the learned public of his time, well beyond his native town of Naples.[8] Born to an impoverished noble family around 1535, Della Porta was taught privately and travelled Europe as a young man, spending time at different Italian and European courts. Later in his life, he lived mainly in Naples, where he died in 1615. Della Porta achieved fame initially with the first edition of *Natural Magic*, which was published in 1558 and quickly reprinted and translated from the Latin original into many other languages.[9] This work conformed to the tradition of books of secrets, which was very popular in Italy at the time. They were collections of short chapters, mostly in recipe form, that reported knowledge originating from ancient and medieval authors and contemporary artisans.[10] In fact, in the Italian language of the time, the term "*secreto*" could also mean a formula or recipe that had previously been transmitted only by word of mouth and was being written down for the first time.[11] Della Porta's *Natural Magic* contained recipes from ancient and medieval sources of various kinds, but also included much original material derived either from Della Porta's own fantasies and experiences, or from those of persons in his circle of friends, who were often mentioned in his writings. The topics treated in the volume were very varied, as we shall see, but they corresponded to Della Porta's broad interests and capabilities. In 1589, he published a second, much expanded version of *Natural Magic*, in which the material from the first edition was reprinted, at times in slightly altered form, and complemented with new experiences, new topics, and some additional learned discussions.[12] While working on the second edition of *Natural Magic*, Della Porta was also experimenting and reading both older and newer literature on many topics. During the late 16th and early 17th centuries, he published a series of monographs devoted to subjects that had been treated in shorter form in *Natural Magic*. For the purposes of this paper, the treatises of interest are those *On Pneumatics*, published in 1601 in Latin and then in 1606 in an expanded Italian edition, and the one on meteorology, *On Transformations of Air*, which appeared in print in 1610.[13]

Before discussing Della Porta's pneumatic experiments, however, a few words must be said about his relationship with another, equally successful literary and philosophical tradition of his age: treatises of "natural magic." Natural magic is a term that had emerged in the 15th century in parallel with the increasing criminalization by Church and state of ritualized practices aimed at enlisting the help of occult powers to produce effects relevant to worldly life.[14] These practices were usually subsumed by those opposing them under the general term "magic," but some of the authors interested in them strove to introduce a distinction between a natural magic that only harnessed the forces of nature and a black, demonic magic that instead relied on the invocation of holy or unholy powers—angels and demons. On this basis, they argued that only the latter should be forbidden, although in practice the distinction remained quite vague. Eventually, the term was appropriated by philosophers, who gave their own diverging characterizations of what "natural magic" was, and to what extent it did or did not involve commerce with demonic powers. In *Natural Magic*, Della Porta combined ideas from earlier authors, especially Marsilio Ficino and Agrippa von Nettesheim, to expound the distinction between allowed and forbidden magic, which was of great relevance for his

[7]For my analysis I have used either the original Latin or Italian text, in the modern critical edition (Della Porta, 1606/2008b; 1610/2000) or, when such an edition does not exist, in early modern ones (Della Porta, 1558/1560; 1589/1597). To my knowledge, no translations exist for the meteorological and pneumatic treatises, and *Natural Magic* was translated into various languages during the 16th and 17th centuries, but the English translations given in this paper are my own.

[8]The following discussion of Della Porta's life and works is based on Eamon (1994; 2017).

[9]Della Porta (1558/1560). On the reception of *Natural Magic* and other works by Giovan Battista Della Porta, see Balbiani (1999).

[10]Eamon (1994).

[11]Byrne (2017, p. 687).

[12]Della Porta (1589/1597).

[13]Della Porta (1601/2008a; 1606/2008b; 1610/2000).

[14]Labouvie & Neugebauer-Wölk (2016).

Sample 3 From *Centaurus*, *62* (2020), p. 409.

Citations are given in footnotes in the author-date format. Unnecessary parentheses are used; for instance, Eamon (1994) could be Eamon 1994.

198 *Electricity in the Seventeenth Century*

Boyle held that a 'good hypothesis' or model must be 'fit and sufficient' to explain the chief phenomena and 'at least consistent with the rest;' whereas an 'excellent hypothesis' must 'enable a skilful Naturalist to foretell future Phaenomena by their Congruity or Incongruity to it; and especially the events of such experiments as are aptly devis'd to examine it, as Things that ought or ought not, to be consequent with it.'[22] In this terminology, physicists early in the century expected only good hypotheses, while those of Boyle's age demanded excellent ones. As for Fabri's sanguine expectations regarding the fit of mechanical models, they are plain enough in the speeches of Antimus and in the *Epistolae* of 1674, where Fabri denied unconvincingly that he followed Descartes, Gassendi and Epicurus.

FABRI AND THE ACCADEMIA DEL CIMENTO

The short-lived Accademia del Cimento (1657–1667) consisted of some ten men supported by Duke Ferdinand and Prince Leopold dei Medici. Its best known and most productive members were G. A. Borelli, Vincenzo Viviani, and Francesco Redi. It appears that Viviani, who liked to style himself the last pupil of Galileo, designed most of the Academy's electrical experiments.[23] The printed record of the Academy's labor, their *Saggi* or *Essays* (1667), do not reveal the extent of their concern with electricity, give no rationale for it, and neither acknowledge Viviani's role in, nor explain the reasons for any of the experiments. All this is characteristic of the *Saggi,* which strive to conceal the thought and preparation behind the Academy's investigations, and avoid identifying individual contributions.[24]

It is said that Magalotti, the principal author of the *Saggi*, chose this impersonal style lest the Church persecute individual members for their opinions about the nature of electricity, magnetism, the vacuum, the hot and the cold, and similar matters fascinating to the Holy Office.[25] That is a serious misinterpretation: the Church no longer cared to martyr physicists, and in any case the opinions of the academicians were no more scandalous than those of Fabri or Lana. Magalotti chose his style for two reasons. First, the peculiar organization of the Academy precluded other forms of publication. The group was Leopold's creation and depended financially upon him and his brother. The Academy tried to act as a whole: individuals proposed experiments; the group decided their importance, feasibility and relevance. The usual, or at least the initial rou-

22. Boyle, 'The Requisites of a Good Hypothesis,' in M. B. Hall, *Boyle* (1965), 134–5. Cf. Boyle, *Excellency Mech. Phil.* (1674), in *Works*, IV, 68–70, 77–8; M. B. Hall, *Boyle* (1965), 189, 191, 195, 207. The changing expectations of the mechanical philosophers have received little attention; see Laudan, *Idea* (1966), 71–8, 89–93, 162–5, and *Ann. Sci.*, 22 (1966), 73–104.

23. Nelli, *Saggio* (1759), 110–111n; Middleton, *Experimenters* (1971), 39.

24. Antinori in Acc. del Cim., *Saggi* (1841^3), 3–133; Targioni-Tozzetti, *Notizie* (1780), *passim;* Abetti in *Celebrazione* (1958), 3–10. For the lesser known members of the Academy, the del Buono brothers, Dati, Rinaldini, Marsili and Oliva, see Antinori in Acc. del Cim., *Saggi* (1841^3), 70–80; Nelli, *Saggio* (1759); and Middleton, *Experimenters* (1971), 26–40.

25. *Ibid.*, 6, 54, 72, 333.

Sample 4 From Heilbron, *Electricity in the 17th and 18th centuries: A study of early modern physics* (Berkeley: The University of California Press, 1979), p. 198.

Citations are in footnotes in abbreviates form. The full reference is found in the appended bibliography.

Ann. Phys. (Berlin) 528, No. 5, 344–349 (2016) / DOI 10.1002/andp.201600105

annalen der physik

THEN & NOW

The renaissance of General Relativity: How and why it happened

Alexander S. Blum, Roberto Lalli, and Jürgen Renn*

Received 14 April 2016, accepted 15 April 2016
Published online 3 May 2016

After an initial burst of excitement about its extraordinary implications for our concept of space and time, the theory of general relativity underwent a thirty-year period of stagnation, during which only a few specialists worked on it, achieving little progress. In the aftermath of World War II, however, general relativity gradually re-entered the mainstream of physics, attracting an increasing number of practitioners and becoming the basis for the current standard theory of gravitation and cosmology-a process Clifford Will baptized the Renaissance of General Relativity. The recent detection of gravitational radiation by the LIGO experiment can be seen as one of the most outstanding achievements in this long-lasting historical process. In the paper, we present a new multifaceted historical perspective on the causes and characteristics of the Renaissance of General Relativity, focusing in particular on the case of gravitational radiation in order to illustrate this complex and far-reaching process.

The year 2015 marked the centenary of Einstein's final formulation of the gravitational equation that bears his name, the cornerstone of the general theory of relativity. Almost exactly one hundred years after the theory's inception, the direct detection of gravitational waves through a large-scale experiment operated by the multinational LIGO Collaboration has confirmed one of its most elusive predictions. With this momentous achievement, general relativity has once again strengthened its position as the standard theory of gravitational phenomena and the basis of cosmological models, in spite of the still unresolved difficulties in reconciling Einstein's theory with quantum mechanics.

The central position the theory continues to hold in our description of the physical world might not seem all that surprising, given its immediate spectacular confirmation (Eddington's solar eclipse expedition of 1919) and Einstein's subsequent rise to scientific superstardom. However, historical investigations have revealed that matters appeared quite different after the initial hype and before the current age of spectacular experimental and observational confirmations. In the early years after its formulation, the epistemic status of general relativity theory was highly uncertain in many respects, from the understanding of its physical implications to the interpretation of the impact of a choice of coordinate system on the result obtained. A perfect example of the uncertainty concerning the epistemic status of the theory is the early debate on the nature and existence of gravitational waves within the framework of general relativity. The 1916 correspondence between Einstein and the astronomers Karl Schwarzschild and Willem de Sitter reveals how confused the connections between the theory and its physical consequences were in the months following the formulation of the theory. Thanks to these epistolary exchanges, Einstein passed from believing that his gravitational theory implied the non-existence of gravitational waves to demonstrating that gravitational wave solutions of the linearized approximation of general relativity exist and carry energy [1]. The formula of gravitational radiation Einstein derived in 1916 was incorrect, though. It permitted the radiation of energy from monopole sources. It was only after reviewing calculations made by the Finnish physicist Gunnar Nördström that Einstein discovered an error in his 1916 derivation of the gravitational radiation and was able to derive the correct quadrupole formula of gravitational radiation in 1918 (apart from a factor of two) [2].

Moreover, only a few years after Eddington's observations of the bending of light general relativity underwent

* Corresponding author E-mail: rlalli@mpiwg-berlin.mpg.de
Max-Planck-Institut für Wissenschaftsgeschichte, Berlin, Germany

References

[1] H. Gutfreund, D. Kormos-Buchwald, and J. Renn, "Gravitational Waves: Ripples in the Fabric of Spacetime" *Huffington Post,* 12 February 2016, http://www.huffingtonpost.com/hanoch-gutfreund/gravitational-waves-ripplb9211978.html

[2] D. Kennefick, Traveling at the Speed of Thoughts: Einstein and the Quest for Gravitational Waves (Princeton: Princeton University Press, 2007), p. 61–65.

[3] J. Eisenstaedt, Arch. Hist. Exact Sci. **35**, 115–185 (1986) J. Eisenstaedt, Arch. Hist. Exact Sci. **37**, 275–357 (1987) J. Eisenstaedt, Einstein and the History of General Relativity, Einstein Studies, Vol. 1, ed. D. Howard and J. Stachel (Boston: Birkhäuser, 1987), p. 277–292 J. Eisenstaedt, The Curious History of Relativity: How Einstein's Theory of Gravity was Lost and Found Again (Princeton: Princeton University Press, 2006).

Sample 5 From *Annalen der Physik, 528* (2016), pp. 344, 349.

Citation appear as numbers in square brackets in the main text, and the references are numbered sequentially.

Eur. Phys. J. H (2021) 46:25
https://doi.org/10.1140/epjh/s13129-021-00024-7

THE EUROPEAN
PHYSICAL JOURNAL H

Check for updates

Regular Article

How to model phase transitions? The changing approaches 1937–1970

Martin Niss[a]

IMFUFA, Department of Science, Roskilde University, Roskilde, Denmark

Received 2 March 2021 / Accepted 28 July 2021 / Published online 7 September 2021

Abstract The paper aims at characterising and documenting a fundamental change in how phase transitions were modelled microscopically in the period 1937–1970. At first, physicists took what will be called *a naturalistic approach* to phase transitions such as the condensation of gases and the Curie point of ferromagnets. Here the purpose was to explain the phenomenon in question, i.e., to show that a model exhibits the same features as the phenomenon. The scope of this approach was broad, as the goal was to account for several aspects of the phenomenon. The employed model should be very realistic and close to the foundational theory, be it classical or quantum mechanics. In the 1960s, the physicists used an alternative approach that they termed *a caricature approach*. This approach not only required explanation in the above sense but also understanding of the physical phenomenon, i.e. insights into why the phenomenon behaves as it does. The scope was limited to certain aspects of the phenomenon, such as the behaviour near the critical point. The caricature approach used a hierarchy of models, ranging from realistic ones over more simplified models to models that were mere caricatures of the system in question. Hence, the two approaches represent very different orientations when it comes to the purpose and scope, the organisation of the resulting theories, and what models are acceptable.

1 Introduction

Modelling is one of the modes that activities in theoretical physics proceed by, according to a report of 1966 by the Physics Survey Committee (appointed by the National Science Foundation). Here, "the emphasis is on the construction of models that isolate from complex physical phenomena certain specific characteristics." [Physics Survey Committee [27], p. 162] The versatile physicist Rudolf Peierls expressed the same point when he said that "Physicists tend to use models of various kinds to aid their understanding of complicated physical situations." [Peierls [26], p. 3] These quotes are indications of the crucial role of models and modelling in what the philosopher R. I. G. Hughes [Hughes [15]] called the theoretical practice of physics in the 20th century, so if we want to understand this practice, we need to understand the models and modelling of physicists. Hughes correctly points out that there is not a single theoretical practice in all of physics. So the practice of solid-state physicists may differ from those of high energy physicists. Here, we study physicists who wanted to give microscopic explanations of macroscopic phenomena. They saw models as crucial; the foundational theory (i.e. quantum mechanics) was in place, but the sheer number of interacting microscopic particles meant that they had to introduce several and substantial simplifications. This fundamental condition introduced some freedom for the individual physicist in the modelling process as to what features of the microscopic constituents were included or left out. In short, there was no canonical way of modelling.

This paper presents a case within this area where a sharp change in the modelling approach took place, namely within the microscopic modelling of phase transitions in the period 1937–1970. While the physicists in this area agreed on the methodology, i.e. how they should use models in the procedure for attaining knowledge [Hon and Goldstein [14]], they chose very different approaches. The change did not involve revolutionary changes in the theoretical basis like the quantum revolution. However, it was still quite profound: it entailed a new view of which models are acceptable and which are not as well as a change in the very idea of what was to be achieved by modelling phase transition phenomena with microscopic models. The purpose of the present paper is to characterise and document this change.

I will argue that there were two general approaches to the modelling of phase transitions in the period. In what I will call *the naturalistic approach*, the purpose was to explain the phenomenon in question, i.e. show that a model exhibits the same features as the phenomenon. The scope of this approach was quite broad as the goal

---

Sample 5bis From *European physical journal H*, *46* (2021), p. 25.

The numbers in square brackets refer to references listed and numbered in alphabetic order at the end of the article.

Olivier Darrigol, Research Director at UMR SPHere, CNRS, Paris; and Research Associate at the Office for History of Science and Technology (OHST) at the University of California, Berkeley. Author of several books on the history and philosophy of physics, mostly in the nineteenth and twentieth century.

Chapter 18
From Hypatia to "Cosmos" and "Agora": The Formation of a Fractal Narrative of Scientific Decline

Michael H. Shank

> *Decline narratives are all unhappy in their own way,*
> *But they can all happily be made like one another.*
>
> *With apologies to Tolstoy*

Abstract This paper examines the evolution of polemical narratives about the life and death of Hypatia of Alexandria. After fifteen centuries, an anecdote from late-antique Alexandrian politics has now become at once a paradigm case of religious hostility to science, a narrative structure that accounts for a millennium of medieval scientific vacuousness, and a terminus a quo marking the onset of the latter. A chronological analysis of key texts (by Toland, Gibbon, and Draper, among other) and recent filmography (Sagan's "Cosmos," Amenábar's "Agora") documents this fractal-like transformation of the microcosmic Hypatia story into a macrocosmic pattern of decline in roughly two phases in which science plays an increasingly prominent role. The first, after sublimating Hypatia and Bishop Cyril into types, reinforced the Protestant periodization of history: Hypatia's death introduces a millennium of repressive Catholicism until the Reformation (and beyond). In a second phase that echoes the small-scale cases of Bruno and Galileo, the large-scale fractal narrative of decline structured the periodization of the history of science itself: so suppressed by religion/Christianity/Catholicism as to be effectively cancelled from Hypatia's death until Copernicus and Kepler.

John Heilbron was on stage when I presented a very preliminary version of this material (University of Notre Dame, April 2017). I will long treasure his friendship, his wit, his criticism of music and scholarship, and his guided tour of Bassano's portrait of Sagredo in the Ashmolean. I thank Stefano Gattei, Nick Jacobson, and Carol Troyer-Shank for their comments and criticisms.

M. H. Shank (✉)
Department of the History of Science, University of Wisconsin-Madison, Madison, WI, USA
e-mail: mhshank@wisc.edu

S. Gattei (ed.), *The Incomparable Historian*, Archimedes 75,
https://doi.org/10.1007/978-3-032-25678-2_18

Keywords Hypatia · Narrative structure · Historiography of science · Decline · Periodization

Readers of John Heilbron's scholarship have long envied the way his writing combines rigorous arguments, novel insights, superb prose, and sharp wit. Both his broad historical judgments and his passing comments have challenged once-secure answers and solved long-term problems thanks to overlooked details. Specifically, they have reshaped this historian of medieval science's obsession with narratives of decline and their role in the periodization of the history of science.[1]

Two Heilbron stimuli—a thesis and a *bon mot*—have informed the questions underlying this chapter. In *The Sun in the Church*, Heilbron debunked a myth that was, and probably still is, deeply entrenched in both popular and scholarly understandings: that, citing Sam Westfall, the Catholic Church made "Copernican astronomy a forbidden topic" for its members for two centuries after Pope Urban VIII forced Galileo to recant his Copernican views and condemned him to house arrest. How then do narratives of scientific suppression and decline become attached to contexts in which they are inappropriate and become favorites in the story-telling repertoire? Later in the same book, Heilbron quipped that if the recent trends in the Catholic Church's rehabilitation of Galileo persisted, he would be canonized within the century, thus bringing his status in line with his self-appreciation.[2] How then do historical episodes turn into legends of sainthood and martyrdom and come to dominate both popular and not-so-popular perceptions of the history of science?

Although narratives of decline and historical legends plague some historical writing, they characterize in particular the popular historiography of science developed by scientists writing as amateur historians. These themes sometimes play such a large role in the long-term periodization of the history of science, both popular and scholarly, that their mutual creation and entrenchment often distort not only the original incident but also the contextual frame into which it is inserted.

Since the literature on Hypatia is too vast to treat comprehensively, this chapter focuses on her role in selected influential narratives drawn, with a few exceptions, from popular Anglo-American historiography of science—what one scholar has called "scientific folklore."[3] Its goal is to highlight key stages in the remarkable transformation of an ancient anecdote into a momentous historical turning-point and a marker of millennial scientific decline. The murder of Hypatia was originally a microhistorical incident in the bloody politics of late-antique Alexandria. A millennium and a half later, popular recastings of this narrative also made her a martyr for science whose death carries world-historical significance for the trajectory of science. Accompanying its introduction into the history of science are

[1] Lindberg and Shank (2013, pp. 1–16).

[2] Westfall (1989, p. 1); Heilbron (1999, pp. 14, 19–20, 211); and Heilbron (2010, pp. 364–65), where sanctification is postponed for 400 years.

[3] Lessl (1999, esp. pp. 146–50).

transformations so striking that the story's later decline narratives function like fractals that replicate their self-similar structure at various scales.

At issue, then, are not the excellent reconstructions of the historical Hypatia's intellectual life, death and context, on which specialists largely agree.[4] I focus instead on the changing narratives about her constructed for widespread (often polemical) consumption and on the way they shape popular understandings of a precipitous millennial decline periodization in the history of science. After a brief introduction, I turn to the most ancient narratives as a reference point for the earliest intents and meanings before sampling their transformation during the subsequent 1500 years.

Hypatia (c. 350–415) is famous today as one of the first women to be recognized as a mathematician, an astronomer, and a philosopher renowned throughout the Eastern Mediterranean. A respected teacher and a leading citizen of Alexandria, she was brutally murdered by a cohort of hotheads led by a minor cleric sympathetic to the bishop of Alexandria. The combination of Hypatia's sex, intellectual prominence, and gruesome murder have made her perennially fascinating. Not surprisingly, story-tellers have tweaked the narratives about her to reflect their changing interests and those of their readers. The sparse or awkward details in the earliest accounts have sometimes led both groups not only to edit, edit out, or supplement them but also to fill the many gaps with appealing embellishments from previous retellings.

Hypatia was the daughter of the mathematician and astronomer Theon of Alexandria, who presumably guided her early education. Theon clearly treated his daughter with reverence. In his commentary on book 3 of the *Almagest*, he calls her "philosopher" and credits her for having *paranagignokein* that section of the book, a verb variously translatable as "checked," "edited," "proof-read," or "revised."[5] Early sources mention her involvement in mathematics generally whereas the *Suda*, a tenth-century Byzantine encyclopedia, attributes to her commentaries on Ptolemy's *Tables*, Apollonius's *Conics*, and Diophantus's *Arithmetic*, for which no other evidence survives. Also reported in the *Suda*, Damascius makes her better than her father because she wanted to become more than a mathematician, i.e. a philosopher. This was an ambiguous compliment since her teacher was Theon, no proper philosopher. Indeed, according to another fragment, Damascius considered her inferior to his own teacher Isidore "not only in that he was a man and she a woman, but in the way a true philosopher differs from a geometer."[6]

The best documentation about her beliefs and accomplishments are the letters of Synesius of Cyrene, who had studied with her in the 390s before becoming bishop of Ptolemaïs. Having died three years before she did, his remarks are untainted by the sensationalism of her death. Synesius found her person, her learning, and her

---

[4] See especially Dzielska (1995) and Watts (2017). Norman and Petkas (2020).

[5] See Acerbi (2008, p. 436).

[6] Photius's fragment of Damascius in Petkas and Norman (2020, p. 248). Deakin (1994) should be tempered with the more cautious evaluations of Acerbi (2008, pp. 435–37) and Berggren (2009, p. 94).

example enthralling. The few letters he addressed to Hypatia are likely genuine even though she evidently did not answer, perhaps in keeping with the strands of Neoplatonism that favored orality and face-to-face contact. Synesius also praised Hypatia in letters to other correspondents.[7]

Hypatia's context suggests that she inclined to one or more varieties of Neoplatonism. The labels "heathen" and "pagan" used to describe her centuries later do not appear in the earliest sources.[8] Throughout her relatively long life (she was likely murdered in her sixties), she was a respected philosopher and an inspiring teacher who probably taught both in public and from home.[9] No early source associates her with an institution, contrary to the popularizations that link her either to the original Museum and Library (not mentioned after Emperor Aurelian's destruction of them in 273) or to the so-called "external" or "daughter library" associated on scant evidence with the Serapeum temple.[10]

According to Damascius, Hypatia's intellectual stature and wisdom drew her into the fractious politics of the city, Indeed, Alexandria had long been plagued by violence between its various political, religious, and ethnic factions. In today's language, she was a public intellectual whose engagement in civic life embroiled her in the political struggles between Bishop Cyril and the imperial prefect Orestes, her close associate. Although both men were Christians, they belonged to political and religious factions with divergent ends, means, and ideas about civic leadership. At stake were the unevenly matched powers of the bishop and imperial prefect in running the city.[11] In Maria Dzielska's words, Hypatia's death was "a political murder provoked by longstanding conflicts in Alexandria."[12]

18.1 The Earliest Sources

The first accounts of Hypatia's death set crucial base lines of detail and emphasis. Without them, one cannot appreciate how markedly the later narratives about her eventually morphed into tales of scientific repression with narrative structures ever more similar to those of Giordano Bruno and Galileo.

The earliest sources are contemporaneous. Socrates Scholasticus (ca. 380–439) was a lawyer in Constantinople who had studied in Alexandria. His *Ecclesiastical*

[7] Harich-Schwarzbauer (2012) and Seng (2020). Based on these letters and other documents, both genuine and fake, scholars have ascribed to her a range of improbable positions from Iamblichean Neoplatonism to Cynicism and even Nestorian Christianity; Dzielska (1995, pp. 20, 62).

[8] John of Nikiu, translated in Petkas and Norman (2020, pp. 251–53).

[9] Damascius, translated in Petkas and Norman (2020, p. 247–48).

[10] For a critical quantitative treatment of these much-mythologized institutions, see Bagnall (2002, esp. p. 358).

[11] Dzielska (1995, p.18) although the discussion of her mathematics is out of date; Wessel (2004, pp. 48 f.).

[12] Dzielska (1995, p.104).

History contrasted the episcopal leadership style in Constantinople (conciliatory) with those in Alexandria and Rome (repressive) and their impacts on each city.[13] His account covered the fourth and early fifth centuries, including Alexandria under Bishop Theophilus and his successor-nephew Cyril, who came to power in a violently contested election (412). Socrates was disturbed by both Bishop Cyril's encroachments on the political realm and his heavy-handed imposition of unity on churches that he deemed schismatic. The most notable were the Novatian purists that his uncle had tolerated and that Socrates favored (Novatians opposed the reintegration into the church of lapsed Christians who under persecution had reneged on their faith or sacrificed to the gods or the emperor).[14]

Socrates brought up Hypatia in the culmination of the "frightfulness of the quarrel between Cyril and Orestes."[15] The daughter of Theon and a Platonist teacher in the Plotinian tradition, she became a widely respected participant in political discussions among men. She nevertheless aroused "envy" and "slander" among the church-going populace that suspected her of discouraging the reconciliation of Orestes with Cyril. Acting on these rumors, some "hot-headed men" led by Peter the Reader took her to the Caesareum church, stripped her, murdered her with potsherds, dismembered her, and burned her remains. Socrates set the crime during Lent, further accentuating the "great opprobrium" that the murder of Hypatia brought upon Cyril and the Alexandrian church.[16]

By Socrates's account, Hypatia's learning and virtue elevated her character and undergirded the propriety of her interactions. The murder followed counterintuitively from political infighting ("It was against this woman [of superlative temperance] that envy armed itself.") The Greek text suggests an organized attack by "men of a hot spirit" with a leader, not the mob of later retellings.[17] Socrates's comparison-contrast of leadership styles in three cities made the murder of Hypatia a horrible, unchristian consequence of Bishop Cyril's political over-reach, one of the many indictments of his governance in Alexandria.

A Byzantine summary of a now-lost *Ecclesiastical History* by Philostorgius, another Arian contemporary, contains two sentences to the effect that Hypatia surpassed her father in astronomy in particular and "was ripped apart by those who championed the homoousian faith."[18] This comment highly praises her competence in mathematics and astronomy and links her murderers with the Trinitarian (anti-Arian) party. Although this does not necessarily imply that she took sides in an intra-Christian debate, it seems reasonable to suppose that a Neoplatonic

[13] Watts (2017, p. 122–23).

[14] Wessel (2004, p. 15); Schaff and Wace (1994, book 7, chap. 15).

[15] Socrates Scholasticus, *Historia Ecclesiastica*, 7.14 translated in Petkas and Norman (2020, p. 249).

[16] Petkas and Norman (2020, pp. 249–50).

[17] Haase (2020a, pp. 92–93), with detailed exegesis in Haase (2020b).

[18] Petkas and Norman (2020, p. 248).

philosopher had opinions about the alleged intermingling (or lack thereof) of human and divine natures in Christ or anyone else.

Written several generations after Hypatia's death, Damascius's *Philosophical History* (also known as the *Life of Isidore*, his teacher) supplemented Socrates's account with memorable details from the perspective of a non-Christian Neoplatonic philosopher. A former student in Alexandria and later head of the Platonic Academy in Athens, Damascius (d. after 538) emphasized that the memory of Hypatia's learning, virtue, and prominence as a philosopher were still vivid in Alexandria in the 480s. She was "endowed with a nobler nature than her father," witness her alleged dissatisfaction with her mathematical education and her achievement of "distinction in the other branches of philosophy." Damascius debunked the "ignorant legend" that Hypatia cured one student's infatuation with her by using music. The passage is obscure with divergent translations: did she not use music at all? or did she try and it failed? Whatever the case may be, she showed him her bloody sanitary napkin, "the symbol of the impurity of birth … [saying] 'This is what you are in love with, young man, and not a thing of beauty.'"[19]

Damascius blamed the murder on Cyril, "the man in charge of the opposite sect," whom he portrays as envious and obsessed with crowd size. A large throng of admirers outside Hypatia's house made the bishop so jealous that he plotted to kill her in horrible fashion, which "a crowd of bestial men" soon did. For Socrates, the murder had offended against Christianity. For Damascius, it "inflict[ed] the greatest pollution and disgrace on the city." After the emperor was bribed to drop the charges against Cyril, justice was delayed but not forever: Socrates claims that the emperor's grandson paid for the offense.[20]

Since later retellings of this story make Hypatia's death the prelude to a scientific and philosophical vacuum, it should be clear that scholarship continued throughout the Eastern Mediterranean in the century after Hypatia despite ongoing intellectual and physical conflicts among many factions, including Christians of various stripes, Jews, so-called pagans and Neoplatonists (sometimes called polytheists). Even after 529, when Emperor Justinian dissolved the Platonic Academy in Athens and Damascius lost his position, philosophy and scientific work continued in Alexandria.

These early sources strongly condemned the murder of Hypatia. In the seventh century, however, the Egyptian bishop John of Nikiu wrote a universal history that was unknown until the late nineteenth century. It praised the pious actions of Peter the lector and his followers for ending paganism in Alexandria, thus concluding Cyril's eradication program (in fact his uncle's, not his). John justifies the murder by charging Hypatia with suspect learning, notably her use of astrolabes and music to manipulate Orestes and others by magic. Reviving the rumor that Damascius had mentioned and tried to quash, John's accusation was perhaps meant to be

[19] Compare the divergent translations in Athanassiadi (1999, pp. 128–31), Leonard (2020, pp. 175–76), and Petkas and Norman (2020, pp. 247–48).

[20] Athanassiadi (1999, p. 131); Gajeri (1992, pp. 18–22).

exculpatory: in some biblical texts and Roman legal pronouncements, magicians deserve death.[21]

Given its late discovery, John of Nikiu's story had no impact on much of the later popular historiography. It is, however, striking from a structural point of view because it pioneers a narrative of decline with a strong triumphalist tone. Hypatia's death marks the demise of paganism, an unmitigated good. That same narrative structure with a negative valence will later frame Hypatia's death in some of the most popular recent accounts of the history of science:

> The subject practices his/her craft/science;
> The Church hierarchy/religion opposes it;
> The subject is condemned and/or executed.
> [Ergo decline follows; an era ends]

Radically different ethical valuations can map onto this same structure. For John, Hypatia's death ushers in the victory of Christianity over paganism in Alexandria. For nearly everyone else before and after John, the decline was a disaster signaling the end of philosophy, and/or science, and/or free thought.

Despite their differences, the early accounts concur that political tensions and personal rivalries between Orestes and Cyril and their supporters motivated Hypatia's death. Socrates Scholasticus and Damascius make the point explicitly, whereas John of Nikiu implies that Orestes' behavior is inexplicable without Hypatia's magic. For Socrates and Damascius, the murder of Hypatia was a local event, albeit one with wider moral lessons about the reprehensible behavior of some Christians (in Socrates) or some citizens (in Damascius). For John of Nikiu, Hypatia's death meant the end of paganism in Alexandria (or so he wished).

18.2 Medieval Interlude: The Invented Martyr Formerly Known as St. Katherine of Alexandria

Although the Hypatia story surfaces in a handful of Byzantine Greek sources, it did not diffuse in the Latin Middle Ages even though Cassiodorus's *Tripartite Ecclesiastical History* had translated Socrates Scholasticus into Latin.[22] Her story somehow found a structural analogue in a new saint's life. The legend of St. Katherine of Alexandria exhibits so many parallels to the earlier Hypatia narratives that some scholars plausibly suspect borrowing in the creation of a martyr from a murder victim. Its earliest form was an eighth-century Greek "passion" that was later translated into Latin. Both versions became very popular. Transcending mere literature, they became the foundation of cults, convents, and churches in both the Orthodox and Roman Catholic traditions. In the Latin inversion of Hypatia, the *Legenda aurea* casts Katherine as a beautiful, virginal, Christian woman from

[21] Petkas and Norman (2020, pp. 251–53); Pharr (1932).

[22] Translated in Petkas and Norman (2020, p. 250).

Alexandria who is supremely learned in all the liberal arts; Emperor Maxentius is a pagan who tries to convert her by making her debate fifty pagan philosophers.[23] His plan goes awry when her arguments convert them to Christianity. After having the reprobates killed, Maxentius proposes marriage to Katherine, who refuses and is condemned to dismemberment with a knife-studded wheel. When the mechanized device shatters, it is deemed a miracle rather than a flawed prototype of circular saw. The emperor thus falls back on time-tested decapitation, which the later iconography of St. Katherine spurned in favor of the more eye-catching broken wheel. After the universities came into being in the late twelfth century, the all-male faculties of arts found in St. Katherine's academic and dialectical brilliance an ideal to emulate and made the new martyr their patron saint.

Despite its alleged origins in Diocletian's third-century persecutions of Christians in Alexandria, the legend of St. Katherine is a literary composite that seems to refract the Hypatia story through a medieval glass darkly. Its creators understood the potential of that narrative structure for martyrology long before the nineteenth century adapted that genre to scientists. Already in the seventeenth century, skepticism about the historicity of St. Katherine caused her removal from the Paris breviary. Her formal demotion throughout Roman Christianity took much longer. In 1969, together with such popular saints as Nicholas and Christopher, she was purged from the Catholic calendar to the distress of Eastern Orthodoxy, which still venerates St. Katherine of Alexandria as a Great Martyr.[24]

18.3 Early-Modern Hypatia Narratives

Meanwhile, in the wake of both Reformation and Counter-Reformation, Hypatia narratives surfaced mostly in polemical ecclesiastical histories such as the multivolume *Annales ecclesiastici* (1588–1607) of Cardinal Caesar Baronius (d. 1607). Although he objected to the blame that Socrates Scholasticus laid on Cyril for his clergy's misdeeds, Baronius used Socrates extensively, supplementing him with details from Synesius and others. His Counter-Reformation stances did not prevent him from praising Hypatia as "the most distinguished of all the philosophers of her era" and a "most chaste virgin" who rejected the advances of an infatuated student.[25] His story fit into a long view of church history with ups and down but largely devoid of sharp breaks. No more than his sources did Baronius emphasize Hypatia's science. For the most part, later ecclesiastical histories followed suit.[26]

By the late seventeenth century, however, some narratives outside church history freed Hypatia from both that genre and the lopsided focus on her murder and its

---

[23] Lewis (2000, pp. 45–48); Voragine (1967, pp. 386–95); Gajeri (1992, pp. 43–44).

[24] Craughwell (2017), Walsh (2007, p. 10); Dzielska (1995, pp. 119, 150–51).

[25] Baronius (1866, pp. 55a–56b).

[26] Tillemont (1732, pp. 274–76).

perpetrators. Even as her horrible death remained irresistibly fascinating, her intellectual accomplishments began to draw attention in their own right. At Jena in 1689, Johannes Andreas Schmidt's "historico-mathematical dissertation" treated Hypatia briefly alongside his compilation of sources about Hipparchus and the two Theons, quoting his sources in Greek and translating them into Latin. Schmidt's booklet was short on both analysis and circulation, but it lifted Hypatia out of ecclesiastical history and gave her a foot-hold in the emerging histories of mathematics and astronomy.[27] The following year, the *érudit* Gilles Ménage, who had tutored some of the most learned women of seventeenth-century Paris, published a 140-page *Historia mulierum philosopharum* (1690). Hypatia got the longest of his brief biographies, twelve pages that highlighted her intellectual work with quotations from the usual sources. The book appeared in English in 1702.[28]

18.4 John Toland's "Hypatia" (1720)

Whether or not Ménage stimulated it, the eponymous essay by John Toland (1670–1722) soon became the most thorough account of Hypatia and remained so until the nineteenth century. Importantly, his post-Reformation retrospective on the preceding 1300 years gave her death a new significance in church history: she effectively stood "on the eve of destruction."[29] The post-Constantinian Church had ushered in centuries of corruption that Toland saw as still endemic in the clergy of Catholicism and High-Church Anglicanism. This enlarged time-frame made Hypatia a signpost of decline.

Toland was a political theorist, an historian, and a deist critic of Anglicanism who undergirded his biographical polemic with ample quotations from Greek primary sources in English translation. Like Ménage,[30] he highlighted the letters of Hypatia's former student Synesius (d. 412) and their details about her philosophical and astronomical learning. In some thirty pages, Toland showcased not just Hypatia's death, but her life. Since he had included all the sources known to him, Toland believed that his biography would make all future accounts redundant. This mistaken judgment did not prevent his work and the controversies it generated from securing Hypatia's fame as an astronomer and philosopher.

Toland first published his biography as the second of four essays in his *Tetradymus* (1720). Its title signaled a polemical work: *Hypatia; or, The History of a most beautiful, most vertuous, most learned, and every way accomplish'd LADY, who was torn to pieces by the Clergy of Alexandria, to gratify the pride, emulation, and cruelty of*

---

[27] Schmidt (1689).

[28] Ménage (1690), Leask (2020, p.1970).

[29] P. F. Sloan (1965) lyrics as sung by Barry McGuire (1965).

[30] Toland (1720, pp. 105–106, 121) paraphrases his Latin with a slight distortion and translates from a modern anthology first cited by Ménage (1690, pp. 3–4, 61–62).

their Archbishop, commonly but undeservedly stil'd St. CYRIL.[31] Born Irish and baptized a Catholic, Toland became an Anglican, then a Dissenter, then a deist and influential radical Whig. His later critics saw him as an atheist wolf in low-church clothing. After training in the classics, he also studied mathematics with David Gregory in Edinburgh and for the ministry in Leiden, where he associated with freethinkers and honed his skills as an historian. A harsh critic of established religion and its political entanglements, Toland touted reason as the road to true religion, attainable by paring away the irrational accretions that had distorted original Christianity and led to its decline (*Christianity Not Mysterious*, 1696).[32] Denounced by the Irish parliament, some of whose members called for Toland to be burned at the stake, the book was burned in Dublin by the public hangman. His personal sympathy for persecuted heretics also drew him to Giordano Bruno, about whose death he republished a rare Latin letter and whose philosophy he summarized and criticized in detail.[33]

Toland was also drawn to the Hypatia story. Although his opponents rightly read his biographical essay as an ecclesiastical polemic, it also marked an interpretative watershed on several fronts. To reach a large British audience, he wrote in English and translated his Greek sources. Toland dwelled at length on Hypatia's beauty and superlative intellectual accomplishments, which made her death all the more shocking. His new emphasis on her scientific and philosophical life was both an artefact of Toland's thoroughness and a reflection of his times. He was keenly aware of both discussions surrounding Newton's recent system of the world and persecutions of individuals associated with new cosmological views.

For Toland, the murder of Hypatia was more than a horrible local incident. Although he did not specifically discuss later incarnations of the Museum and library, he linked her accomplishments to "the Alexandrian School." Her death thus marked a dramatic institutional decline with long-term implications. Toland may have been a deist, but his irredeemably Protestant sensibilities shaped his readings of world history in the heyday of Newtonian science. After two centuries of Reformation polemics and more than a millennium of foreshortened hindsight, Hypatia's death roughly coincided with the dramatic break in Protestant historical periodization: the Church's fall after Constantine legalized Christianity. Like some Reformers, Toland believed that the Church's corruption began when it colluded with imperial power. Cyril's violent tactics and clerical jealousy of Hypatia's learning and influence were more than the personal flaws of one individual;[34] they characterized post-Constantinian Christianity down to his own day.

Beyond the historical interests of his essay, Toland's preface framed her story as follows: "Excesses of a nature beyond those I have hitherto related [in the essay

[31] Toland (1720, pp. 101–36), later reissued as a separate book.

[32] Hudson (2009, pp. 79–81, 84–91; Daniel (1984, pp. 22, 37, 53); Jacob (1969, esp. p. 310).

[33] Nicholl (1965, p. 54); Toland (1727, pp. 304–49).

[34] Cyril was "one of those clergymen who will neither acknowledge nor bear the superiority of any Laymen in this respect [learning];" Toland (1720, p. 131).

"Cliophorus" that preceded "Hypatia"], as endeavors to mob, if not even to worry men of learning and virtue, on account of their dissent from others in speculative opinions, made me write the life of HYPATIA." Toland took the Hypatia narrative in a new direction by linking the mob's actions to her "speculative opinions." Although vague, the relevance of her ideas to her death became the occasion for Toland to address the persecution of a contemporary for his ideas:

> I was mov'd to undertake this work in a particular manner, by the unlawful and unchristian usage that Mr. Whiston receiv'd from Dr. Sacheverell, who wou'd needs forcibly thrust him out of the Church in the midst of divine Service; tho the one fail'd of his attemt, and the other came off with no more harm, than being call'd *Rogue* by the pious old women, said to look like one by the sage Alms-men, and hooted (as I heard) by some boys instead of learning their Catechism.[35]

Toland's association of the two incidents demonstrates his awareness of structural similarities between the two narratives. The recent incident had occurred soon after January 23, 1719.[36] Since Toland's essay contained a wealth of material irrelevant to his analogy, the Whiston kerfuffle was surely not his primary motive for writing. It nevertheless offered ammunition for his two-decade-long politico-religious feud with Henry Sacheverell, a Tory and strident High-Church Anglican divine. Sacheverell defended kingship by divine right and opposed the toleration of dissent from the Church of England, views that Whigs saw as "a direct attack upon the Revolution settlement." Parliament eventually impeached Sacheverell, whose raucous popular support ("the mob" in one account) drew him a three-year ban on preaching.[37]

For his part, William Whiston was one of the most prominent mathematically proficient clergyman of the day. For twenty years he had been closer to Newton than most, after which, like most, he was also pushed away. His mastery of the *Principia* was so impressive that Newton had convinced Trinity College to appoint Whiston his successor in the Lucasian Chair of Mathematics (1702). Whiston later gave the prestigious Boyle lectures endowed to defend orthodox Christianity on natural philosophical principles. His life took an awkward turn in 1708 when, using Newton's biblical hermeneutics, Whiston became convinced that the New Testament offered no warrant for trinitarian doctrine. Unlike Newton, however, Whiston published his views. His *Sermons and Essays* (1709) precipitated his expulsion from Cambridge (1710), which he immediately protested in his four-volume *Primitive Christianity Reviv'd* (1710).[38] To the secretive Newton, Whiston soon became a liability for trying too hard to get his idol to "come out" as anti-trinitarian. Newton responded by blocking Whiston's nomination to the Royal Society (1716) and cutting off contact

[35] Toland (1720, pp. vii–viii).

[36] A good summary appears in Watts (2020).

[37] Harkness (1957, pp. 60–62); Force (1985, pp. 88).

[38] Force (1985, pp. 106–107).

while safely remaining master of the Mint, president of the Royal Society, and a member of Parliament.[39]

When Whiston was tried for heresy, he fought the charges for several years and unsuccessfully demanded a hearing in which to defend his conclusions on the evidence.[40] By 1713, he judged his case procedurally analogous to Galileo's. The church authorities had deprived both men of due process by refusing to consider evidence (in theology and philosophy, respectively). Referring to himself in the third-person, he wrote:

> Nor, methinks, did the Papists proceed much more absurdly, in not examining, but prosecuting, the famous Galilaeo, for his Assertion about the Motion of the Earth; which was then a notorious Heresy in Philosophy, and forcing him to recant it, than the Protestants in not examining, but prosecuting Mr. Whiston for his pretended Heretical Assertions now, and under the utmost Penalties the Law can inflict, to oblige them [him and Galileo] to recant the same. But, however, as notwithstanding that foolish Prosecution of Galilaeo, his Doctrine prevail'd by its own native Evidence; and the Observations afterwards confirm'd it, and is now the universal Belief of Astronomers; while the former Philosophick Orthodoxy is exploded for ever: so I believe will these [Unitarian] Doctrines, which are now discourag'd, prevail by their own original Evidence, and the farther Enquiries of the Learned, till the present Errors, which are unsupported by real Evidence, will gradually wear away, and disappear out of the Christian World.[41]

Whiston's analogies to his own life pioneered a loose nexus between Hypatia and Galileo that later martyrologists of science would tighten. Fortified by Galileo's triumph, Whiston for several years fought with Sacheverell, who was also the rector of his parish church. When Sacheverell ordered the former Cambridge don expelled from services, Whiston refused to obey without a lawful order. Denied a seat, he was forced to stand in the aisles with the servants.[42]

Referring to this *cause célèbre*, Toland's preface to the *Tetradymos* continues:

> But Mr. Whiston may (next to God's providence) thank the good nature of the English people for his preservation; for in some other parts of the world he had upon much fewer threats and remonstrances from two Priests, not to speak of the pulling and haling of their officer, been torn to pieces for all his gown and his innocence. For the rest, there never was a story better attested than that of Hypatia.[43]

Despite Toland's sympathetic allusions to Whiston's troubles, the two men strongly disagreed about many topics from biblical interpretation to the uses of natural philosophy. Indeed, Whiston's "Boyle Lectures" (1707) had sought to combat precisely the views of deists like Toland. The famous "General Scholium" that Newton added to the second edition of his *Principia* (1713) addressed the perceived "materialist" uses to which the likes of Toland had put his natural philosophy.[44] Like Newton,

---

[39] Westfall (1980, pp. 647–53). See also Snobelen (1999); Snobelen (2004); Feingold (2017).

[40] Force (1985, pp. 18–19, 109–10).

[41] Whiston (1713, pp. 7–8).

[42] Snobelen and Stewart (2003, pp. 137–56); Watts (2020, pp. 193–207).

[43] Toland (1720, pp. vii–viii).

[44] Snobelen (2001, pp. 174–75, 187).

Whiston was something of a biblical literalist who rejected Toland's rationalist depreciation of Christian revelation. Indeed Whiston had defended his anti-Trinitarianism on biblical grounds, not rationalist ones.[45] However much he may have appreciated Toland's defense of him, he wanted no association with the latter's ideas.[46]

Toland obviously knew his analogy between the fates of Hypatia and Whiston was hyperbolic, but it was too good to pass up. The Hypatia story was an extreme version of the acrimonious debates that intertwined contemporary politics, religion, philosophy, and science with the role of women in Restoration Britain. More importantly, Toland's analogy also helped to foreground Hypatia's scientific work. Although her knowledge and wisdom were irrelevant to her death in the earliest sources, Toland quoted Hesychius of Miletus's comment from the *Suda* that her dismemberment "*befell her on the score of her extraordinary wisdom, and especially her skill in astronomy*" (Toland's italics signify quotation).[47] Toland thus cast Hypatia as persecuted also for her learning, including her astronomy, rather than solely for her political actions or associations. Both episodes concerned individuals incidentally engaged in science and illustrated persecution for their beliefs, which Toland defined as "suffering of any degree or kind on the account of mere opinion."[48]

Earlier sources had shamed Hypatia's murderers for conduct unbecoming Christians (Socrates) or Alexandrian citizens (Damascius). Toland now broadened the list to include men. His essay opened by claiming that Hypatia would "ever continue the glory of her own sex and the disgrace of ours," a proto-feminist outlook that echoed Ménage's *Historia mulierum philosopharum*.[49] Following Philostorgius, he appraised Hypatia's competence in mathematics and astronomy as surpassing her father's.[50] He presented philosophy ("the highest perfection") as a proper object of study for women and identified Hypatia as the pre-eminent philosopher of her day, defensively insisting that her scientific competence did not detract from her excellence as a philosopher. As Toland had defended the intellectual capabilities of women in earlier writings, this motivation for writing his "Hypatia" long antedated the Whiston-Sacheverell episode. Around 1720, however, this theme was even timelier for he was courting.[51]

Toland's essay emphasized another thread that would resonate long after him: the linkage between science and long-term historical decline within Christianity. To his ingrained Protestant sensibilities, Cyril and the Alexandrian clergy were

---

[45] Force (1985, pp. 6, 100, 113); Champion (1999, p. 91); Force (1994, p. 181).

[46] Whiston lumped Toland and his rationalist approach to the Bible with "comparatively ignorant and unlearned Writers ... [who] may have proceeded [much farther] in their grosser Degrees of Infidelity." Whiston (1749, p. 109).

[47] Toland (1720, p. 131); compare with Petkas and Norman (2020, p. 248).

[48] Toland (1720, p. xviii).

[49] Toland (1720, p. 103); Ménage (1690, pp. 52–63); Wyles (2016); Leask (2020); Watts (2020, pp. 199–200).

[50] Toland (1720, pp. 105–06).

[51] Hudson (2009, pp. 94–95); Champion (2003, pp. 52–4).

emblematic of the post-Constantinian church, whose persecution of Hypatia signaled a historic decay: "of genuin [sic] Christianity, there remain'd very little at that time unless ... in the bare name and profession."[52] The legalization of Christianity in the fourth century had allied it with imperial politics, a collusion that explained the decadent strands of traditional Christianity that Toland criticized as needing reform in the Roman and Anglican Churches of his day.[53] When he fitted his emphases on Hypatia's science and the persecution of her ideas into the framework of church-historical decline, he was echoing the narrative structure of the Galileo Affair.

Toland concluded his essay by attacking the post-Constantinian proliferation of illegitimate saints, which included Cyril: "the epithet Saint, which peculiarly belong'd to Piety and Innocence, was thus pompously bestowed on Vice and Impiety," prompting a "deluge of Ignorance, Superstition, and Tyranny, which overwhelmed almost the whole Christian world." Persecutions of learning and virtue followed. Toland stopped short of calling Hypatia a saint, merely hinting at it with stark ethical and intellectual contrasts between her and the clergy: she was "farr more learned than themselves, not to be endured in the laity."[54] Likewise, without discussing a widespread scientific decline, he implied as much. Having earlier placed Hypatia at the head of the "Platonic school at Alexandria," Toland had her preside over its culmination: "the Alexandrian School never flourish'd more."[55] This institutional embellishment made the subsequent decline sharper as he conflated Hypatia's death with that of "the Alexandrian School." Later storytellers would tighten this association even further.

18.5 Telltale Attacks on Toland's "Hypatia"

The firestorm that Toland's essay prompted illustrates how he was read. The Anglican priest Thomas Lewis's *History of Hypatia* (1721) attacked Toland for using his essay to "blast the Reputation of the Venerable St. Cyril, a strenuous Assertor of Orthodoxy and Church Discipline, and of Dr. Sacheverell his faithful successor in Zeal, against the Modern Novatians and the Arian Heresie." Lewis conceded Toland's analogy between Cyril and Sacheverell but reversed its valuation. He also endorsed that between Hypatia and Whiston, which he used to denigrate both their astronomical interests and their unorthodox beliefs: "the Characters of these Persecuted two Philosophers run exactly parallel, both of them affecting to be thought Astronomers, and alike professed enemies to the Divinity of Christ." Thirteen-hundred years later, trinitarian debates were still raging and Hypatia was on the wrong side. After slurring her as a "She Philosopher," Lewis also impugned

[52] Toland (1720, p. 133).

[53] Sullivan (1982, pp. 129–30, 142).

[54] Toland (1720, pp. 129, 136).

[55] Toland (1720, chap. 2–5, esp. pp. 107–108).

her modesty. She had visited Orestes alone and used her bloody sanitary napkin to teach a disgusting lesson "which I believe the most common prostitute in Venice would blush at." As his reading also shows, he correctly understood Toland's emphasis on her science: Lewis deemed it preposterous that Cyril might have been jealous of Hypatia's astronomical knowledge.[56]

Toland had attempted a definitive biography of Hypatia that reflected his own post-Reformation critiques and his post-*Principia* elation. As his critics recognized, it was also a polemical broadside in both ancient and modern ecclesiastical history. Although the earliest histories had chastised the behaviors of local individuals and the Alexandrian church hierarchy, Toland's critique of post-Constantinian Christendom condemned the Christianity of a millennium. Blame for the murder of the astronomer-philosopher Hypatia and incidentally the demise of the "school of Alexandria" rested with Cyril, who represented the corrupt and ignorant version of Christianity that constituted medieval and early modern Catholicism. With his analogies and use of the Hypatia story to frame his millennial polemics, Toland expanded at once its temporal baseline, its scale, and its significance. He intuitively understood the fractal character of its narrative structure, which was transposable both on the small scale (Hypatia-Whiston, Cyril-Sacheverell) and on the large scale of the decline narrative that he exploited for his polemics in church history. Although he dropped hints of a parallel large-scale decline in intellectual life and science, he did not develop them.

18.6 Echoes Across the Channel

Toland had at least one notable impact in France. A young woman who had heard about Hypatia asked a certain M.G. to write an essay about her. The latter complied with a long critical *Dissertation sur Hypacie* that rivaled Toland's in size and, like him, attended to her mathematical and astronomical work before turning to the murder and its perpetrators. The *Dissertation* took most existing sources into account, including Ménage, Fabricius, de Tillemont, and Toland. It devoted much of his essay to Hypatia's accomplishments, but implicitly took to task both Toland ("*les Protestants … [mais] pas … tous*") and his sources—Socrates the Arian heretic, Damascius the pagan philosopher—for tendentiously pronouncing St. Cyril guilty of murder.[57]

M.G. was exceptional in France for adopting Toland's emphasis on Hypatia's work in mathematics and astronomy. When the *philosophes* treated her, they cast her learning as broadly philosophical. In 1751, when the French mathematician Jean Le Rond d'Alembert published his *Discours préliminaire* to the *Encyclopédie,* he took the long view of variable progress in the arts and sciences. The despotism of

[56] Lewis (1721, pp. 3, 9, 13, 31).

[57] M.G. (1728, pp. 182–83); Watts (2020, pp. 202–205).

"a few powerful theologians" (*quelques Théologiens peu nombreux, mais puissans*") over philosophy, including science, had retarded progress. Later he alluded to only two cases: Pope Zachary's condemnation of a bishop for believing in the antipodes before Columbus discovered them, and the Inquisition's condemnation of Galileo.[58] Hypatia did not come up.

In the *Encyclopédie* itself, Hypatia appears as a philosopher in the article "Éclectisme." To twenty-first century eyes, this seems an odd rubric. For Diderot, however, the term encapsulated the ideal of non-dogmatic free thought: Eclecticism was not a philosophical sect because it opposed all sectarianism. Eclectics shared with Skeptics the Horatian motto *Nullius addictus jurare in verba magistri* ("committed to swearing upon the words of no master," *Epistles* I.1). After identifying Hypatia as one of the female "disciples of Eclecticism," the article made her death the end of not philosophy or science, but the ancient "eclectic sect" (!), which was only revived in the sixteenth century. The article mentioned Hypatia's education in geometry and astronomy, rounding out her universal genius in the arts, sciences, philosophy (Plato, Aristotle), and oratory.[59]

For the *Encyclopédie*, then, Hypatia symbolized the end not of Greek science, but of the philosophical "non-sect" of eclecticism, left nameless until the late sixteenth century. Diderot listed seven names among its early members, but "after more careful consideration" he redistributed all but one under non-eclectic headings (Peripatetics, Platonists, and Christians). In retrospect "Eclecticism" looks like a convenient rubric for handling the uncertainty and many competing claims about Hypatia's philosophical views.

Soon after Toland published "Hypatia" (1720) and died (1722), his younger contemporary Voltaire exiled himself to England for three years. Although conversant in English, Voltaire seems not to have read Toland.[60] When he later diffused the Hypatia story in France, his few pithy sentences relied on other sources. In his (initially anonymous) *L'examen important de Milord Bolingbroke...*, Voltaire used the Hypatia story as an anticlerical vignette.[61] At the end of chapter thirty-seven ("On the Christians to Theodosius"), he lashed out at the cowardice of Cyril's priests toward Hypatia, "famous for her beauty and 'spirit'… praised as much for her knowledge (*science*) as for her behavior but she was pagan." She was persecuted not for her natural science, but for her religion. Voltaire's brief retelling drew few distinctions and was otherwise sloppy. He made Theon and Hypatia generic philosophers who held official chairs. Skipping the political tensions in Alexandria, he blamed the murder directly on Cyril's fanatical "tonsured mastiffs" and indirectly

58 D'Alembert (1751, pp. xxii–xxiv); see also Finocchiaro (2005, pp. 120–21).

59 Diderot (1755, pp. 273, 282–83).

60 Torrey (1930, pp. 4, 12–13, 19, 201).

61 Voltaire later republished the *Examen* as *Le tombeau du fanatisme*.

on the latter and Emperor Theodosius I (who was dead) for failing to condemn this "excess of inhumanity."[62]

The first editions of Voltaire's anonymous *Dictionnaire encyclopédique* (1764, 1765) did not mention Hypatia, but he added to the augmented Geneva edition (1772) an article "Hypathie" that referred to the *Encyclopédie* article.[63] Paralleling Toland's contemporary analogies, Voltaire updated Hypatia's story to *fin-de-siècle* Paris in clear awareness of its transposable narrative structure. He imagined the archbishop of Paris murdering Anne Dacier (d. 1720) in the cathedral of Notre Dame after accusing the brilliant translator of Homer (and reputedly the most beautiful woman of her day) of valuing Homer more than monkish poetry. Dacier was indeed a famous intellectual. Ménard had dedicated his book on women philosophers to her, proclaiming her the most learned of all women ever.[64] Voltaire's transposition also had an implicit religious dimension: Dacier had been raised Protestant.[65] In sharp contrast to Toland's feminist sensibilities, Voltaire ended his story nastily by remarking that there were better uses for naked women than murder victim.[66] Since he judged his analogy to Dacier to be "precisely the story of Hypatia," he saw her scientific activities as irrelevant to her fate. Had they mattered, Voltaire should have given the role to someone like his former companion, Madame du Châtelet, who among other accomplishments in physics and mathematics, had translated and commented on Newton's *Principia*.

18.7 Gibbon's *Decline and Fall* (1776–1787)

Writing from Lausanne a half-century after Toland and some forty miles from Voltaire's château near the French border, Edward Gibbon surprisingly called his *History of the Decline and Fall of the Roman Empire* "my ecclesiastical history,"[67] the *genus classicum* for Hypatia narratives. His treatment was short—one page in the long forty-seventh chapter in the fourth volume (of six). It nevertheless deserves a brief discussion because many later writers used it and extrapolated from it. Writing under the heading "His tyranny," Gibbon's Hypatia story was one indictment of many in his extensive prosecution of Cyril of Alexandria.[68]

---

62 [Voltaire] (1776, pp. 185–86). Voltaire later modified the conclusion. Having learned, perhaps from the *Encyclopédie*, that Theodosius I no longer ruled in 415, he blamed "Theodosius the Younger and his devout sister Pulcheria," while salvaging part of his critique of the father; see chap. 34 in the later version of the work; Voltaire (1961, p. 1108).

63 Voltaire (1772, pp. 138–140); Voltaire (1879, pp. 392–93).

64 Ménage (1690, pp. 3–5).

65 Wyles (2016, p. 65).

66 Watts (2017, p. 137).

67 Imbruglia (1997, p. 76); see also Chadwick (1976, p. 111).

68 Gibbon (1788, Table of Contents and p. 549).

Gibbon's account is also important because his framework and fame embedded Hypatia and her death in the centuries-long picture of *Decline and Fall* in his title, an historical arc with a radical Protestant structure.[69] More obviously than in Toland, Cyril was but one actor in a debased Christianity that shared the blame for ruining classical culture. In keeping with his "ecclesiastical history," the backdrop for his Hypatia story was a dense account of competing doctrines of the Incarnation and their supporters' ugly political battles. Gibbon used Baronius extensively elsewhere in his work, but his treatment of Hypatia drew heavily on LeNain de Tillemont's massive church history to excoriate Cyril.[70] The bishop had "usurped the state and authority of a civil magistrate," repressed the "most innocent" Novatian sect, destroyed the synagogues, and expelled the large Jewish community. Moreover, Cyril had honored as a martyr his disciple Ammonius, the leader of the Nitrian monks, whom the Roman prefect Orestes had executed for trying to assassinate him. That execution in turn incited Cyril's followers to take revenge on Orestes by murdering Hypatia.

Gibbon echoed Toland's sacrificial imagery, bitingly given Christian opposition to pagan sacrifices: "[Cyril] soon prompted, or accepted, the sacrifice of a virgin, who professed the religion of the Greeks, and cultivated the friendship of Orestes."[71] The liability of Hypatia's pagan religion preceded that of her political associations. Although Gibbon mentioned her commentaries on Apollonius and Diophantus before her teaching of Plato and Aristotle, her science figured in her death only because her popularity as a teacher had aroused Cyril's jealousy and murderous fantasies. Precipitating her death was the "rumour among the Christians that the daughter of Theon was the only obstacle to the reconciliation of the praefect with the archbishop; and that obstacle was speedily removed" by Peter the reader and his followers. In keeping with his sacrificial motif, Gibbon enhanced the horror of the murder with garish, idiosyncratic details: "her flesh was scraped from her bones with sharp oyster shells, and her quivering limbs were delivered to the flames." One historian reads this curious rendering of *ostraka* (properly "tiles," or "potsherds") as Gibbon "maliciously using the language of Christian martyr texts."[72] Gibbon concluded that "the murder of Hypatia has imprinted an indelible stain on the character and religion of Cyril," a harsh paraphrase of Socrates. Toland had limited his criticisms of Cyril to the frame of Hypatia's life but in Gibbon's saga, the worst of Cyril's behavior was yet to come.[73] In both cases, Cyril's actions stood for more than those of one man; they represented an entire era of Christendom.

---

[69] Chadwick (1976, pp. 113–16).

[70] Tillemont (1732, pp. 274–76). Besides Socrates Scholasticus, Renaudot, and Suidas [the *Suda*], his sources included Fabricius (1717, pp. 219–21). For a clear-eyed evaluation of Gibbon's sources and interpretations, see Chadwick (1976).

[71] Gibbon (1952, II, pp. 138–39). Some scholars also see sacrificial connotations in the vocabularies of Socrates, Philostorgius, and Hesychius (Haase 2020b, p. 269).

[72] Cameron (2013, pp. 67–68).

[73] Gibbon (1952, II, p. 139); Chadwick (1976, pp. 112, 122).

Despite his overarching thesis, Gibbon did not suggest ecclesiastical opposition to Hypatia's scientific work. To the contrary, Cyril's jealousy of Hypatia's mathematical and philosophical knowledge implicitly acknowledged their appeal. Like Toland before him, Gibbon collected, integrated, and diffused widely in his brief narrative the details that later, large-scale science-versus-religion narratives would pick up. Indeed, he had no desire to highlight the persecution of science for he resented the neglect of letters and erudition that he associated with the intellectual imperialism of mathematics and physical science found in D'Alembert's *Discours préliminaire* among other encyclopedists. Against this trend, which he distrusted, his youthful "Essai sur l'étude de la littérature" called geometry "the imperious queen who is not satisfied with reigning: she must also banish her sisters, declaring all reasoning that does not run to lines and numbers unworthy of that name." Or again: "Physics and mathematics are now on the throne. They see all their sisters prostrate before them, chained to their chariot or at most engaged in adorning their triumph. Perhaps their fall is not too distant."[74]

Gibbon's account assembled the elements from which later narrators could construct the proposition "churchman opposes scientist," but he did not do so. His quarry was Cyril, less because he was a Christian than because he was a tyrant. It by-passed the personal and political reasons for the murder offered in the early accounts. Most importantly, Gibbon placed the Hypatia incident in a sweeping narrative of centuries-long political fall. Despite their different sensibilities and purposes, for Gibbon as for Toland, the blame for her death lay with unreformed Christianity, which Gibbon's work took into the sixteenth century.

Strikingly, Gibbon's Hypatia story leaves no hint of the later synecdoche according to which her death stands for that of ancient learning. Unlike many later authors, he did not link her story to the demolition of the temple to Serapis under Theophilus and Emperor Theodosius (several hundred pages earlier in Chap. 28). There, his attention went primarily to the destruction of the building and the dismemberment of the giant statue of the god. He dealt with the associated library in one laconic, doubt-filled disjunction: "The valuable library of Alexandria was pillaged or destroyed." To this he added Orosius's eye-witness account of empty shelves 20 years later and a comment bemoaning the archbishop's failure to separate the attack on idolatry from the gratuitous loss to posterity of "the compositions of ancient genius."[75]

---

[74] Section XLV. My translation adapts Pococke's based on the original French cited in Pococke (2004, vol. 1, pp. 215–16, 227–28; see also 225–26, 239; p. 71 about Alexandrian Neoplatonists and magic (chap. 13).

[75] Gibbon (1952, I, p. 462).

18.8 Categorical Digression: Brewster's *Martyrs* (1841) and Barni's *Les Martyrs* (1862)

The sacrificial language in Toland and Gibbon merely hinted at Hypatia's martyrdom, the main theme in the parallel legend of St. Catherine of Alexandria. For very different reasons, that motif became explicit in both *The Martyrs of Science, or The Lives of Galileo, Tycho Brahe, and Kepler* (1841), the bizarrely titled book by Sir David Brewster, the physicist and Newton biographer, and the philosopher Jules Barni's *Les Martyrs de la pensée libre* (1862).

Destined for the ministry, Brewster had studied divinity at Edinburgh but hated public speaking.[76] Against many career odds and financial difficulties, he became a prominent physicist who specialized in optics and co-founded the British Association for the Advancement of Science.[77] Chided even by his family for his title, Brewster used "martyr" loosely. Although his three heroes died unaided, they all faced hardships with which he sympathized. As J. R. R. Christie has noted, Brewster identified with "the civil martyrdom inflicted upon men of science by governments, kings, and emperors, who for much of history had constantly failed to provide support for men of science and to grant them the recognition, status and state salary they deserved."[78] Equally odd, "martyr" appeared in the text only in two florid sentences in which Brewster bemoaned Galileo's refusal to die for his convictions:

> Why did this master-spirit of the age … reject the crown of martyrdom which he had himself coveted, and which, plaited with immortal laurels, was about to descend upon his head? […] in the love of truth he found a miserable substitute for the hopes of the martyr. Galileo cowered under the fear of man, and his submission was the salvation of the church.[79]

Brewster paired his insight into Galileo's psychology with a charge of cowardice that expressed his own quasi-religious commitment to scientific truth.[80] As a man of science, Galileo had effectively apostatized. A more insightful and compelling account of Galileo's state of mind and behavior appears in Heilbron's *Galileo*.[81] Given Brewster's wish for a Galilean *auto-da-fé*, it is revealing that he worked neither Hypatia nor Giordano Bruno into his book: their deaths had not yet become popular scientific lore. Indeed, in 1837 William Whewell's *History of the Inductive Sciences* had specifically dissociated Bruno's speculative astronomical opinions from his trial for heresy (he did not discuss Hypatia).[82]

---

[76] Cochran (1984, p. 13); Finocchiaro (2005, pp. 223–25).

[77] Morrell (1984).

[78] Christie (1984, p. 53).

[79] Brewster (1841, pp. 94–95). Whewell, without naming him, criticized Brewster's designation of Galileo as a martyr; Whewell (1875, 1, pp. 281–82); see also Baxter (1984, pp. 48–50).

[80] Christie (1984, p. 54).

[81] Heilbron (2010, chap. 8).

[82] Whewell (1837, vol. 1, p. 384).

Jules Barni’s *Martyrs de la libre pensée* (1862) confirms that the connections between the deaths of Hypatia and Bruno and science were not obvious. A Frenchman, Barni was a Kantian professor of philosophy whose democratic activism during the Second Empire led to his exile in Geneva, where he taught and delivered eight public lectures on persecuted free thinkers from Socrates to Madame de Staël. His subject allowed him to cast a wide net. He devoted his third lecture to Hypatia as a philosopher and to her Alexandrian context. Even though he used Synesius’s letters, he barely mentioned her mathematical and astronomical training. Although he did not specify the ideas for which she was persecuted, he counted her death as martyrdom for free thought. By contrast, Barni crammed four Italians into the eighth lecture: Bruno and Lucilio Vanini (both executed), Tommaso Campanella (severely tortured) and Galileo. Bruno got the most space, followed by Galileo. Barni’s quotation of the Inquisition’s reasons for demanding Bruno’s extradition from the Venetian Republic for heresy did not include his cosmic views. For Galileo, Barni quoted Jean-Baptiste Biot’s 1816 account of the trial, including the latter’s designation of Galileo as a martyr for the moral torture that his abjuration caused.[83] These cases of martyrdom for free thought paid little attention to scientific beliefs. Within a few decades, however, popularizers of the history of science would firmly inscribe the names of Hypatia, Bruno, and Galileo in a growing scientific martyrology.[84]

18.9 Draper’s New Periodization of the History of Science (1874)

A leading promoter of this development was one of Brewster’s younger colleagues in scientific photography, the wide-ranging Anglo-American scientist John W. Draper (1811–82).[85] In two popular works of Western intellectual history, Draper presented a compelling *longue durée* narrative of the history of science in which Hypatia’s death marked an inflection of millennial significance. In his equation of the history of science with the story of its conflict with religion, the murder of Hypatia marked the onset of a catastrophic scientific decline. In Draper’s retelling, the microstructure of her story has a large-scale counterpart in his periodization of the history of science. This new fractal narrative about scientific rise and fall parallels the large-scale church-historical narrative of Christianity’s spread and corruption that Toland and Gibbon shared. This parallelism is no accident. In a developmental account that identifies the history of science with its conflicts with religion (construed as Catholic Christianity), the template that structures the history

[83] Barni (1862, pp. 234–35); see also Finocchiaro (2005, p. 234).

[84] See the 126 names (not Hypatia, however) in Tissandier (1879, p. [335]); and Slater (1880), who comments on Tissandier.

[85] Gillespie (2012, pp. 248–50).

of science will in part reflect the arc of Christian history, in this case from a Protestant point of view. Hypatia thus entered scientific lore in a position more pivotal than that of her modest presence in earlier histories of mathematics.[86]

The son of an English Methodist minister, Draper had studied chemistry at the University of London. After emigrating to the U.S. and earning an M.D. at the University of Pennsylvania, he became a major figure in American science, medicine, and technology. His polymathy included pioneering photographic techniques, establishing the thermal limit of black-body radiation (1847), co-founding New York University's medical school and serving as the first president of the American Chemical Society (1876).[87] Draper also had serious amateur historical interests with Comtean underpinnings: a search for the laws underlying not only natural phenomena but also the "career of nations" and the "seasons" of civilization.[88] Indeed, in 1860, at the Oxford meeting of the British Association for the Advancement of Science, Draper had delivered before a packed hall a long paper entitled "The Intellectual Development of Europe (considered with reference to the views of Mr. Darwin and others) that the Progression of Organisms is Determined by Law." The discussions that followed Bishop Wilberforce's official response to his paper became the occasion for the interchange that was mythologized two decades later into "the Huxley-Wilberforce debate."[89] Sketched in his paper, Draper's quest for lawlike historical patterns induced him to write two works of intellectual history, between which he published a remarkably early history of the U.S. Civil War (1867).

Draper initially included the Hypatia story in his unfootnoted two-volume *History of the Intellectual Development of Europe* (1863). His Oxford paper was the abstract of his first major historical venture. This first account of Hypatia and her context are important because Draper would condense its details and complexity to different effect a decade later. Subtitled "The Rise of Science," Chapter six of *Intellectual Development* highlighted the glory of Alexandrian science from its early days to Ptolemy, after whom Draper judged the city in intellectual decline three centuries before Hypatia. The next chapter was subtitled "The Death of Greek Philosophy." Chapter ten ("The European Age of Faith") returned to Alexandria under Emperor Theodosius during whose reign "the pagan religion and pagan knowledge were together destroyed." Paralleling Gibbon's account, the purple prose of his relatively nuanced account painted the Serapeum temple as both the majestic site for the remains of the Alexandrian library and a center of pagan worship. After Bishop Theophilus insulted their cultic symbols, the pagans led by a philosopher rioted, killed several Christians, forced others into making pagan sacrifices, and retreated into the Serapeum. News of the drawn-out disturbance led Emperor

[86] Montucla (1758, pp. 259, 326).
[87] Draper (1847); Gillespie (2012).
[88] Ungureanu (2019, pp. 31–39).
[89] Ungureanu (2016); James (2017, esp. pp. 176–182).

Theodosius to order the demolition of the temple. Like Gibbon, Draper cited Orosius's claim that the library building survived but with shelves devoid of books.[90]

Draper's *Intellectual Development* used that episode to set up the story of Hypatia, the "beautiful young woman" to whom he devoted three pages. His account pioneered a close association between the destruction of the Serapeum's library and her death. "Suppression of Science/in Alexandria" was the running head for the chapter's last two pages, which did not distinguish between science and philosophy: "in the 414th year of our era, the position of philosophy in the intellectual metropolis of the world was determined; henceforth science will no longer be tolerated … from this period for some centuries it altogether disappeared." This is counterfactual, as we shall see. Draper nevertheless had science resurface in Alexandria thanks to the "Arabian conquerors" under whom astronomers measured the earth's circumference, the motion of the sun's apogee, and the third lunar inequality. Since Draper, like a few before him and many after him, wrongly believed that medieval Christendom took the earth to be flat, the contrast with the Muslim astronomers was stark indeed.[91]

Draper's celebration of astronomy under Islam would soon clash with the title of his most famous book, *A History of the Conflict between Religion and Science* (1874). The generality of the word "religion" was slightly misleading. It cloaked an anti-Catholic polemic triggered by Pius IX's recent aggressive pronouncements and by the upheavals in war, politics, and religion brewing during Draper's first trip to the European continent (1870–1871). In 1870, the Franco-Prussian War broke out, Italian troops annexed Rome and dissolved the papal states, which the papacy ("the prisoner in the Vatican") refused to recognize for 60 years. In the "Syllabus of Errors (1864), the pope had condemned freedom of religion and the separation of church and state, among other subversive modern ideas. More recently, the First Vatican Council (1869–70) had made papal infallibility a dogma and neo-Thomism the Roman Church's official theology. Meanwhile, 1871 inaugurated the *Kulturkampf* between Bismarck and Pope Pius IX about the appropriate boundaries of church and state.

These developments were fresh in Draper's mind when in 1873 his publisher commissioned *Conflict* for a series of books by eminent scientists. Draper briskly recycled earlier articles and much content from *Intellectual Development* while abridging earlier arguments and updating his polemic in light of current developments. He finished within the year. *Conflict* appeared in 1874 with three concluding chapters that attacked the pope's recent pronouncements and political meddling.[92]

In his preface, Draper made clear that the book was no antiquarian project but a historical polemic that responded to recent events. The aggressive modern papacy colored the long arc of the history of science because it insisted "on a political

---

[90] Draper (1864, I, pp. 307, 309–10); Gibbon (1952, I, pp. 462, 819).

[91] Draper (1864, I, pp. 314–15; II, pp. 154–55); Draper (1874, pp. 159, 294); Cormack (2009, pp. 28–29).

[92] Fleming (1950, pp. 122–30); Principe (2018, 12–21); Draper (1874, chap. 10–12).

supremacy in accordance with its claims to a divine origin and mission, and a restoration of the mediaeval order of things, loudly declaring that it will accept no reconciliation with modern civilization." After touting his impartiality, Draper offered the following extraordinary definition of what eventually became our discipline:

> the history of Science is not a mere record of isolated discoveries; it is a narrative of the conflict between two contending powers, the expansive force of the human intellect on one side and the compression arising from traditionary faith on the other.[93]

By defining the history of science as a long conflict between science and religion, Draper shaped the popular understanding of the discipline before it was professionalized. Thanks to the large audiences that his book eventually draw, he left an indelible imprint on both popular and learned historiography. Although he defined neither "traditionary faith" nor "religion," he shrank the generality of his title: the book would say little about either moderate views or the Protestant and Greek Churches, which had welcomed science since its "restoration" around the sixteenth century. Draper intentionally avoided a statistically balanced account: "in conflicts of this kind it is not with the moderates but with the extremists that the impartial reader is mainly concerned. Their movements determine the issue." By "Christianity," he generally meant "Catholicism" because of its large membership, its pretentious demands, and its enforcement of the latter with civil power.[94] This last point deserves emphasis: religion's use of political power was particularly nefarious.

Draper's preface is crucial to understanding Hypatia's paradoxical place in his overall argument. He stated that "its [Christianity's] incompatibility with science caused it to suppress forcibly the Schools of Alexandria. It was constrained to this by political necessities of its position." Surprisingly, however, the events in turn-of-the-century Alexandria did not count as one of Draper's conflicts. Instead, they introduced "the parties to the conflict," which unfolded in four to five stages long after the Hypatia episode.[95]

Draper's conflicts are remarkable for the theological themes that drove them. The "first, or Southern Reformation" concerned "the nature of God" as the Muslim conquests imposed on the Near-East and North Africa their radical monotheism and "fatalism:" unlike the interventionist miracles of Christendom, the conviction that immutable law governed the world would become a prelude to "the restoration of science."[96] Islam's rejection of anthropomorphism led to "a second conflict, that respecting the nature of the soul" with the rise of Averroism (identified with the "active intellect"). Since the late-medieval church had condemned Averroism and the recent Vatican Council had anathemized it again, Draper treated this conflict as ongoing. In the more famous third conflict about the place of the Earth, Galileo had taken the side of science, leading to the eventual "overthrow of the Church on the question in dispute." Under this rubric, Draper also included the conflict about the

93 Draper (1874, p. vi).

94 Draper (1874, pp. vi, x–xi).

95 Draper (1874, pp. xi-xiii; see also 336–39, 343).

96 Draper (1874, pp. 106–110).

age of the Earth, which the Church also lost. Draper's fourth conflict was out of order chronologically and jarring thematically: "known to us as the Reformation—the second or Northern Reformation," it turned, curiously enough, on whether the Church or the Bible was the criterion of truth—not exactly a burning scientific issue. Draper optimistically considered that conflict settled thanks to Luther by "reason, or intellectual freedom" and resulting in Roman Christianity's loss of Northern Europe—not exactly a scientific outcome. Finally, as the new conflict unfolding in his own day, Draper identified the question whether the world is governed "by incessant divine intervention or by the operation of primordial and unchangeable law" in the matters of "Evolution, Creation, Development."[97] Despite this emphasis and his personal involvement in the incident, Draper did not mention the 1860 meeting of the B.A.A.S. The "Huxley-Wilberforce debate" had not yet been turned into an iconic rival of the Galileo Affair.[98]

As Draper's outline shows, his framework for the history of science bore a deep church-historical imprint, which extended from its Protestant perspective on rise-and-decline, through its Reformation vocabulary, to the key issues that organized his four-to-five conflicts.

In this schema, the Hypatia episode was an outlier, and not simply because it preceded Draper's canonical conflicts. In every one of the latter, he identified a long struggle about a specific intellectual issue. The brief Hypatia story suggested none. Although her story fell outside his schema, it involved an abuse of ecclesiastical power gripping enough to include. Draper thus gave it two other significant roles: it introduced the "parties to the conflict," and it marked a sharp break in his periodization of the history of science.

Writing before the decipherment of cuneiform astronomical tablets, Draper may be excused for starting *Conflict* with the Greeks. He devoted half of the first chapter ("The Origin of Science") to Alexandria, whose Museum he made "the birthplace of modern science" because, unlike the rest of ancient science, it relied on "observation, experiment, and mathematical discussion instead of mere speculation." In Alexandria, he saw "the first example of that system, which in the hands of modern physicists, has led to such wonderful results." He made the great Library stupendous: 400,000 volumes with an annex of 300,000 more in the temple of Serapis.[99]

Although he might have concluded the first chapter with Hypatia, he saved her story for the next one, which treated the rise of Christianity and its corruption after Constantine. When discussing earlier religious conflicts in Alexandria, Draper wrote empathetically about persecuted Christianity. Once the persecutions ended, however, Christianity's collusion with imperial power "brought it into conflict with science." The early tensions came less from its original beliefs than from its self-arrogation of political power. For Draper as for some sixteenth-century Reformers, the church's involvement in secular power after Constantine initiated a long decline

[97] Draper (1874, pp. xiii–xv).
[98] James (2017, p. 187).
[99] Draper (1874, pp. xiii, 19, 27).

caused by the politically expedient "amalgamation of Christianity with paganism" (e.g., the Trinity, the cult of Mary, some rituals, the veneration of relics).[100] Thus tainted and empowered, the Church set itself up as the sole arbiter of permissible belief and knowledge: "She thus took a course which determined her whole future career: she became a stumbling-block in the intellectual advancement of Europe for more than a thousand years,"[101] a sweeping claim that still circulates in misinformed circles.

Bishop Theophilus had been under imperial orders to destroy the Serapeum temple after which "the great library, which been collected by the Ptolemies and had escaped the fire of Julius Caesar was by that fanatic dispersed." Draper took care to disperse rather than destroy the remnants of the earlier library he associated with the Serapeum: manuscripts were needed to explain the revival of Greek science under Islam (the "first Reformation"). Within this large frame, his one-page story of Hypatia solidified the modern popularized association of her death with a precipitous intellectual decline. Lumping together philosophy, science, profane knowledge, and free thought, Draper used the episode to set up the polarity needed to establish the "parties to the conflict" in his title:

> Hypatia and Cyril! Philosophy and bigotry. They cannot exist together. So Cyril felt and on that feeling he acted. [An account of the assault and murder follows] So ended Greek philosophy in Alexandria … The "Daughter Library," that of the Serapeum, had been dispersed. The fate of Hypatia was a warning to all who would cultivate profane knowledge. Henceforth there was to be no freedom for human thought.[102]

In a change of scale reminiscent of Toland, Draper jumped from two individuals to two general features that they embodied. Incarnating the post-Constantinian Church's lust for power, Cyril was the bigoted opponent of the free thought and "profane knowledge" that Hypatia the philosopher embodied. Here was also one of the least convincing motives for murder: Cyril's conviction that bigotry and philosophy cannot co-exist. Crucial to Draper's construction of these antitheses, the Cyril type kills the Hypatia type, a pattern that is extremely difficult to find in the history of science.

In his preface, Draper had claimed that Christianity had suppressed the schools of Alexandria. Yet he evidently had enough scruples about the political causes of Hypatia's death not to make her fate an instance of the conflict between science and religion. Instead, he claimed, her story served to introduce the *dramatis personae* in *later* struggles between religion and science. Draper's cautious qualification in the preface was surely overridden by the book's title. In any case, his equivocation did not prevent the murder of Hypatia from marking a dramatic fall in the grand narrative of the history of science, which he identified with the history of its conflict with religion understood as Catholicism.

---

[100] Draper (1874, pp. 45–7, 52).

[101] Draper (1874, p. 52).

[102] Draper (1874, pp. 54–55).

Draper constructed the sudden collapse of Alexandrian learning that Hypatia's death preceded by conflating institutions and teaching and by overlooking the decades of later intellectual activity in the city. According to Plutarch and others, the Museum library had suffered earlier assaults. Julius Caesar's troops in 48 B.C. had "destroyed the great library" whereas Emperor Aurelian had laid waste the palace quarter that housed the Museum complex when he suppressed an insurgency around 273. After that date, no source mentions the Museum.[103] *Pace* Draper, the demise of buildings and even damage to libraries rarely ends all learning, as attested by the long careers of Theon and Hypatia themselves and of many others after similar conflagrations.

Contrary to Draper's account, for more than a century after Hypatia, scholars moved back and forth between Alexandria and Athens. Some tensions between Christians and pagans occasionally flared into violence, but compromises and truces followed during which philosophers taught in Alexandria even after most no longer did in Athens. In the fifth century, Proclus and Damascius had studied in Alexandria before going to Athens, where the latter headed the Platonic Academy until Emperor Justinian suppressed it (529). Other pagans and Christians taught philosophy and mathematics in Alexandria at least until Olympiodorus (d. later sixth century), the last pagan known to have done so.[104] A century and a half after Hypatia, two intellectual opponents who studied in Alexandria were the pagan Simplicius (d. 560) and the Christian John Philoponus (d. 570), whose writings demonstrate their access to many texts. Both exerted great influence on subsequent natural philosophy (i.e., their works also survived). Philoponus not only wrote the earliest surviving treatise on the astrolabe but also fought the harmonizing tendencies of contemporary philosophy by restoring a more critical, adversarial approach to philosophical argumentation. The results were impressive. He refuted Aristotle's theory of motion, replacing it with a ground-breaking "impetus theory" that later echoed in the writings of John Buridan, Copernicus, and Galileo because it made the earth's motion conceivable instead of impossible, as it had been for Aristotle and Ptolemy.[105]

Draper's book excluded the Hypatia incident from his numbered conflicts, but it was much more than a bloody local incident in the nasty political fights of Alexandria. Although unrelated to her scientific work, her death two decades after the destruction of the Serapeum marked a momentous turning point in the history of science. The modern science that had begun at the Museum of Alexandria ended in Christendom for a millennium during which the Catholic Church used "the civil power" to suppress freedom of thought.

---

[103] Bagnall urges great caution about the library generally and skepticism about the mythical size of its holdings. His quantitative and contextual arguments show that ancient reports of hundreds of thousands of manuscripts are literally unbelievable. Bagnall (2002). See also Casson (2001, p. 47) and Barnes (2000, p. 73).

[104] Athanassiadi (1999, pp. 20–21); Dzielska (1995, pp. 105, 117–19); Neugebauer (1975, pp. 1031–54); Watts (2006, chap. 8–9) and the relevant entries in the Stanford Encyclopedia of Philosophy (https://plato.stanford.edu).

[105] Sorabji (2014).

For Draper, that decline was not global, however. Thanks to the Muslim conquests, darkness in Alexandria lasted only two centuries, as he argued while weakening the force of his general title. Even though he abhorred the entanglements of religion with political power, he let pass those of Islam. Instead, he praised the positive scientific role of its uncompromising monotheism and its alleged "fatalism." Seemingly oblivious to Islamic occasionalism's emphasis on God's will, he associated the latter trait with an unshakable belief in a universe governed by laws, which he contrasted with Christian belief in miracles and prayers for divine intervention into the natural course of events. He highlighted Islamic accomplishments in astronomy, which he also contrasted sharply with those of Christendom before the sixteenth century: "Christianity had been in existence fifteen hundred years, and had not produced a single astronomer."[106] With this sweeping claim, Draper created a timeline devoid of astronomy during the late-antique and medieval Greco-Latin eras.

Draper had read widely in the literature on Galileo. Indeed, despite occasional howlers, his *Intellectual Development* gave a detailed and passable account of Galileo's interaction with the Church authorities.[107] In *Conflict*, he therefore could have discussed Galileo at much greater length than he did Hypatia (Chapter Six, "Conflict Respecting the Nature of the World") but did not. After making Galileo's telescopic discoveries more decisive for Copernicanism than they were, he mischaracterized the latter's private 1616 interview (with Cardinal Bellarmine, not mentioned). It became a summons before the Inquisition in which he pledged not to publish or defend the Copernican theory: "Knowing well that Truth has no need of martyrs, he assented to the required recantation …" He thus sided with Whewell against Brewster, but the abjuration was premature. Galileo neither appeared before the Inquisition nor abjured before his 1633 trial, which Draper discussed by misleadingly claiming that he was "forced by the threat of death to deny facts which his judges as well as himself knew to be true."[108] This interpretation again overstates the cogency of Galileo's arguments, which were inductive and dependent on his highly problematic theory of the tides. As Heilbron has shown, those arguments drew such skepticism from Urban VIII on logical and philosophical grounds that the pope had Galileo change his book title to remove its mention of the tides.[109]

Draper strikingly climaxed this chapter anti-chronologically—and off-topically—with Giordano Bruno (d. 1600), whom he had treated chronologically in *Intellectual Development*.[110] His inversion is as notable as his equivocation. In the text, he called Bruno's musings about the cosmos "rather speculations than facts" even though "Scientific Ideas of Bruno" was the running head on the next folio.

---

106 Draper (1874, pp.107, 157–58); Yalcinkaya (2011, p. 172).

107 Draper (1864, II, pp. 252–57).

108 Draper (1874, pp. 171–72); see Finocchiaro (2009).

109 Heilbron (2010, e.g. pp. 115–16, 216–259-61, 298).

110 Draper (1864, II, pp. 249–50) claimed that Bruno had fled Italy for teaching that the earth rotated, and that his capital offence was teaching that space is infinite and "filled with self-luminous and opaque worlds, many of them inhabited." He implied that Francis Bacon (pp. 250–51) and Galileo (p. 254) were cowed by Bruno's execution.

Draper's book powerfully stimulated the later polemical arguments that either connected the fates of Bruno and Galileo or made their cases similar by glossing over the different reasons for their very different condemnations.[111] Unlike many a successor, Draper had a modicum of justification for including Bruno in his account. The main thread in his second conflict followed the fate of Averroes's doctrine of the soul, which Draper also associated with Bruno's ("intermediate between Averroes and Spinoza"). The Inquisition condemned Bruno as a "heresiarch" for "philosophical opinions" that included the unicity of the intellect and the plurality of worlds.[112] Draper thus had a semi-plausible case for treating Bruno's "speculations" as scientific conflicts with religion. Not least from a dramatic point of view, Bruno's life ended with his execution at the stake, a death akin to Hypatia's because of both its horror and its association with Church authorities. Executions for scientific beliefs being exceedingly rare, Draper found a way to make the narratives of both Bruno and Hypatia count.

Draper concluded the Bruno episode with an agenda: he suggested that posterity "offer an expiation for this great ecclesiastical crime" by erecting a statue of Bruno under the dome of St. Peter's Basilica. Had he heard this idea during his trip to Rome? Had he thought of it himself? Either way, Draper had clout in Italy. He had been elected to Galileo's old academy, the prestigious Accademia dei Lincei in Rome.[113] In addition, the Italian translation of *Conflict* (1876) resonated in the democratic and anti-clerical circles of Risorgimento Italy. In 1883, in language nearly identical to Draper's, the co-founder of the progressive paper "La Lega della democrazia" advocated erecting statues of both Bruno and Galileo under the dome of St. Peter's. In 1885, leading European and American intellectuals started an international subscription campaign for a statue of Bruno. In 1889, the monument was duly inaugurated at the very spot of Bruno's execution in the Campo de' Fiori.[114] Given the structure of his Hypatia narrative, it is perhaps surprising that Draper did not conclude from these two ecclesiastical condemnations that Italy suffered a scientific decline. Others would do so later.[115]

Draper's last chapter ("The Impending Crisis") was a frontal attack on Pius IX's initiatives. After his "Syllabus of Errors" (1864), the pope had secured the Vatican Council's contentious approval of papal infallibility (1870) even as he attacked Protestants as atheists, pressed the papacy's territorial claims, and sought "control over the intellectual development of the nations professing Christianity."[116] Although Draper criticized Catholicism for belief in miracles and other violations of natural law, he was more concerned with its use of political power to restrict belief and action, its intrusions in the secular realm, and its attacks on Reason and Nature. By

[111] Shackelford (2009).

[112] Draper (1874, p. 179–80).

[113] Barker (1886, p. 379).

[114] Bucciantini (2015, pp. 100, 120–24, 242, 329).

[115] Tarrant (2019).

[116] Draper (1874, p. 343).

contrast, "Science … has never sought to ally herself to civil power," a dubious claim contradicted by Draper's own discussions of the Alexandrian Museum, patronage, and science in Islam, among other examples.[117]

Several pages quoting from the "Dogmatic Constitutions of the Catholic Faith" justified Draper's argument that scientists should oppose the militant Catholicism of his day: "Roman Christianity and Science … cannot exist together; one must yield to the other; mankind must make its choice—it cannot have both."[118] This stance again undercut the generality of "religion" in the title of his book. According to this former Methodist, the minority status of Protestantism came with far less power, making it reconcilable with science.

Thanks to twenty printings in its first decade alone and many later translations, the framework and vignettes of Draper's *Conflict* proved highly influential. His narrative eventually shaped the dominant "big picture" of the history of science for not only the general public but also many an academic. The murder of Hypatia now inaugurated a precipitous intellectual decline as science and philosophy allegedly disappeared in Alexandria under Cyril. In Draper's account, science effectively ended for a millennium in Western Christendom. The scale of his fractal narrative was not universal, however: two centuries after Hypatia, the Muslims revived science (especially astronomy) in Alexandria and the other territories they had conquered.

Despite that qualification, the wide diffusion of Draper's book gave his total negation of science in medieval Christendom an enduring foothold in popularized histories of science and of astronomy in particular. His generalization remains a common-place of historical misinformation even as hundreds of pre-Copernican Latin astronomical manuscripts have yet to be studied.

18.10 The Bowdlerization of Draper: Sagan's "Cosmos" Series (1980)

More than a century after its publication, Draper's periodization of the history of early science underwent dramatic simplification and enjoyed exponentially greater diffusion as the medieval decline of science became global. Carl Sagan's "Cosmos: A Personal Voyage" (1980; reissued 2009), the epic Public Broadcasting Service television series, was reportedly seen by more than half a billion viewers in the following decades.[119] In thirteen episodes, Sagan offered an up-to-date introduction to astronomy interwoven with interpretations of its history that, thanks to Gibbon and Draper, were sometimes out-of-date by more than a century. In a series that featured very few women, Sagan gave Hypatia great prominence: a summary of her story

[117] Draper (1874, pp. xi, 18ff, 111ff).

[118] Draper (1874, p. 363).

[119] Hollinger (1999).

bracketed the entire series. Although it was ostensibly about the beginnings of astronomy, Sagan reinforced the myth about its Greek origins by saying nothing about the predictive mathematical astronomy of the Babylonians.[120] Instead, he devoted the first episode to a long segment about Alexandria, including a memorable re-creation of Eratosthenes's measurement of the Earth's radius. Using the fabulous figure of 400,000 books, he also graphically re-imagined the Museum and Library on a lavish scale. He gave these institutions a seven-century lifespan from Euclid to Hypatia, "the last light of the library whose martyrdom is bound up with the destruction of this place." He did not yet say how, or for what, she died. Despite his vague language, Sagan made explicit the hints of martyrdom in Toland, Gibbon, and Draper. As these sources had given the Library and Museum an afterlife in the Serapeum of Hypatia's day, Sagan associated her with both institutions, with the library holding "at its peak [unspecified] nearly one million scrolls," an impossible number.[121] Leaving the details of Hypatia's death in suspense, he offered a timeline of the history of astronomy that merged Draper's periodization with echoes of Burckhartian language. Europe allegedly fell into a deep "sleep" for a thousand years. Skipping the medieval vacuum, Sagan's narrative picked up with the reawakening during the Renaissance of the fifteenth century and its rediscovery of Alexandrian "tools, books, and concepts." After a millennium, these led linearly through Kepler (the direct heir of Aristarchus of Samos) and Newton to late-twentieth-century astronomy and space travel.[122]

In the final episode, Sagan concluded both "Cosmos" and his Hypatia narrative by tying her scientific work to the Library:

> Let me tell you about the end. It's a story about the last scientist to work in this place. A mathematician, astronomer, physicist and head of the Neoplatonic school of philosophy in Alexandria ... The growing Christian Church was consolidating its power and attempting to eradicate pagan influence and culture ... Cyril, the bishop of Alexandria, despised her in part because of her close friendship with the Roman governor, but also because she was a symbol of learning and science, which were largely identified by the early Church with paganism.[123]

His retelling re-echoes vignettes from Toland, Gibbon, and Draper while leaving their occasional shades of gray on the cutting-room floor. By omitting important details (e.g., Orestes's conversion to Christianity, Hypatia's Christian students), Sagan side-stepped the complexities of elite fifth-century Alexandrian culture, including the intense intra-religious political enmity between the imperial prefect and the bishop. His reading of Draper also overlooked the latter's limited use of the

---

[120] Neugebauer ([1951] 1969), chap. 2–3, 5.

[121] Draper (1874, pp. 54–55).

[122] Transcription of Sagan's Episode 1 at https://www.organism.earth/library/document/cosmos-1; see also Burckhardt (1990, pp. 97–98).

[123] https://www.organism.earth/library/document/cosmos-13 (accessed 19 December 2024). Darin Hayton's imaginative interview with Hypatia (accessed 5 April 2017) formerly included a link to the soundtrack of the thirteenth *Cosmos* episode; http://dhayton.haverford.edu/blog/2015/03/09/an-interview-with-hypatia/#fn:6015

Hypatia story to introduce the parties to *later* conflicts between science and religion. Instead, Sagan made her a martyr (implicitly of science) at the hands of Christianity writ large. Describing the murder, Sagan embellished Gibbon's idiosyncratic translation of *ostraka* (tiles or potsherds) with greater specificity: "they flayed her flesh from her bones with abalone shells. Her remains were burned, her works obliterated, her name forgotten. Cyril was made a saint."[124] These short, juxtaposed phrases are misleading. For works to be obliterated, they must first exist and it is far from clear that they did.[125] Her name was obviously not forgotten. Like Toland, Sagan concluded with Cyril's canonization—true but centuries later and disingenuously hinting at a reward for his association with the murderers.

More enduring and damaging than these problematic details was Sagan's large-scale periodization, which radically simplified Draper's schema. In the book accompanying the television series, Sagan summarized the first episode's narrative into a timeline of astronomy from antiquity to the present. Omitting more than a millennium of Babylonian data and mathematical astronomy, it started with the Greeks from Thales to Hypatia and left blank the centuries between Hypatia and Leonardo: "The millennium gap in the middle of the diagram represents a poignant lost opportunity for the human species." The language recalls Draper's post-Hypatian precipitous decline, albeit shorn of its restriction to Europe. Sagan unaccountably eliminated the latter's praise of the "Mohammedan astronomy" that limited the Alexandrian darkness to two centuries. By 1980, research on astronomy in Islamic civilization had been enriched by a century of new scholarship highly relevant to the "Cosmos" series.[126] Sagan's generalized medieval gap reached back beyond Draper to an old Enlightenment historiography that was already out-of-date in 1819, when Delambre published his *Histoire de l'astronomie du Moyen Âge*, more than six hundred pages documenting medieval theories, tables, observations, and observatories—to say nothing of the earlier summary under "Astronomie" in the *Encyclopédie*.[127]

[124] Haase (2020b, pp. 271–75).

[125] Acerbi (2008, p. 436).

[126] Sagan (1980, p. 335); Draper (1874, p. 52); and the pre-1980 literature cited in Saliba (2007).

[127] Delambre (1819); Diderot et al. 1751–65, vol. 1, pp. 788–89); see also Sayili (1981) and Heilbron (1999, pp. 70–71).

18.11 "Cosmos" Goes to the Movies: Amenábar's "Agora" (2009) and the Millennial Delay of the Scientific Revolution

One of the 500 million viewers of "Cosmos" was Alejandro Amenábar, who credited Sagan's series for inspiring "Agora," the 2009 film he directed and co-wrote with Mateo Gil.[128] "Agora" vastly enlarged Sagan's compact Hypatia narrative with many details from the science and history of later millennia. The script apparently did not take advantage of Maria Dzielska's sober assessment of Hypatia scholarship (1995) or even of all of his advisor Justin Pollard's detailed knowledge of Alexandrian politics (judging from his 2006 history of the city). The reason was perhaps that Amenábar had more urgent goals than historical accuracy.[129] In the aftermath of the September 11, 2001, attacks on New York's Twin Towers, the title and ambitions of "Agora" transcended a mere biography of Hypatia. Amenábar's target was purportedly "every kind of fundamentalism."[130] In the Greek city-state, the agora was the all-purpose public space in which citizens, among other activities, debated politics and civic issues. Combined with Pollard's historical advice, that focus restored a much-needed complexity to the politics behind the Hypatia story.

As the context of Hypatia's life, the film depicts the breakdown of civil discourse and violent altercations between the various parties in Alexandria: pagans, one of the largest Jewish communities of the time, and a social spectrum of adherents to an ascendant Christianity. Fanaticism, threats of violence, and actual violence eventually engulf all camps, including the scholarly Theon to his later regret.[131] True to Socrates Scholasticus's account, Hypatia is the conciliatory exception. When she is not teaching or engrossed in her astronomical research, she moderates various disagreements that range from the intellectual ones in her classroom (all her students are to be as brothers) to much tenser ones in the agora between ethnic and religious parties on the verge of violence.

Presumably thanks to Pollard's advice, the political and social complexity of Amenábar's film greatly improved on Sagan's simplistic account thanks to its technical excellence, brilliant cast, documentary conventions, and striking special effects.[132] This is all to the good. Unfortunately, these qualities also serve to reinforce Sagan's disturbingly anachronistic interpretations of the history of science, and even to amplify them by creating new myths.

With visually impressive recreations of ancient Alexandria, the first half of "Agora" is set around 391–92 up to the spectacular destruction of the Serapeum (both temple and library). The second, around 415, presents Hypatia's political

---

[128] An excellent summary and brilliant analysis appear in Scheidegger-Laemmle (2020); see also García (2015, pp. 116–117).

[129] Pollard and Reid (2006).

[130] Eco (2010).

[131] Scheidegger-Laemmle (2020, pp. 218–20).

[132] Scheidegger-Laemmle (2020, pp. 224–27).

activities and attempts at mediation, which often interrupt her work. Her research is taking her to the threshold of early modern science for she proposes that Earth is in an elliptical orbit around a stationary Sun). The film climaxes in a bizarre reimagination of Hypatia's death in which the love interests of her fictional former slave predominate. This scene and its antecedents at least make clear that she did not die for her science.

As the opening credits roll against stars slowly rising almost vertically in the night sky, an introductory text describes a world in decline à la Gibbon: the Roman Empire is "on the verge of collapse," an old common-place that modern specialists reject.[133] The lone exception is Alexandria. As the panorama zooms out, the edge of the spherical Earth becomes the Blue Ball, which exits left as stars fill the screen. While a woman's voice lectures about the perfection of the stars' circular courses, the Moon traces a quick arc and vanishes. Then the Sun's dazzling circle shrinks into the "O" in ΛGORΛ (sic!) and turns the surrounding darkness into the bright Alexandrian sky. Reinforcing the light metaphors are the text's allusions to the city's wondrous lighthouse and "the greatest library on earth."[134]

The camera now shows only her hand holding a handkerchief that falls to the ground before a small class of male students. Hypatia is illustrating the contrast between the eternal circular motion of the heavens and the temporary rectilinear motion of sublunar objects. The room is impressively reminiscent of the sixth-century, post-Hypatian classroom uncovered in a 2004 excavation.[135] Set around 391–92, the scene introduces Hypatia as a teacher and some of her students, a microcosm of tensions in the city. Amenábar implausibly casts Synesius as a doctrinaire Christian and Orestes as a pagan whom he identifies with the anonymous love-sick student whose advances, according to Damascius, Hypatia rebuffed with her menstrual rag. After Orestes converts to Christianity and becomes imperial prefect, he and Hypatia enjoy a (neo)-Platonic intellectual and political friendship, but Orestes's weakness fails to stop attacks on the ancient Jewish community (driven out in a later scene) or to stop Cyril's encroachments on the imperial power that he represents.

Amenábar's spectacular rendition of the attack on the Serapeum owes much to Gibbon's details and favors Draper's moderate interpretation over Sagan's totalizing one. During the assault on the temple and library (depicted inside one fortified complex), Hypatia and her students desperately save some manuscripts. (She will have two more decades of teaching in a more peaceful Alexandria acknowledged by an on-screen script.) After the emperor allows Christians to take over the premises, the

133 Bowersock (1996).

134 "By the end of the 4th century A.D., the Roman Empire was on the verge of collapse. Yet Alexandria, in the Roman province of Egypt, still retained much of its splendor. It boasted one of the seven wonders of the ancient world, the legendary lighthouse, as well as the greatest library on earth. The library was not only a cultural symbol, but also a religious one, a place where the pagans worshipped their ancestral gods. The city's long-established pagan cult was now challenged by the Jewish faith and a rapidly spreading religion until recently banned: Christianity."

135 Sorabji (2014).

empty library houses livestock. Amenábar will eventually transform the former temple into the Caesareum church in which Socrates Scholasticus located the murder of Hypatia.

During the second half of the film (circa 415), Hypatia's astronomical research begins to criticize Ptolemy's theory and devices. After one character mentions Aristarchus's sun-centered hypothesis from a work lost when the "mother library" was destroyed centuries earlier, Hypatia explores heliocentric possibilities. She quickly dismisses an epicyclic arrangement for a circumsolar Earth (ironically, Copernicus's approach). While searching for another solution, she offers striking visual demonstrations of the conic sections (according to a late Byzantine source, she commented on Apollonius). Using a taut rope and two flaming torches as foci (a nice visual pun) in a sandpit, she constructs an ellipse thanks to which she gives the Earth an elliptical orbit, anticipating Kepler's radical post-Copernican heliocentrism.

Amenábar's great cinema and striking demonstrations flesh out the theses of both Draper and Sagan with mythical amplifications that wreak havoc on the history of science. Imagining that Hypatia discovered the Earth's elliptical orbit by contemplating conic sections promotes a gross scientific anachronism. Having dismissed (Copernicus's!) epicycle-based heliocentric theory, Amenábar makes her leap directly from her sliced cone to Keplerian ellipses without explaining how a sphereless planet might keep to a mathematical curve.

Kepler's discovery of Mars's elliptical orbit, however, was one-of-a-kind. Unlike Hypatia, he had access to the excellent data from Tycho Brahe's late-sixteenth-century observations. He needed more than their accuracy, however: his own hyper-idiosyncratic *physical* theory, which postulated a force that could keep planets in motion if spheres no longer carried them and also justified his selection of the ellipse from all the curves that could fit the new data. No one else had reason to consider elliptical orbits, let alone to "see" them (and certainly not by studying the Earth's orbit first). Only after his many laborious calculations showed that all existing theories, (including Copernicus's) failed to fit the new data did Kepler consider an elliptical trajectory. Indeed, contrary to the exaggerated ellipses depicted in the film and many textbooks, most planetary orbits around the Sun are so slightly elliptical that they appear circular both to the naked eye and to computations from data less accurate than a few minutes of arc.[136] Hypatia was not about to scoop Kepler.

Consistent with Draper's criteria for the modernity of Alexandrian science, Amenábar's Hypatia also experiments. In another fine anachronistic scene, she performs the thought experiment from Galileo's *Dialogo* by dropping a sack from the mast of Orestes's moving ship.[137] Having incorrectly predicted that the weight would fall behind the mast, she is both puzzled and delighted to be wrong. She cannot explain the result but sees that it can nevertheless answer objections to a moving earth. This ahistorical foreknowledge further reinforces the Draper-Sagan notion of

---

[136] Stephenson (1994, e.g., p. 125; chap. 5); Gingerich (1994, p. 15).

[137] Heilbron (2010, p. 271).

a post-Hypatian darkness that stalls the progress of ancient science for 1200 years. Ironically, the source-based history of science in fifth- and sixth-century Alexandria contradicts this picture by offering a solution to Hypatia's conundrum. Unaware of living in a scientific vacuum or of contravening his Christian beliefs by doing so, the Alexandrian natural philosopher John Philoponus refuted Aristotle's theory of motion in the later fifth century. His new, alternative impetus theory could explain the puzzling result of Hypatia's fictional experiment.[138] Indeed not only John Buridan in his fourteenth-century vacuum but also Copernicus and Galileo used impetus theory to establish the possibility of the Earth's rotation.[139]

At the end of "Cosmos," Sagan had alluded to the scourge of slavery in Alexandria, which "Agora" depicts in several characters. Most central is the fictional youth Davus, who goes through a dizzying series of religious, political and emotional crises. He too has developed a crush on Hypatia while serving her every need, from teaching assistant through informal student to bath attendant. After secretly becoming a Christian, he assaults Hypatia, who forgives him instantly and frees him. He then joins the *parabalani* (literally "risk-takers") who bury the often-contagious dead, care for the sick and poor, and also serve as the bishop's vigilantes.[140] Davus now rejects his former learning, destroying his model of the Ptolemaic system. While Hypatia is becoming a Keplerian, he professes ignorance of the celestial order when Davus's fellow *parabalani* defend a cosmos with a flat earth and an ark-shaped heavens. Hypatia's former students Synesius and Orestes now put pressure on her to convert to Christianity. The timing of the scene suggests that she will die for refusing to become a Christian, an unsatisfying explanation since confessing Christianity was no guarantee of safety in the deadly intra-Christian altercations between the followers of Cyril and Orestes.

Meanwhile, Davus has overheard his colleagues plotting to kill Hypatia. After remorsefully throwing down his weapons, he finds Hypatia as his fellow *parabalani* are leading her to the church in the repurposed Serapeum. In a strange twist of poetic license, the film rewrites the murder. It climaxes in a painful scene that fleshes out Sagan's "martyrdom" vignette from "Cosmos" without, however, implying that she was killed for her scientific work. Having stripped Hypatia before the altar, the *parabalani* want to kill her with knives but Davus convinces them not to spill "impure blood" (pagan? female? belonging to a witch, as someone in the crowd calls her?). To lapidate her, they rush out to gather stones, leaving Hypatia alone with Davus. Instead of fleeing, she stands disturbingly motionless before the altar, apparently resigned to her death. Davus approaches her from behind. With his hands on her neck, his mind replays the scenes associated with his past desires. Her gaze turns to the circular aperture in the dome, elliptical for the viewer, as he clasps his hand over her mouth and nose. Whether dead or unconscious, she sinks to the floor. Davus tells the returning *parabalani* that she has fainted and they proceed to stone

[138] Wildberg (1999).

[139] Wolff (1987).

[140] Beers (2020, pp. 79–81).

her. The political and personal motives for the murder, the consensus about the role of Peter the lector and his associates in Hypatia's death and dismemberment move into the background, replaced by a fictional character whose strange mercy-killing doubles as an eroticized asphyxiation as Hypatia's vision of Keplerian ellipses dies with her.

Like a documentarian, Amenábar skips the gory aftermath described in the historical sources. Instead, the view zooms out above the old Serapeum. Peering through the circular *oculus* in its dome, the viewer sees Davus walk away as the shift in perspective again transforms the circle into an ellipse. Ever wider vistas fill the screen—the Nile delta, the Middle East, the Earth—as an on-screen paragraph concludes in classic documentary style:

> Hypatia's mutilated body was dragged through the streets and burnt on a pyre. Orestes disappeared, never to be seen again. Cyril seized power of (sic) Alexandria. Much later Cyril was declared a saint and doctor of the Church. Although none of Hypatia's works survived, it is known that she was an outstanding astronomer and renowned for her mathematical studies on conic curves. 1200 years later, in the 17th century, the astronomer Johannes Kepler discovered that one of these curves, the ellipse, governs the motion of the planets.

The last sentence is very confusing. Viewers have just watched Hypatia give the Earth an elliptical orbit around the Sun. Is Amenábar retracting his plot and giving Kepler credit for Hypatia's work?

Given its roots in Draper and Sagan, "Agora" not surprisingly brings into sharp relief the fractal character of its Hypatia narrative for the history of science. At the local scale, it portrays a life of ground-breaking astronomical research cut short by Christian fanatics. But that narrative structure also has an even more important large-scale counterpart. Hypatia's prescient Keplerian astronomy and Galilean physics put her at the apex of Alexandrian mathematical and experimental science, which she symbolizes. She not only brings to life the skeletal proposition in Draper and Sagan that Alexandria was the birthplace of modern science, but also marks its high point. Her death at the hands of the *parabalani* is much more than a personal and local tragedy. It is also the linchpin of a much larger decline narrative for it abruptly ends the eight-century upward trend of Alexandrian science and inaugurates the Draper-Sagan millennium of medieval nothingness. Bishop Cyril's murderous flat-earthers have effectively retarded the incipient Scientific Revolution by 1200 years. The film reinforces its prose graphically. After the bright light has gone out in Alexandria, the starry dark sky that opened the film also ends it. Astronomy and the viewer are back where they started. It was all for naught. After 1200 years, Kepler had to start over.

18.12 Coda: Hypatia's Retreat in deGrasse Tyson's New "Cosmos" (2014)

Did the historical criticisms that followed Sagan's efforts and the more political rationale for Hypatia's death in Amenábar's "Agora" bear some fruit? Perhaps. When Neil deGrasse Tyson reworked Sagan's "Cosmos" for Fox Broadcasting Company ("Cosmos: A Spacetime Odyssey," 2014), he parted ways with his mentor by leaving Hypatia out of the opening episode.

The influence of Draper's *Conflict* was still strong, however. The first episode avoided traditional astronomical pioneers to introduce Giordano Bruno, to whom deGrasse Tyson devoted fully one quarter of the first episode. As he conceded, Bruno's speculations on an infinite universe and a plurality of worlds and his execution for heresy had little to do with astronomy: "Bruno was no scientist. His vision of the cosmos was a lucky guess, because he had no evidence to support it." The lure of the Inquisition's fiery judicial murder that loosely connected heresy with cosmic speculation was enough to shift attention away from Copernicus and Galileo, each only mentioned twice, primarily to introduce or connect with Bruno. For example, "Ten years after Bruno's martyrdom, Galileo first looked through a telescope, realizing that Bruno had been right all along," a fanciful sentence for an astrophysicist to paraphrase from A. D. White's abridged *Warfare of Science with Theology*.[141] Neither infinity, nor a plurality of worlds, nor Copernicanism can be seen through a telescope.

Hypatia nevertheless appeared in the series when deGrasse Tyson, like Sagan, alluded to the Alexandrian Library in the final thirteenth episode. He named Hypatia just once: between Pythagoras (!) and Euclid, in a short list of names that he associated with "the greatest library on Earth." His point was that many of us now have at our fingertips vastly more information than the Alexandrian Library ever contained. By contrast, its "knowledge belonged to the elite. So in the fourth century AD, when the mob came to destroy the library and the genius of classical civilization, there were not enough people to defend it."[142] As this equivocal sentence identified neither the institution nor the mob with its abstract goal, viewers were evidently expected to fill in the story with "common knowledge" gleaned from other sources—Sagan's "Cosmos," perhaps "Agora" (a box-office failure outside of Spain, however), or more likely the internet, the Alexandrian library's huge modern counterpart. In the new "Cosmos," Hypatia all but disappeared.[143]

---

[141] deGrasse Tyson (2014a); White (1965, p. 125).

[142] deGrasse Tyson (2014b).

[143] Some missed Hypatia; see Adam Lee, "Cosmic Terror," https://www.salon.com/2014/03/27/cosmic_terror_why_neil_degrasse_tyson_has_religious_fundamentalists_so_freaked_partner/ accessed 1 October 2014; https://www.the-numbers.com/movie/Agora#tab=summary accessed 18 December 2024.

18.13 Summation

Of the many variants of the Hypatia narrative since the fifth century, the preceding pages document but one strand relevant to the popularized history of science. Juxtaposing the earliest historical sources with the most recent narratives shows the magnitude of the transformations. The story began as a moralizing church history anecdote about a deadly confrontation in fifth-century Alexandrian politics. Nineteenth- to twenty-first-century Hypatia narratives retold for mass consumption assigned it the global significance of inaugurating a millennial decline in the history of science. However problematically, her death is now enshrined in the popular martyrology of science alongside the execution of Giordano Bruno and the condemnation of Galileo.

The earliest sources recounted a dramatic incident in the perennially bloody power politics of fifth-century Alexandria. This particular brutal murder stood out, however: the victim was not only a highly learned woman but also a widely respected teacher and citizen. Not least, the perpetrators were the bishop's men. Although the motives ascribed to the murderers were generally political, they ranged more specifically from Cyril's personal resentment (jealousy of Hypatia's popularity) through collateral revenge (to punish the imperial prefect Orestes, her associate) to political expediency (to end her alleged opposition to reconciling Cyril and Orestes as they sought control of the city.) In the fifth- and sixth-century accounts, Hypatia's scientific and philosophical activities marked her as uniquely admirable but played a negligible role in her death.

Almost 1600 years later, Amenábar's post-9/11 film "Agora" turned Hypatia's life into a parable about the doomsday consequences of fanaticism. The violence that follows the breakdown of civic discourse in Alexandria leads to devastating long-term consequences for science. In Amenábar's rewriting of the history of science, Hypatia's experiments glimpse Galilean physics while her mastery of conic sections lets her see the ellipses of Keplerian heliocentrism. Before she can tidy up these discoveries, Cyril's *parabalani* kill her. Her death thus interrupts the linear progress of astronomy until Kepler 1200 years later.

In the centuries between these strikingly different messages, Hypatia narratives change in both scale and structure. Only after the sixth-century do her knowledge and practices became significant for her death. John Nikiu's seventh-century account applauds Hypatia's murderers for eliminating a magician who bewitches Orestes with astrolabes and music, thus ending paganism in Alexandria. The legend of St. Katherine, a likely inversion of the Hypatia story, exploits its potential for creating a new martyr whose remarkable knowledge and skills directly affect her fate. Despite the very different aims and values in these two stories, the protagonist's beliefs and practices play a role in her death.

The long retrospective of Toland's *Hypatia* marks a crucial turning point in the evolution of the narrative. Because he synthesized all available sources, promoted the intellectual capabilities of women, and lived in the age of Newton, he highlighted Hypatia's accomplishments in mathematics and astronomy. Even more important was his retrospective. Looking back on fifth-century Alexandria from early-eighteenth-century Britain, Toland's post-Reformation perspective enlarged the story's frame and significance. The polarities between Cyril and Hypatia in the earliest sources were largely individual, but Toland saw in them clashes between enduring types of authority. Within his big-picture, Protestant time-line of church history, Cyril's cruelty and injustice were also features of the 1300-year-old, degenerate, post-Constantinian, unreformed Christendom of his day, as he knew from personal experience. Conversely, Toland's Hypatia, although not a Christian, embodied the good (her ethics), the true (her scientific and philosophical brilliance), and the beautiful (her physique), which made her more deserving of sainthood than St. Cyril.

Since the structure of the Hypatia narrative transcended its fifth-century particulars, Toland's polemical biography doubled as a parable with contemporary analogies. Toland's minoritarian outlook and personal history sensitized him to the persecution of individuals for their beliefs. Anyone persecuted for their opinions by church authorities was akin to Hypatia. He noted one analogue in the recent Whiston-Sacheverell altercation, less deadly by far but suited to his longstanding feud with Sacheverell. Whiston in turn saw his persecution for anti-trinitarianism as paralleling Galileo's for Copernicanism. With their analogies, Toland and Whiston loosely brought together elements that were poised to transform the Hypatia story into a deadly version of the stereotypical Galileo Affair.

Deep inside hundreds of pages of *Decline and Fall*, Gibbon's brief account of Hypatia's death was on a different scale from Toland's. Like the latter, however, he was still writing an ecclesiastical history periodized by the Protestant conviction that traced more than a millennium of fallen Christianity back to Constantine. As he documented the Church's nefarious role in politics and culture up to the sixteenth century, Gibbon used the Hypatia story to prosecute Cyril, a leading reprobate among the many he assailed. The Hypatia story dramatically illustrated the fall of Christendom on a small scale. Whereas Toland had generalized it, the scope of Gibbon's work made it only one of the many grievances in his case against post-Constantinian Christianity. Although he mentioned Hypatia's scientific activities in passing, he was far more concerned with Cyril's role and the "indelible stain" that her death left on his "character and religion."

By contrast, Draper's *History of the Conflict between Religion and Science* decisively changed the significance of the Hypatia story by making it a key episode in his popularized history of science. Draper shared with Toland and Gibbon the Protestant perspective on the first millennium and a half of Christianity. In his predecessors' accounts, this rise-and-fall framework structured narratives that intertwined events drawn mostly from ecclesiastical and political history. Draper, however, used it to organize his large-scale narrative of the history of science, which he equated with its conflict with religion (read: Catholicism). Thus framed, Hypatia entered scientific lore.

In Draper's book, the role of the Hypatia narrative was both ambiguous and momentous. Although he claimed that Christianity had suppressed Alexandrian science, he omitted the Hypatia episode when he subdivided the history of science into his four conflicts. Instead, her story served to introduce the "parties" to those later conflicts. Beyond representing themselves, Cyril and Hypatia were once again types. Beyond this broad function, Hypatia's death as a specific historical event also carried epochal significance: it closed a glorious era and opened a dismal one. For Draper, the science of the Alexandrian Library and Museum was experimental and mathematical. It therefore marked the beginning of modern science. Although Draper's earlier work had identified a decline after Ptolemy (second century), in *Conflict* he extended the city's vibrant intellectual life by several centuries until the destruction of the Serapeum and the murder of Hypatia by assorted Christians. His juxtaposition of the attacks on the Serapeum library with Hypatia's death became a feature of later polemical accounts.

Draper's typology resulted in a contorted abstract explanation for the motive behind the murder: Cyril acted on his feeling that bigotry (he) and philosophy (she) could not co-exist. Her death at the hands of his mob thus ended philosophy, "profane knowledge," and freedom of thought not only in Alexandria but far beyond. Here was the most significant aspect of her story for Draper's long-term argument. In Europe, the precipitous decline that followed her death allegedly lasted a millennium or more (no astronomers for 1500 years).

Draper's formal and temporal parallelisms between intellectual and ecclesiastical declines were not coincidental. A fallen Church had caused both, and both ended when it lost control during the Reformation—the "second Reformation" in Draper's four-conflict schema. The "first Reformation" had happened in Alexandria, where science disappeared for only two centuries, because the conquests of Islam helped to revive it. Draper liked Islam's alleged "fatalism," which he oddly equated with the belief that deterministic laws governed the world. Once properly harnessed, that outlook led to the first scientific renewal, a remarkable outcome in a book on the conflict between religion and science.

When Draper reached the canonical topic of the Earth's status (his third conflict), he curiously treated Galileo in less detail and with more distortions than in his earlier work. More surprising, he followed Galileo's abjuration (1633) with the story of Bruno and his execution by the Inquisition (1600). Indeed, he ended the chapter by proposing that the Vatican expiate its crime by memorializing Bruno with a statue in St. Peter's. Draper's anti-chronological outline calls for an explanation. Both Galileo and Bruno belonged to the "party" that Draper identified with Hypatia, and Galileo was clearly the more important scientific figure. Why then displace Galileo to make Bruno's "speculations," execution, and death the culmination of his argument? In a book about conflicts with religion (read: Catholicism), the likely answer lies in the relative polemical value of their contrasting sentences and deaths. In brutality, the Inquisition's condemnation of Bruno to the stake dwarfed its imposition of house arrest on Galileo, who died in bed. Bruno was the closest Draper could come to finding a Cyril type killing a Hypatia type. His case therefore clamored for inclusion in the book despite the hazy connection between his death and his cosmological views. The dramatic horror of Bruno's death thus earned him his place at the climax of the chapter.

Draper's unexpected inversion of the Galileo and Bruno stories may shed light on his transformation of the Hypatia narrative. It is a near-certainty that Draper, like most of us, was first exposed to potted versions of the Galileo Affair before ever learning about Hypatia or even Bruno. The reasons for Galileo's house arrest ultimately traced back to his scientific beliefs. This was not the case for the deaths of either the Hypatia or the Bruno of history. Conversely, both deaths incriminated Church authorities. Consistent with the types implicit in his "parties to the conflict" and his search for law-like historical patterns in history, Draper tweaked the structures of the three narratives to make them more analogous than the sources suggest. Without explicitly stating that Hypatia and Bruno died for their scientific beliefs, both Draper's typology and his inclusion of them alongside Galileo in the book's argument suggested as much. His cagey qualifications were largely lost on later readers eager to make Hypatia and Bruno martyrs of science (they were already among Barni's martyrs of free thought).

In the popular historiography of science, Sagan's concise treatment of Hypatia in "Cosmos" offers the most extreme version of her story in both his summary of the murder and its consequences. With problematic generalizations from particulars and an uncritical echo of Draper's conclusions, Sagan's bipolar narrative makes Hypatia the last remnant of the Alexandrian Library's scientific greatness and a martyr. While conceding that Cyril hated her for her friendship with the imperial governor, Sagan made the overarching motive for the murder institutional and religious. The early Christian church associated science with paganism, which it sought to eradicate; with the help of his mob, Cyril therefore eradicated Hypatia.

In Sagan's narrative as in Draper's, the consequences of her murder (and the alleged suppression of her works) reached far beyond Alexandria in 415. Preceded by the destruction of the library, Hypatia's death did more than end a seven-century run associated with the greatest names in the ancient mathematical sciences; it also opened onto a scientific void. The timeline for Sagan's popular history of astronomy echoed Draper's sweeping conclusions while omitting his qualifications. Unlike Draper's, Sagan's mythical thousand-year medieval vacuum was not limited to Christendom; it featured no one, not even the Muslim astronomers that earned Draper's praise.

Although Amenábar's inspiration for "Agora" came from the Hypatia narrative in Sagan's "Cosmos," his film abandoned the latter's glib account of the Alexandrian context. All parties showed divisions and some members of each were prone to violence and engaged in it. This richer and more complex historical account made it impossible to keep Sagan's simplistic implication that Christianity persecuted Hypatia for her scientific work. Unfortunately, Amenábar did more than retain Sagan's discredited vision of a millennial vacuum in the history of astronomy. He effectively expanded it into a larger myth with new fictional elaborations even more seductive for an uncritical viewer. Hypatia is portrayed as a critic of Ptolemy who anticipates both Galileo's thought experiment about free fall on a moving ship and the elliptical orbits of Kepler's post-Copernican astronomy. In short, her work represents the culmination of Alexandria's mathematical and experimental tradition, which Draper had equated with the "birth of modern science." The fanatical

Christian *parabalani* who abruptly make her a martyr end not only her ground-breaking work but the entire trajectory of Alexandrian science. As in Draper and Sagan, the fractal-like small-scale Hypatian decline narrative both expands and lends its structure to the large-scale periodization of the history of science. After Alexandrian science dies, a millennium of darkness follows. Only 1200 years later, according to Amenábar, Kepler will pick up where the fictional Hypatia left off and move the Scientific Revolution forward.

For better or worse, movies are memorable, round numbers appealing, and sharp breaks seductive. In his recent *Women of Genius in Science* (2023), the highly educated Lars Jaeger devoted the first chapter to Hypatia. He cared enough about evidence to offer detailed documentation for most of his claims, with multiple footnoted quotations from primary sources. Yet he also made the following unfootnoted comment: "According to tradition, Hypatia was fascinated by the ellipse, a figure that results when a plane is made to cut through a cone. Twelve hundred years before Johannes Kepler, she tried to explain the irregular orbits of the planets in terms of this geometric figure."[144] One surmises that the tradition in question is not ancient and oral tradition, but recent, visual, and available on DVD. With the small fractal clearly sketched, can the large one be far behind? The penultimate paragraph of the chapter ends as follows: "Today's historians see her death as a turning point that marked the transition from the ancient to the Christian age—and thus the beginning of a millennium of efforts to restrict education and oppose scientific thinking."[145] Published by a respected scientific press, this astonishing statement refers to the millennium that witnessed the first two great scientific translation movements, from Greek and Syriac into Arabic and Persian (from the eighth century) and from Greek and Arabic into Latin (from the eleventh century). In the late twelfth century, the first universities emerged and grew so successfully that, by the late Middle Ages, *per capita* matriculations were some five times greater than in the late nineteenth century. The medieval arts faculties devoted a greater portion of their curricula to scientific requirements (natural philosophy and the mathematical sciences) than do most modern universities.[146] No matter. The thousand years of darkness will live on, bolstered by the seductive new myth of Hypatia's aborted elliptical astronomy.

References

Acerbi, F. 2008. Hypatia. In *New Dictionary of Scientific Biography*, ed. N. Koertge, vol. 3, 435–437. Detroit: Thompson Gale.

Athanassiadi, P. 1999. *Damascius, The Philosophical History, text with translation and notes*. Athens: Apamea.

[144] Jaeger (2023, p. 7).

[145] Jaeger (2023, p. 12).

[146] Sayili (160, chap. 3); Shank (2013, pp. 228–230), Grant (1984, esp. pp. 68–71).

Bagnall, R. S. 2002. Alexandria: Library of Dreams. *Proceedings of the American Philosophical Society* 146 (4): 348–362.

Barker, G. 1886. *A Memoir of John William Draper, 1811–1888*, 349–388. Washington, DC: National Academy of Sciences.

Barnes, R. 2000. Cloistered Bookworks in the Chicken-Coop of the Muses: The Ancient Library of Alexandria. In T*he Library of Alexandria, Centre of Learning in the Ancient World*, ed. R. MacLeod, 61–77.

Barni, J. 1862. *Les Martyrs de la libre pensée. Cours public professé dans la salle du Grand Conseil de Genève*. Geneva: n.p.

Baronius, C. 1866. *Annales ecclesiastici Cardinalis Baronii*, ed. Augustin, vol. 7. Bar-le-Duc: Louis Guérin.

Baxter, P. 1984. Brewster, Evangelism and the Disruption of the Church of Scotland. In *'Martyr of Science': Sir David Brewster 1781-1868*, ed. A. D. Morrison-Low and J. R. R. Christie, 45–50.

Beers, W. F. 2020. Bloody Iuvenalia: Hypatia, Pulcheria Augusta, and the Beginnings of Cyril of Alexandria's Episcopate. In *Hypatia of Alexandria: Her Context and Legacy*, ed. D. L. Norman and A. Petkas, 67–86.

Berggren, J. L. 2009. The Life and Death of Hypatia. *Metascience* 18:93–97.

Bowersock, G. W. 1996. The Vanishing Paradigm of the Fall of Rome. *Bulletin of the American Academy of Arts and Sciences* 49 (8): 29–43.

Brewster, D. 1841. *The Martyrs of Science, or The Lives of Galileo, Tycho Brahe, and Kepler*. London: John Murray.

Bucciantini, M. 2015. *Campo dei Fiori: Storia di un monumento maledetto*. Turin: Einaudi.

Burckhardt, J. 1990. *The Civilization of the Renaissance in Italy*. S.G.C. Middelmore (trs.). London: Penguin Books.

Cameron, A. 2013. The Life, Death, and Work of Hypatia. In *Le voyage des légendes: Hommage à Pierre Chuvin*, ed. M. Tardieu and D. Lauritzen, 65–82.

Casson, L. 2001. *Libraries in the Ancient World*. New Haven: Yale University Press.

Chadwick, O. 1976. Gibbon and the Church Historians. *Daedalus* 105 (3): 111–123.

Champion, J. A. I. 1999. 'Acceptable to Inquisitive Men': Some Simonian Contexts for Newton's Biblical Criticism, 1680–1692. In *'Newton and Religion: Context, Nature, and Influence*, ed. J. E. Force and R. Popkin, 77–96.

Champion, J. 2003. *Republican Learning: John Toland and the Crisis of Christian Culture, 1696–1722*. Manchester: Manchester University Press.

Christie, J. R. R. 1984. Sir David Brewster as an Historian of Science. In *'Martyr of Science': Sir David Brewster 1781–1868*, ed. A. D. Morrison-Low and J. R. R. Christie, 53–56.

Cochran, W. 1984. Sir David Brewster: An Outline Biography. In *'Martyr of Science': Sir David Brewster 1781–1868*, ed. A. D. Morrison-Low and J. R. R. Christie, 11–14.

Cormack, L. B. 2009. Myth 3: That Medieval Christians Taught that the Earth was Flat. In *Galileo Goes to Jail and Other Myths about Science and Religion*, ed. R. L. Numbers, 28–49.

Craughwell, T. 2017. Why Pope John Paul II Restored this Saint to the Calendar. *The National Catholic Register*. https://www.ncregister.com/blog/why-john-paul-ii-restored-this-saint-to-the-calendar. Accessed 28 June 2023.

D'Alembert, J. L. 1751. Discours préliminaire des éditeurs. In *Encyclopédie ou dictionnaire raisonné des sciences, des arts, et des métiers...*, ed. D. Diderot et al. Paris: Briasson, etc.

Daniel, S. 1984. *John Toland: His Methods, Manners and Mind*. Montreal: McGill-Queen's University Press.

Deakin, M. A. B. 1994. Hypatia and Her Mathematics. *American Mathematical Monthly* 101:234–243.

deGrasse Tyson, N. 2014a. *Cosmos: A Space-Time Odyssey*. Transcript of Episode 1: Standing Up in the Milky Way. https://docs.google.com/document/d/1h9CPur0qYZ2NDK78bLViX9ogfaOAifv61w15KgZGYhU/edit?tab=t.0

deGrasse Tyson, N. 2014b. *Cosmos: A Space-Time Odyssey*. Transcript of Episode 13: Unafraid of the Dark. https://docs.google.com/document/d/1jFu4QiGfNtU1SZJ-jOF64lKFMjpDGXE6BqckcIwyfa0/edit?tab=t.0
Delambre, J. B. J. 1819. *Histoire de l'astronomie du Moyen Âge*. Paris: Veuve Courcier.
Diderot, D., et al., eds. 1751–65. *Encyclopédie ou dictionnaire raisonné des sciences, des arts, et des métiers...* Paris: Briasson et al.
Diderot, D. 1755. Éclectisme. In *Encyclopédie, ou Dictionnaire raisonné des sciences, des arts et des métiers*, vol. 5, 270–293.
Draper, J. W. 1847. On the production of light by heat. *The London, Edinburgh, and Dublin Philosophical Magazine and Journal of Science* 30 (202): 345–360.
Draper, J. W. 1864. *History of the Intellectual Development of Europe*. Vol. 1. London: Bell and Daldy.
Draper, J. W. 1874. *History of the Conflict between Religion and Science*. London: Henry S. King and Company.
Dzielska, M. 1995. *Hypatia of Alexandria*. Cambridge, MA: Harvard University Press.
Eco, U. 2010. La bustina di Minerva: Ipazziammo. *L'Espresso* (29 April 2010). https://lespresso.it/c/opinioni/2010/4/29/ipazziammo/26137. Accessed 14 October 2023.
Fabricius, J. A. 1717. *Bibliotheca graeca*. Vol. 8. Hamburg: Christian Liebezeit.
Feingold, M. 2017. Isaac Newton, Heretic? Some Eighteenth-Century Perceptions. In *Reading Newton in Early Modern Europe*, ed. E. A. Boran and M. Feingold, 328–345.
Finocchiaro, M. A. 2005. *Retrying Galileo, 1633–1992*. Berkeley: University of California Press.
Finocchiaro, M. A. 2009. Myth 8: That Galileo Was Imprisoned and Tortured for Advancing Copernicanism. In *Galileo Goes to Jail and Other Myths about Science and Religion*, ed. R. L. Numbers, 68–78.
Fleming, D. 1950. *John William Draper and the Religion of Science* . Philadelphia: University of Pennsylvania Press.
Force, J. E. 1985. *William Whiston, Honest Newtonian*. Cambridge: Cambridge University Press.
Force, J. E. 1994. The God of Abraham and Isaac (Newton). In *The Books of Nature and Scripture: Recent Essay on Natural Philosophy, Theology, and Biblical Criticism in the Netherlands of Spinoza's Time and the British Isles of Newton's Time* , ed. J. E. Force and R. Popkin, 179–200.
Gajeri, E. 1992. *Ipazia: Un mito letterario*. Rome: La Meridiana Editori.
García, Nacho. 2015. City of Lights: Ancient Alexandria in Cinema and Modern Imagination. In *Imagining Ancient Cities in Film: From Babylon to Cinecittà*, ed. Marta García Morcillo, P. Hainesworth, and Ó. Lapeña Marchena, 13–34.
Gibbon, E. 1788. *The History of the Decline and Fall of the Roman Empire*. Vol. 4. London: A. Strahan and T. Cadell.
Gibbon, E. 1952. *The Decline and Fall of the Roman Empire*. Great Books of the Western World, 40–41. 2 vol. Chicago: Encyclopedia Britannica.
Gillespie, S. K. 2012. John William Draper and the Reception of Early Scientific Photography. *History of Photography* 36:241–254.
Gingerich, O. J. 1994. Circles of the Gods: Copernicus, Kepler, and the Ellipse. *Bulletin of the American Academy of Arts and Sciences* 47 (4): 15–27.
Grant, E. 1984. Science and the Medieval University. In *Rebirth, Reform, and Resilience: Universities in Transition, 1300-1700*, ed. J. M. Kittelson and P. J. Transue, 68–102.
Harich-Schwarzbauer, H. 2012. Hypatie d'Alexandrie. *Clio* 35:201–214.
Harkness, D. 1957. *Bolingbroke: The Man and his Career*. London: Staples Press.
Haase, M. 2020a. The Shattered Icon: An Alternative Reading of Hypatia's Killing (Socrates, Hist. eccl. 7.15.5-7, John of Nikiu, Chron. 84.100-103. and Rufinus, Hist. eccl. 11.23). In *Hypatia of Alexandria*, ed. D.L. Norman and A. Petkas, pp. 87–117.
Haase, M. 2020b. Appendix B: Hyatia's Death According to Socrates, Hist. eccl. 7.15: A Textual Commentary. In *Hypatia of Alexandria* , ed. D. L. Norman and A. Petkas, 255–284.
Heilbron, J. L. 1999. *The Sun in the Church: Cathedrals as Solar Observatories*. Cambridge, MA: Harvard University Press.

Heilbron, J. L. 2010. *Galileo*. Oxford: Oxford University Press.
Hollinger, D. 1999. Star Power. *The New York Times Book Review* (28 November 1999) https://archive.nytimes.com/www.nytimes.com/books/99/11/28/reviews/991128.28holingt.html. Accessed 17 October 2023.
Hudson, W. 2009. *The English Deists: Studies in Early Enlightenment*. London and New York: Routledge.
Imbruglia, G. 1997. 'My Ecclesiastical History:' Gibbon between Hume and Raynal. In *Edward Gibbon: Bicentenary Essays*, ed. D. Womersly, 73–102.
Jacob, M. 1969. John Toland and the Newtonian Ideology. *Journal of the Warburg and Courtauld Institutes* 32:307–331.
Jaeger, L. 2023. *Women of Genius in Science Whose Frequently Overlooked Contributions Changed the World*. Zürich: Springer.
James, Frank A. J. L. 2017. In An *'Open Clash between Science and the Church'?: Wilberforce, Huxley and Hooker on Darwin at the British Association, Oxford, 1860* , ed. D. M. Knight and M. D. Eddy, 171–193.
Leask, I. 2020. From Serena to Hypatia: John Toland's Women. *Royal Institute of Philosophy Supplements* 88:195–214.
Leonard, V. 2020. The Ideal (Bleeding?) Female: Hypatia of Alexandria and Distorting Patriarchal Narratives. In *Hypatia of Alexandria*, ed. D. L. Norman and A. Petkas, 171–192.
Lessl, T. 1999. The Galileo Legend as Scientific Folklore. *Quarterly Journal of Speech* 85:146–168.
Lewis, K. 2000. *The Cult of Saint Katherine of Alexandria in Late Medieval England*. Woodbridge: Boydell and Brewer.
Lewis, T. 1721. *The History of Hypatia, A most Impudent School-Mistress of Alexandria: murder'd and torn to pieces by the populace, in defence of Saint Cyril and the Alexandrian Clergy. From the aspersions of Mr. Toland*. London: T. Bickerton, at the Crown in Pater-noster-Row.
Lindberg, D., and M. H. Shank, eds. 2013. *The Cambridge History of Science*, vol. 2: *Medieval Science*. New York: Cambridge University Press.
Ménage, G. 1690. *Historia mulierum philosopharum*. Lyon: Anissonios, Posuel and Rigaud.
M.G. 1728. Dissertation sur Hypacie où l'on justifie Saint Cyrille d'Alexandrie sur la mort de cette Sçavante. *Continuation des Mémoires de Littérature et d'Histoire*. Vol. 5, partie 1, 139–186. Paris: Simart.
Montucla, J. E. 1758. *Histoire des mathématiques*. Vol. 1. Paris: Jombert.
Morrell, J. B. 1984. Brewster and the early British Association for the Advancement of Science. In *'Martyr of Science': Sir David Brewster 1781–1868*, ed. A. D. Morrison-Low and J. R. R. Christie, 25–29.
Neugebauer, O. [1951] 1969. *The Exact Sciences in Antiquity*. 2nd edition. New York: Dover.
Neugebauer, O. 1975. *A History of Ancient Mathematical Astronomy*. Berlin: Springer.
Nicholl, H. F. 1965. John Toland: Religion Without Mystery. *Hermathena*, No. 100 (Summer 1965), 54–65.
Norman, D. L., and A. Petkas, eds. 2020. *Hypatia of Alexandria: Her Context and Legacy*. Tübingen: Mohr-Siebeck.
Petkas, A., and D. L. Norman. 2020. Appendix A: The Principal Ancient Sources, Translated. In *Hypatia of Alexandria*, ed. D. L. Norman and A. Petkas, 247–248.
Pharr, C. 1932. The Interdiction of Magic in Roman Law. *Transactions and Proceedings of the American Philological Association* 63:269–295.
Pococke, J. G. A. 2004. *Barbarism and Religion: The Enlightenments of Edward Gibbon, 1737–1764*. Vol. 1. Cambridge: Cambridge University Press.
Pollard, J., and H. Reid. 2006. *The Rise and Fall of Alexandria: Birthplace of the Modern Mind*. New York: Viking.
Principe, L. 2018. The Warfare Thesis. In *The Warfare between Science and Religion: The Idea That Wouldn't Die*, ed. J. Hardin, R. L. Numbers, and R. A. Binzley, 6–26.
Sagan, C. 1980. *Cosmos*. New York: Random House.

Saliba, G. 2007. *Islamic Science and the Making of the European Renaissance*. Cambridge, MA: MIT Press.

Sayili, A. [1960] 1981. The Observatory in Islam. New York: Arno Press.

Schaff, P. and H. Wace (Eds). 1994. *Socrates: Church History from A.D. 305-438*. Nicene and Post-Nicene Fathers, ser. 2, vol. 2. Peabody, MA: Hendrickson Publishers.

Scheidegger-Laemmle, C. 2020. In *Starring Hypatia: Amenábar's Agora and the Tropology of Reception*, ed. D. L. Norman and A. Petkas, 209–237.

Schmidt, J. A. 1689. *Hipparchum Theonas, doctamque Hypatiam in mathesi celebres. Dissertatione historico-mathematica... exhibet*. Jena: Krebs.

Seng, H. 2020. In *Desire and Despair: Synesius, Hypatia, and No Consolation of Philosophy*, ed. D. L. Norman and A. Petkas, 29–49.

Shackelford, J. 2009. Myth 7: That Giordano Bruno was the First Martyr of Modern Science. In *Myth 7: Galileo Goes to Jail and Other Myths About Science and Religion*, ed. R. L. Numbers, 59–67.

Shank, M. H. 2013. In *Schools and Universities in Medieval Latin Science*, ed. D. C. Lindberg and M. H. Shank, 207–239.

Slater, J. W. 1880. The Martyrdom of Science. *Popular Science Monthly* 17:71–77. [reprinted from *The Journal of Science* 1880].

Snobelen, S. D. 1999. Isaac Newton, Heretic: The Strategies of a Nicodemite. *British Journal for the History of Science* 32:381–419.

Snobelen, S. D. 2001. 'God of Gods, and Lord of Lords': The Theology of Isaac Newton's General Scholium to the *Principia. Osiris* 16:169–208.

Snobelen, S. D., and L. Stewart. 2003. Making Newton Easy: William Whiston in Cambridge and London, In *From Newton to Hawlking*, ed. K. C. Knox and R. Noakes, 135–170.

Snobelen, S. D. 2004. William Whiston, Isaac Newton and the Crisis of Publicity. *Studies in the History and Philosophy of Science* 35:573–603.

Sorabji, R. 2014. The Alexandrian Classrooms Excavated and Sixth-Century Philosophy Teaching. In P. Remes and S. Slaveva-Griffin (eds). Accessed on 17 Oct 2023.

Stephenson, Bruce. [1987] 1994. *Kepler's Physical Astronomy. Princeton*: Princeton University Press.

Sullivan, R. 1982. *John Toland and the Deist Controversy*. Cambridge, MA: Harvard University Press.

Tarrant, N. 2019. Science, Religion, and Italy's Seventeenth-Century Decline: From Francesco de Sanctis to Benedetto Croce. *Zygon: Journal of Religion and Science* 54:1125–1144.

de Tillemont, S.L. [1693–1712] 1732. Mémoires pour servir à l'Histoire ecclésiastique des six premiers siècles. 2nd ed. Vol. 14. Venice: François Pitteri.

Tissandier, G. 1879. *Les Martyrs de la science*. Paris: Maurice Dreyfous.

Toland, J. 1720. *Tetradymus* .... London: n. p.

Toland, J. 1727. *A collection of several pieces of Mr. John Toland now first publish'd from his original manuscripts*. London: J. Peele.

Torrey, N. L. 1930. *Voltaire and the English Deists*. New Haven: Yale University Press.

Ungureanu, J. C. 2016. A Yankee at Oxford: John William Draper at the British Association for the Advancement of Science at Oxford, 30 June 1860. *Notes and Records of the Royal Society of London* 70:135–150.

Ungureanu, J. C. 2019. *Science, Religion, and the Protestant Tradition: Retracing the Origins of Conflict*. Pittsburgh: University of Pittsburgh Press.

Voltaire. 1772. *Dictionnaire philosophique par des amateurs*, 9e partie. [Geneva]: n.p.

Voltaire. 1776. *L'examen important de Milord Bolingbroke, écrit sur la fin de 1736 et accompagné des notes de M...., éditeur de ses ouvrages*. 10th edition corrected and expanded. London: n.p.

Voltaire. 1879. *Œuvres complètes*. Vol. 19. Paris: Garnier.

Voltaire. 1961. *Mélanges de Voltaire*. Bibliothèque de La Pléiade, 152. Paris: Gallimard.

de Voragine, J.. 1967. *La Légende dorée*. Trans. J.-B. M. Roze. 2 vol. Paris: Flammarion, Vol. 2.

Walsh, C. 2007. *The Cult of St. Catherine of Alexandria in Early Medieval Europe*. Aldershot: Ashgate.

Watts, E. J. 2006. *City and School in Late-Antique Athens and Alexandria.* Berkeley: University of California Press.
Watts, E. J. 2017. *Hypatia: The Life and Legend of an Ancient Philosopher.* Oxford: Oxford University Press.
Watts, E. J. 2020. Hypatia and her Eighteenth-Century Reception. In *Hypatia of Alexandria*, ed. D. L. Norman and A. Petkas, 193–207.
Wessel, S. 2004. *Cyril of Alexandria and the Nestorian Controversy: The Making of a Saint and of a Heretic.* Oxford: Oxford University Press.
Westfall, R. S. 1980. *Never at Rest: A Biography of Isaac Newton.* Cambridge: Cambridge University Press.
Westfall, R. S. 1989. *Essays on the Trial of Galileo.* Vatican City: Vatican Observatory.
Wildberg, C. 1999. Impetus Theory and the Hermeneutics of Science in Simplicius and Philoponus. *Hyperboreus* 5:107–124.
Whewell, W. 1837. *History of the Inductive Sciences from the Earliest to the Present Time.* Vol. 3 vol. London: John Parker.
Whiston, W. 1713. *Reasons for not Proceeding against Mr. Whiston by the Court of Delegates in a letter to the Reverend Dr. Pelling.* London: A. Baldwin.
Whiston, W. 1749. *Memoirs of the Life and Writings of Mr. William Whiston...* London: Printed for J. Whiston and B. White.
Whewell, W. 1875. *History of the Inductive Sciences from the Earliest to the Present Time.* 3rd ed. 2 vol. New York: Appleton
White, A.D. [1896] 1965. *A History of the Warfare of Science with Theology in Christendom.* Abridged by B. Mazlish. New York: Free Press.
Wolff, M. 1987. Impetus Mechanics as a Physical Argument for Copernicanism: Copernicus, Benedetti, Galileo. *Science in Context* 1:215–256.
Wyles, R. 2016. Ménage's Learned Ladies: Anne Dacier (1647–1720) and Anna Maria van Schurman (1607–1678). In *Women Classical Scholars: Unsealing the Fountain from the Renaissance to Jacqueline de Romilly*, ed. R. Wyles and E. Hall, 61–77.
Yalcinkaya, M. A. 2011. Science as an Ally of Religion: A Muslim Appropriation of 'the Conflict Thesis. *British Journal for the History of Science* 44:161–181.

Michael H. Shank, Professor Emeritus in the Department of the History of Science (now merged with History), University of Wisconsin-Madison, he researches late-medieval astronomy, natural philosophy, and university history. Recent work includes "The Medieval University", in A. Amaral and A. Magalhães (eds.), *Handbook on Higher Education Management and Governance* (Cheltenham, 2023); and "The Monsters of Astronomy before Copernicus: Regiomontanus's Criticisms of Ptolemaic Models in the *Defensio Theonis contra Georgium Trapezuntium*", *Studia Copernicana* 46 (in press).

GPSR Compliance
The European Union's (EU) General Product Safety Regulation (GPSR) is a set of rules that requires consumer products to be safe and our obligations to ensure this.

If you have any concerns about our products, you can contact us on

ProductSafety@springernature.com

In case Publisher is established outside the EU, the EU authorized representative is:

Springer Nature Customer Service Center GmbH
Europaplatz 3
69115 Heidelberg, Germany

www.ingramcontent.com/pod-product-compliance
Ingram Content Group UK Ltd.
Pitfield, Milton Keynes, MK11 3LW, UK
UKHW020933180926
475692UK00003B/18
* 9 7 8 3 0 3 2 2 5 6 7 7 5 *